# MACROMOLECULAR DESIGN: CONCEPT AND PRACTICE

## *"Advanced Polymers Via Macromolecular Engineering" Series*

1. Macromolecular Design: Concept and Practice

*Additional Volumes in Preparation.*

# MACROMOLECULAR DESIGN: CONCEPT AND PRACTICE

## Macromonomers
## Macroinitiators
## Macroiniferters
### Macroinimers
### Macroinifers
### Macroiniters

**EDITED BY**

**Munmaya K. Mishra**
Polymers Research
Texaco, R & D Department
P. O. Box 509
Beacon, New York 12508

Polymer Frontiers International, Inc. Hopewell Jct.(New York)
New Delhi

**ISBN 0-9639138-0-8**

Library of Congress 92-85448 CIP



The authors have made effort to get permission from the copyright holders for any borrowed material. If they have inadvertently overlooked any, they will be pleased to make the necessary arrangements at the first opportunity.

First published in 1994 10 9 8 7 6 5 4 3 2

Printed in the United States of America

# PREFACE

The field of polymer science, particularly the area of *macromolecular engineering*, has grown very large indeed, and it would scarcely be possible in a single volume to do justice to all the excellent research in various branches of this subject. For this reason several volumes under the series title "Advanced Polymers Via Macromolecular Engineering" will be published.

The present volume, "Macromolecular Design: Concept and Practice," discusses macromolecular design and applications with special emphasis on macromonomers, macroinitiators, and macroiniferters. This book is divided into four parts with chapters describing various design strategies, applications of new polymer products, and challenges facing polymer scientists in *macromolecular engineering*.

This book is written for academic and industrial scientists interested in designing new polymers and polymeric materials for various applications, as well as for the student who is searching for working relations between abstract ideas, contemporary research in polymer science, and ultimately, some of the today's important technologies. For all those involved in designing polymer and *macromolecular engineering*, this book represents a vital source of information. As this book describes polymeric materials for various applications of technological importance, industrial scientists, will find the contents particularly useful and challenging.

It would not have been possible to complete a project like this without the help, understanding and participation of numerous individuals. I gratefully acknowledge with thanks all the authors (Profs./Drs. S. Kobayashi, H. Uyama, K. Ishizu, K. Hatada, T. Kitayama, K. Ito, Y. Tsukahara, Y. Yagci, A. Ueda, S. Nagai, O. Nuyken, B. Voit, M. Okada, C. P. R. Nair, P. Chaumont, G. Clouet, T. Otsu, A. Matsumoto, R. C. Kumar, R. D. Dueltgen, and M. H. Andrus, Jr.) whose contributions made this book possible and for their patiences, understanding and suggestions through out the preparation of this volume. The thoughtful comments of Prof. Y. Yagci, Prof. K. Hatada, Dr. H. Chafetz, Dr. B. L. Papke, and Dr. E. T. Hessell are highly appreciated. Also, I would like to point out that the tragic death of Prof. G. Clouet by an air-accident during the preparation of this book not only shocked me, but also the polymer community around the world. His enthusiasm and passion for research has been responsible for most of the results described in Chapter 11 by his coauthors. Last, but not the least, it is with love and appreciation I acknowledge my wife *Minu*, without whose help and support this project would never have started or ended.

Beacon, New York, January 1994 **Munmaya K. Mishra**

# Contents

# 1. INTRODUCTION

This article overviews recent advances of macromonomer synthesis and its applications. Macromonomer is abbreviation of macromolecular monomer. A macromonomer is generally referred to a linear macromolecule possessing a polymerizable group at the chain end. In most cases, the polymerizable group is of vinyl type; typically styryl, (meth)acryl, or a vinylester group. Cyclic polymerizable groups such as an oxirane and a cyclic imino ether and bifunctional groups such as a glycol and a dicarboxylate have been reported as other polymerizable groups of macromonomers for ring-opening polymerization and step-growth polymerization, respectively.

Recently, block and graft copolymers have received much attention as novel polymeric materials with multi-components. For the preparation of graft copolymers, several methods are known; typically, initiation from a polymer backbone and deactivation of a living polymer by a polymer backbone. However, these methods are difficult to control the defined structure of the resulting graft copolymer. In the middle of 1970's, Milkovich of CPC International Inc. presented the concept "Macromer®" (a trade mark of macromonomer). He synthesized various macromonomers by the end capping of anionic living polymers with electrophiles possessing a polymerizable group [1]. Then, much attention has been devoted to macromonomer techniques as new convenient and powerful methodology for the preparation of graft copolymers with well-defined structure.

A pioneering work on the synthesis of polymers having a polymerizable group may be dated back to 1958. However, the work was not fully recognized as a useful technique to prepare graft copolymers until Milkovich. First, vinyl-terminated poly(methyl methacrylate) was formed by disproportion in the radical polymerization and its copolymerizability with acrylonitrile was investigated [2]. Second, vinyl and allyl-terminated polystyrene macromonomers were obtained by the anionic polymerization using corresponding lithium compounds as an initiator [3,4]. Greber et al. synthesized polysiloxane (**1**) and polystyrene (**2**) macromonomers; the former was synthesized by the reaction of $\alpha$-chloropolysiloxane with a Grignard compound of p-chlorostyrene [5] and the latter by using dimethyl(p-styryl)silyl chloride as an end-capping agent for living end of anionic polystyrene [6]. A chain-transfer agent of the radical polymerization was used for the preparation of macromonomer in 1967; the polymerization of MMA in the presence of 2-mercaptoethanol produced $\alpha$-hydroxyl PMMA, which was converted into methacryl-type PMMA macromonomer **3** by the reaction with methacryrloyl chloride [7].

The first application of macromonomers in the technical field was made by the ICI; a graft copolymer from a lipophilic macromonomer was used as a stabilizer for the dispersion polymerization in the hydrocarbon media [8]. The graft copolymer consisted of the grafted segment from the macromonomer soluble in the media and the main chain insoluble in the media. The steric repulsion of the grafted polymer

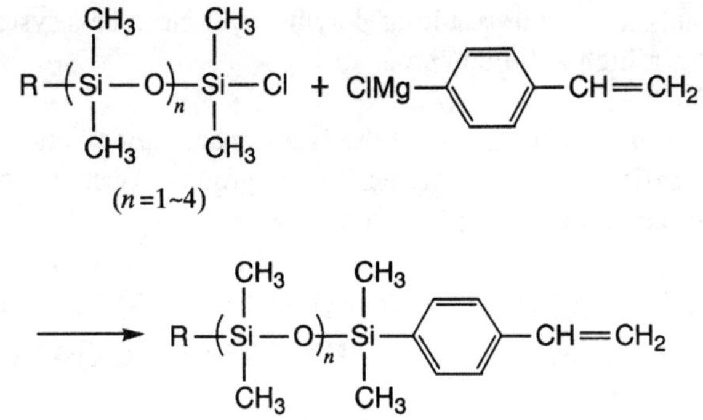

$$R\!-\!\left(\underset{CH_3}{\overset{CH_3}{Si}}\!-\!O\right)_n\!\!-\underset{CH_3}{\overset{CH_3}{Si}}\!-\!Cl \;+\; ClMg\!-\!C_6H_4\!-\!CH{=}CH_2$$

$(n = 1\text{~}4)$

$$\longrightarrow R\!-\!\left(\underset{CH_3}{\overset{CH_3}{Si}}\!-\!O\right)_n\!\!-\underset{CH_3}{\overset{CH_3}{Si}}\!-\!C_6H_4\!-\!CH{=}CH_2$$

**1**

$$\text{(Polystyrene)}\!-\!CH_2\!-\!\underset{C_6H_5}{\bar{C}H}\;Na^+ \;+\; Cl\!-\!\underset{CH_3}{\overset{CH_3}{Si}}\!-\!C_6H_4\!-\!CH{=}CH_2$$

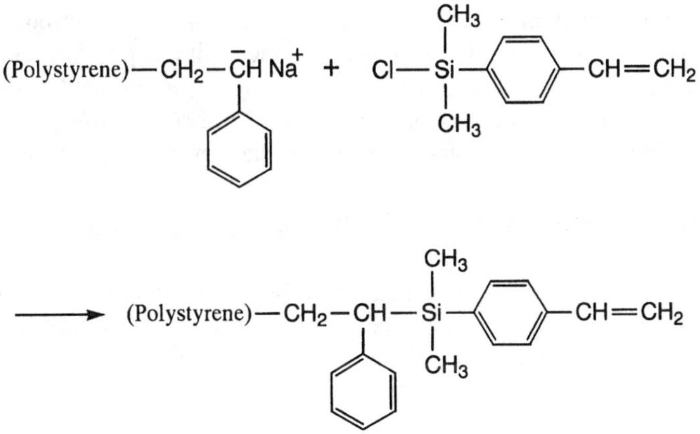

$$\longrightarrow \text{(Polystyrene)}\!-\!CH_2\!-\!\underset{C_6H_5}{CH}\!-\!\underset{CH_3}{\overset{CH_3}{Si}}\!-\!C_6H_4\!-\!CH{=}CH_2$$

**2**

$$n\; CH_2{=}C(CH_3)CO_2CH_3 \;+\; HOCH_2CH_2SH$$

$$\xrightarrow{R\cdot} HOCH_2CH_2S\!-\!\left(CH_2\!-\!\underset{CO_2CH_3}{\overset{CH_3}{C}}\right)_n\!\!-H$$

$$\xrightarrow{CH_2=C(CH_3)COCl} CH_2{=}C(CH_3)CO_2CH_2CH_2S\!-\!\left(CH_2\!-\!\underset{CO_2CH_3}{\overset{CH_3}{C}}\right)_n\!\!-H$$

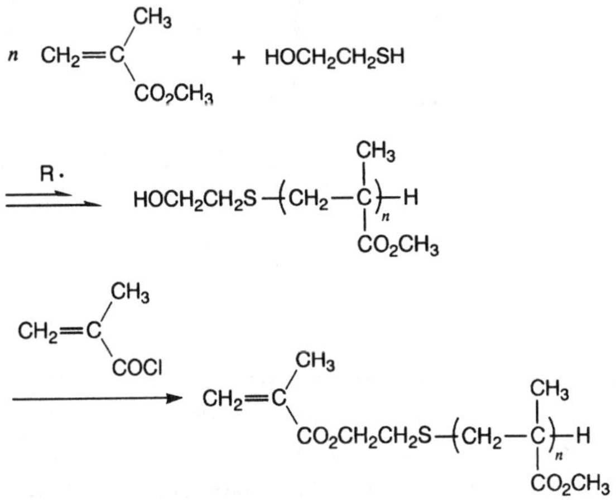

**3**

chain prevented coagulation of the resulting polymer particles. This system gave stable polymer dispersion in a high solid content.

In the first part of this article the synthesis of macromonomers are described according to the classification of polymerizable groups. Second, several recent applications of macromonomers are presented.

## 2. SYNTHESIS OF MACROMONOMERS: CLASSIFICATION OF MACROMONOMERS ACCORDING TO POLYMERIZABLE GROUPS

Generally, macromonomer synthesis is carried out by four methods; (I) initiation by a reagent possessing a polymerizable group ("Initiator Method"); (II) direct termination of a living polymer with a reagent having a polymerizable group ("Terminator Method"); (III) reaction of $\alpha$-functional polymer with a polymerizable compound (2-step synthesis); and (IV) polyaddition using a bis-unsaturated monomer as one component. In this part, macromonomers are classified according to polymerizable groups and typical methods for the synthesis of these macromonomers are described.

### 2.1 Macromonomers Having Styryl Group

Compounds possessing styryl or $\alpha$-methylstyryl group were used as an initiator for the synthesis of styryl-type macromonomers. Polymerization of ethylene oxide (EO) initiated by potassium p-isopropenylbenzyloxide gave polyEO macromonomer **4** with narrow molecular weight distribution [9].

$$n\ \text{(ethylene oxide)} + CH_2{=}C(CH_3){-}C_6H_4{-}CH_2O^{-}K^{+} \longrightarrow CH_2{=}C(CH_3){-}C_6H_4{-}CH_2O{+}CH_2CH_2O{+}_n$$

**4**

p- and m-Vinylbenzylmagnesium chlorides (VBzMgCl) were employed as an initiator for polymerizations of styrene and MMA to produce styryl-type macromonomers [10]. The polymerization of MMA with m-VBzMgCl in THF at -110 °C gave highly syndiotactic (s>96%) PMMA of low polydispersity (Mw/Mn=1.19). Polystyrene macromonomer was obtained from p-VBzMgCl, however, m-BzMgCl did not induce the polymerization.

In the aldol group-transfer polymerization of silylenol ether catalyzed by a Lewis acid, the use of p-formylstyrene as an initiator gave poly(silyl vinyl ether) macromonomer **5**, [11] which would be converted into polyvinylalcohol macromonomer by hydrolysis.

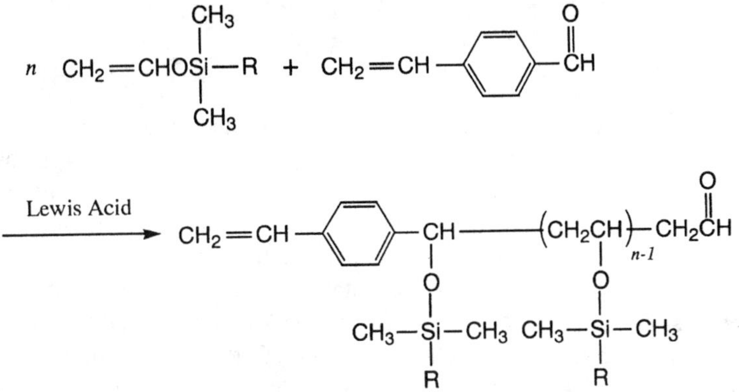

$$n\ CH_2{=}CHOSi(CH_3)_2{-}R \ + \ CH_2{=}CH{-}C_6H_4{-}CHO \xrightarrow{\text{Lewis Acid}} CH_2{=}CH{-}C_6H_4{-}CH(OSi(CH_3)_2R){-}(CH_2CH(OSi(CH_3)_2R))_{n-1}{-}CH_2CHO$$

**5**

Macromonomer syntheses via initiator method in the ring-opening polymerization have been reported. Polymerization of $\epsilon$-caprolactone ($\epsilon$-CL) using lithium isopropenylbezyloxide as an initiator produced poly($\epsilon$-CL) macromonomer [12]. Poly(2-oxazoline) macromonomer **6** was synthesized by cationic polymerization of 2-oxazoline using vinylbenzyl iodide [13,14]. A mixture of vinylbenzyl chloride and silver hexafluorophosphate or silver perchlorate was also an effective initiator for preparation of poly(tetrahydrofuran) (THF) macromonomer [15]. Polymerization of $\epsilon$-benzyl-L-glutamate-N-carboxyanhydride initiated by N-methyl-N-(2-(p-styryl)ethyl)ethylene diamine (**7**) gave poly(amino acid) type macromonomer **8** [16].

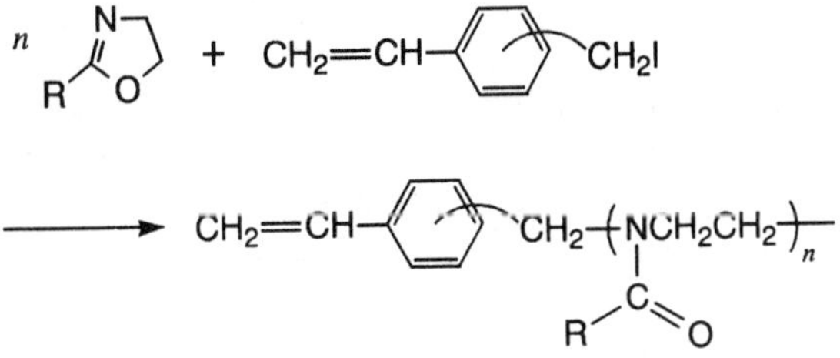

$$n\ \text{(2-R-2-oxazoline)} \ + \ CH_2{=}CH{-}C_6H_4{-}CH_2I \longrightarrow CH_2{=}CH{-}C_6H_4{-}CH_2{-}(N(COR)CH_2CH_2)_n{-}$$

**6**

$$n\ \text{(R-substituted N-carboxyanhydride)} \ + \ CH_2{=}CH{-}C_6H_4{-}CH_2CH_2N(CH_3)CH_2CH_2NH_2 \longrightarrow$$

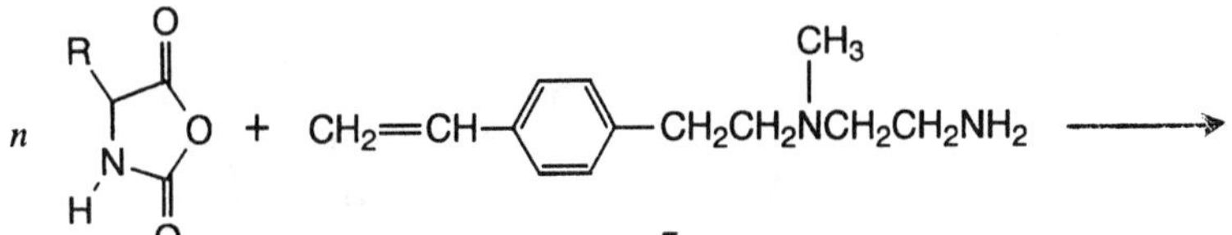

**7**

$(R{=}CH_2CH_2CO_2CH_2C_6H_5)$

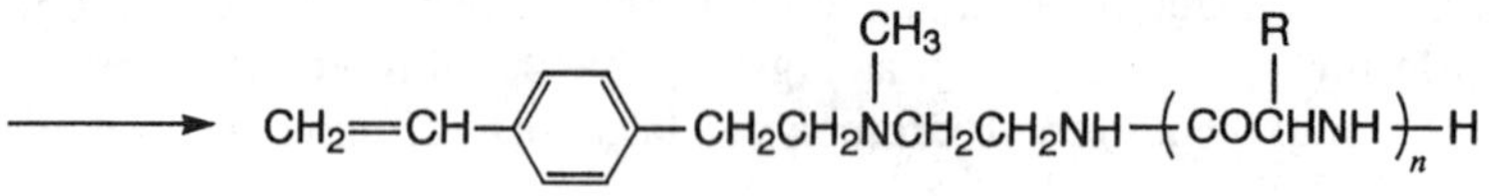

**8**

Milkovich prepared macromonomers by end-capping of an anionic living polymer with a terminator possessing a polymerizable group [1]. In case of styryl-type polystyrene macromonomer, the direct coupling of the living carbanion end with benzyl halide competed with a side reaction involving attack of the carbanion at the double bond of p-benzyl halide. In order to suppress the side reaction, the terminal living end of polystyrene anion was converted to the alcoholate by the reaction with ethylene oxide, and then the resulting end was reacted with the p-vinylbenzyl halide to produce a styryl-type macromonomer. The alcoholate decreased the nucleophilicity of the end group. 1,1-Diphenylethylene was also employed to decrease the reactivity of the living end [17].

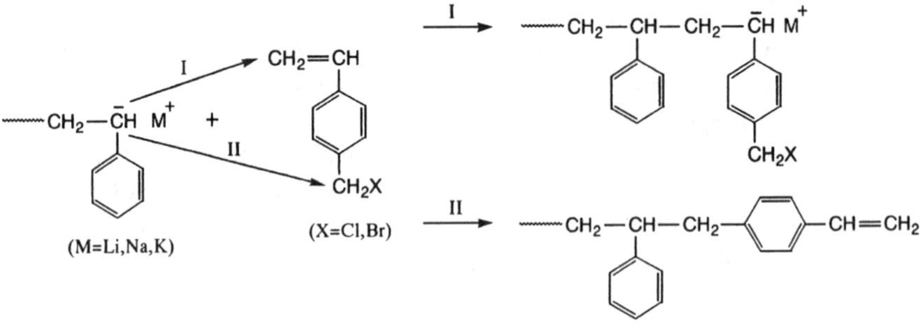

(M=Li,Na,K)

(X=Cl,Br)

The direct coupling of the polystyrene living end with p-benzyl halide was studied by Asami [18]. The solvent effect on the suppression of the side reaction was pronounced. The coupling reaction of polystyrene lithium with excess p-vinylbenzyl chloride in a THF-containing solvent prevented the side reaction exclusively.

The termination of anionic living end of other polymers from alkyl methacrylates [17,19], ethylene oxide [17,20], and 2-vinylpyridine [17,21] with a vinylbenzyl (or isopropenylbenzyl) halide gave the styryl-type macromonomers. In all cases, the direct termination gave quantitative introduction of the polymerizable function at the chain end.

Styryl-type poly(siloxane) macromonomer **9** was synthesized by the termination of dimethyl(p-vinylbenzyl)silyl chloride with the anionic living end of poly(siloxane), prepared by ring-opening polymerization of cyclic trimer of dimethylsiloxane ($D_3$) [22]. Ring-opening polymerization of a bicyclic oxalactam followed by the reaction with p-vinylbenzylamine produced polyamide macromonomer **10** [23].

$$n\ \left[\!\left(\mathrm{Si(CH_3)_2}-\mathrm{O}\right)_3\!\right] + \mathrm{CH_3-Si(CH_3)_2-OLi} \longrightarrow \mathrm{CH_3-Si(CH_3)_2-O}\!\left(\mathrm{Si(CH_3)_2-O}\right)_{3n}\!\mathrm{-Li}$$

$$\xrightarrow{\mathrm{Cl-Si(CH_3)_2-C_6H_4-CH{=}CH_2}} \mathrm{CH_3-Si(CH_3)_2-O}\!\left(\mathrm{Si(CH_3)_2-O}\right)_{3n}\!\mathrm{-Si(CH_3)_2-C_6H_4-CH{=}CH_2}$$

**9**

$$\mathrm{CH_2{=}CH-C_6H_4-CH_2-NH}\!\left(\mathrm{C(=O)-C_5H_8O-NH}\right)_n\!\mathrm{-C(=O)-C_6H_5}$$

**10**

The termination of cationic living end with a nucleophile having a polymerizable group is a versatile method to prepare macromonomers. Ring-opening polymerization of THF using triethyloxonium tetrafluoroborate ($Et_3OBF_4$) as an initiator followed by the end-capping with sodium p-vinylbenzyloxide [24] or sodium p-vinylphenoxide [25] produced styryl-type polyTHF macromonomer **11**. Similarly, the use of potassium p-isopropenylbenzyloxide as a terminator gave polyTHF macromonomer possessing $\alpha$-methylstyryl group [26]. Styryl-type poly(2-oxazoline) was synthesized by the termination of the living end with **7** [13].

Another route to synthesize macromonomers is the reaction of a polymer having a functional group at the chain end with a compound having polymerizable group. The reaction of $\alpha$-hydroxyl terminated polyEO with vinylbenzyl chloride gave styryl-type polyEO macromonomer [27]. Aromatic polyether sulfone [28] and poly(2,6-dimethyl-1,4-phenylene oxide) [29] having phenol group at the chain end were converted into corresponding macromonomers by the reaction with vinylbenzyl chloride. N-Acetylchito oligosaccharide macromonomer **12** was prepared by the reaction of oligosaccharide possessing lactone moiety at the chain end with p-vinylbenzylamine [30].

$$n\ \mathrm{THF} \xrightarrow{\mathrm{Et_3\overset{+}{O}B\overset{-}{F_4}}} \mathrm{Et}\!\left(\mathrm{OCH_2CH_2CH_2CH_2}\right)_{n-1}\!\mathrm{-\overset{+}{O}(CH_2)_4}\ \mathrm{BF_4^-}$$

$$\xrightarrow[\mathrm{(X=OCH_2,\ O)}]{\mathrm{Na-X-C_6H_4-CH{=}CH_2}} \mathrm{Et}\!\left(\mathrm{OCH_2CH_2CH_2CH_2}\right)_n\!\mathrm{-X-C_6H_4-CH{=}CH_2}$$

**11**

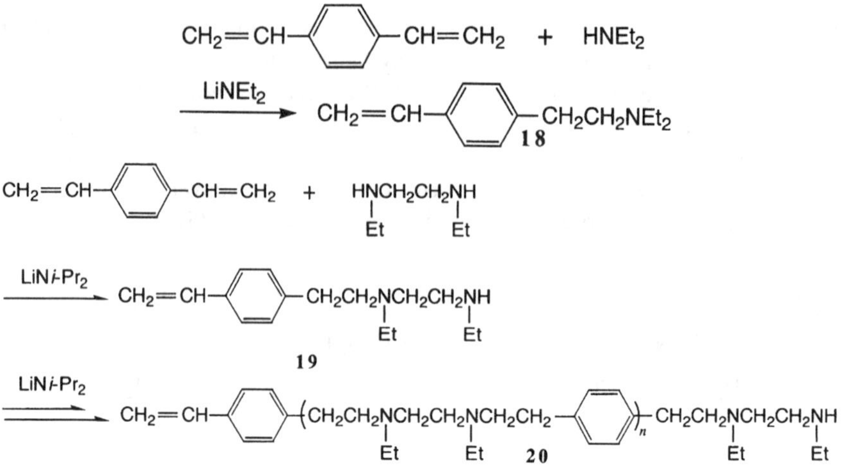

$CH_2{=}CH{-}C_6H_4{-}CH{=}CH_2$ + $HNEt_2$ $\xrightarrow{LiNEt_2}$ $CH_2{=}CH{-}C_6H_4{-}CH_2CH_2NEt_2$

**18**

$CH_2{=}CH{-}C_6H_4{-}CH{=}CH_2$ + $HN(Et)CH_2CH_2NH(Et)$

$\xrightarrow{LiN\textit{i}\text{-}Pr_2}$ $CH_2{=}CH{-}C_6H_4{-}CH_2CH_2N(Et)CH_2CH_2NH(Et)$

**19**

$\xrightarrow{LiN\textit{i}\text{-}Pr_2}$ $CH_2{=}CH{-}C_6H_4{-}(CH_2CH_2N(Et)CH_2CH_2N(Et)CH_2CH_2{-}C_6H_4)_n{-}CH_2CH_2N(Et)CH_2CH_2NH(Et)$

**20**

## 2.2 Macromonomers Having (Meth)acryl Group

Compounds having a (meth)acryl function have been used as an initiator for preparation of macromonomers. Oxocarbenium salts are well known to be efficient initiators for cationic polymerization of THF. The polymerization using methacryloyl hexafluoroantimonate, produced by reaction of methacrloyl chloride and silverhexafluoroantimonate gave methacryl-type polyTHF macromonomer with narrow molecular weight distribution [40].

(Meth)acryloyl hexafluorophosphate or hexafluoroantimonate induced the polymerization of a 1,4-anhydro sugar, 1,4-anhydro-2,3-bis[O-t-butyldimethylsilyl)-$\alpha$-D-ribopyranose **21** to produce polysaccharide macromonomer **22**. The resulting polymer consisted of (1$\rightarrow$5)-$\alpha$-D-ribofuranan and (1$\rightarrow$4)-$\beta$-D-ribopyranan units. The polymerization of **21** initiated acryloyl hexafluorophosphate at -40 °C gave the macromonomer with high $\alpha$-stereoregularity. On the other hand, mainly $\beta$-pyranose structure was formed from hexafluoroantimonate salts [41].

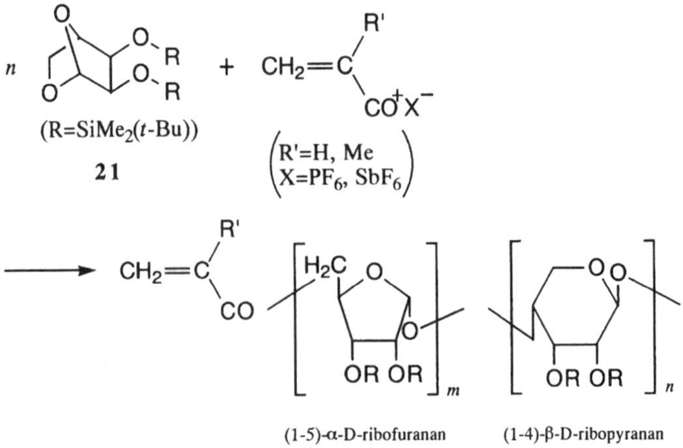

*n* **21** (R=SiMe$_2$($t$-Bu)) + $CH_2{=}C(R')CO^+X^-$ (R'=H, Me; X=$PF_6$, $SbF_6$)

$\longrightarrow$ $CH_2{=}C(R')CO{-}$[(1-5)-α-D-ribofuranan]$_m$[(1-4)-β-D-ribopyranan]$_n$ (OR OR)

(1-5)-α-D-ribofuranan (1-4)-β-D-ribopyranan

**22**

Higashimura et al. [42] found that cationic polymerization of vinyl ether initiated by hydrogen iodide/iodine ($HI/I_2$) proceeded through a highly living system. Monodisperse poly(vinyl ether) macromonomer **24** was obtained by the living polymerization using the adduct **23** between hydrogen iodide and vinyl ether having a methacryl group [42]. Cyclopolymerization of divinyl ether, 1,2-bis[2-(2-vinyloxyethoxy)ethoxy]benzene (**25**) using **23** and iodine as initiator produced methacryl-terminated poly(benzo-19-crown-6) with narrow molecular weight distribution [43].

It is well known that aluminum isopropoxide is a very effective initiator for $\epsilon$-CL polymerization, where the propagation step is a living system that leads to a monodisperse (uniform) polymer. An aluminum alkoxide bearing a double bond of the methacrylic acid **26**, prepared by the equimolar reaction of triethylaluminum and 2-hydroxylethylmethacrylate (HEMA) [44] was used for the preparation of poly($\epsilon$-CL) macromonomer via initiator method. The polymerization was of highly living nature and the molecular weight distribution of the resulting macromonomer was narrow. Furthermore, $\alpha,\omega$-dimethacrylic poly($\epsilon$-CL) was obtained by the termination of the living end with methacryloyl chloride.

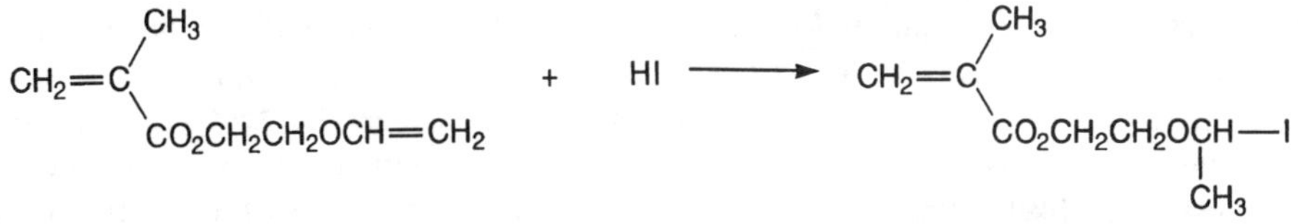

$$CH_2{=}C(CH_3)CO_2CH_2CH_2OCH{=}CH_2 + HI \longrightarrow CH_2{=}C(CH_3)CO_2CH_2CH_2OCH(CH_3){-}I$$

**23**

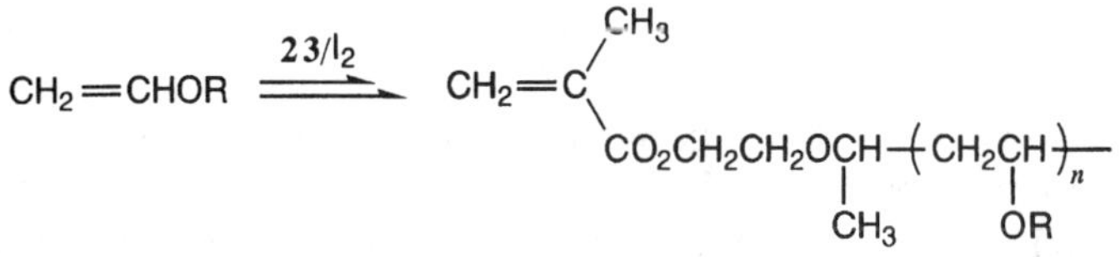

$$CH_2{=}CHOR \xrightarrow{23/I_2} CH_2{=}C(CH_3)CO_2CH_2CH_2OCH(CH_3){+}CH_2CH(OR){+}_n$$

**24**

$$\xrightarrow{23/I_2} CH_2{=}C(CH_3)CO_2CH_2CH_2OCH(CH_3){-}[\ \ ]_n$$

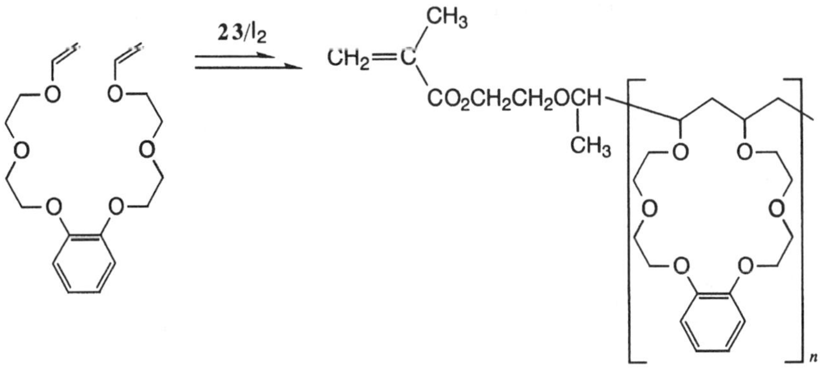

**25**

$$CH_2{=}C(CH_3)CO_2CH_2CH_2OAlEt_2$$

**26**

Preparation of methacryl-type macromonomers by terminator method in the living anionic polymerization was started by Milkovich [1]; first anionic polymer end was converted into alcoholate by the reaction of EO, followed by the end-capping with methacryloyl chloride. In a similar manner, methacryl-type macromonomers from styrene [17,45]and 2-vinylpyridine[17,21]were prepared. In case of 4-vinylpyridine[46] and EO [9,20],the direct termination by methacryloyl chloride produced corresponding macromonomers. Methacryl-type polysiloxane macromonomer was synthesized by the termination of the anionic living end with (methacryloxyalkyl)dimethylsilyl chloride [22,47].

In living cationic polymerization, (meth)acrylic acids or their salts are effective terminators for introduction of (meth)acryl group at the chain end. Methacryl-type polyamine macromonomer was prepared by cationic ring-opening polymerization of 1-t-butylaziridine followed by the termination with methacrylic acid [48]. Sodium methacrylate was used as a terminator for the preparation of polyTHF macromonomer [49]. Quantitative termination of living poly(2-oxazoline) end with triethylammonium[50] and 2,6-lutidinium[51](meth)acylates produced (meth)acryl-type macromonomers.

Recently, it was reported that cationic polymerization initiated by a strong acid in the presence of a nucleophile proceeded in a living nature. The nucleophile acted as a stabilizer of the active chain end. A mixture of trifluoromethanesulfonic acid and thiolane induced the living polymerization of vinyl ether, followed by termination with 2-(dimethylamino)ethyl methacrylate to give poly(vinyl ether) macromonomer **27** [52].

$$CH_2{=}CHOR \xrightarrow{CF_3SO_3H\ /\ \text{thiolane}} H{+}CH_2CH(OR){+}_{n-1}CH_2CH(OR){-}S^{+}(\text{thiolane})\ CF_3SO_3^{-}$$

$$\xrightarrow{(CH_3)_2NCH_2CH_2O_2C(CH_3)C{=}CH_2} H{+}CH_2CH(OR){+}_{n}\overset{+}{N}(CH_3)_2CH_2CH_2O_2C(CH_3)C{=}CH_2\ \ CF_3SO_3^{-}$$

**27**

A polymer having hydroxyl group at the chain end is readily converted into methacyl-type macromonomer by the reaction with methacryloyl chloride in the presence of amine. Methacryl-type macromonomers of aromatic polyether sulfone [28], poly(2-oxazoline)[50] and poly(isobutene)[53,54] were prepared from corresponding $\alpha$-hydroxyl prepolymer.

Methacryl-type macromonomers of the carboxyl-terminated (prepared using a chain transfer process) prepolymers of styrene[55] and N-vinylpyrrolidone[33]were prepared by the treatment with glycidylmethacrylate. $\alpha$-Amino polystyrene **16** was converted into the macromonomer by the reaction with glycidyl methacrylate[35]. Isocyanoethyl methacrylate was also an effective agent for the preparation of polyMMA[56] and poly($\epsilon$-CL) [12] macromonomers from the corresponding hydroxyl-terminated prepolymers.

Acryl-type poly(vinyl chloride) (CV) and poly(vinylidene chloride) ($CV_2$) macromonomers were prepared from their telomers. The polymerization in the presence of carbon tetrachloride produced the telomer **28**. Then **28** was reacted with allyl acetate in the presence of $CuCl_2$ followed by hydrolysis with $H_2SO_4$ to produce the hydroxyl-terminated polymer **29**. Then the reaction of **29** with acrylic acid produced the acryl-type polyCV or polyCV$_2$ macromonomer **30** [57]. Methacryl-type polyCV macromonomer was synthesized from the prepolymer having carboxyl and hydroxyl groups, which was prepared using thioglycolic acid and 2-mercaptoethanol, respectively [58].

$$n\ CH_2{=}CRCl + CCl_4 \xrightarrow{R\cdot} CCl_3{-}\!\left(CH_2{-}CRCl\right)_n\!{-}Cl$$

(R=H, Cl) **28**

$$\xrightarrow{CH_2{=}CHCH_2O_2CCH_3} \xrightarrow{H_2SO_4} CCl_3{-}\!\left(CH_2{-}CRCl\right)_n\!{-}CH_2CHClCH_2OH$$

**29**

$$\xrightarrow{CH_2{=}CHCO_2H} CCl_3{-}\!\left(CH_2{-}CRCl\right)_n\!{-}CH_2CHClCH_2O_2CCH{=}CH_2$$

**30**

Polystyrene prepared by using **14** as a functional initiator possesses an amidino group at the chain end. This prepolymer was transferred into the methacryl-type macromonomer by the reaction with glycidylmethacrylate[59].

$\alpha,\omega$-Dimethacryloyl poly(epoxide-amine) macromonomer **31** was synthesized by reaction of bis-phenol A diglycidyl ether and methacrylic acid followed by addition polymerization of unreacted epoxide groups with primary amines[60].

$$CH_2{=}C(CH_3){-}CO_2CH_2CH(OH)CH_2{-}\left[O{-}C_6H_4{-}C(CH_3)_2{-}C_6H_4{-}OCH_2CH(OH)CH_2N(C_6H_5)CH_2CH(OH)CH_2\right]_n{-}O{-}C_6H_4{-}C(CH_3)_2{-}C_6H_4{-}OCH_2CH(OH)CH_2O_2C{-}C(CH_3){=}CH_2$$

**31**

Michael-type polyaddition reactions using compounds possessing a diacrylamide function gave an acrylamide-type macromonomer[61,62]. For example, the polyaddition of 1,4-bis-acryloylpiperazine and N,N'-dimethylethylenediamine produced poly(amide-amine) macromonomer **32** [61].

$$CH_2{=}CH{-}C({=}O){-}N\langle\rangle N{-}C({=}O){-}CH{=}CH_2 + HN(CH_3)CH_2CH_2NH(CH_3)$$

$$\longrightarrow CH_2{=}CH{-}C({=}O){-}N\langle\rangle N{-}C({=}O){-}CH_2CH_2{-}\left[N(CH_3)CH_2CH_2N(CH_3)CH_2CH_2{-}C({=}O){-}N\langle\rangle N{-}C({=}O){-}CH_2CH_2\right]_n{-}N(CH_3)CH_2CH_2NH(CH_3)$$

**32**

## 2.3 Macromonomers Having Other Vinyl Groups

The polymerizable groups of macromonomers reported are mainly styryl and (meth)acryl types. This section deals with macromonomers bearing other polymerizable vinyl groups.

Vinyl monomers are divided into conjugated and non-conjugated ones in the stand point of radical polymerization. Typical monomers of the former are styrene and (meth)acryl compounds. These monomers are readily copolymerized radically with macromonomers having styryl and (meth)acryl groups. The latter examples are ethylene, propylene, vinyl chloride, and vinyl acetate. Such monomers have different reactivities in the radical copolymerization compared with conjugated monomers, and hence, are hardly copolymerized with the macromonomers as described above. Therefore, macromonomers having non-conjugated polymerizable groups such as 1-alkenyl, vinyl ester, vinyl silane, etc. are necessary for the copolymerization of non-conjugated monomers in the preparation of graft copolymers of a well-defined structure by the macromonomer technique.

For the preparation of macromonomers possessing alkenyl group, a polymerizable moiety was introduced by the initiation with alkenyl lithium or by the termination of the anionic living end with alkenyl halide. Initiators used for the preparation of the polystyrene macromonomer were vinyl and allyl lithium[2,3]. Allyl halide was employed as a terminator for the anionic living end of poly(styrene)[1] and polyMMA[19]. Dimethyl(1-hexenyl)silyl chloride was also an effective terminator for the living end of $D_3$ [63].

Butadiene-type macromonomer **33** was first synthesized by ring-opening polymerization of 2-oxazoline initiated by allylic dihalide, 2,3-bis(bromomethyl)-1,3-butadiene [64]. The resultant macromonomer possessed the polymerizable group in the main chain. Polystyrene macromonomer bearing a vinyl ether group was prepared by the termination of polystyrene living end with 2-chloroethyl vinyl ether [1]. The reaction

$$\text{2-R-oxazoline} \xrightarrow{\text{BrCH}_2\text{C}(=\text{CH}_2)\text{C}(=\text{CH}_2)\text{CH}_2\text{Br}} -\!\!(\text{CH}_2\text{CH}_2\text{N}(\text{COR}))_m\!-\!\text{CH}_2\text{C}(=\text{CH}_2)\text{C}(=\text{CH}_2)\text{CH}_2\!-\!(\text{N}(\text{COR})\text{CH}_2\text{CH}_2)_n\!-$$

**33**

of potassium alcoholate-terminated polyisobutene with 2-chloroethyl vinyl ether gave polyisobutene macromonomer[65]. $\alpha,\omega$-Divinyl ether of polyEO was prepared by the reaction of potassium alcoholate of polyEO with acetylene at 180°C [66]. The complete conversion of hydroxyethyl to vinylether group was confirmed by NMR analysis. These vinyl ether type macromonomers are used as a co-monomer for cationic vinyl polymerization.

Functional initiator **35** prepared from 2-(allyloxy)ethyl vinyl ether (**34**), induced living polymerization of vinyl ether in the presence of iodine to produce macromonomer **36** having allyl ether group[67]. The termination of living poly(vinyl ether) with allyl alcohol also produced allyl-ether type macromonomer[52]. PolyTHF macromonomer bearing allyl-ether was prepared by ring-opening polymerization of THF initiated by a mixture of allyl halide and silver hexafluorophosphate[15].

$$\underset{\textbf{34}}{CH_2{=}CHCH_2OCH_2CH_2OCH{=}CH_2} + HI$$

$$\longrightarrow \underset{\textbf{35}}{CH_2{=}CHCH_2OCH_2CH_2OCH(CH_3){-}I}$$

$$CH_2{=}CHOR \xrightarrow{\textbf{35}/I_2} \underset{\textbf{36}}{CH_2{=}CHCH_2OCH_2CH_2OCH(CH_3){-}\!\!(CH_2CH(OR))_n\!-}$$

Allyl glycidyl ether is reactive for the introduction of allyl ether at the polymer end. Amino-[35] and amidino-[59] terminated polystyrenes were converted into the allyl-ether type macromonomer by the reaction with allyl glycidyl ether.

Vinylsilane-type macromonomers were prepared by the terminator method. Cationic living polymerization of THF followed by the termination with sodium 3-(dimethylvinylsilyl)propyloxide produced vinylsilyl-terminated polyTHF **37**[68]. In case of anionic living polymerization of $D_3$, the termination of 3-(dimethylvinylsilyl)propyl chloride yielded vinyl-silane type macromonomer[69].

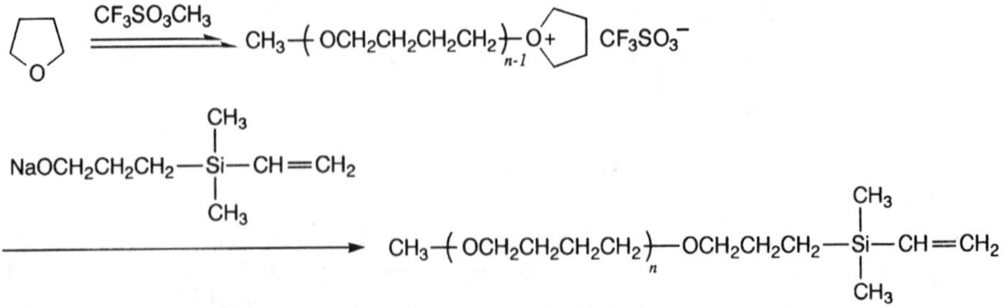

$$\text{THF} \xrightarrow{CF_3SO_3CH_3} CH_3\text{-}(OCH_2CH_2CH_2CH_2)_{n-1}\text{-}O^{+}(\text{THF ring})\ CF_3SO_3^-$$

$$\xrightarrow{NaOCH_2CH_2CH_2\text{-}Si(CH_3)_2\text{-}CH{=}CH_2} CH_3\text{-}(OCH_2CH_2CH_2CH_2)_n\text{-}OCH_2CH_2CH_2\text{-}Si(CH_3)_2\text{-}CH{=}CH_2$$

**37**

Macromonomers possessing vinyl ester group were prepared by termination of the living end of polystyrene with vinyl chloroacetate[1] or initiation of 2-oxazoline with vinyl iodoacetate[70]. Hydroxyl-terminated polyVC was transferred into the corresponding vinyl-ester type macromonomer by the reaction with vinyl chloroformate[58].

Polyisobutene macromonomer **38** having dicyclopentadiene group was prepared by the carbo-cationic polymerization initiated by 1-chlorocyclopentadiene in the presence of chlorodiethylaluminum[71].

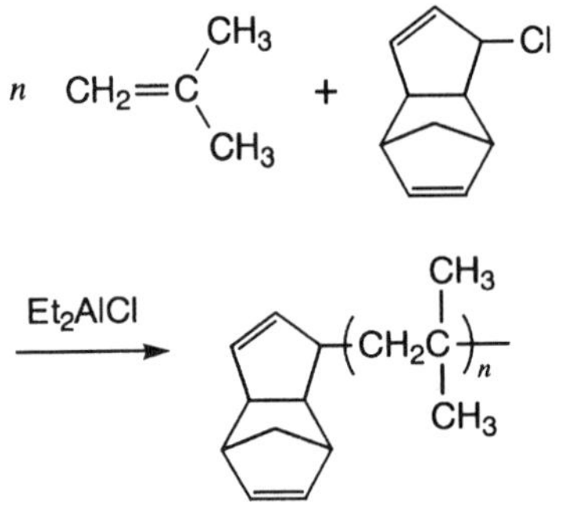

$$n\ CH_2{=}C(CH_3)_2 + \text{Cl-dicyclopentadiene} \xrightarrow{Et_2AlCl} \text{dicyclopentadienyl}\text{-}(CH_2C(CH_3)_2)_n\text{-}$$

**38**

Radical polymerization of MMA in the presence of methyl $\alpha$-(bromomethyl)acrylate yielded polyMMA macromonomer **39** bearing the 2-methoxycarbonylallyl end group through chain reaction involving bimolecular termination[72].

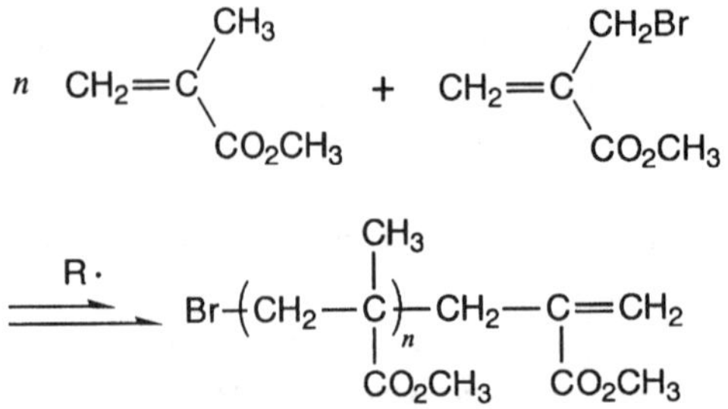

$$n\ CH_2{=}C(CH_3)(CO_2CH_3) + CH_2{=}C(CH_2Br)(CO_2CH_3) \xrightarrow{R\cdot} Br\text{-}(CH_2\text{-}C(CH_3)(CO_2CH_3))_n\text{-}CH_2\text{-}C(CO_2CH_3){=}CH_2$$

**39**

PolyEO macromonomer having fumaric half-ester was prepared from hydroxyl-terminated polyEO and maleic anhydride in the presence of 4-dimethylaminopyridine[73]. The half-ester macromonomer was converted into fumaric diester polyEO macromonomer by esterification with monoether of polyEO.

Polyguanamine macromonomer **41** was synthesized by self-addition reaction of 2-amino-4-(m-amino-N-alkylanilino)-6-isopropenyl-1,3,5-triazine (**40**)inN,N-dimethyl-2 acetamide at 120°C for ca. 300h [74]. The molecualr weight was 3000-4000 estimated by GPC.

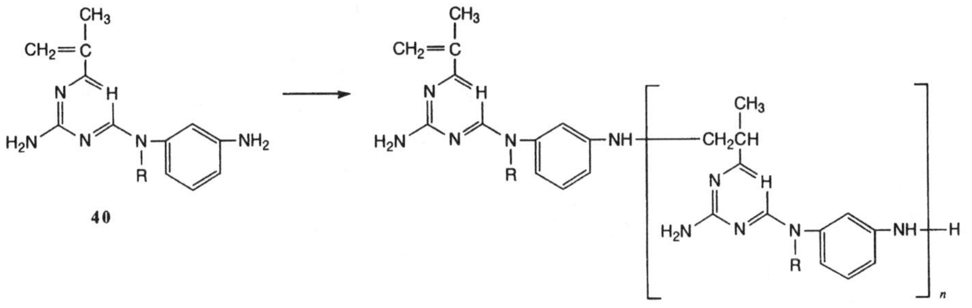

40

41

## 2.4 Macromonomers Having Ring-Opening Polymerizable Group

Polymerizable groups other than vinyl functions, such as cyclic groups for ring-opening polymerization and bifunctional groups at one polymer end for step-growth polymerization are known. So far, macromonomers having oxirane, 2-oxazoline, and norbornene groups capable of ring-opening polymerization have been reported.

$\alpha$-Glycidyl ether polyisobutylene was prepared by quantitative reaction between phenol-capped polyisobutylene and epichlorohydrin in the presence of tetra-n-butylammonium hydrogen sulfate[75]. Amidino-terminated polystyrene was converted into $\alpha$-glycidylether macromonomer by the reaction with diglycidylether[59]. Terminal allyl ether of poly(vinyl ether) was transferred into glycidyl ether group by the oxidation using a peracid[67].

Anionic ring-opening polymerization of EO by lithium salt of 2-(hydroxyphenyl)-2-oxazoline produced polyEO macromonomers **42** bearing polymerizable 2-oxazoline group[76]. The molecular weight was controlled by varying the feed ratio. Carbanion species **43** derived by 2-methyl-2-oxazoline initiated the polymerization of methyl acrylate to produce 2-oxazoline-terminated macromonomer **44**[77].

Norbornene is well known to be subjected to ring-opening olefin methathesis polymerization. Polystyrene macromonomers bearing norbornene group was prepared by the termination of anionic living end with 5-norbornene-2-carbonyl chloride [78].

Monochloro functionalized polydimethylsiloxane was converted into the corresponding norbornene-type macromonomer by the reaction with 2-metalated 5-norbornene-2-carbonitrile[79].

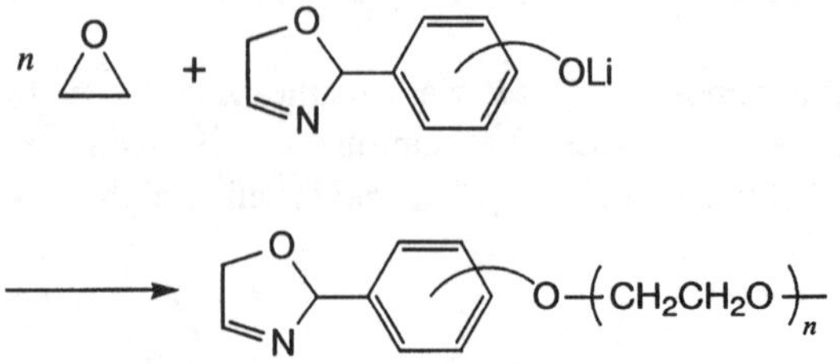

*n* (ethylene oxide) + (oxazoline)–C₆H₄–OLi → (oxazoline)–C₆H₄–O–(CH₂CH₂O)ₙ–

**42**

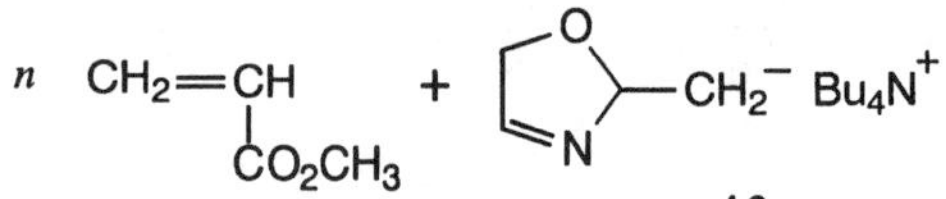

*n* $CH_2{=}CH(CO_2CH_3)$ + (oxazoline)–$CH_2^-$ $Bu_4N^+$

**43**

→ (oxazoline)–$CH_2$–$(CH_2CH(CO_2CH_3))_n$–

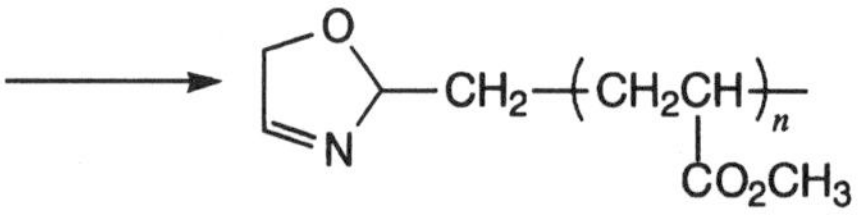

**44**

## 2.5 Macromonomers Having Other Polymerizable Groups

Syntheses of dicarboxyl, dihydroxyl, and diamino-terminated macromonomers have been reported as a comonomer of step-growth copolymerization. Radical polymerizations of (meth)acryl esters such as MMA and HEMA in the presence of thiomalic acid produced dicarboxyl-terminated macromonomers **45**[80,82]. For the preparation of aromatic polyamide or polyester graft copolymers, aromatic dicarboxyl terminated polyMMA macromonomer **46** was prepared;[83] hydroxyl-terminated prepolymer, prepared by radical polymerization of MMA in the presence of 2-mercaptoethanol, was reacted with trimellitic anhydride to form macromonomer **46**.

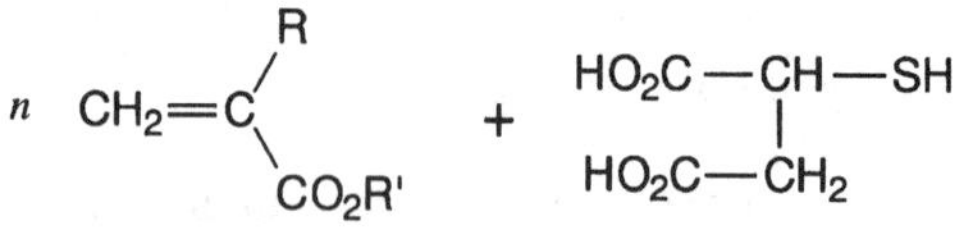

*n* $CH_2{=}C(R)(CO_2R')$ + $HO_2C{-}CH(SH){-}CH_2{-}CO_2H$

(R=H, $CH_3$)

$\xrightarrow{\text{AIBN}}$ $HO_2C{-}CH(CH_2CO_2H){-}S{-}(CH_2{-}C(R)(CO_2R'))_n{-}H$

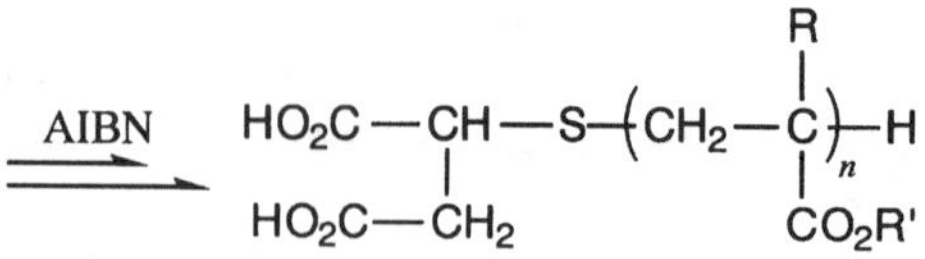

**45**

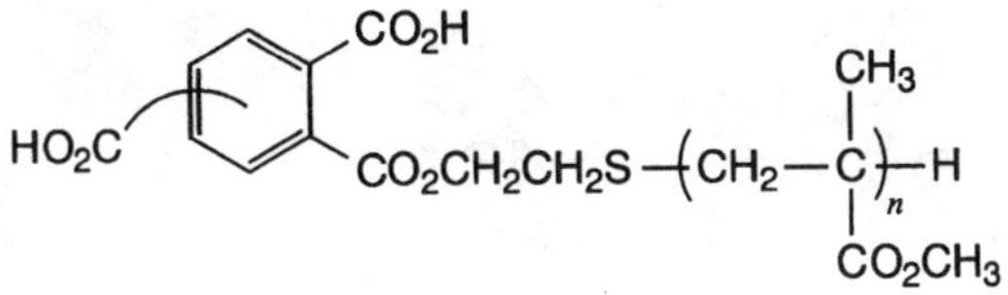

$$HO_2C-C_6H_3(CO_2H)-CO_2CH_2CH_2S\text{-}(CH_2-C(CH_3)(CO_2CH_3))_n\text{-}H$$

**46**

Glycol-type polyMMA macromonomer **47** was synthesized by radical polymerization using $\alpha$-thioglycerol as a chain transfer agent[84]. Living cationic ring-opening polymerization of 2-oxazoline followed by quantitative termination with diethanolamine yielded macromonomer bearing diol end-group **48**[85]. Similarly, dihydroxyl-terminated polyTHF or poly(N-t-butylaziridine) macromonomer was obtained by end-capping with diethanolamine[86]. Monodisperse glycol-type poly(dimethylsiloxane) macromonomer was prepared by hydrosilylation of poly(dimethylsiloxane) containing a silane group at the chain end with 3-allyloxy-1,2-propanediol[87].

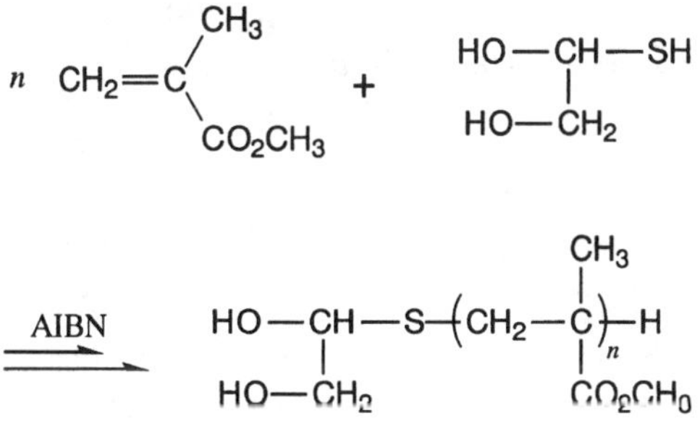

$$n\ CH_2=C(CH_3)CO_2CH_3 + HO-CH(CH_2OH)-SH \xrightarrow{AIBN} HO-CH(CH_2OH)-S\text{-}(CH_2-C(CH_3)(CO_2CH_3))_n\text{-}H$$

**47**

$$\text{2-R-2-oxazoline} \xrightarrow{CH_3OTs} CH_3\text{-}(N(COR)CH_2CH_2)_{n-1}\text{-}N^+(\text{oxazolinium, R}) \ TsO^-$$

$$\xrightarrow[-TsOH]{HN(CH_2CH_2OH)_2} CH_3\text{-}(N(COR)CH_2CH_2)_n\text{-}N(CH_2CH_2OH)_2$$

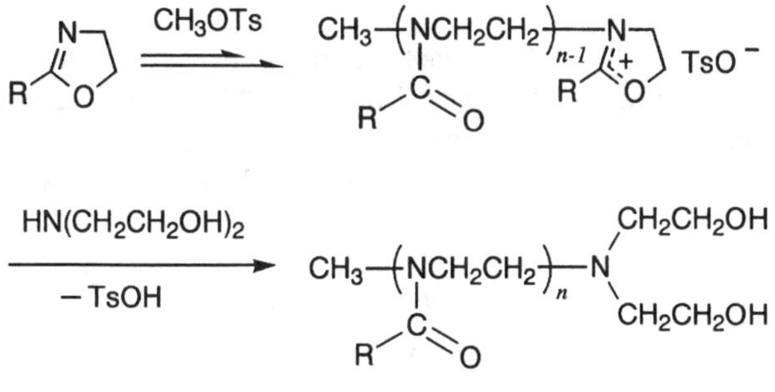

**48**

Hydroquinone-type polystyrene macromonomer **49** was synthesized by utilizing anionic living polymerization; polystyrene living end was reacted with 1,1-diphenylethylene followed by the treatment with 2,5-dimethoxybenzyl bromide and ether cleavage with boron tribromide resulted in the formation of hydroquinone-type macromonomer **49** [88]. Direct termination of polystyrene lithium 1,1-bis(4-methoxyphenyl)ethylene followed by de-protection with boron tribromide produced hydroquinone-type macromonomer.

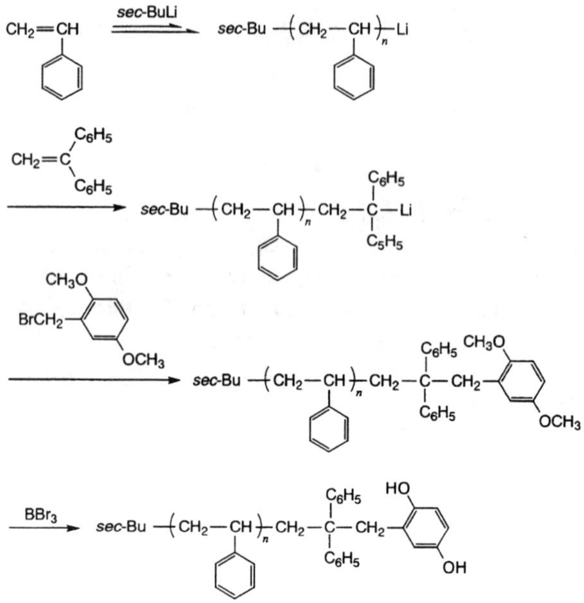

49

Ullmann reaction of 1-bromo-2,5-dimethoxybenzene and potassium 4-bromophenolate produced methyl-group-protected poly(oxy-1,4-phenylene)prepolymer. This prepolymer was converted into hydroquinone-type poly(oxy-1,4-phenylene) macromonomer **50** by the reaction with trimethylsilyl iodide[89].

Methyl 1-trimethylsiloxy-2-[2,5-bis(trimethylsiloxy)phenyl]vinylether (**51**) was used as an initiator for the group-transfer polymerization of MMA to give macromonomer with aromatic dihydroxyl group[90]. Quantitative removal of the silyl-protective group was achieved by fluoride-catalyzed cleavage to give the deprotected macromonomer **52**.

3,5-Diaminophenyl-terminated polydimethylsiloxane macromonomer **53** was synthesized by quantitative hydrosilylation of poly(dimethylsiloxane) having a silane group at the chain end with 3,5-dinitrostyrene, followed by the reduction of nitro groups[91].

## 3. APPLICATION OF MACROMONOMERS

Preparation of graft copolymers by macromonomer technique possesses several advangates compared with that by conventional methods; molecular weight and number of grafted polymer chain are readily controlled and combinations between backbone and graft polymer chains can be selected for desired purposes. Recently, design and synthesis of functional polymers via macromonomer technique have been extensively

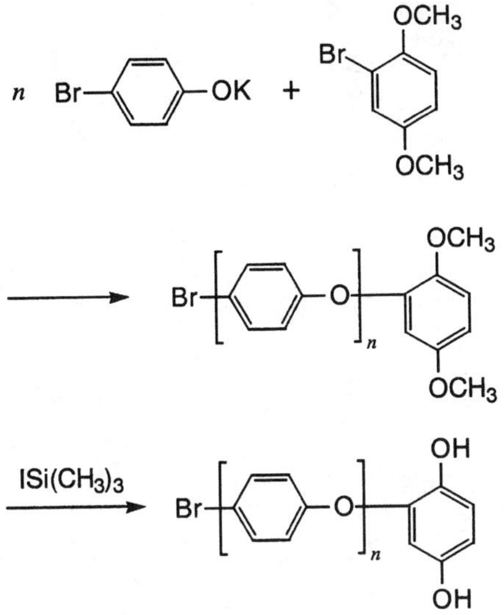

$n$ Br–C6H4–OK + 2-bromo-1,4-dimethoxybenzene

$\xrightarrow{}$ $Br{-}[C_6H_4{-}O]_n{-}C_6H_3(OCH_3)_2$

$\xrightarrow{ISi(CH_3)_3}$ $Br{-}[C_6H_4{-}O]_n{-}C_6H_3(OH)_2$

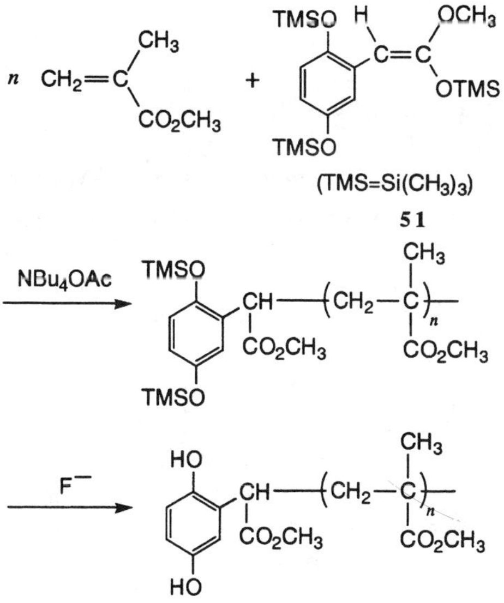

**50**

$n$ $CH_2{=}C(CH_3)CO_2CH_3$ + 2,5-(TMSO)$_2C_6H_3$–CH=C(OCH$_3$)(OTMS)

(TMS=Si(CH$_3$)$_3$)

**51**

$\xrightarrow{NBu_4OAc}$ 2,5-(TMSO)$_2C_6H_3$–CH(CO$_2$CH$_3$)–$(CH_2{-}C(CH_3)(CO_2CH_3))_n$–

$\xrightarrow{F^-}$ 2,5-(HO)$_2C_6H_3$–CH(CO$_2$CH$_3$)–$(CH_2{-}C(CH_3)(CO_2CH_3))_n$–

**52**

investigated. In this chapter, some applications of graft copolymers prepared using macromonomers are described.

$$\text{3,5-}(O_2N)_2C_6H_3\text{-}CH{=}CH_2 + H\text{-}(Si(CH_3)_2\text{-}O)_n\text{-}Si(CH_3)_2\text{-}CH_3$$

$$\xrightarrow{H_2PtCl_6} \text{3,5-}(O_2N)_2C_6H_3\text{-}CH_2CH_2\text{-}(Si(CH_3)_2\text{-}O)_n\text{-}Si(CH_3)_2\text{-}CH_3$$

$$\xrightarrow{H_2/Pd\text{-}C} \text{3,5-}(H_2N)_2C_6H_3\text{-}CH_2CH_2\text{-}(Si(CH_3)_2\text{-}O)_n\text{-}Si(CH_3)_2\text{-}CH_3$$

**53**

## 3.1 Graft Copolymers for Polymeric Surfactants and Surface Modifiers

Amphiphilic graft copolymers are conveniently synthesized by copolymerization of a hydrophobic monomer with a hydrophilic macromonomer and vice versa. The resulting copolymers are expected to show surfactant properties.

An amphiphilic graft copolymer synthesized by HEMA with methacryl-terminated polystyrene macromonomer was found to form micellar aggregate in methanol by $^1H$ NMR analysis; spectrum of HEMA-rich graft copolymer (85 wt% of polyHEMA content) in $CD_3OD$ showed only peaks due to polyHEMA segments[92]. By the addition of $CDCl_3$ to the solution of the graft copolymer, peaks ascribed to polystyrene segments appeared. These data clearly indicate that the polystyrene segments insoluble in methanol constitute a micelle core which behaves as a solid on the NMR time scale, consequently their peaks were too broad to be observed. The reverse phenomenon was observed in the case of polystyrene-rich graft copolymer (72 wt% of polystyrene content); in the $^1H$ NMR spectrum measured in $CDCl_3$, peaks due to the polystyrene segments were only observed, indicating the formation of a micelle with a polyHEMA core surrounded by polystyrene segments.

Two kinds of amphiphilic graft copolymers were prepared by replacing the backbone and the branched components by macromonomer method and their colloidal properties to form micellar aggregates were examined[93]. The amphiphilic macromonomers

**54** and **55** consisted of hydrophobic polyMMA chain and hydrophilic polymer having quaternary ammounium group in the side chain. In order to examine the effect of molecular structure of **54** and **55**, $^1$H NMR spectra were measured in a mixed solvent of water and acetone. In both cases, the integrals of peaks due to polyMMA decreased comparable with those ascribed to dimethylammonium group as the water content of the mixed solvent increased. These data imply that the graft copolymers formed a micelle with a hydrophobic polyMMA core in a water-rich solvent. By changing the composition of the medium, the integral ratio between carbomethoxy protons of polyMMA and dimethylammonium protons in **55** changed largely, compared with that in **54**. This is because **55** possesses a hydrophobic polyMMA chain in the main chain, and hence, forms a more rigid micelle in a water-rich solvent as illustrated in Figure 1.

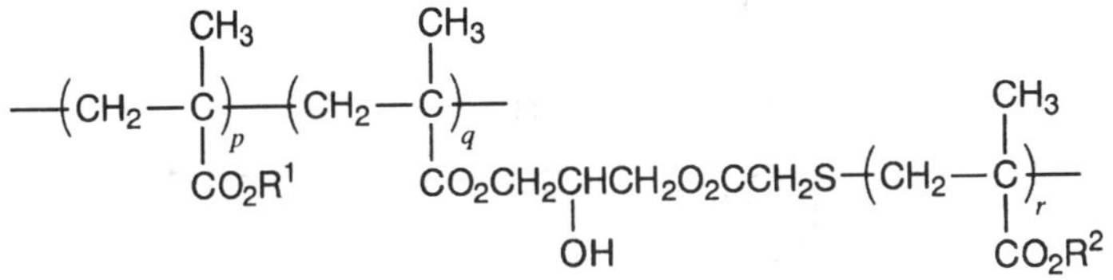

$$-\!\left(CH_2-\underset{CO_2R^1}{\overset{CH_3}{C}}\right)_p\!\!-\!\left(CH_2-\underset{CO_2CH_2\underset{OH}{C}HCH_2O_2CCH_2S-\left(CH_2-\underset{CO_2R^2}{\overset{CH_3}{C}}\right)_r-}{\overset{CH_3}{C}}\right)_q\!\!-$$

**54** : $R^1=CH_2CH_2\overset{+}{N}(CH_3)_2C_2H_5\,Br^-$, $R_2=CH_3$

**55** : $R^1=CH_3$, $R^2=CH_2CH_2\overset{+}{N}(CH_3)_2C_2H_5\,Br^-$

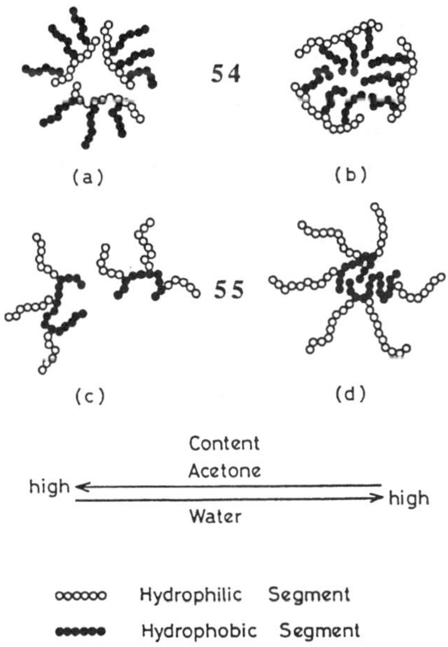

54

(a) (b)

55

(c) (d)

Content

high ← Acetone

Water → high

Hydrophilic Segment

Hydrophobic Segment

***Figure 1*** *Schematic models for the formation of micelle from the copolymers* ***54*** *and* ***55*** *in water-acetone mixed solvent.*

used as a steric stabilizer. Instead of using a highly polar polymer, hydrophilic poly(2-oxazoline) macromonomer **6a** was used as stabilizer as well as comonomer in the dispersion copolymerization of MMA in an aqueous methanol solution. The copolymerization using **6a** gave monodisperse polyMMA particles in the micron size and their surface was smooth (Figure 2)[100, 101].

$$CH_2{=}CH-C_6H_4-CH_2-\left(N(C(=O)R)CH_2CH_2\right)_n-$$

**6a** : R=Me (Mw=3900)
**6b** : R=Me (Mw=1200)
**6c** : R=Me (Mw=700)
**6d** : R=Et (Mw=3500)
**6e** : R=Et (Mw=1200)
**6f** : R=*n*-Pr (Mw=4200)
**6g** : R=*i*-Pr (Mw=4200)

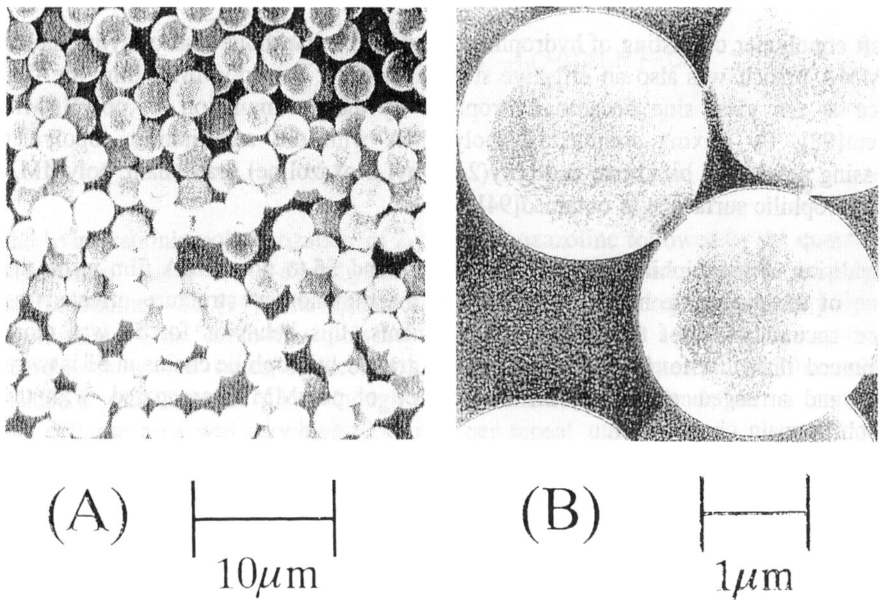


*Figure 2 SEM photographs (A and B) of polyMMA particles prepared by dispersion copolymerization of MMA with **6a** of 0.2 wt%.*

Effects of molecular weight and substituent of the macromonomer on the particle formation were examined. Macromonomers **6b** and **6d** as well as **6a** produced monodisperse particles. In each case, the particle size increased with decreasing the macromonomer concentration (Figure 3). In the same concentration of the macromonomer, the particle size by **6a** was smaller than those by **6b** and **6d**. In the copolymerizations using **6c**, **6e**, and **6f**, polydisperse particles were obtained. These data indicate that a macromonomer of higher hydrophilicity or of higher molecular weight stabilizes polymer particles more effectively. Furthermore, poly(2-oxazoline) chains were found to be locally present on the surface from ESCA analysis.

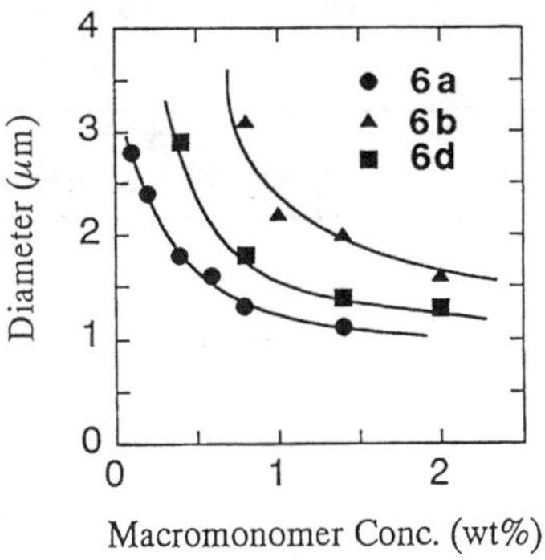


*Figure 3 Effect of the macromonomer concentration on the particle size in the dispersion copolymerization of MMA with* ***6****.*

In case of dispersion polymerization of MMA using a highly polar homopolymer as a stabilizer, more than 10 wt% of the stabilizer of higher molecular weight (Mw$\geq$ 4x$10^4$) was always necessary to obtain micron-size particles with narrow size distribution (Table 1). On the other hand, the present method provided that the small concentration of **6** with relatively low molecular weight is enough to produce such polymer particles, i.e., the macromonomer acted as a very efficient stabilizer for the preparation of monodisperse micron-size polymer particles. Especially, **6a** was efficient with a very small concentration ($<$ 1 wt% for the total monomers), which is of one-order's magnitude less compared with that in conventional dispersion polymerizations.

The dispersion copolymerization of MMA with butadiene-type poly(2-oxazoline) macromonomer **33** (R=$CH_3$, Mw=2300) also produced micron-size particles with narrow size distribution[64]. The concentration of **33** necessary for preparation of monodisperse particles was higher than that of **6a** or **6b**, indicating that the copolymerizability of **33** is lower than that of MMA or of the styryl-type macromonomer.

In the dispersion copolymerization of styrene with poly(2-oxazoline) macromonomer **6** in aqueous ethanol solution, **6a, 6b, 6d** and **6g** were effective for the preparation of monodisperse polystyrene particles in the micron range[102]. The size of the particles by macromonomers with molecular weight of ca. 4000 (**6a**, **6d**, **6g**) was almost the same, indicating that the substituent of the macromonomer does not affect the particle size.

Very recently, monodisperse particles of highly hydrophilic polymers such as poly(acrylamide) and poly(N-vinylformamide) were prepared in the dispersion

**Table 1.** *Comparison of ability to stabilize polymer particles between 2-oxazoline homopolymer and poly(2-oxazoline) macromonomer in the dispersion polymerization of MMA*

| Stabilizer | Mw | Conc. of stabilizer | Dn(μm)[a] | Dw/Dn[a] |
|---|---|---|---|---|
| **6a** | 3,900 | 0.2[b] | 2.4 | 1.01 |
| **6b** | 1,200 | 0.8[b] | 3.1 | 1.03 |
| **6d** | 3,500 | 0.8[b] | 3.1 | 1.03 |
| PEtOZO[c] | 50,000 | 30[d] | 3.7 | 1.03 |
| PEtOZO[c] | 200,000 | 10[d] | 4.1 | 1.22 |
| PEtOZO[c] | 200,000 | 20[d] | 3.8 | 1.03 |
| PEtOZO[c] | 500,000 | 5[d] | 4.5 | 1.15 |
| PEtOZO[c] | 500,000 | 10[d] | 3.9 | 1.02 |

a) Determined by SEM.
b) Weight % for the total monomer. The concentration is the lowest necessary for the preparation of relatively monodisperse PMMA particles.
c) Poly(2-ethyl-2-oxazoline)
d) Weight % for MMA.

polymerization in polar media. The preparation of monodisperse polymer particles by the dispersion copolymerization using hydrophilic macromonomer was applied to that of such particles[103]. In this system, the concentration of **6** necessary for the preparation of monodisperse particles was lower than that of homopolymer-type stabilizer, indicating that the macromonomer is also an efficient stabilizer for the preparation of hydrophilic polymer particles.

Akashi et al.[33] prepared sub-micron size polystyrene or polyMMA particles by the dispersion copolymerization using much amount of styryl- or methacryl-type poly(N-vinylpyrrolidone) macromonomer in ethanol. The resulting dispersion of the particles was stable for a long period. Similarly, the dispersion copolymerization of MMA with methacryl-terminated polyEO in aqueous ethanol solution gave the stable dispersion[104]. The resulting particles were in sub-micron range. As a possible application, polymeric drugs dispersed stably in water were prepared[105]. The dispersion copolymerizations of vinyl monomers containing pharmaceutically active purine and pyrimidine derivatives with styryl-type poly(N-vinylpyrrolidone) macromonomer in methanol gave stable dispersion of sub-micron size polymer particles. The vinyl monomers employed were 7-(2-methacryloyloxy)ethyltheophylline and N-(2-methacryloyloxyethyl)-5-fluorouracil. The resulting particles dispersed in aqueous solution gradually released the corresponding drug components by the hydrolysis. This water dispersed polymeric drugs are expected to be of large advantage in medical treatment compared with insoluble ones.

Monodisperse particles of less than 1$\mu$ possessing hydrophilic polymer segments on the surface were prepared by soap-free emulsion and emulsion copolymerizations. The soap-free emulsion copolymerization of styrene with methacryl-terminated polyEO macromonomer was investigated[106,107]. The polymerization using the macromonomer with short polyEO chains resulted in the formation of uneven-shaped particles. On the other hand, the macromonomer with long polyEO chains served to form smooth and spherical copolymer particles. The size of dry particles measured by transmission electron micrograph was smaller than the hydrodynamic size and this tendency was pronounced in case of the particles having long polyEO chains. These data indicate that hydrophilic polyEO chains are locally present on the surface and the particles are highly hydrated.

Hydrophilic styryl-type poly(2-oxazoline) macromonomer **6** was soap-free copolymerized with styrene to produce monodisperse polymer particles with narrow size distribution[108]. The particle size was smaller than that of pure polystyrene particles prepared in a similar way. From the analysis of the surface composition of the particles by ESCA, poly(2-oxazoline) chains were enriched on the surface.

Polymerizations of 2-oxazoline monomers proceed through a highly living nature. By utilizing this nature, a block-type amphiphilic macromonomer **58** having a polymerizable styryl group was synthesized by one-pot two-stage procedure[109]. The hydrophilic segment comes from polymerization of 2-methyl-2-oxazoline. 2-n-Butyl-2-oxazoline was used for the construction of the hydrophobic segment. Macromonomers **58a** and **58b** possess each segment of the almost same chain length, the position of the segments being reversed in **58a** and **58b**. These macromonomers exhibited good surfactant properties evaluated by the measurement of the surface tension in water. Emulsion copolymerization of styrene with macromonomer **58a** possessing the polymerizable group attached to the hydrophobic segment gave monodisperse polymer particles in sub-micron range (Figure 4). The macromonomer acted as polymeric surfactant as well as comonomer. In case of the emulsion copolymerization using **58b**, on the other hand, the resulting particle was larger than that by **58a** and the size distribution was broader. These results indicate that the position of styryl group affects the present copolymerization. Schematic representation of the macromonomers in micelle is shown in Figure 5. In the micelle of amphiphilic macromonomer **58a** before the copolymerization, the styryl group of **58a** is inside the micelle (Figure 5(a)). Thus, the copolymerization of styrene with the macromonomer readily proceeds inside the micelle to produce the copolymer particles covalently stabilized by the hydrophilic polymer segment. On the other hand, the polymerizable group of **58b** is assumed to be located outside the micelle (Figure 5(b)), and hence, is hard to be copolymerized with styrene in micelle, in the early stage of the copolymerization. Therefore, the weak stabilization by the physically adsorbed macromonomer causes coalescence between the particles to produce larger polydisperse particles.

Emulsion copolymerization of vinyl acetate with block-type amphiphilic poly(2-oxazoline) macromonomer bearing a vinyl ester group gave stable latex, in

$CH_2{=}CH{-}C_6H_4{-}CH_2{-}(N(C(=O)R^1)CH_2CH_2)_m{-}(N(C(=O)R^2)CH_2CH_2)_n{-}OH$

**58a**: $R^1$=*n*-Bu, $R^2$=Me, $m$=3.2, $n$=6.4

**58b**: $R^1$=Me, $R^2$=*n*-Bu, $m$=5.8, $n$=3.2

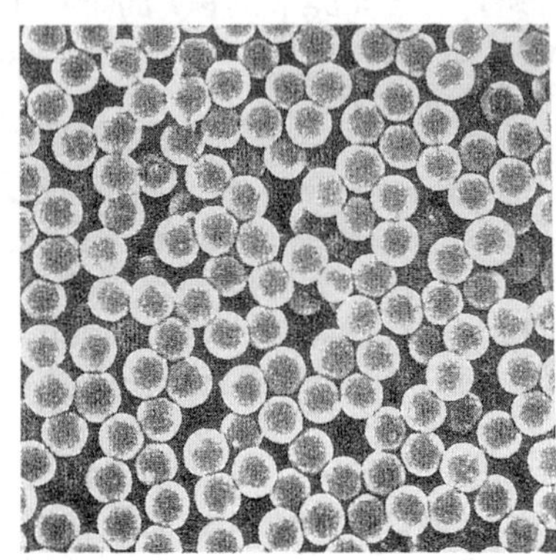

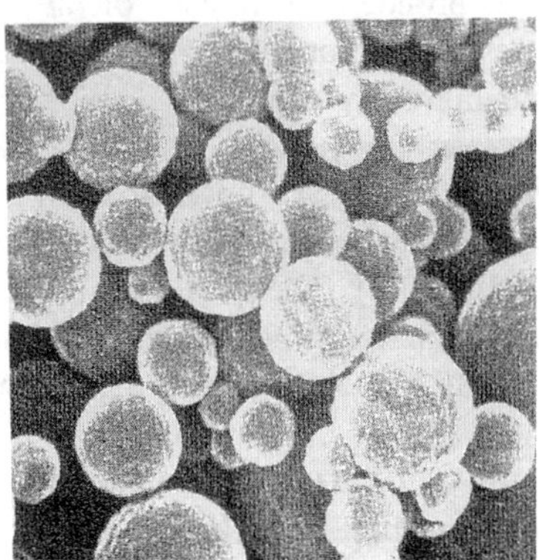

***Figure 4*** *SEM photographs of copolymer particles by emulsion copolymerization of styrene with (A)* ***58a****; (B)* ***58b*** *of 5 wt%.*

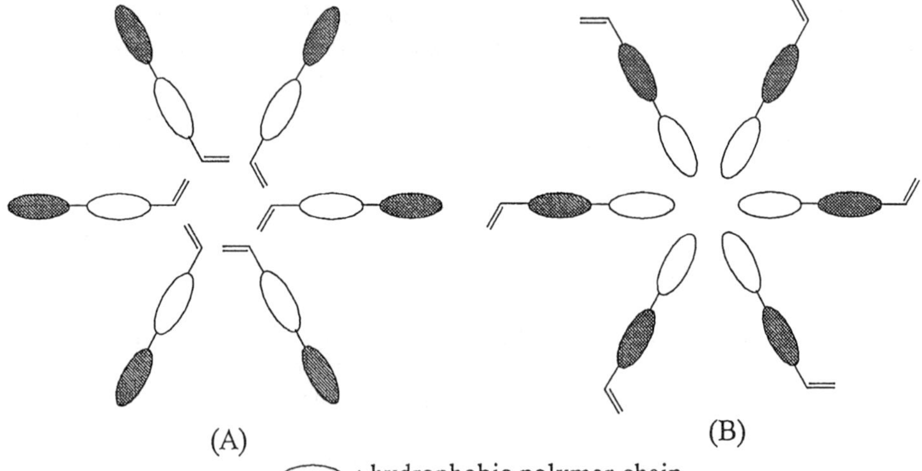


: hydrophobic polymer chain

: hydrophilic polymer chain

: polymerizable group

***Figure 5*** *Schematic representation of macromonomer in the micelle: (A) macromonomer* ***58a****; (B) macromonomer* ***58b****.*

the sub-micron range[110]. When the macromonomer possessed its polymerizable group bonded to the hydrophobic polymer chain, the conversion rate was larger and the particle size became smaller, as compared with experiments carried out with a macromonomer in which vinyl ester group was linked to the hydrophilic segment.

## 3.3 Compatibilizers for Polymer Blends

Since demands of various functions and properties for polymer materials are increasing, it has become difficult to satisfy such demands by a single polymer. Recently composites of more than two polymers have been investigated in order to achieve such demands. In blending systems of polymer materials, compatibility between the respective polymers becomes very important because it affects the properties of the resulting polymer blend. In immiscible cases, block and graft copolymers are often used as a compatibilizer.

Poly(styrene-g-MMA) prepared by macromonomer technique was an effective compatibilizer for blending between polystyrene and polyVC; elongation and tensile strength of the composite containing this graft copolymer are superior to those without the compatibilizer[111]. In case of the blend between polystyrene with nylon 6, the dispersity was improved by the addition of graft copolymer prepared by the copolymerization of styrene, maleic anhydride, and polystyrene macromonomer.

Recently, graft-type reactive compatibilizers have been developed[112]. The compatibilizers possess an acid or epoxy group in the main chain, and hence, can be reacted with nylon, polycarbonate, poly(ethylene terephthalate), etc. The branched component is polyMMA or polystyrene, which is compatible with ABS, poly(phenylene oxide), polyVC, etc. Therefore, the graft copolymer is expected to be a novel type of compatibilizers between the former and the latter polymer groups, which are immiscible with each other.

## 3.4 Adsorbents for Chromatographic Separation

Cell separations have been extensively investigated because of the growing demand for pure cell population in the field of biomedical science and biotechnology. Various techniques such as density gradient centrifugation, electrophoresis, and affinity panning have been developed for cell separation based on different cell properties.

Tsuruta et al.[113-115] developed new polymeric adsorbent for adsorption chromatography of B and T cells, two major subpopulations of lymphocytes. They used graft copolymers prepared by the copolymerization of styrene or HEMA with polyamine macromonomer **20** as the adsorbent polymer. In these graft copolymer films, the polyamine segment formed island-shaped micro-domains in a continuous phase of polystyrene or polyHEMA determined by TEM analysis. The diameter of the polyamine island are in sub-micron range, which is much smaller than that of rat lymphocytes.

copolymers via macromonomer methods, the length and number of graft chains are readily controlled, and hence, design and synthesis of new functional polymers become possible.

Amphiphilic graft copolymers in which the backbone is hydrophobic and the graft is hydrophilic or *vice versa* show interesting properties such as micelle formation and surface modification. Macromonomers are effective as stabilizers for the preparation of monodisperse polymer particles having chemically-bonded stabilizer chains on the surface.

In the future, research in the macromonomer area will continue to develop. It is further hoped that syntheses of new macromonomers as well as improvement of the existing methods of macromonomer preparations will be performed. For preparations of functional graft copolymers with higher functions, it will be necessary to understand relationships between structure and properties of the copolymers. We expect that further exciting investigations based on the macromonomer technique contribute to basic polymer science as well as to various practical applications.

## REFERENCES

1. R. Milkovich and M. T. Chiang, U.S. Patents, 3788116,3842050,3842057, 3842058,3842059(1974),3862098,3862101,3862102,3862267(1945); R. Milkovich, ACS Polym. Prepr., **21**, 40(1980).
2. C. H. Bamford and E. F. T. White, *Trans. Faraday Soc.*, **54**, 268(1958).
3. R. Waack and M. A. Doran, *Polymer*, **2**, 365(1961).
4. R. Waack and M. A. Doran, *J. Org. Chem.*, **32**, 3395(1967).
5. G. Greber and E. Reese, *Makromol. Chem.*, **55**, 96(1962).
6. G. Greber and A. Balciunas, *Makromol. Chem.*, **69**, 193(1963).
7. K. F. Gillman and E. Senogles, *Polym. Lett.*, **5**, 477(1967).
8. K. E. Barrett, Ed., "Dispersion Polymerization in Organic Media", John Wiley & Sons, Inc., New York, 1975.
9. P. Masson, G. Beinert, E. Franta, and P, Rempp, *Polym. Bull.*, **7**, 17(1982).
10. K. Hatada, H. Nakanishi, K. Ute, and T. Kitayama, *Polym. J.*, **18**, 581(1986).
11. Y. Kawakami, T. Aoki, and Y. Yamashita, *Polym. Bull.*, **18**, 473(1987).
12. Y. Gnanou and P. Rempp, *Makromol. Chem.*, **188**, 2267(1987).
13. S. Kobayashi, M. Kaku, S. Sawada, and T. Saegusa, *Polym. Bull.*, **13**, 447(1985).
14. R. C. Schulz and E. Schwarzenbach, *Makromol. Chem., Macromol. Symp.*, **3/14**, 495(1988).
15. F. J. Burgess, A. V. Cunliffe, D. H. Richards, and D. Thompson, *Polymer*, **19**, 334 (1978).
16. M. Maeda and S. Inoue, *Makromol. Chem., Rapid Commun.*, **2**, 357(1981).
17. P. Rempp, P. Lutz, P. Masson, and E. Franta, *Makromol. Chem.*, Suppl., **8**, 3(1984).
18. R. Asami, M. Takaki, and H. Hanahata, *Macromolecules,* **16**, 628(1983).

19. B. C. Anderson, G. D. Andrews, P. Arthur, Jr., H. W. Jacobson, L. R. Melby, A. J. Playtis, and W. H. Sharkey, *Macromolecules,* **14**, 1599(1981).
20. K. Ito, H. Tsuchida, A. Hayashi, T. Kitano, E. Yamada, and T. Matsumoto, *Polym. J.*, **17**, 827(1985).
21. P. R. Rao, P. Masson, P. Lutz, G. Beinert, and P. Rempp, *Polym. Bull.*, **11**, 115(1984).
22. Y. Kawakami, Y. Miki, T. Tsuda, R. A. N. Murthy, and Y. Yamashita, *Polym. J.*, **14**, 913(1982).
23. K. Hashimoto, H. Sumitomo, and M. Kawasumi, *Polym. Bull.,* **11**, 121(1984).
24. R. Asami, M. Takaki, K. Kita, and E. Asakura, *Polym. Bull.*, **2**, 713(1980).
25. R. Asami, M. Takaki, K. Kyuda, and E. Asakura, *Polym. J.*, **15**, 139(1984).
26. J. Sierra-Vargas, P. Masson, G. Beinert, P. Rempp, and E. Franta, *Polym. Bull.*, **7**, 277(1982).
27. T. Hamaide, N. Mariaggi, J. L. Foureys, P. Le Perchec, and A. Guyot, *J. Polym. Sci., Polym. Chem. Ed.*, **22**, 3091(1984).
28. V. Percec, P. L. Rinaldi, and B. C. Auman, *Polym. Bull.*, **10**, 215(1983).
29. V. Percec, P. K. Rinaldi, and B. C. Auman, *Polym. Bull.*, **10**, 397(1983).
30. K. Kobayashi, K. Aoki, H. Sumitomo, and T. Akaike, *Makromol. Chem., Rapid Commun.*, **11**, 577(1990).
31. Y. Gnanou and P. Rempp, *Makromol. Chem.*, **188**, 2111(1987).
32. M. Akashi, I. Kirikihira, and N. Miyauchi, *Angew. Makromol. Chem.,* **132**, 81(1985).
33. M. Akashi, T. Yanagi, E. Yashima, and N. Miyauchi, *J. Polym. Sci., Polym. Chem. Ed.,* **27**, 3521(1989).
34. K. Ishizu, K. Mitsutani, and T. Fukutomi, *J. Polym. Sci., Polym. Lett. Ed.*, **25**, 287(1987).
35. K. Ishizu, *J. Polym. Sci., Polym. Chem. Ed.*, **28**, 1887(1990).
36. T. Tsuruta, T. Narita, Y. Nitadori, and T. Irie, *Makromol. Chem.*, **177**, 3255(1976).
37. Y. Nitadori and T. Tsuruta, *Makromol. Chem.*, **180**, 1877(1979).
38. Y. Nabeshima, A. Maruyama, T. Tsuruta, and K. Kataoka, *Polym. J.*, **19**, 593(1987).
39. T. Nishimura, M. Maeda, Y. Nitadori, and T. Tsuruta, *Makromol. Chem.*, **183**, 29(1981).
40. J. S. Vergas, J. G. Zilliox, P. Rempp, and E. Franta, *Polym. Bull.*, **3,** 83(1980).
41. T. Uryu, M. Yamanaka, M. Date, M. Ogawa, and K. Hatanaka, *Macromolecules,* **21**, 1916(1986).
42. S. Aoshima, K. Ebara, and T. Higashimura, *Polym. Bull.*, **14**, 425(1985).
43. T. Kakuchi, and O. Kobayashi, D. Nakaya, and K. Yokota, *Polym. J.*, **21**, 649(1989).
44. Ph.Dubois, R. Jèrôme, and Ph. Teyssiè, *Macromolecules*, **24**, 977(1991).
45. R. Asami, M. Takaki, and Y. Moriyama, *Polym. Bull.*, **16**, 125(1986).
46. F. Severini, M. Pegoraro, and L. Saija, *Angew. Makromol. Chem.*, **133**, 111(1985).

47. T. Suzuki and P. Y. Lo, *Macromolecules,* **24**, 460(1991).
48. E. J. Goethals and M. A. Vlegels, *Polym. Bull.*, **4**, 521(1981).
49. M. Takaki, R. Asami, and T. Kuwabara, *Polym. Bull.*, **7**, 521(1992).
50. S. Kobayashi, E. Masuda, S. Shoda, and Y. Shimano, *Macromolecules*, **22**, 2878(1989).
51. M. Miyamoto, K. Naka, M. Tokumizu, and T. Saegusa, *Macromolecules*, **22**, 1604(1989).
52. E. J. Goethals, N. H. Haucourt, A. M. Berheyen, and J. Habimana, *Makromol. Chem., Rapid Commun.*, **11**, 623(1990).
53. J. P. Kennedy and M. Hiza, *J. Polym. Sci., Polym. Chem. Ed.*, **21**, 1033(1983).
54. T. Kitayama, S. Kishiro, and K. Hatada, *Polym. Bull.*, **25**, 161(1991).
55. Y. Yamashita, K. Ito, H. Mizuno, and K. Okada, *Polym. J.*, **14**, 255(1982).
56. G. F. Chen and F. N. Jones, *Macromolecules*, **24**, 2151(1991).
57. B. Boutevin, Y. Piètrasanta, M. Taha, and T. E. Sarraf, *Polym. Bull.*, **10**, 157(1983).
58. B. Boutevin, Y. Piètrasanta, M. Taha, and T. E. Sarraf, *Makromol. Chem.*, **184**, 2401(1983).
59. K. Ishizu and T. Fukutomi, *J. Polym. Sci., Polym. Lett. Ed.*, **25**, 169(1987).
60. J. Klee, F. Claussen, H. H. Hörhold, and J. Raddatz, *Polym. Bull.*, **27**, 511(1992).
61. P. Ferruti, D. Arnoldi, M. A. Marchiso, E. Martuscelli, M. Palma, F. Riva, and L. Provenzale, *J. Polym. Sci., Polym. Chem. Ed.*, **15**, 2151(1977).
62. P. Ferruti and E. Ranucci, *Polym. J.*, **23**, 541(1991).
63. J. M. DeSimone, G. A. York, J. E. McGrath, A. S. Gozdz, and M. J. Bowden, *Macromolecules,* **24**, 5330(1991).
64. S. Kobayashi, H. Uyama, and Y. Narita, *Makromol. Chem., Rapid Commun.*, **13**, 337(1992).
65. S. Nemes, T. Pernecker, J. P. Kennedy, *Polym. Bull.*, **25**, 633(1991).
66. L. J. Mathias, J. B. Canterberry, and M. South, *J. Polym. Sci., Polym. Lett. Ed.*, **20**, 473(1982).
67. S. Sawamoto, S. Aoshima, and T. Higashimura, *Makromol. Chem., Macromol. Symp.*, **13/14**, 513(1988).
68. Y. Tezuka, A. Okabayashi, and K. Imai, *Makromol. Chem.*, **190**, 753(1989).
69. Y. Tezuka, A. Fukushima, and K. Imai, *Makromol. Chem.*, **186**, 685(1985).
70. H.Uyama and S. Kobayashi, *Macromolecules*, **24**, 614(1991).
71. M. F. Farona and J. P. Kennedy, *Polym. Bull.*, **11**, 359(1984).
72. B. Yamada, E. Kato, S. Kobatake, and T. Otsu, *Polym. Bull.*, **25**, 423(1991).
73. I. V. Berlinova and I. M. Panayotov, *Makromol. Chem.*, **188**, 2141(1987).
74. H. Kunisada, Y. Yuki, S. Kondo, Y. Nishimori, and A. Masuyama, *Polym. J.*, **23**, 1455(1991).
75. J. P. Kennedy and J. D. Carter, *Macromolecules,* **23**, 1238(1990).
76. S. Kobayashi, T. Mizutani, and T. Saegusa, *Makromol. Chem.*, **185**, 441(1984).
77. S. Sivaram and R. S. Khisti, *Makromol. Chem., Rapid Commun.*, **12**, 435(1991).
78. R. L. Norton and T. J. McCarthy, *Macromolecules*, **22**, 1022(1989).

79. Y. Kawakami, H. Toda, M. Higashino, and Y. Yamashita, *Polym. J.*, **20**, 285(1988).
80. Y. Yamashita, Y. Chujo, H. Kobayashi, and Y. Kawakami, *Polym. Bull.*, **5**, 361(1981).
81. Y. Chujo, H. Kobayashi, and Y. Yamashita, *Polym. Commun.*, **25**, 278(1984).
82. Y. Chujo, A. Hiraiwa, H. Kobayashi, and Y. Yamashita, *J. Polym. Sci., Polym. Chem. Ed.*, **26**, 2991(1988).
83. Y. Chujo, H. Kobayashi, and Y. Yamashita, *J. Polym. Sci., Polym. Chem. Ed.*, **27**, 2007(1989).
84. Y. Chujo, T. Tatsuda, and Y. Yamashita, *Polym. Bull.*, **8**, 239(1982).
85. S. Kobayashi, H. Uyama, and H. Shirasaka, *Makromol. Chem., Rapid Commun.*, **11**, 11(1990).
86. H. Kazama, M. Hoshi, H. Nakajima, D. Horak, Y. Tezuka, and K. Imai, *Polymer*, **31**, 2207(1990).
87. H. Kazama, T. Ono, Y. Tezuka, and K. Imai, *Polymer*, **30**, 553(1989).
88. T. Heitz and H. Hocker, *Makromol. Chem.*, **189**, 777(1988).
89. N. Niessner and W. Heitz, *Makromol. Chem.*, **191**, 1463(1990).
90. T. Heitz and O. W. Webster, *Makromol. Chem.*, **192**, 2463(1991).
91. Y. Nagase, S. Mori, M. Egawa, and K. Matsui, *Makromol. Chem., Rapid Commum.*, **11**, 185(1990).
92. K. Ito, Y. Masuda, T. Shintani, T. Kitano, and Y. Yamashita, *Polym. J.*, **15**, 443(1983).
93. Y. Chujo, T. Shishino, Y. Tsukahara, and Y. Yamashita, *Polym. J.*, **17**, 133(1985).
94. S. Shoda, E. Masuda, M. Furukawa, and S. Kobayashi, *J. Polym. Sci., Polym. Chem. Ed.*, **30**, 1489(1992).
95. Y. Yamashita, Y. Tsukahara, K. Ito, K. Okada, and Y. Tajima, *Polym. Bull.*, **5**, 335(1981).
96. Y. Chujo, A. Hiraiwa, H. Kobayashi, and Y. Yamashita, *J. Polym. Sci., Polym. Chem. Ed.*, **26**, 2991(1988).
97. Y. Kawakami, R. Aswatha, N. Murthy, and Y. Yamashita, *Makromol. Chem.*, **185**, 9(1984).
98. Y. Yamashita, Y. Tsukahara, and H. Ito, *Polym. Bull.*, **7**, 289(1982).
99. S. Kobayashi and H. Uyama, *Kobunshi Ronbunshu*, **50**, 209 (1993).
100. S. Kobayashi, H. Uyama, J. H. Choi, and Y. Matsumoto, *Proc. Japan Acad., Ser. B*, **67**, 140(1991).
101. S. Kobayashi, H. Uyama, J. H. Choi, and Y. Matsumoto, *Polym. International*, **30**, 265(1993).
102. H. Uyama, S. W. Lee, Y. Matsumoto, and S. Kobayashi, *Polym. Prepr. Jpn.*, **41**, 1908(1992).
103. H. Uyama, H. Kato, and S. Kobayashi, *Chem. Lett.*, 261 (1993).
104. M. Akashi, D. Chao, E. Yashima, and N. Miyauchi, *J. Appl. Polym. Sci.*, **39**, 2027(1990).
105. M. Akashi, M. Wada, S. Yanase, and N. Miyauchi, *J. Polym. Sci., Polym. Lett. Ed.*, **27**, 377(1989).

106. F. Hoshino, M. Sakai, H. Kawaguchi, and Y. Ohtsuka, *Polym. J.*, **19**, 383(1987).
107. G. Levesque, V. Moitie, B. Bacle, and P. Depraètère, *Polymer,* **29**, 2271(1988).
108. S. Kobayashi, H. Uyama, and I. Yamamoto, *Makromol. Chem.*, **191**, 3115(1990).
109. H. Uyama, Y. Matsumoto, and S. Kobayashi, *Chem. Lett.*, 2401(1992).
110. H. Uyama, Y. Honda, and S. Kobayashi, *J. Polym. Sci., Polym. Chem. Ed.*, **31**, 123(1993).
111. K. Azuma and T. Tsuda, *Kino Zairyo*, No.10, pp5(1987).
112. Y. Maeda and K. Azuma, *Kobunshi Kako*, **40**, 182(1991).
113. K. Kataoka, Y. Sakurai, and T. Tsuruta, *Makromol. Chem., Suppl.*, **9**, 53(1985).
114. T. Tsuruta, Makromol. Chem., *Macromol. Symp.*, **12**, 323(1987).
115. K. Kataoka, T. Okano, Y. Sakurai, T. Nishimura, M. Maeda, S. Inoue, T. Watanabe, and T. Tsuruta, *Makromol. Chem., Rapid Commun.*, **3**, 275(1982).
116. K. Kataoka, T. Okano, Y. Sakurai, T. Nishimura, S. Inoue, A. Maruyama, T. Tsuruta, *Eur. Polym. J.*, **19**, 979(1983).
117. A. Maruyama, E. Senda, N. Yoshino, T. Tsuruta, K. Kataoka, T. Okano, Y. Sakurai, T. Nishimura, and S. Inoue, Jpn. *J. Artif. Organs.*, **13**, 1201(1984).
118. Y. Kawakami, H. Karasawa, T. Aoki, Y. Yamamura, H. Hisada, and Y.Yamashita, *Polym. J.*, **17**, 1159(1985).

# Macromonomers Chapter 2

## REACTION OF MACROMONOMERS AND DEVELOPMENT OF HIGH-PERFORMANCE MATERIALS

**Koji Ishizu**

Department of Polymer Science,
Tokyo Institute of Technology,
2-12, Ookayama, Meguro-ku, Tokyo 152
JAPAN





Macromolecular Design: Concept and Practice (Macromonomers, Macroinitiators, Macroiniferters, Macroinimers, Macroiniters, and Macroinifers) Pages 39 - 84

initiators for living polymerization of methacrylic acid esters. So, we checked the coupling reaction of living PMMA initiated by triphenylmethyl sodium or lithium with p-CMS at -78$^0$ C in THF or THF/toluene mixture [16]. As a result, the couling of living PMMA/Li$^+$ ($M_w/M_n$=1.17) with p-CMS took place quantitatively in

TABLE 1 Macromonomers prepared by anionic polymerization

| Polymer | Functional Groups | References |
|---|---|---|
| Polystyrene | $-CH_2-C_6H_4-CH=CH_2$ | [1,4] |
| | $-OC(=O)C(CH_3)=CH_2$ | [1-3,5-7] |
| Poly(α-methylstyrene) | $-CH_2-C_6H_4-CH=CH_2$ | [8] |
| Polyisoprene | $-CH_2-C_6H_4-CH=CH_2$ | [9] |
| Poly(2-vinylpyridine) | $-CH_2-C_6H_4-C(CH_3)=CH_2$ | [10] |
| | $-CH_2CH_2OC(=O)C(CH_3)=CH_2$ | [11] |
| | $-CH_2-C_6H_4-CH=CH_2$ | [12] |
| Poly(4-vinylpyridine) | $-C(=O)-C(CH_3)=CH_2$ | [11,13] |
| Poly(methyl methacrylate) | $-CH_2-C_6H_4-CH=CH_2$ | [14-18] |
| | $-CH_2-CH=CH_2$ | [15] |
| Poly(t-butyl methacrylate) | $-CH_2-C_6H_4-CH=CH_2$ | [19] |
| Poly(ethylene oxide) | $-OCH_2-C_6H_4-C(CH_3)=CH_2$ | [20] |
| | $-C(=O)-C(CH_3)=CH_2$ | [11,20-24] |
| | $-CH_2-C_6H_4-CH=CH_2$ | [20-22,25] |
| | $-CH=CH_2$ | [26,27] |

| Polymer | Functional Groups | References |
|---|---|---|
| Poly(dimethyl siloxane) | $-Si(CH_3)_2-C_6H_4-CH=CH_2$ | [28] |
| | $-Si(CH_3)_2(CH_2)_5OC(=O)C(CH_3)=CH_2$ | [29,30] |
| | $-Si(CH_3)_2(CH_2)_3OC(=O)C(CH_3)=CH_2$ | [31] |
| | $-Si(CH_3)_2CH=CH_2$ | [32] |

THF/toluene mixture. This result was not agreed with that reported by Anderson et al. [15]. A precise answer to the cause can not be obtained. The purification method of vinylbenzylchloride may be the direct cause. This macromonomer showed relatively high triad syndiotacticity (74%). On the other hand, Hatada et al. [17] synthesized the PMMA macromonomer by anionic polymerization of MMA with functional Grignard reagent as an initiator according to following scheme.

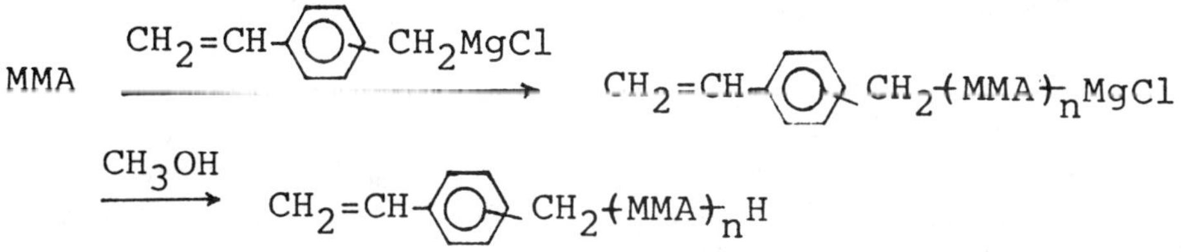

$$\text{MMA} \xrightarrow{CH_2=CH-C_6H_4-CH_2MgCl} CH_2=CH-C_6H_4-CH_2\text{-(MMA)}_n\text{-}MgCl$$

$$\xrightarrow{CH_3OH} CH_2=CH-C_6H_4-CH_2\text{-(MMA)}_n\text{-}H$$

The syndiotacticity increased with a decrease in the polymerization temperature, and the polymer prepared with *m*-vinylbenzylmagnesium chloride at -110$^0$ C showed extremely high syndiotacticity (96.6% in triads) and narrow molecular weight distribution ($M_w/M_n$=1.19).

Poly(ethylene oxide) macromonomers were synthesized by several methods shown in Table 1. This macromonomer is even more interesting, as intermediates for the synthesis of amphiphilic graft copolymers, because poly(ethylene oxide) is a non-ionic hydrophilic polymer.

Poly(dimethyl siloxane) (PDMS) macromonomers [28] were synthesized by ring-opening polymerization of hexamethylcyclotrisiloxane ($D_3$) initiated by lithium trimethylsilanolate according to the following scheme.

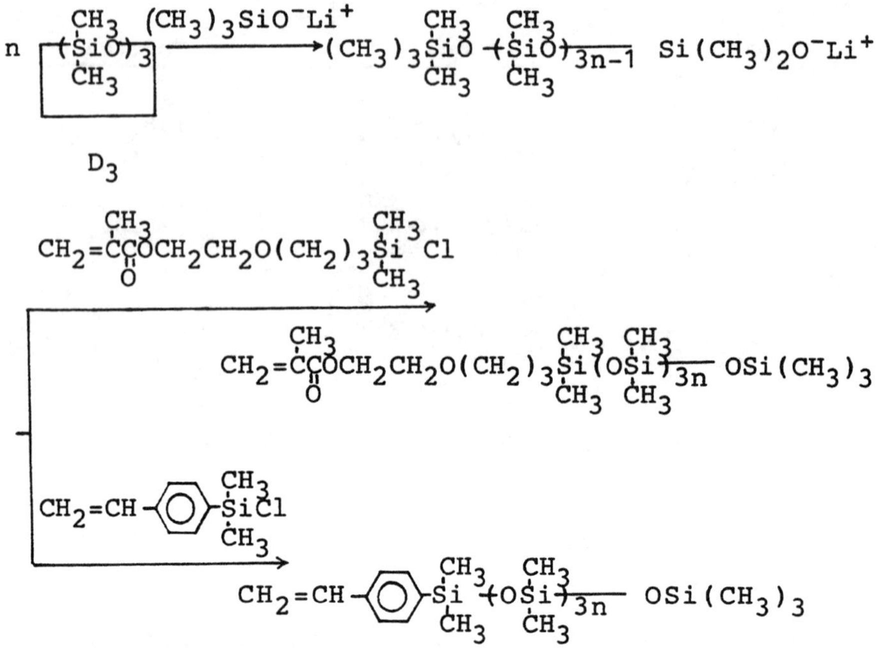

$$n\ \boxed{\left(\begin{matrix}CH_3\\ -SiO-\\ CH_3\end{matrix}\right)_3} \xrightarrow{(CH_3)_3SiO^-Li^+} (CH_3)_3\overset{CH_3}{\underset{CH_3}{Si}}O\left(\overset{CH_3}{\underset{CH_3}{Si}}O\right)_{3n-1} Si(CH_3)_2O^-Li^+$$

$D_3$

$$\xrightarrow{CH_2=\overset{CH_3}{C}\underset{O}{\overset{\|}{C}}OCH_2CH_2O(CH_2)_3\overset{CH_3}{\underset{CH_3}{Si}}Cl} CH_2=\overset{CH_3}{C}\underset{O}{\overset{\|}{C}}OCH_2CH_2O(CH_2)_3\overset{CH_3}{\underset{CH_3}{Si}}\left(O\overset{CH_3}{\underset{CH_3}{Si}}\right)_{3n} OSi(CH_3)_3$$

$$\xrightarrow{CH_2=CH-C_6H_4-\overset{CH_3}{\underset{CH_3}{Si}}Cl} CH_2=CH-C_6H_4-\overset{CH_3}{\underset{CH_3}{Si}}\left(O\overset{CH_3}{\underset{CH_3}{Si}}\right)_{3n} OSi(CH_3)_3$$

Methacryloyl-terminated PDMS macromonomer can be also synthesized by ring-opening polymerization of $D_3$ initiated by n-BuLi and termination of living PDMS with 3-methacryloyloxypropyl(dimethyl chlorosilane) [31]. Silicones are noteworthy for their high permeability toward gases.

There have been reports of attempts to synthesize the macromonomers by group transfer polymerization (GTP) and immortal polymerization. It is well known that the GTP was discovered by researchers at du Pont [34]. Asami et al.[35] prepared the vinylbenzyl-terminated PMMA macromonomer by GTP with unsaturated initiators. Following scheme shows PMMA macromonomer synthesis by using vinylphenylketene methyl trimethylsilyl acetal as an initiator.

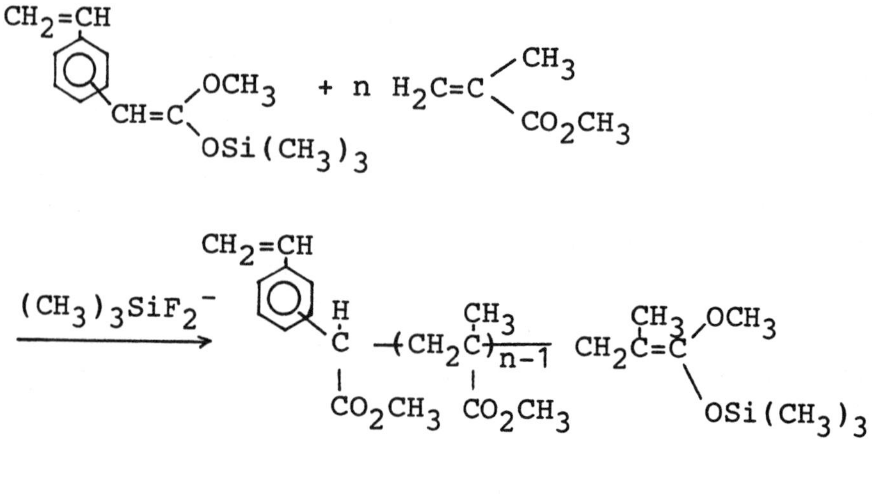

$$CH_2=CH-C_6H_4-CH=C\begin{matrix}OCH_3\\ OSi(CH_3)_3\end{matrix} + n\ H_2C=C\begin{matrix}CH_3\\ CO_2CH_3\end{matrix}$$

$$\xrightarrow{(CH_3)_3SiF_2^-} CH_2=CH-C_6H_4-\underset{CO_2CH_3}{\overset{H}{C}}\left(CH_2\underset{CO_2CH_3}{\overset{CH_3}{C}}\right)_{n-1} CH_2\overset{CH_3}{C}=C\begin{matrix}OCH_3\\ OSi(CH_3)_3\end{matrix}$$

The vinyl groups are introduced quantitatively to the end of PMMA molecule with narrow molecular weight distribution ($M_w/M_n=1.10$). Trimethylsilylated vinylbenzyl cyanide (1) can also be used as an effective initiator for GTP of MMA, producing PMMA macromonomer with high functionality and narrow molecular weight distribution.

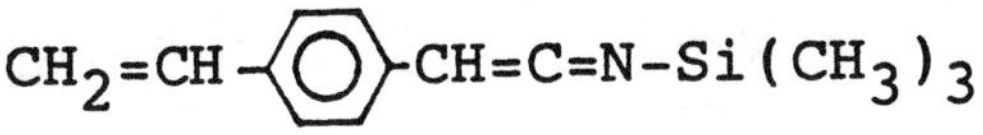

$$CH_2=CH-C_6H_4-CH=C=N-Si(CH_3)_3 \qquad \underline{1}$$

Inoue et al. have reported that some aluminum porphyrins (2; TPPAlX) exhibit a 100 % efficiency as initiator for the ring-opening polymerization of epoxide [36] and β-lactone [37], to produce living polyether and polyester with narrow molecular weight distribution, respectively. They named this system as "immortal" polymerization and

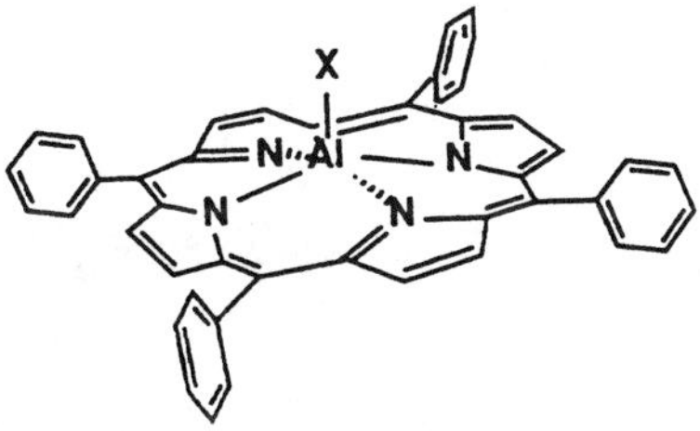

2 TPPAlX

found that it also initiates the living polymerization of methacrylic ester [38]. By using this immortal polymerization, they prepared polyester macromonomer by ring-opening polymerization of β-butyrolactone with tetraphenylporphinatoaluminum acrylate $TPPAlOCOCH=CH_2$ as an initiator by following scheme [39].

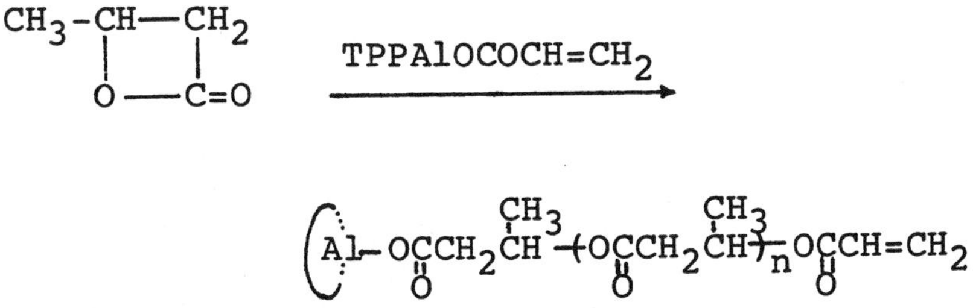

$$\underset{O\text{———}C=O}{CH_3-CH\text{—}CH_2} \xrightarrow{TPPAlOCOCH=CH_2}$$

$$(Al)-OC(=O)CH_2CH(CH_3)-\!\!\left(OC(=O)CH_2CH(CH_3)\right)_n\!-OC(=O)CH=CH_2$$

## 2.2 Macromonomer by Cationic Polymerization

Kennedy et al. [40-42] synthesized the chlorine-terminated poly isobutylene (PIB) prepolymer by "inifer"method of cationic polymerization and prepared various type of PIB macromonomers by end-capping reactions. In contrast with macromonomer synthesis by anionic polymerization, the macromonomers by cationic polymerization can be generally prepared from the coupling reaction of living polycation with functional nucleophile. Table 2 lists the macromonomers prepared by cationic polymerization. Rempp et al. [45] carried out the ring-opening polymerization of THF with an unsaturated initiator for the preparation of PTHF macromonomer. Also Rempp's [46] and Asami's groups [43,44,47,48] have prepared successfully the well-defined PTHF macromonomers by the end-capping method of living PTHF with an unsaturated nucleophile. Four types of PTHF macromonomers prepared by Asami et al. having styryl or acryloyl function at chain end are shown in the following scheme.

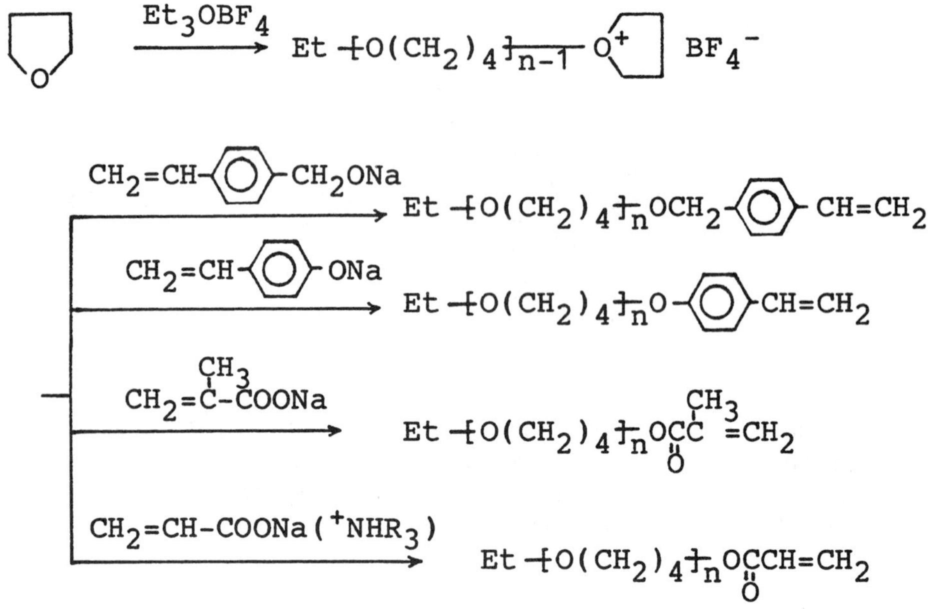

$$\text{THF} \xrightarrow{Et_3OBF_4} Et{-}\!\!\left[O(CH_2)_4\right]_{n-1}\!{-}O^{+}\!\!\left\langle\right. \quad BF_4^-$$

$$\xrightarrow{CH_2{=}CH{-}C_6H_4{-}CH_2ONa} Et{-}\!\left[O(CH_2)_4\right]_n OCH_2{-}C_6H_4{-}CH{=}CH_2$$

$$\xrightarrow{CH_2{=}CH{-}C_6H_4{-}ONa} Et{-}\!\left[O(CH_2)_4\right]_n O{-}C_6H_4{-}CH{=}CH_2$$

$$\xrightarrow{CH_2{=}C(CH_3){-}COONa} Et{-}\!\left[O(CH_2)_4\right]_n OC({=}O)C(CH_3){=}CH_2$$

$$\xrightarrow{CH_2{=}CH{-}COONa\,(^{+}NHR_3)} Et{-}\!\left[O(CH_2)_4\right]_n OC({=}O)CH{=}CH_2$$

These macromonomers have high functionality (more than f=0.9) and narrow molecular weight distribution ($M_w/M_n=1.10$).

Aliphatic polyamine macromonomers can be synthesized by ring-opening polymerizations. Desactivation of the active species (aziridinium ion) of living poly(t-butyl aziridine) by methacrylic acid leads to the corresponding methacrylate ester-terminated polyamine macromonomer [49]. On the other hand, Kobayashi et al.[50] have performed the synthesis of poly(2-methyl-2-oxazoline) (PMeOZO) macromonomers by following two methods. The first one is based on the living nature of the MeOZO polymerization; the living cationic propagating species of PMeOZO is

TABLE 2 Macromonomers prepared by cationic polymerization

| Polymer | Functional Groups | References |
|---|---|---|
| Poly(isobutylene) | $-OC(=O)C(CH_3)=CH_2$ | [40] |
| | $-C(CH_3)_2-CH=CH_2$ | [40,41] |
| | 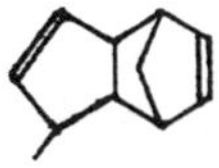 (methyl-substituted dicyclopentadienyl group, drawn as a structure) | [42] |
| Poly(tetrahydrofuran) | $-OCH_2-C_6H_4-CH=CH_2$ | [43] |
| | $-O-C_6H_4-CH=CH_2$ | [44] |
| | $-OC(=O)C(CH_3)=CH_2$ | [45-47] |
| | $-OC(=O)CH=CH_2$ | [48] |
| Poly $-(N(^tBu)CH_2CH_2)_n-$ | $-OC(=O)C(CH_3)=CH_2$ | [49] |
| Poly(2-methyl-2-oxazoline) | $-CH_2-C_6H_4-CH=CH_2$ | [50] |
| Poly $-(CH_2CH(OR))_n-$ | $-CH(CH_3)OCH_2CH_2OC(=O)C(CH_3)=CH_2$ | [51,52] |
| | $-CH(CH_3)OCH_2CH_2OCH=CH_2$ | [52] |
| | $-CH_2CH_2OCH=CH_2$ | [53] |

terminated with an amine having a styryl groups (terminator method) according to the following scheme.

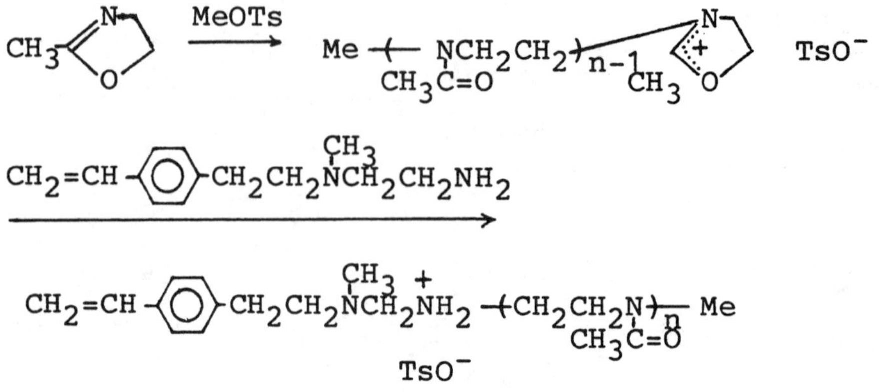

$$CH_3\text{-oxazoline} \xrightarrow{MeOTs} Me\!-\!\!\left(\underset{CH_3C=O}{N}CH_2CH_2\right)_{n-1}\text{-oxazolinium}(CH_3)^{+}\quad TsO^-$$

$$\xrightarrow{CH_2=CH-C_6H_4-CH_2CH_2\overset{CH_3}{N}CH_2CH_2NH_2}$$

$$CH_2=CH-C_6H_4-CH_2CH_2\overset{CH_3}{N}CH_2\overset{+}{N}H_2-\!\left(CH_2CH_2\underset{CH_3C=O}{N}\right)_n\!-Me\quad TsO^-$$

The second one is based on the use of an initiator having a styryl group to induce the polymerization of MeOZO (initiator method) according following scheme.

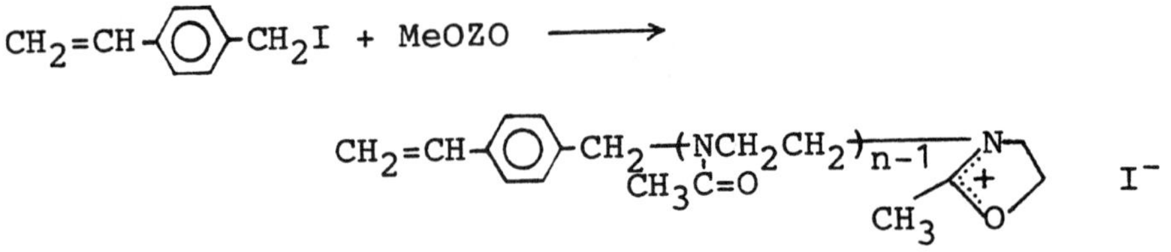

$$CH_2=CH-C_6H_4-CH_2I + MeOZO \longrightarrow$$

$$CH_2=CH-C_6H_4-CH_2-\!\left(\underset{CH_3C=O}{N}CH_2CH_2\right)_{n-1}\text{-oxazolinium}(CH_3)^{+}\quad I^-$$

These PMeOZO macromonomers can be converted to poly(ethylene imine) macromonomers by alkaline hydrolysis.

Higashimura et al.[51,52] have reported the synthesis of poly(vinyl ether) macromonomer by following scheme, where the adduct between hydrogen iodide and vinyl ether was employed as a functional initiator that bears a methacrylate group. They also prepared poly(vinylether) macromonomer by end-capping method according to following scheme [53]. The living growing end, generated by $HI/I_2$, is so stable and hence selective that it combines with an added functional nucleophile :Nu-X without undergoing $\beta$-proton expulsion and other side reactions.

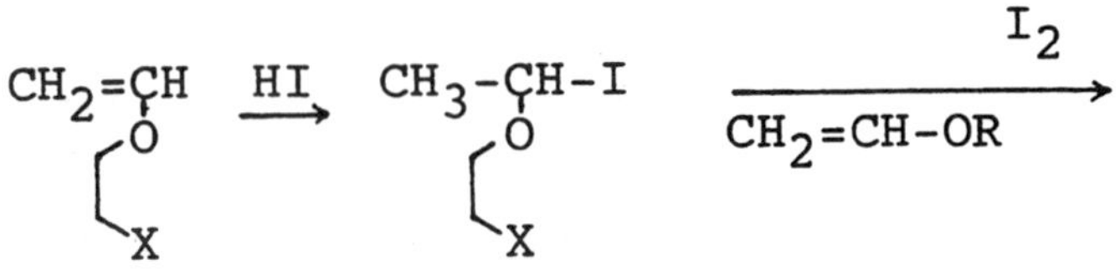

$$\underset{\underset{X}{\frown}}{CH_2=\underset{|}{CH}-O} \xrightarrow{HI} \underset{\underset{X}{\frown}}{CH_3-\underset{|}{CH}-I\;(O)} \xrightarrow[CH_2=CH-OR]{I_2}$$

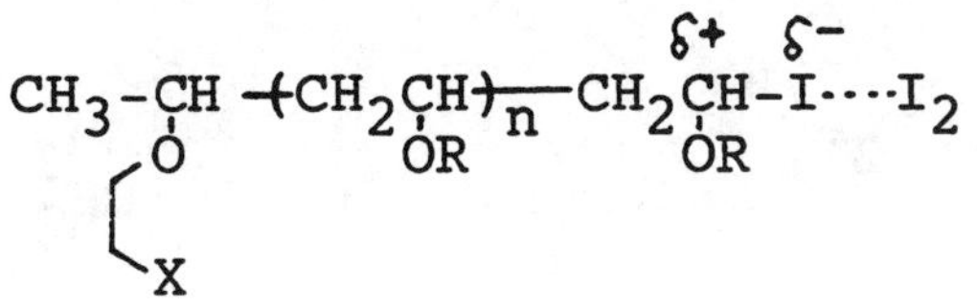

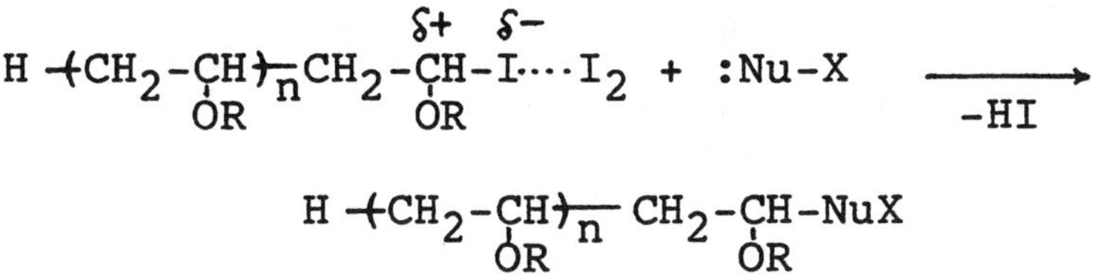

$$CH_3-\underset{\underset{\underset{X}{(}}{O}}{CH}-(CH_2\underset{OR}{CH})_n-CH_2\underset{OR}{CH}-\overset{\delta+}{I}\cdots\overset{\delta-}{I_2}$$

$$H-(CH_2-\underset{OR}{CH})_n CH_2-\underset{OR}{CH}-\overset{\delta+}{I}\cdots\overset{\delta-}{I_2} + :Nu-X \xrightarrow[-HI]{}$$

$$H-(CH_2-\underset{OR}{CH})_n CH_2-\underset{OR}{CH}-NuX$$

## 2.3 Macromonomer by Radical Polymerization

Living polymerization systems provide us with the most elegant tool available for the synthesis of tailor-made macromonomers, such as a predictable molecular weight and very narrow molecular weight distribution. However, the limited monomers are used in living polymerization. Free radical polymerization is the most commonly used method for the preparation of polymers from a wide variety of vinyl and diene monomers. Free radicals may be terminated by a variety of mechanisms both first and second-order in reactive species. Second-order processes are by far the most important where the two principal modes of termination are combination and disproportionation. In order to obtain the macromonomers by way of free radical polymerization, it is necessary to control the mechanism of polymerization.

One of the most convenient method for the preparation of macromonomers by way of free radical polymerization may be as follows. The mechanical treatment of polymer-vinyl monomer mixtures under suitable conditions leads to addition polymerization from free radicals produced by mechanochemical scission as initiator. Ishizu et al. [59,64] reported a method of preparing macromonomers by mechanical scission reaction of the main chain in the presence of divinylbenzene or triallylcyanurate. By this method it is possible to prepare telechelic macromonomers with various types of main chain and controlled number of unsaturated end groups.

Yamashita et al.[56] have also prepared poly(methacrylic acid ester) by way of free radical polymerization according to the method described by Walbridge and Waite [54,55]. Various steps for the synthesis are shown in following scheme.

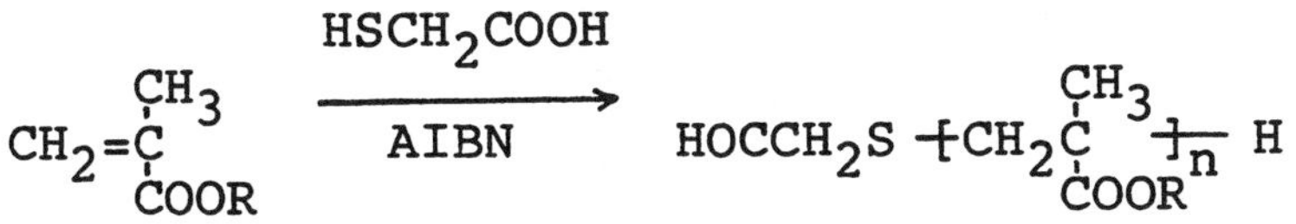

$$CH_2=\overset{\overset{CH_3}{|}}{\underset{\underset{COOR}{|}}{C}} \xrightarrow[AIBN]{HSCH_2COOH} HOCCH_2S-(CH_2\overset{\overset{CH_3}{|}}{\underset{\underset{COOR}{|}}{C}})_n-H$$

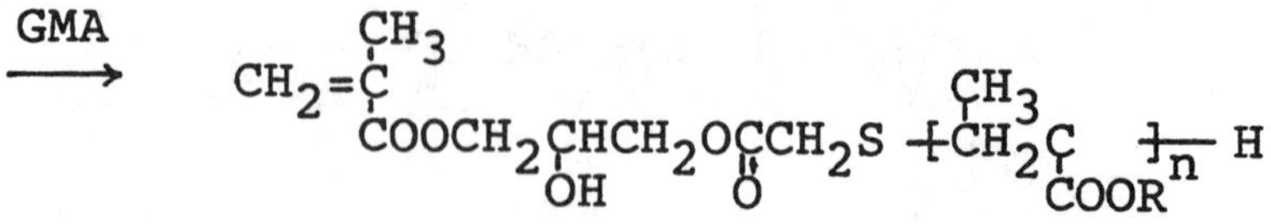

$$\xrightarrow{\text{GMA}} CH_2{=}\underset{|}{\overset{CH_3}{|}}{C}\text{-}COOCH_2\underset{OH}{CH}CH_2O\underset{O}{\overset{\|}{C}}CH_2S\text{-}(CH_2\overset{CH_3}{\underset{COOR}{C}})_n\text{-}H$$

GMA: glycidyl methacrylate

The carboxyl-terminated PMMA prepolymer was synthesized by free radical polymerization of MMA in the presence of thioglycolic acid as a transfer agent, which was then reacted with glycidyl methacrylate (GMA) to give PMMA macromonomer having methacryl group terminal. The synthesis method of carboxyl-terminated PMMA prepolymer corresponds to telomerization. Table 3 lists macromonomers prepared by radical polymerization.

TABLE 3 Macromonomers prepared by radical polymerization

| Polymer | Functional Groups | References |
|---|---|---|
| $-(\overset{CH_3}{\underset{COOR}{C}}-CH_2)_n-$ <br> $R=CH_3, C_{12}H_{25}, C_{18}H_{37}, CH_2CH_2NEt_2$ | $-SCH_2\overset{O}{C}OCH_2\underset{OH}{CH}CH_2O\overset{O}{C}\overset{CH_3}{C}=CH_2$ | [54-57] |
| $-(\overset{CH_3}{\underset{COOC_{12}H_{25}}{C}}-CH_2)_n-$ | $-SCH_2CH_2O\overset{O}{C}CH=CH_2$ | [58] |
| $-(\overset{CH_3}{\underset{COOCH_3}{C}}-CH_2)_n-$ | $-(CH_2\underset{C_6H_4CH=CH_2}{CH})_n-$ | [59] |
| Poly(t-butyl methacrylate) | $-CH_2-C_6H_4-CH=CH_2$ | [60] |
| Polystyrene | $-CH_2\overset{O}{C}OCH_2\underset{OH}{CH}CH_2O\overset{O}{C}\overset{CH_3}{C}=CH_2$ | [61] |
| | $-CH_2-C_6H_4-CH=CH_2$ | [62] |
| | $-CH_2\underset{OH}{CH}CH_2OCH_2CH=CH_2$ | [63] |

| Polymer | Functional Groups | References |
|---|---|---|
| | $-CH_2CH(OH)CH_2OC(=O)C(CH_3)=CH_2$ | [63] |
| | $-(CH_2CH=CH_2)_n$ | [64] |
| Poly(vinyl-chloride) | $-SCH_2C(=O)OCH_2CH(OH)CH_2OC(=O)C(CH_3)=CH_2$ | [65] |
| | $-SCH_2CH_2OC(=O)C(CH_3)=CH_2$ | [66] |
| Poly(acryl-amide) | $-SCH_2CH_2C(=O)OCH_2-C_6H_4-CH=CH_2$ | [67] |
| Poly(vinyl-pyrrolidone) | $-SCH_2CH_2C(=O)OCH_2-C_6H_4-CH=CH_2$ | [67] |
| Poly(vinyl-acetate) | $-CH_2-C_6H_4-CH=CH_2$ | [68] |
| | $-C(=O)C(CH_3)=CH_2$ | [69] |

Poly(vinylchloride) [65], poly(acrylamide) [67], and poly(vinylpyrrolidone) [67] macromonomers were also synthesized according to the method developed by Yamashita et al.

Ishizu et al. have synthesized polystyrene (PS) [62], poly(t-butyl methacrylate) [60], and poly(vinyl acetate) [68] macromonomers by radical polymerization of each monomer initiated by a functional initiator (2,2'-azobis(N,N'-dimethyleneisobutyramidine):

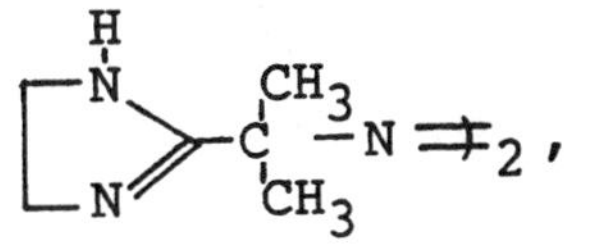

$$\left(\text{imidazolin-2-yl}-C(CH_3)_2-N=\right)_2,\ \text{VA-061}$$

in the presence of allylmalonicacid diethylester as a degradative chain transfer agent, followed by an end-capping reaction with CMS. In these reaction conditions, the polymerization of styrene is regarded to proceed entirely with the unimolecular termination mechanism. Because allyl compound leads to degradative chain transfer, the resulting allyl radical is quite stable. This preparation method of the imidazol-terminated prepolymers at one end is applicable to common vinylmonomers.

This terminal imidazol group of PS prepolymer can be converted into other various functional groups with chemical modifications according to following scheme [63].

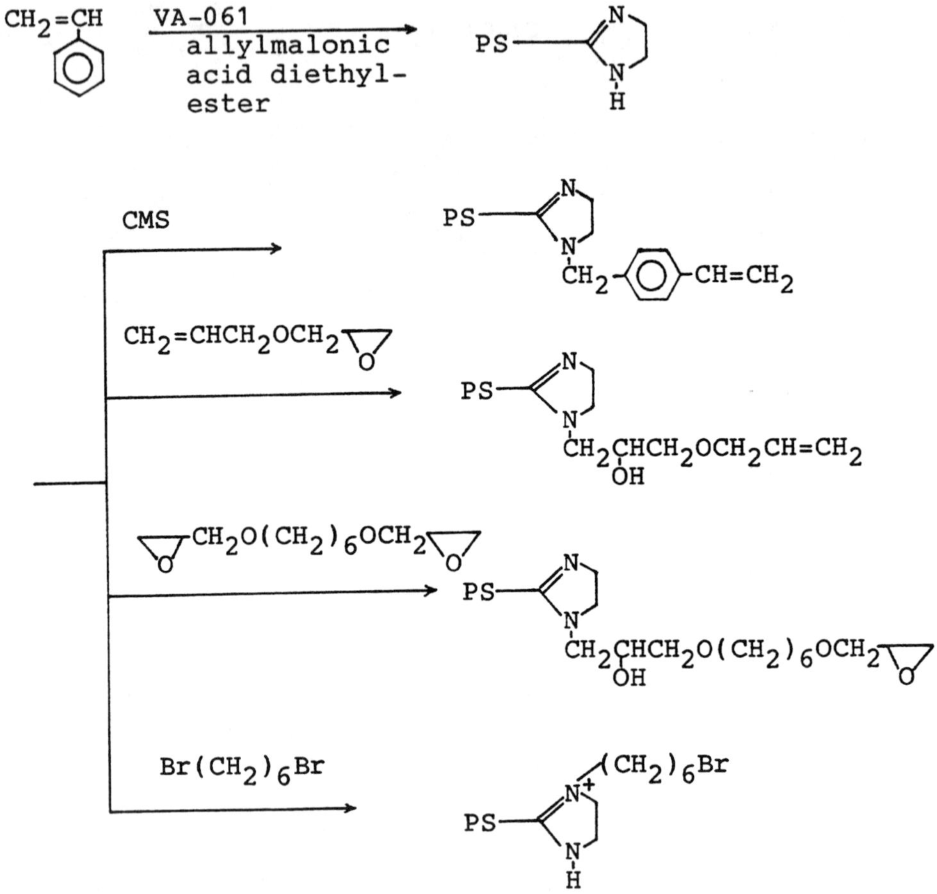

$CH_2{=}CH(C_6H_5)$ —(VA-061, allylmalonic acid diethyl-ester)→ PS–(2-imidazolinyl, N–H)

CMS → PS–imidazolinyl–N–$CH_2$–$C_6H_4$–$CH{=}CH_2$

$CH_2{=}CHCH_2OCH_2$–(epoxide) → PS–imidazolinyl–N–$CH_2CH(OH)CH_2OCH_2CH{=}CH_2$

(epoxide)–$CH_2O(CH_2)_6OCH_2$–(epoxide) → PS–imidazolinyl–N–$CH_2CH(OH)CH_2O(CH_2)_6OCH_2$–(epoxide)

$Br(CH_2)_6Br$ → PS–imidazolinium ($N^+$–$(CH_2)_6Br$, N–H)

Macromonomers obtained from radical polymerization have broad molecular weight distribution. So, we can get macromonomers with relatively narrow molecular weight distribution by fractionation of raw macromonomers.

Macromonomers are synthesized by way of other polymerizations. Ferruti et al.[70,71] have synthesized poly(amidoamine) macromonomers (3) from the polyaddition of N,N'-dimethylethylenediamine to 1,4-bis-acryloyl-piperazine.

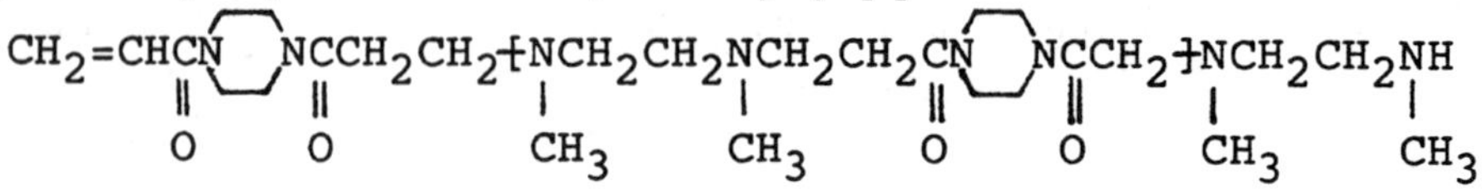

$$CH_2{=}CHC({=}O)N\langle piperazine\rangle NC({=}O)CH_2CH_2{+}N(CH_3)CH_2CH_2N(CH_3)CH_2CH_2C({=}O)N\langle piperazine\rangle NC({=}O)CH_2{+}N(CH_3)CH_2CH_2NH(CH_3)$$

3

A poly(2,6-dimethyl-1,4-phenyleneoxide) (PPO) macromonomer (4) containing styrene as a polymerizable group was prepared from a PPO containing phenol and p-bromophenoxy groups as the two chain ends [72].

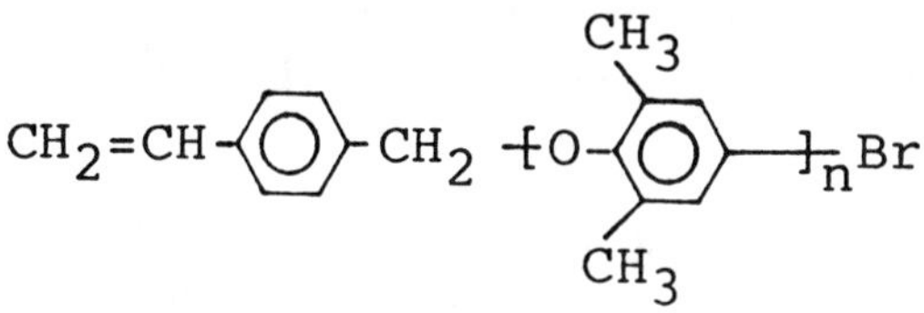

$$CH_2{=}CH{-}C_6H_4{-}CH_2{-}\!\!\left[O{-}C_6H_2(CH_3)_2\right]_n\!Br$$

4

The main reaction consists of the quantitative esterification of the phenol chain end with CMS in the presence of tetrabutylammonium hydrogen sulfate as a phase transfer catalyst. Thus, the macromonomers are possible to prepare from various types of polymerizations.

## 3. APPLICATION FOR MULTI-COMPONENT POLYMERS

Macromonomer techniques have made possible the preparation of a variety of graft copolymers of well-defined structure. In the preparation process, the relevant copolymerization parameters are very important in determining the trunk-branch distribution in the resulting graft copolymers [22]. On the other hand, homopolymerization system of macromonomers exhibits the following features in comparison with polymerization systems of small monomers [73,74]:(i) high viscosity of the polymerization media from the beginning of the polymerization, (ii) low concentration of the polymerizable reactive end group, (iii) the propagating step is a repeat of the polymer-polymer reaction, and (iv) high segment density or multi-branched structure around the propagating radical site. It is important to solve these problems described above to obtain the well-defined graft copolymers and comb-like branched polymers. Copolymerization and homopolymerization reactivities of macromonomers will be mentioned by Ito [Poly(ethylene oxide) Macromonomers] and Tsukahara [Macromonomers: Characterization, Polymerization, Reactivity and Application] in other chapters. In this section, four examples will be given to clarify the application of macromonomers for high-performance materials.

### 3.1 Graft Copolymers

We have often adopted one of two classes as the method of graft copolymer synthesis: (i) growth of a branch from an active site on a polymer chain, and (ii) coupling of an already-formed polymer chain with functional groups located on the backbone polymer chain. It is clear that both of these methods are also available to other mechanisms of polymerization. The anionic nonterminating systems offer the unique advantage that the branches, whether grown out from the main chain or coupled to the main chain, can be of remarkable uniformity in length. The design of graft copolymers by macromonomer method also exhibits similar advantage to anionic nonterminating systems. That is to say, the macromonomer system can be of uniformity in branch length. Moreover, this system can be applied for copolymerization with a wide variety of vinyl monomers.

The solution and solid behaviors of graft copolymers are very similar to those of block copolymers. Generally, graft and block copolymers form micelles in a selective solvent. Both copolymers exhibit unique physical properties as a result of the microphase separation of the constituent polymers. The thermodynamic approach to the problem of microdomain formation for block copolymers has been worked out [75-79] by demonstrating that the morphology of the domain structure - such as equilibrium size, shape, and interfacial thickness - can be described in terms of a balance of physical factors. The morphology of block copolymers such as sphere, cylinder, and lamella has been found to depend, according to Molau's rule [80], on the volume fractions of the phases.

Kawai et al.[81] indicated from a quasi-equilibrium phenomenon of micelle formation that two types of rodlike domains, usually found for block copolymers as the intermediates between the spherical and lamellar domains, were missing in graft copolymers. Ishizu et al.[62] synthesized poly[4-vinylpyridine(4VP)-g-S] graft copolymers by macromonomer method. Where vinylbenzyl-terminated PS macromonomers were prepared by way of radical polymerization. Table 4 lists the copolymerization condition of PS macromonomer with 4VP and characteristics of graft copolymer C2.

Figure 1 shows transmission electron micrograph (TEM) of graft copolymer C2 specimen cast from 1,1,2-trichloroethane (TCE) solution. Lamellar arrangement structure was detected for this specimen. In an earlier work [82], the monodispersed poly(S-b-4VP) diblock copolymer (prepared by anionic living polymerization) specimens having PS 30-35 mol% cast from TCE showed the texture of PS rods (cylinders) in a P4VP matrix. So, graft copolymers are found to have a wide range of alternating lamellar domains in comparison with block copolymers. The average domain identify period of the C2 specimen (PS; 22 nm, P4VP; 13 nm) is about half of the values of those obtained in monodispersed poly(S-b-4VP) diblock copolymer having the same molecular weight. The main reason is due to the low volume fraction of grafted segments. The distribution of microdomain sizes for the C2 specimen is relatively uniform as a result of precipitate fractionation of PS macromonomer ($M_w/M_n$ =1.38). Hashimoto et al.[83] have reported that the uniformity of the microdomain sizes of block copolymers is much higher than the uniformity of molecular weights of the block copolymers comprising the microdomains. Then, even in microphase separation of graft copolymers with relatively broad molecular weight distribution, the molecules are packed in the domain space so as to compensate their molecular weight distributions.

TABLE 4 Results of copolymerization of PS macromonomer with 4-vinylpyridine[a]

| Exptl. No. | PS macromonomer[b] $10^{-4}\bar{M}_n$ | (g) | $[4VP]/[end$ $styryl$ $group]$ (mol/mol) | Graft copolymer $10^{-5}\bar{M}_n$ [c] | composition of PS[d] (mol%) | number of graft (per chain) |
|---|---|---|---|---|---|---|
| C2 | 2.8 | 2.3 | 505 | 1.82 | 31.6 | 2.9 |

[a] Copolymerization in benzene/methanol [10/3 (v/v)] initiated by 2,2'-azobisisobutyronitrile (AIBN; 50mg) at 60°C for 60 h.
[b] End styryl group=0.8 per chain.
[c] Determined by osmometry.
[d] Determined by IR spectra.

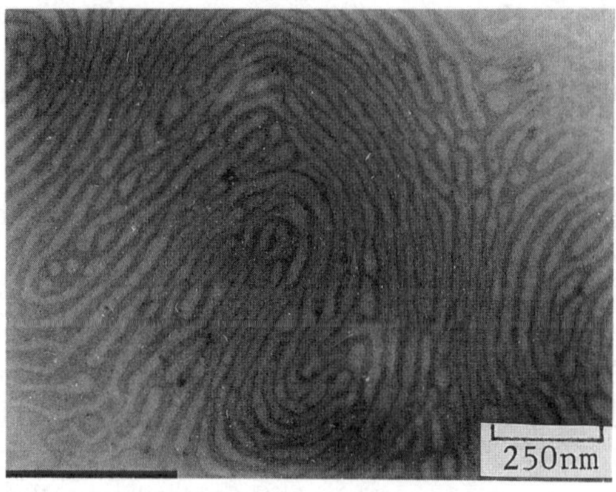


*Fig. 1. Electron micrograph of poly(4VP-g-S) graft copolymer C2 specimen cast from TCE. Taken from Ref. 62.*

Thus two-phase morphology is much more commonly observed in graft copolymers. This nature can be used to advantage in controlling properties such as modulus and permeability. Two-phase morphology of graft copolymers is also reflected in their thermal transition behavior. These graft copolymers display two distinct glass transition temperature similar to the incompatible physical blends. One important characteristics of two-phase graft copolymers in their facility for blending well with their respective homopolymers. Another way to utilize this property is to employ graft copolymers as interfacial emulsifying or compatibilizing agents in polymer blends. Several books [79, 84-87] give a good survey of the graft copolymers described above. The field of conducting polymers, selective transport membranes, polymeric biomaterials, and various types of polymeric catalysts have become increasingly important for the target of application of block and graft copolymers.

## 3.2 Star-Block Copolymers

Theoretical models based on confined-chain and mean-field principles have proven particularly useful in elucidating fundamental relationships between microstructural dimensions and molecular characteristics in microphase-separated block copolymers in the strong-segregation limit [75,79,88]. In the case of the weak-interaction limit, the free energy of the system, near the order-disorder transition (ODT), has been modeled utilizing de Gennes' random phase approximation [89] in a theory originally pioneered by Leibler [90]. He has pointed out that a simple 50:50 diblock AB copolymer containing N monomer units has a large critical value:

$$(\chi N)_c = 10.5 \tag{1}$$

where $\chi$ is Flory-Huggins interaction parameter.
This implies that phase separation is more difficult in diblock copolymers than in analogous homopolymer systems $[(\chi N)_c=4]$. Recently, de la Cruz and Sanchez [91] have derived the phase stability criteria and static structure factors near the ODT for a simple AB graft copolymer, for star copolymers with equal numbers of A and B arms ($A_nB_n$ star) and n-arm star diblock copolymers [$(AB)_n$ star]. The schematic representation of simple AB graft, $A_nB_n$ star, and $(AB)_n$ star copolymers are shown in Figure 2.

Figure 3 shows the variation of $(\chi N)_c$ with composition and arm number for $A_nB_n$ star copolymers. This type of star copolymer can be constructed by bringing n identical diblock copolymers together and joining them at their A-B junction points. There is a very surprising result for this type of star copolymer; $(\chi N)_c$ is independent of n and equals 10.5. The critical temperature does not change with the number of arms in the star. On the other hand, Figure 4 shows the variation of $(\chi N)_c$ with composition and arm number for $(AB)_n$ star copolymers. The $(AB)_2$ star corresponds to triblock copolymer. The $(\chi N)_c$ curves are asymmetric. As n gets large, the $(AB)_n$ star begins to develop a "core and shell" type structure. The core will be rich in A monomer and the shell will be rich in B monomer even in the disordered state. This self-segregation or self-micellarization tends to create significant concentration fluctuations at the core-shell interface. Thus, this theory predicts that phase separation is easier in star copolymers than in analogous block copolymers.

There have been attempts [92-99] to prepare $A_nB_n$ star block (heteroarm star) copolymers. For example, Berlinova and Panayotov [92] have prepared amphiphilic diblock macromonomers possessing a central unsaturated group. In the radical cis-trans isomerization of these diblock macromonomers, four- to eight-arm amphiphilic $A_nB_n$ star-shaped copolymers are formed. Rempp et al.[93,94]have prepared $A_nB_n$ star copolymers by using a three-step anionic process. A living precursor polymer was made first, and used to initiate the polymerization of a small amount of a bis-unsaturated monomer, so as to build star molecules. Each of the resulting cores was linked to the precursor chains which had contributed to its initiation. This living star polymer was used subsequently to initiate the polymerization of another monomer, implying that new branches were growing out from the core. However, it does not seem that the initiation step involves the living star polymer is at least as rapid as the propagation step, due to high segment density around the core anions.

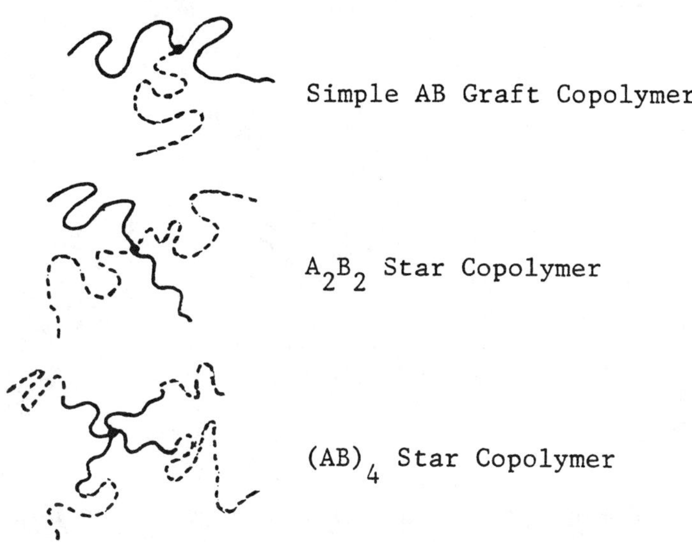


*Fig. 2 Schematic representation of three types of copolymers*

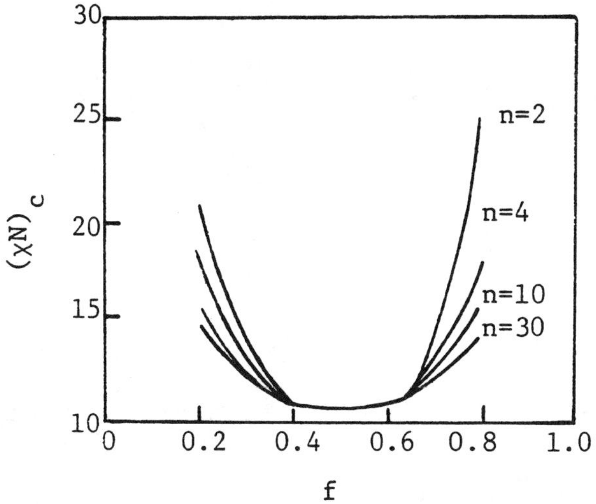


*Fig. 3 Variation of $(\chi N)_c$ with composition and arm number for $A_nB_n$ star copolymers. Taken from Ref. 91.*

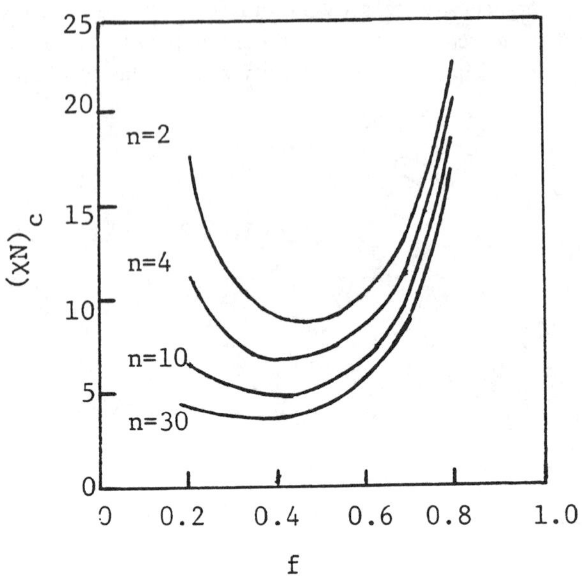


*Fig. 4 Variation of $(\chi N)_c$ with composition and arm number for $(AB)_n$ star copolymers. Taken from Ref. 91.*

Ishizu et al.[100,101]have reported the synthetic method for diblock macromonomers possessing central vinylbenzyl groups by anionic addition in a three-stage process using styrene, 1,4-divinylbenzene, and t-butyl methacrylate (BMA) monomers (5).

$$\text{n-Bu}\!-\!(CH_2\!-\!\underset{C_6H_5}{CH})_l\!-\!(CH_2\!-\!\underset{C_6H_4\!-\!CH=CH_2}{CH})_m\!-\!(CH_2\!-\!\underset{C=O,\ OC(CH_3)_3}{\overset{CH_3}{C}})_n\!-\!H \quad (l \text{ and } n > m)$$

5

They also suggested a novel architecture for $A_nB_n$ star block copolymers by means of the organized-polymerization by diblock macromonomers possessing central polymerizable groups. In this environment, the polymerizable groups at the position of block junction should be oriented regularly at the domain interfaces. The interface is narrow (ca. 2.0 nm) for block copolymers possessing narrow molecular weight distributions regardless of molecular weight [102 -104]. Table 5 lists the characteristics of the poly(S-b-BMA) diblock macromonomer BP2 possessing central vinylbenzyl groups.

TABLE 5 Characteristics of poly(S-b-BMA) diblock macromonomer possessing central vinylbenzyl groups

| Sample | PS precursor $10^{-4}\bar{M}_n$ [a] | Diblock macromonomer $10^{-4}\bar{M}_w$ [a] | $\bar{M}_w/\bar{M}_n$ [a] | Vinylbenzyl groups[b] (number /1-polymer) | PS blocks[c] (wt%) |
|---|---|---|---|---|---|
| BP2 | 0.40 | 1.92 | 1.09 | 2.5 | 20.8 |

[a] Determined by gel permeation chromatography (GPC).
[b] Determined by GPC using double detectors (RI and UV).
[c] Determined by NMR.
Taken from Ref. 101.

Figure 5 shows TEM micrograph of the diblock macromonomer BP2 (PS blocks 20.8 wt%).The dark areas are the selectively stained PS blocks with $RuO_4$. The morphology of BP2 is a microstructure of PS spherical domains dispersed in a PBMA matrix. The central vinylbenzylgroups at the position of block junction should be oriented regularly at the domain interface. Photosensitive polymerization of diblock macromonmer BP2 films was carried out in various conditions. $A_nB_n$ star block copolymer was formed by the polymerization system CP22 by adding both tetramethylthiuram disulphite (photosensitizer) and AIBN irradiated with an UV beam in a nitrogen atmosphere. Figure 6 shows GPC profiles of CP22 and diblock macromonomer BP2 taken with RI and light scattering (LS) detectors. The GPC distribution of CP22 is bimodal on an RI chart. The elution peak on the high molecular weight side corresponds to star block copolymer ($A_nB_n$-1). Table 6 lists the characteristics of $A_nB_n$ star block copolymer. The conversion of $A_nB_n$-1 is 23.4wt%. The molecular weight distribution is very narrow ($M_w/M_n$=1.08). This star block copolymer exhibits an $A_3B_3$ type molecular structure.

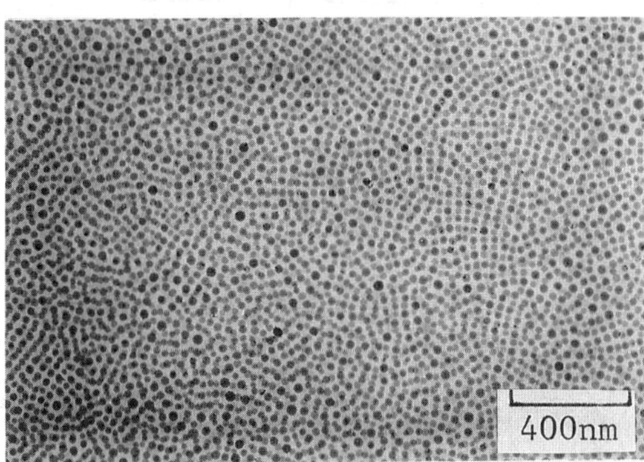


*Fig. 5 TEM micrograph of diblock macromonomer BP2 cast from toluene. Taken from Ref. 101.*

More recently, the functional diblock macromonomers possessing central isoprene groups have been also synthesized by anionic addition in three-stage process using styrene, isoprene, and 2VP monomers [105]. These functional diblock copolymers formed microphase-separated structure in the solid state. The P2VP blocks of the cast film (60 μm thick) were converted into P2VP.HCl salts by exposing to HCl vapor. Crosslinking of the isoprene groups at the domain interfaces was carried out by soaking in n-heptane/ chloroform: 8/2 (v/v) mixture of $S_2Cl_2$ at room temperature, varying $S_2Cl_2$ concentration and reaction time. As a result, micro-gelation (addition condensation) of poly(S-b-2VP) diblock macromonomers formed $A_nB_n$ star block copolymers having the arm number n=3-19. Thus, $A_nB_n$ star block copolymers can be synthesized by organized-polymerization of diblock macromonomers possessing central polymerizable groups.

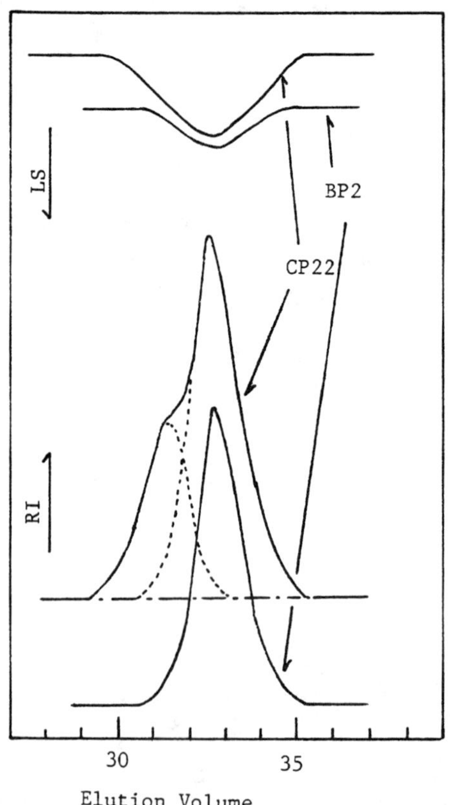


*Fig. 6 GPC profiles of polymerization product CP22 and diblock macromonomer BP2. Taken from Ref. 101.*

The formation of star polymers of monocarbanionic chains with chlorosilane compounds as linking agents is well known [106]. Roovers and co-workers [107-109] synthesized and studied a series of 4- and 6-armed star polymers using corresponding chlorosilane. The reaction of the silicon-chlorine groups with carbon-lithium chain end yields a process free from deleterious side reactions. The only potential difficulty encountered is that due to the steric hindrance. Fetters et al.[110-113] applied this technique for the synthesis of $(AB)_n$ star block copolymers by the condensation of AB diblock anions with chlorosilane compound. On the other hand, Rempp et al.[114-117] reported the formation of star polymers by crosslinking monocarbanionic chains with divinyl benzene. Fetters et al. [111,118,119]also reported that an extention of the forgoing procedures led to the synthesis of $(AB)_n$ star block copolymers. At the same time, they investigated in detail the microphase separation of $(AB)_n$ star block copolymer films and found out a new morphology such as the ordered bicontinuous structure [118,119].

TABLE 6 Characteristics of $A_nB_n$ star block copolymers

| No. | $A_nB_n$ star block copolymer | | | |
|---|---|---|---|---|
| | Conversion[a] (%) | $10^{-4}\bar{M}_w$[b] | $\bar{M}_w/\bar{M}_n$ | Arm number |
| $A_nB_n$-1 | 23.4 | 5.38 | 1.08 | 2.8 |

[a] Calculated from the areas of GPC distributions.
[b] Determined by GPC using double detectors (RI and LS)
Taken from Ref. 101.

Ishizu et al.[120] have made clear that the direct coupling between poly[ $\alpha$-methylstyrene(MS)-b-isoprene (I)] diblock living anions end-capped with MS units and p-chloromethylstyrene (CMS) formed the vinylbenzyl-terminated poly(MS-b-I) diblock macromonomer having both high functionality and narrow polydispersity. The $S_N$ reaction of carbanions with chloromethyl groups is much faster than the addition reaction of carbanions to vinyl groups. So, if we choose the reaction condition that a large excess of poly(MS-b-I) diblock anions (6) is reacted with CMS, the AB diblock macromonomers equivalent to the feed mole of CMS forms at the first stage of the reaction. Subsequently the anionic polymerization of residual AB diblock anions with AB diblock macromonomers proceeds in these reaction systems by the following scheme.

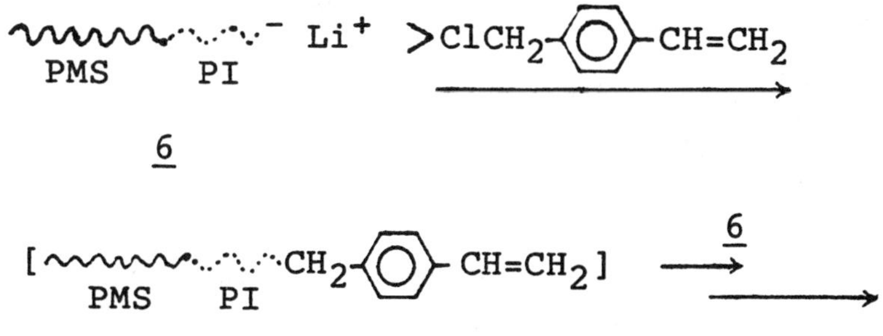

$$\underset{\text{PMS}\quad\text{PI}}{\sim\sim\sim\cdots\cdots^{-}}\ Li^{+}\ \xrightarrow{>ClCH_2-C_6H_4-CH=CH_2}$$

6

$$[\underset{\text{PMS}\quad\text{PI}}{\sim\sim\sim\cdots\cdots}CH_2-C_6H_4-CH=CH_2]\ \xrightarrow{6}\ \longrightarrow$$

Comb-Shaped $(AB)_n$ Copolymer

According to the above design, the comb-shaped $(AB)_n$ copolymers (n=4-15) were formed by means of the one-shot anionic polymerization of poly(MS-b-I) macromonomers with CMS [121]. This primary structure is very similar to that of $(AB)_n$ star block copolymer. More recently, Ishizu et al.[122]have suggested a novel architecture of $(AB)_n$ star block copolymers by means of micro-gelation in micelle between AB diblock macromonomers and low molecular weight difunctional unsaturated compounds as linking agents. Figure 7 shows the schematic representation on conformation of AB diblock macromonomers in micelle. In general, block copolymers form the micelle or mesomorphic structure in solution. The terminal polymerizable groups of AB diblock macromonomers are located with the condensed concentration within a micelle. So, free radical polymerization of AB diblock macromonomers seem to proceed in the organized-field, such as micellar domains. Where the block junctions are located regularly at the interface of micelles.

Vinylbenzyl-terminated poly(S-b-I) AB type diblock macromonomers were prepared by the direct coupling of corresponding diblock anions with a large excess of CMS [122]. Table 7 lists characteristics of AB diblock macromonomers. The end group functionality of these diblock macromonomers was close to unity. Diblock macromonomers showed a narrow polydispersity ($M_w/M_n$=1.04-1.09).

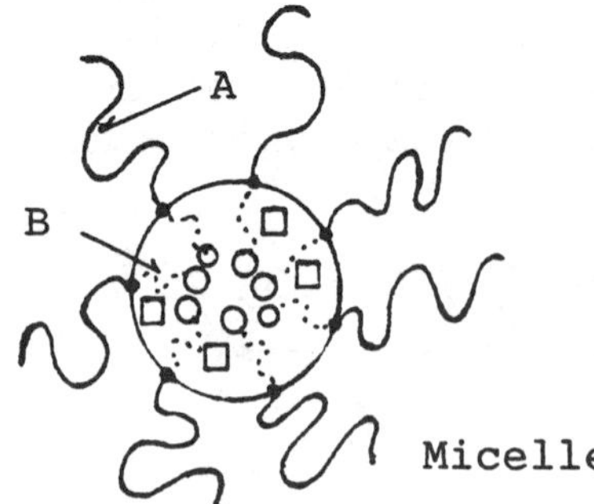


O; Polymerizable group of AB diblock macromonomer end
•; Block junction
□; Crosslinking agent

*Fig. 7 Schematic representation on conformation of AB diblock macromonomers in micelle.*

TABLE 7 Characteristics of poly(styrene-b-isoprene) AB diblock macromonomers

| | Diblock macromonomer | | | |
|---|---|---|---|---|
| Code | $10^{-4} \cdot \bar{M}_n$ [a] | $\bar{M}_w/\bar{M}_n$ [b] | PS[c] blocks (wt%) | Vinylbenzyl[d] groups (per chain) |
| SIM-1 | 3.70 | 1.09 | 80.0 | 0.95 |
| SIM-2 | 2.50 | 1.04 | 51.2 | 0.97 |

[a] Determined by universal calibration on a GPC.
[b] Determined by GPC.
[c] Determined by $^1$H-NMR in $CDCl_3$.
[d] Determined by the area ratio of RI and UV (292 nm) intensities in a GPC chart.

Free radical polymerization of diblock macromonomer SIM-2 (PS block 51.2wt%) was carried out in benzene initiated by AIBN, varying the feed SIM-2 concentration [M]. This polymerization system will form the comb-shaped $(AB)_n$ copolymer as a primary structure. Figure 8 shows the plots of arm number (n) /or polydispersity $M_w/M_n$ of comb-shaped copolymers against the feed SIM-2 concentration [M]. It can be seen from this figure that no homopolymerization occurs at [M] less than $0.2 \times 10^{-2}$mol/L. Beyond this concentration, both arm number and $M_w/M_n$ increase gradually with an increase of [M]. However, the arm number is small (n=3-5) and its polydispersity is very narrow ($M_w/M_n$–1.1-1.3). In general, block copolymers form the micelle or mesomorphic structure in solution. So, free radical polymerizations of these experimental series seem to proceed in the organized-field, such as micellar domains.

Subsequently, micro-gelation of SIM-2 diblock macromonomer was carried out with the addition of ethylene glycol dimethacrylate (EGDM) into above polymerization systems. The EGDM is compatible with PI but not so compatible with PS segments. So, the feed EGDM may be omnipresent in PI domains of micelles formed by diblock macromonomer. This system will form $(AB)_n$ star block copolymers as a primary structure. Macro-gelation was observed only in the polymerization condition: [M]=$0.96 \times 10^{-2}$mol/L, [EGDM]/[M]=0.10 mol/mol. Other experimental series were maintained the solution state during copolymerization. This phenomenon suggests that the copolymerization of SIM-2 diblock macromonomer with EGDM leads to micro-gelation in PI domains of micelles formed by diblock macromonomers. Figure 9 shows the plots of the arm number /or $M_w/M_n$ of $(AB)_n$ star block copolymers against the feed SIM-2 concentration, in the case of the same mole ratio of [EGDM] to [M] ([EGDM]/[M]=0.04 mol/mol). Beyond [M]=$0.2 \times 10^{-2}$mol/L, the arm number increases rapidly with an increase of [M]. The maximum of n appears at ca. [M]=$2.0 \times 10^{-2}$mol/L (polymer concentration=50.0 wt%). And beyond [M]=$2.0 \times 10^{-2}$ mol/L, the arm number decreases rapidly with an increase of [M]. On the other hand, the polydispersity is almost constant ($M_w/M_n$=ca. 1.5) in the range of [M]=$0.2 \times 10^{-2}$

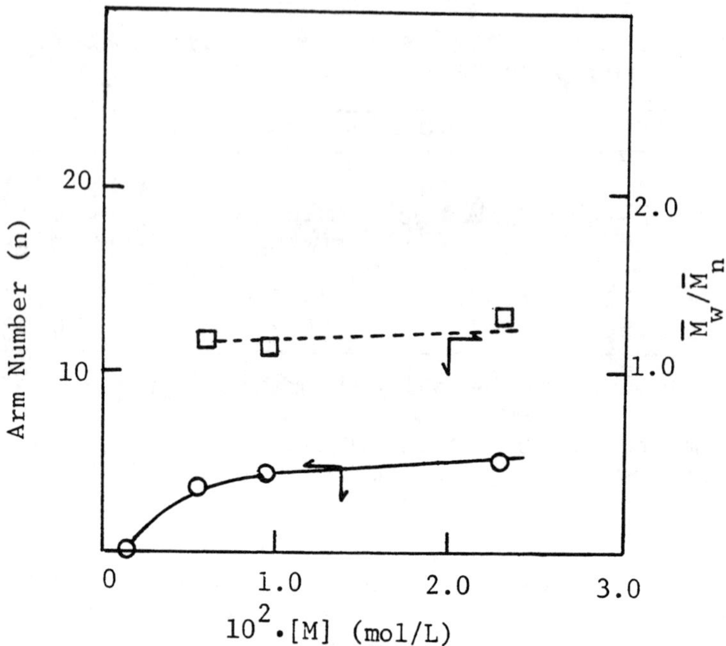


*Fig. 8* *Plots of arm number or polydispersity of comb-shaped $(AB)_n$ copolymers against the feed SIM-2 concentration.*

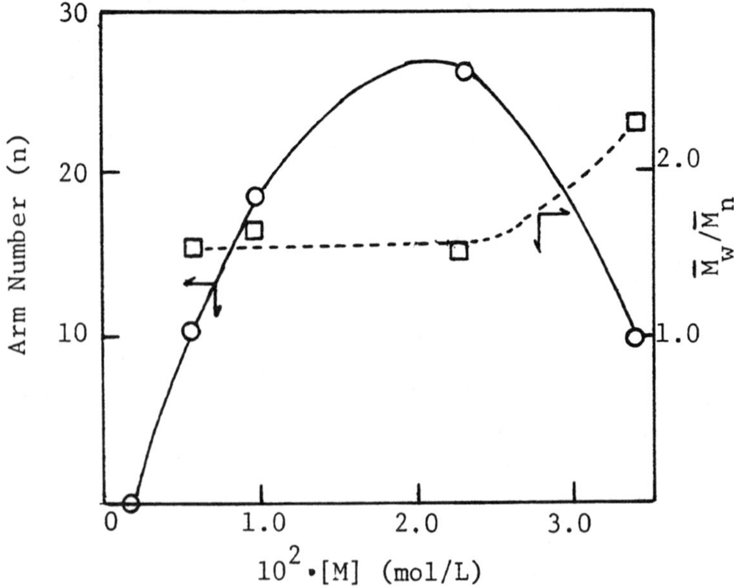


*Fig. 9* *Plots of arm number or $M_w/M_n$ of $(AB)_n$ star block copolymers against the feed SIM-2 concentration, in the case of [EGDM]/[M]=0.04 mol/mol.*

-2.0x$10^{-2}$mol/L. However, its value increases rapidly beyond [M]=2.0x$10^{-2}$mol/L. In the region of this polymer concentration (85.0 wt%; [M] = 3.40x$10^{-2}$mol/L), the mobility of diblock copolymer radicals in micellar domains seems to decrease extremely due to high viscosity of polymer solution. Thus the micro-gelation system is more advantageous than the free radical polymerization to control the arm number. In micro-gelation system, the radical species are activated by copolymerization of SIM-2 with EGDM. The arm number depends strongly on the feed SIM-2 macromonomer concentration.

It is suggested from above results that the micro-gelation proceeds in PI domains formed by diblock macromonomers. We tested to observe directly the TEM image of cross-section of the freezing dry film. Figure 10a shows the cross-sectional TEM micrograph of SIM-2 freezing dry film from 37.5 wt% of polymer concentration. The dark portions are the selectively stained PI blocks with $OsO_4$. The texture shows lamellar-like morphlogy. It is found from this micrograph that the SIM-2 diblock macromonomer exhibits such organized-structure in benzene solution. The terminal vinylbenzyl groups of diblock macromonomers are located at the inside of PI domains. The crosslinking reaction of terminal vinylbenzyl groups with EGDM is able to take place in the localized field, such as PI lamellar domains.

Next, we focused on the dependence of the arm number on the feed mole ratio of [EGDM] to [M]. Figure 11 shows the plots of the arm number /or $M_w/M_n$ of $(AB)_n$ star block copolymers against [EGDM]/[M], in the condition that the feed diblock macromonomer concentration is constant ([M]=0.96x$10^{-2}$mol/L). The arm number increases rapidly with an increase of [EGDM]/[M]. On the other hand, the $M_w/M_n$ increases gradually with an increase of [EGDM]/[M]. These copolymerization systems lead to macro-gelation at near [EGDM]/[M]=0.10 mol/mol. It is indicates from these results that the arm number is possible to control by changing not only the feed diblock macromonomer concentration [M] but also the feed ratio of [EGDM]/[M].

Micro-gelation of SIM-1 diblock macromonomer (PS block 80.0wt%) was also carried out in benzene. Fig. 10b shows the cross-sectional TEM micrograph of SIM-1 freezing dry film from 55.5 wt% of polymer solution ([M]=1.5x$10^{-2}$mol/L). The texture shows the morphology of the island (PI) - sea structure. So, the terminal vinylbenzyl groups of SIM-1 are located at the inside of PI spherical domains. Figure 12 shows the plots of the arm number /or $M_w/M_n$ of $(AB)_n$ star block copolymers against the feed SIM-1 concentration. Where the ratio value (0.04 mol/mol) of [EGDM]/[M] is constant for all experimental series. Beyond 0.10x$10^{-2}$mol/L of macromonomer concentration, the arm number increases in rough proportion to the feed diblock macromonomer concentration. The polydispersity is very narrow ($M_w/M_n$=1.1-1.2) for all experimental series. It is seen that the micro-gelation of diblock macromonomers proceeds effectively in the spherical domains compared with lamellar domains. Crosslinking reaction within the spherical domains is subject to restriction of the reaction-field compared with that within lamellar domains due to the small curvature of the domain interface.

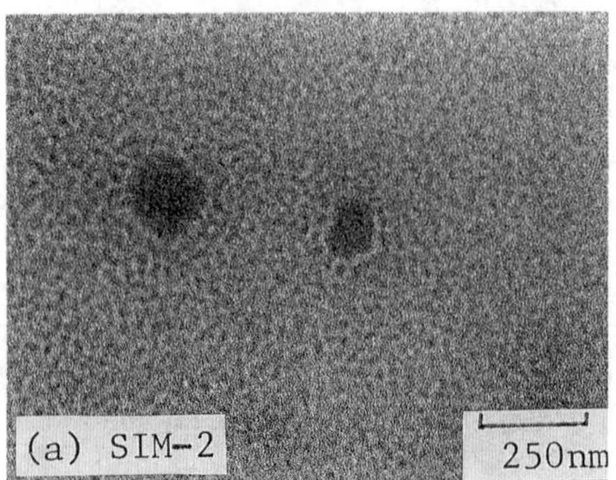


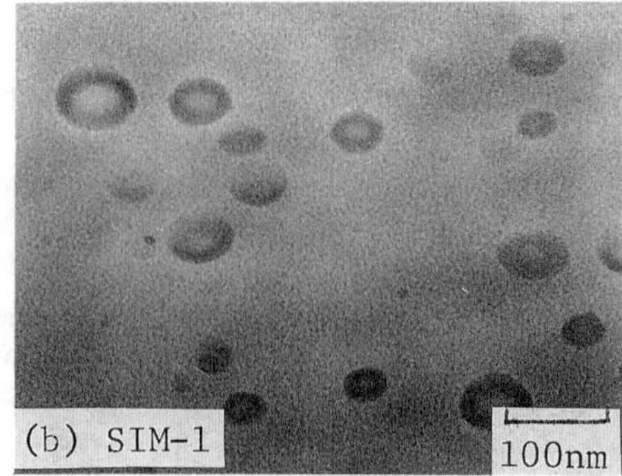


*Fig. 10 Cross-sectional TEM micrographs of micellar structures of diblock macromonomers: (a), SIM-2 freezing dry film from 37.5 wt% of polymer solution; (b), SIM-1 freezing dry film from 55.5 wt% of polymer solution.*

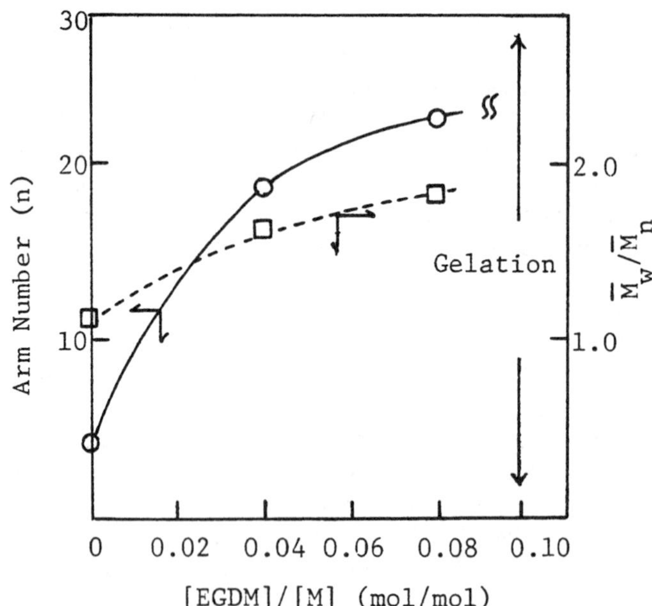


*Fig. 11 Plots of arm number or $M_w/M_n$ of $(AB)_n$ star block copolymers against the feed ratio of [EGDM]/[M], in the condition that the feed SIM-2 concentration is constant ([M]=0.96x10$^{-2}$ mol/L).*

In conclusion, $A_nB_n$ and $(AB)_n$ type star block copolymers can be synthesized by means of organized-polymerization of block macromonomers.

## 3.3 Microspheres

It was mentioned previously that block and graft copolymers formed micelles in a selective solvent. Poly[S-b-butadiene(Bu)-b-S] block copolymer micelles with cores of PBu blocks in dilute solution were stabilized by crosslinking of the chains in the micellar cores by UV irradiation in presence of a photoinitiator and by fast electrons [123-125]. The irradiated, stabilized micelles, examined by light scattering, sedimentation, and GPC, did not decompose upon heating or in good solvents for both blocks. Fukutomi et al.[126-132] have also obtained such core-shell type polymer microspheres by crosslinking of core domains in micelles formed in selective solvent for several block and graft copolymers.

More recently, we have established a preparative method for core-shell type microspheres by the crosslinking reaction of diblock copolymer films with microphase-separated spherical cores with crosslinking reagents in the solid state. As a matter of fact, the graft copolymers prepared by macromonomer methods also form spherical microdomains in the solid state. So, this method of preparation for core-shell type microspheres is possible to apply for graft copolymer films with

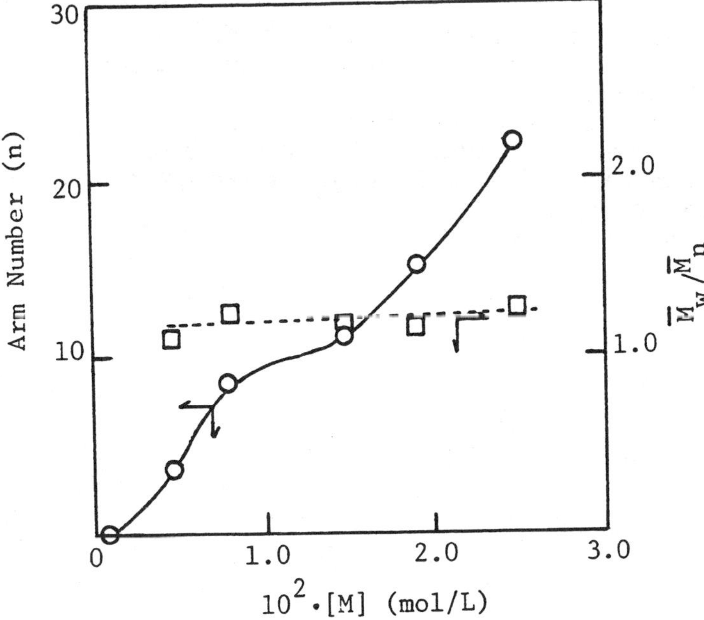


*Fig. 12 Plots of arm number or $M_w/M_n$ of $(AB)_n$ star block copolymers against the feed SIM-1 concentration.*

microphase-separated spherical cores. In this section, the author outlines a few of recent studies concerning the synthesis method for core-shell type microspheres and their properties in solution and solid.

The well-defined poly(S-b-4VP) diblock copolymer was prepared by the usual anionic addition polymerization using n-BuLi as an initiator in THF at -78° C [133]. Table 8 lists the characteristics of the monodisperse diblock copolymer SV1 and the microdomain spacing of the specimen cast from 1,1,2-trichloroethane (TCE). Figure 13 shows the schematic representation of the core-shell microspheres in solution prepared from poly[S-b(g)-4VP] block or graft copolymer films. This microphase-separated structure is more stable in thermal equilibrium than the micelle formed in selective solvent. The representation of the core-shell microspheres indicates the structure in which the P4VP chains in core are crosslinked in the state of maintaining spherical microdomains.

TABLE 8 Characteristics of poly(styrene-b-4-vinylpyridine) diblock copolymer and domain spacing

| Specimen code | $10^{-4}\cdot\overline{M}_n$ Block copolymer[a] | $10^{-4}\cdot\overline{M}_n$ P4VP block[b] | P4VP block[c] (wt%) | Shape of P4VP domain | Domain radius $\overline{R}_{P4VP}$[d] (nm) |
|---|---|---|---|---|---|
| SV1 | 11.2 | 2.7 | 24 | Sphere | 17 |

[a] Measured by osmometry in TCE at 37°C.
[b] Estimated from the $\overline{M}_n$ data of the diblock copolymer and PS precursor.
[c] Determined from infrared spectra.
[d] Determined by means of electron microscopy of specimen cast from TCE. $\overline{R}_{P4VP}$=average radius of P4VP sphere structure.
Taken from Ref. 133.

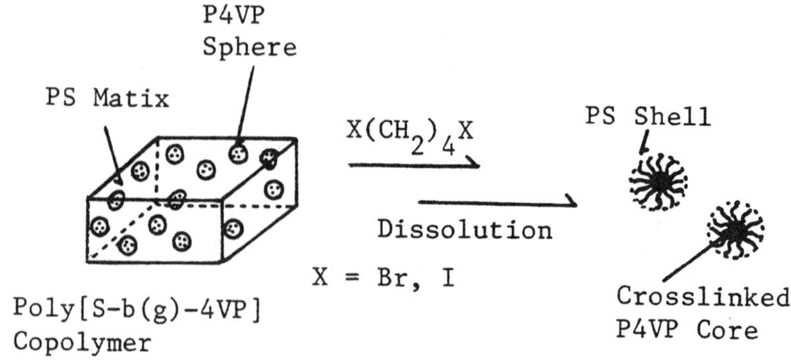


*Fig. 13 Schematic presentation of core-shell type polymer microspheres prepared from poly[S-b(g)-4VP] diblock or graft copolymer film with P4VP spherical microdomains.*

Figure 14 shows TEM micrograph of a diblock copolymer SV1 specimen cast from TCE. The dark portions are the selectively stained P4VP blocks. The microdomain structure in SV1 specimen (PS 24 wt%) shows the texture of discrete P4VP spheres in a PS matrix. The absolute value of the radius of the P4VP spheres, $R_{P4VP}$ is far below the theoretical value (32 nm) [133]. The TCE used is a good solvent for PS and a somewhat poor one for P4VP blocks, judging from the Hildebrand parameter. This morphology is considered to demonstrate the nonequilibrium state of the spherical domain system. These segregated P4VP chains in a sphere were crosslinked by using quaternization with 1,4-dibromobutane (DBB) vapor in the solid state.

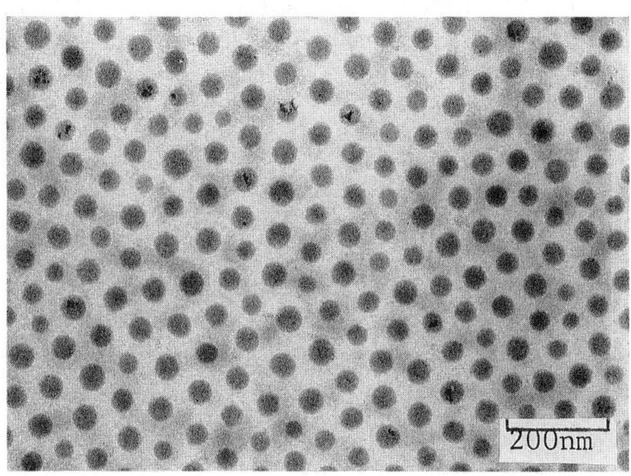


*Fig. 14 TEM micrograph of SV1 diblock copolymer specimen cast from TCE. Taken from Ref. 133.*

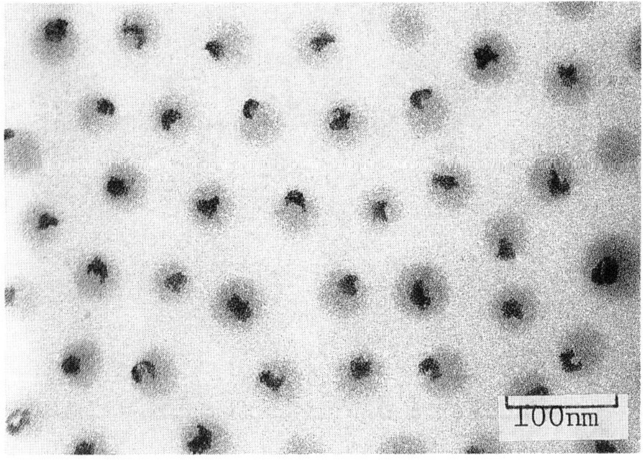


*Fig. 15 TEM micrograph of core-shell type polymer microsphere SV1M. Taken from Ref. 134.*

Figure 15 is a TEM micrograph of crosslinked product SV1M cast from 0.1 wt% TCE/nitrobenzene: 10/1 (v/v) solution [134]. In the preparation conditions of this cast specimen, the multimolecular micelles are frozen with a structure like the core-shell type polymer microspheres. The dark, grey, and white portions indicate the crosslinked P4VP cores, shell of PS chains, and carbon support, respectively. This micrograph shows clearly the structure of the core-shell type polymer microspheres. It is clear that these microsphere particles have only a narrow size distribution.

The turbidity curves of the core-shell microsphere SV1M and ionic block copolymer are shown in Figure 16. Where the corresponding ionic block copolymer was prepared from the quaternization of SV1 with methyl iodide, the ionic block copolymer of SV1 shows such behavior. It was observed that the appearance of new micelles was accompanied by the sharp maximum of the turbidity (water fraction=0.17). This behavior is explained as follows: At first, the micelles with a quaternized P4VP core and a shell of solvated PS chains are formed in THF solution, and this micellar form changes into the reversed micelles with a PS core and shell of ionized P4VP chains at the water fraction of about 0.17. These reversed micelles were stabilized until the water fraction of about 0.6, due to the strong ionic strength of quaternized P4VP shell. However, the core-shell microsphere SV1M prepared by crosslinking of the segregated P4VP chains in core are very stable in THF/water mixture. This result is explained by the formation of stabilized core-shell type polymer microspheres with not only a crosslinked but also a quaternized P4VP core and a shell of PS chains.

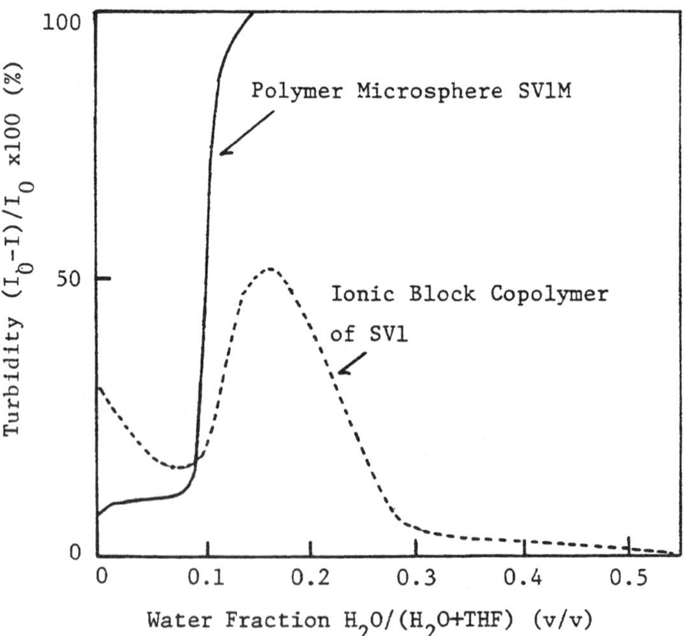


*Fig. 16 Turbidity curves of the core-shell type polymer microsphere and ionic block copolymer. Taken from Ref. 133.*

The particle size distribution of core-shell microsphere SV1M was measured in benzene or in THF dispersion by dynamic light scattering. The SV1M microsphere had narrow particle size distribution in both dispersions. The average-diameters (D) of SV1M were 159.9 (in benzene) and 172.2 nm (in THF), respectively.

Solution behavior of core-shell microspheres was made more clear by studies on the microsphere with polyisoprene (PI) core and PS shell. The preparation method of core-shell type polymer microspheres from poly[S-b-isoprene(I)] diblock copolymer films is described [135]. The monodisperse poly(S-b-I) diblock copolymers were prepared by the usual sequential anionic addition in benzene using sec-BuLi as the initiator. The microstructure of PI blocks obtained from this method contained 50% 3,4 and 50% 1,4 structures (60% cis, 40% trans) [136]. The results of polymerization reactions are shown in Table 9.

TABLE 9 Characterization of poly(S-b-I) diblock copolymers and microdomain size

| | $10^{-4} \cdot \bar{M}_n$ | | | Domain size[d] |
|---|---|---|---|---|
| Sample code | PS[a] block | Block[b] copolymer | Wt% of PI[c] blocks | $\bar{R}$ (nm) |
| SI1 | 0.99 | 1.08 | 8 | - (1.2) |

[a] Determined by GPC.
[b] Determined by $\bar{M}_n$ of PS precursor and composition data.
[c] Determined by $^1$H-NMR.
[d] $\bar{R}$ is the average radius of PI spheres. Parenthese shows the equilibrium value calculated from Helfand's theory [79].
Taken from Ref. 135.

Crosslinking of the microphase-separated PI chains in the core was carried out by soaking in n-hexane solution of sulfur monochloride ($S_2Cl_2$) at room temperature, varying the reaction time. It is well known that unsaturated allyl groups of PI units are crosslinked by addition with $S_2Cl_2$ [137]. Intermolecular crosslinking of PI units (1,4 structure) in a core with $S_2Cl_2$ is represented by the following scheme.

$$-CH_2-\overset{CH_3}{\overset{|}{C}}=CH-CH_2- \xrightarrow{S_2Cl_2} \begin{array}{c} -CH_2-\overset{CH_3}{\overset{|}{\underset{Cl}{\underset{|}{C}}}}-CH-CH_2- \\ \qquad\quad | \\ \qquad\quad S_2 \\ \qquad\quad | \\ -CH_2-\overset{Cl}{\overset{|}{\underset{CH_3}{\underset{|}{C}}}}-CH-CH_2- \end{array}$$

Figure 17 shows GPC profiles of crosslinked products SI1M1-SI1M3 (starting material: SI1 film 50 m thickness). The GPCs of SI1M1-SI1M3 show a bimodal pattern. The GPC elution peaks at the side of high molecular weight for SI1M1-SI1M3 are shifted to the side of high molecular weight as compared with that of SI1 diblock copolymer as starting material. On the other hand, the elution peaks appeared at the side of low molecular weight are identical with that of SI1 diblock copolymer. In the crosslinking condition of SI1M1 (n-hexane solution of 1 wt% $S_2Cl_2$, reaction time 6 h), the segregated PI chains in core are crosslinked partially with $S_2Cl_2$ due to heterogeneous reaction in the solid state. However, it is found from SI1M2 (3 wt% $S_2Cl_2$, 48 h) and SI1M3 (20 wt% $S_2Cl_2$, 3 h) profiles that these segregated PI chains in the core are fixed almost completely by crosslinking reaction under the conditions of long reaction time (SI1M2) or high concentration of crosslinking reagent (SI1M3).

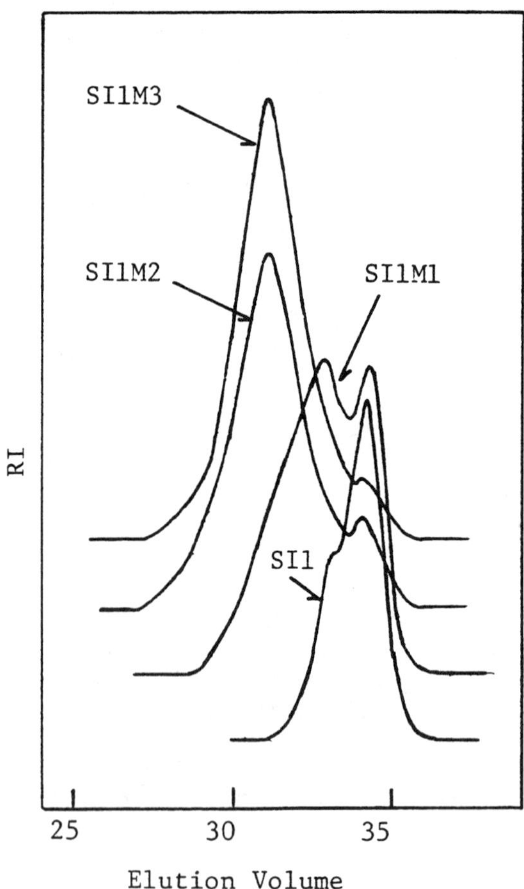


*Fig. 17 GPC profiles of microspheres SI1M1-SI1M3: SI1M1: 1 wt% $S_2Cl_2$, reaction time 6 h; SI1M2: 3 wt% $S_2Cl_2$, 48 h; SI1M3: 20 wt% $S_2Cl_2$, 3 h.*

Figure 18 shows the typical particle size distribution of crosslinked products SI1M1-SI1M3 in benzene by means of dynamic light scattering. The SI1M1 has two kinds of particle species (particle size: 13 and 66 nm). The result corresponds to the GPC pattern of SI1M1 (see Fig. 17). Dynamic light scattering data support the assumption that the crosslinked SI1M1 is a mixture of the unfixed SI1 diblock copolymer and polymer microspheres in which PI chains in core is partially fixed. The SI1M2 and SI1M3 polymer microspheres show almost the same average particle size (95 nm), with narrow distribution. Nevertheless, polymer particles are dispersed homogeneously in solution. Figure 19 shows typical $^1$H-NMR spectra of the SI1 diblock copolymer, and of the SI1M1 and SI1M2 core-shell microspheres in $CDCl_3$. In NMR spectrum of SI1 diblock copolymer, peaks due to PS segments appear at 6.5-7.4 ppm (aromatic protons) and peaks due to the PI segments appear at 5.2 (1,4 structure of double bond) and 4.7 ppm (3,4 structure of double bond), and 1.6 ppm (methyl protons), respectively. In the spectrum of SI1M1, these peaks due to PI segments decrease as compared with those of SI1 diblock copolymer. The spectrum of SI1M2 polymer microsphere shows only these peaks due to PS segments, but SI1M2 is apparently soluble in $CDCl_3$, giving a clear solution. It was found from IR spectrum of SI1M2 that the conversion of addition reaction of double bond with $S_2Cl_2$ did not reach 100%. Nevertheless, the segregated PI chains in the core are fixed effectively with intermolecular crosslinking. So, the NMR data indicate clearly that the crosslinked PI segments constitute a micellar core which behaves as a solid on the time scale of segment relaxation. The solubility of the core-shell type polymer microspheres depends strongly on the solubility of block chains constructing shell parts. The micelle of core-shell microspheres moves like a pseudolatex.

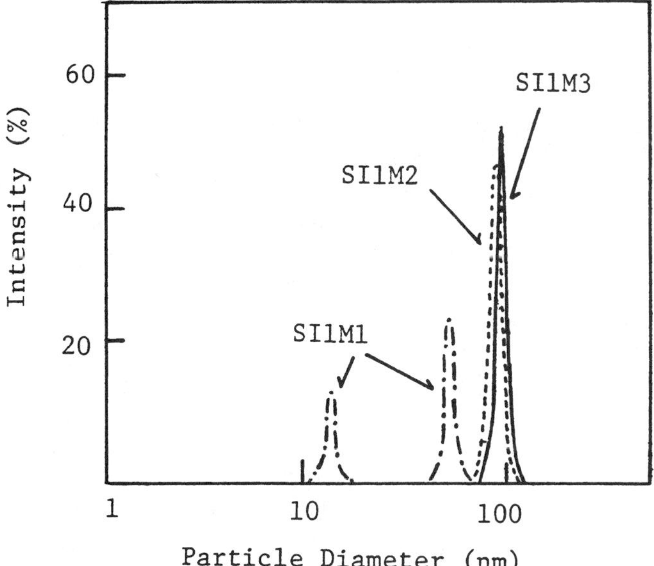


*Fig. 18 Particle size distributions of core-shell type polymer microspheres SI1M1-SI1M3. Taken from Ref. 135.*

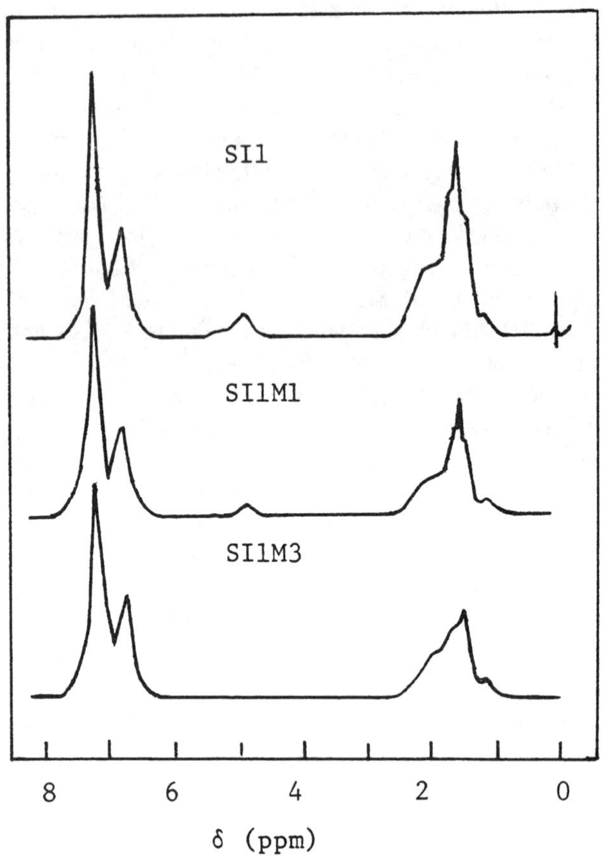


*Fig. 19 ¹H-NMR spectra of SI1 diblock copolymer and polymer microspheres SI1M1 and SI1M2 in $CDCl_3$. Taken from Ref. 135.*

The structure of core-shell microspheres is very similar to that of star polymers. Daoud and Cotton [138] were the first to study the conformation and dimensions of stars by means of scaling ideas. The picture presented for a star polymer consists of three regions: a central core, a shell with a semidilute segment density in which the arms have the unperturbed chain conformation, and an outer shell in which the arms of the star assume a self-avoiding conformation. Stars with many arms (the critical number of arms is estimated to be of the order $10^2$) are expected to form a crystalline array in near the overlap concentration ($C^*$) [139]. Star polymers with many arms may also be good models for by grafted polymers. Indeed, in the Daoud-Cotton picture of star polymers, the arms of a star are contained within an outwardly widening cone, and a similar picture is used to describe colloidal stabilization by graft polymers [140, 141].

The micelle of core-shell type polymer microspheres prepared in our works moves like a pseudolatex. So, we have investigated the possibility of macrocrystal ordering in core-shell type microsphere solution [142]. In this work, the core-shell microsphere is employed the structure of quaternized P4VP core and PS shell (SV1M microsphere). We calculated the overlap concentration $C^*$ of core-shell type polymer microsphere SV1M as follows.

The molecular weight of segregated P4VP cores [$M_{w(c)}$] is given by eq. 2.

$$M_{w(c)} = (4\pi /3)R_{P4VP}{}^{3}\ \rho_{P4VP}\ N_A \qquad (2)$$

where $R_{P4VP}$, $\rho_{P4VP}$, and $N_A$ are the radius of P4VP core, density of P4VP, and Avogadro number ($6.023\ x10^{-23}\ mol^{-1}$), respectively. So, the $R_{P4VP}$ and $\rho_{P4VP}$ are adopted 17 nm (see Fig. 15) and the density ($11.4\ x10^{3}\ mol\ m^{-3}$) of P2VP [143], respectively.

The aggregation number N of SV1M microsphere is given by eq. 3.

$$N = M_{w(c)}/M_{P4VP} \qquad (3)$$

where $M_{P4VP}$ is the molecular weight of P4VP block in diblock copolymer SV1. The aggregation number N of SV1M microsphere can be calculated to be 550 by using eq. 2 and 3.

The overlap concentration $C^*$ is given by eq. 4.

$$C^* = 3M_{w(m)}/(4\ \pi\ <S^2>^{3/2}N_A) \qquad (4)$$

where $<S^2>$ is the mean square radius of gyration of core-shell microsphere. In this work, the $<S^2>^{1/2}$ is assumed to be equal to 1/(2D) (D. the average-diameter determined by dynamic light scattering; D=159.9 nm in benzene for SV1M microsphere). The $M_{w(m)}$ ($6.16x10^{7}$) is the molecular weight of SV1M microsphere. So, the $C^*$ can be estimated to be $4.77x10^{-2}$g/mL in benzene. It is expected that the structure of crystalline array for SV1M microsphere may appear at ca. 5 wt% polymer concentration of benzene dispersion.

We have studied the arrangement of core-shell microsphere SV1M from its dispersion on the carbon substrate. Figure 20 shows scanning electron micrograph (SEM) of SV1M cast from 0.3 wt% dispersion of benzene. It is clear that particles of SV1Mmicrospheres are frozen on carbon substrate and these particles have only a narrow size distribution. The average-diameter of particles (D=24 nm) obtained from the SEM image is the same as that obtained from TEM (see Fig. 15). Moreover, the particles of core-shell microspheres are well-aligned in a hexagonal array on carbon substrate.The mechanism on ordered arrangement of SV1M microspheres is speculated as follows. The SV1M specimen cast from 0.3wt% dispersion of benzene reaches near the $C^*$ concentration during solvent evaporation. The structure of macrocrystalline array appears with the form of dispersed monolayer at this concentration. In order to make clear precisely the existence of macrocrystalline solution, it is necessary to measure the small angle X-ray scattering, etc. of core-shell microspheres in near the $C^*$ concentration.

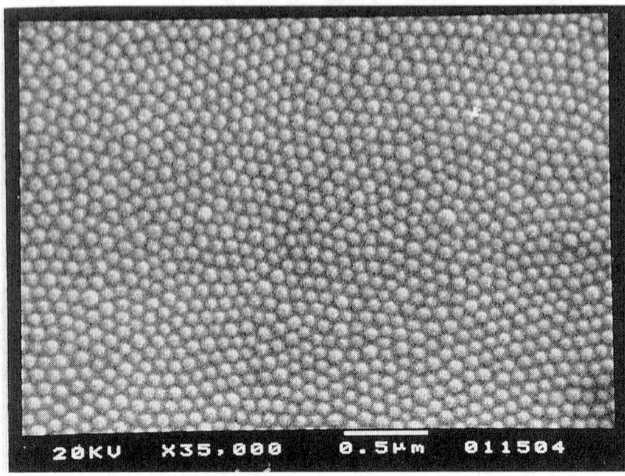


Fig. 20 SEM micrograph of SV1M microsphere specimen cast from 0.3 wt% dispersion of benzene. Taken from Ref. 142.

More recently, there have been a few studies reported on microspheres prepared from dispersion or emulsion copolymerizations of hydrophilic macromonomer with hydrophobic vinyl monomers [144-146]. In these systems, grafted chains (hydrophilic macromonomer) play an important part of the colloidal stabilization for hydrophobic core parts.

## 3.4 Dendritic Macromolecules

Dendritic macromolecules are hyperbranched fractal-like structures that emanate from a central core and contain a large number of terminal groups. Up .to now, two synthetic approaches have been reported for the preparation of these macromolecules: a divergent approach which is characterized by the "starburst" and "arborol" structures of Tomalia [147,148]and Newkome [149],respectively, and more recently a "convergent growth" approach [150-153]. One limitation encountered in the synthesis of starburst structures is that, because the molecules used as building blocks have a low molecular weight, the rate of increase in molecular weight per generation is relatively low. Many synthetic steps are therefore necessary to produce high molecular weight materials. The divergent approach also has the disadvantage of being intrinsically very sensitive to incomplete or side reactions leading to structure imperfections. On the other hand, the convergent approach is characterized by reactions occurring at a single functional group at the focal point of the growing dendritic macromolecules. The presence of a single functional group at the focal point increases the ease of purification and the degree of control over the architecture of the final dendritic macromolecules.

More recently, Hawker and Frechet [154] have synthesized polyether macromonomer in which the macromolecular part is a hyperbranched dendritic moiety. Typical [G-4]-styrene macromonomer (8) was prepared from the reaction of hydroxymethyl-terminated [G-4]-OH polyether macromolecules (7) with p-chloromethyl styrene according to following scheme.

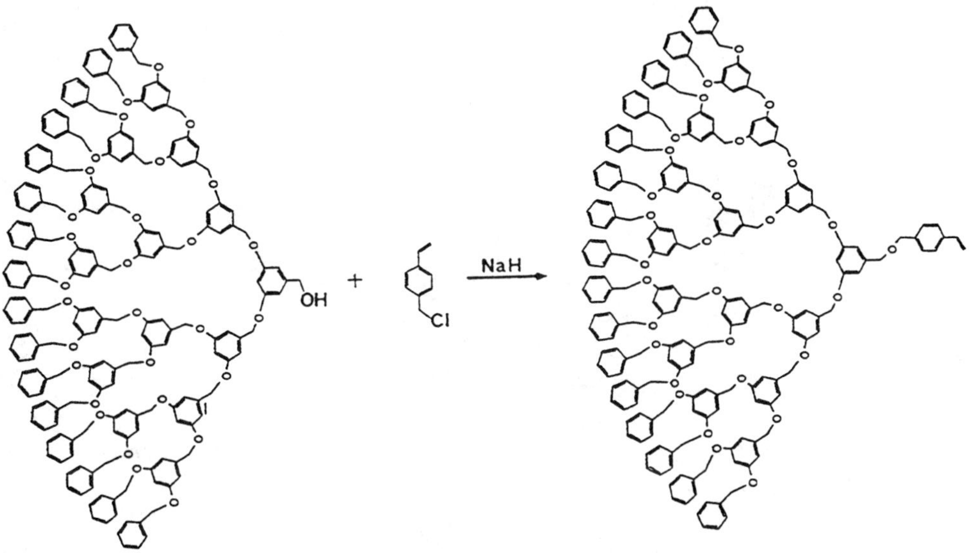

OH + Cl NaH

[G-4]-OH
7

[G-4]-styrene
8

Subsequently, several novel copolymers were prepared from copolymerization of the macromonomers with styrene by free radical polymerization. To avoid the synthetic problems, the macromonomer with hyperbranched dendritic moiety may be one of the most useful materials for the preparation of the dendritic macromolecules.

Mourey et al. [155] have reported the interesting characterization study of dendrimers based on benzyl ether linkages prepared by the convergent growth approach. Dendrimers obtained by coupling monodendron wedges to a trifunctional core 1,1,1-tris(4'-hydroxyphenyl)ethane, are referred to as tridendrons $[G\text{-}4]_3$-[C] (Figure 21). The accompanying designation $[G\text{-}4]_3$-[C] refers to three fourth-generation wedges attached to a core.

The intrinsic viscosity [$\eta$ ], refractive index, and hydrodynamic radius of dendrimers were obtained from size exclusion chromatography and low-angle laser light scattering measurements. Figure 22 shows the relation of the intrinsic viscosity of dendrimers with dendrimer generation g. The intrinsic viscosity of both monodendrons and

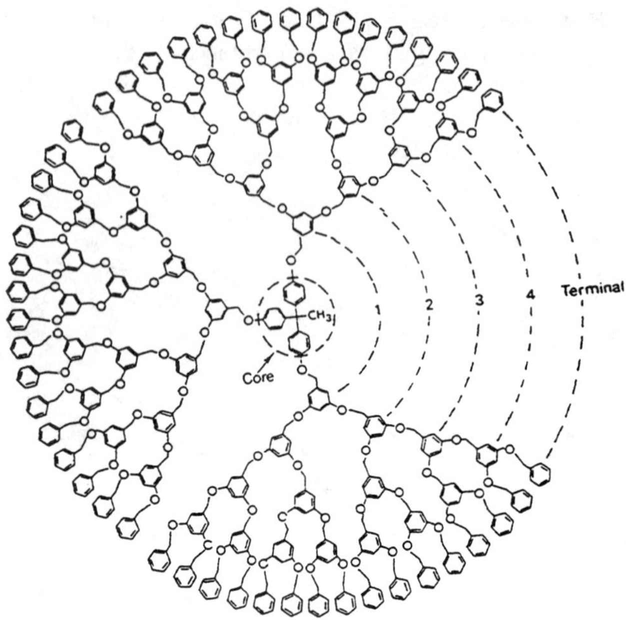


*Fig. 21 Representation of $[G\text{-}4]_3\text{-}[C]$ tridendron obtained by convergent growth. Taken from Ref. 155.*

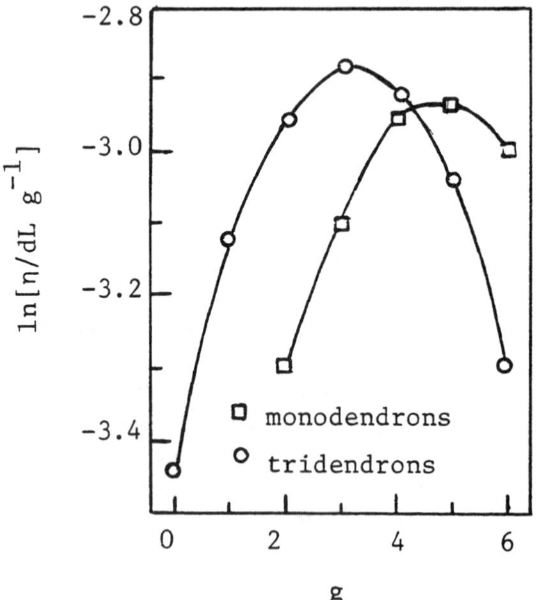


*Fig. 22 Intrinsic viscosity in THF of convergent growth dendrimers. Taken from Ref. 155.*

tridendrons passes through a maximum as a function of generation number g. While the refractive index increment appears to pass through a minimum. Hydrodynamic radii, calculated from intrinsic viscosity data, increase approximately linearly with dendrimer generation. These properties distinguish these dendrimers from completely collapsed, globular structures.

A model developed by Lescanec and Muthukumar [156] predicts inward folding of branch units, a density maximum in the center, and a distribution of terminal groups throughout the structure. The model is based on a self-avoiding walk algorithm that simulates kinetic growth of dendrimers. The experimental data are in agreement with simulations of Lescanec and Muthukumar. This suggests, but does not directly prove, that the density of these dendrimers decreases monotonically from the center.

## 4. CONCLUSION

Macromonomers were synthesized not only by ionic and radical polymerizations, but also by polycondensation and polyaddition reactions. So, we have a great variety to choose both chemical component and terminal functional group of macromonomers. Homopolymerization and copolymerization of macromonomers form comb-like and graft copolymers, respectively. Especially in copolymerization, the reactivity ratios of macromonomers play an important part for the speculation of the primary structure of graft copolymers.

These graft copolymers form the micelles in selective solvent and the microphase-separated structure in the solid state, respectively. The micelles can be used as a carrier of drug delivery systems. Two-phase morphology can be used to advantage in controlling properties such as modulus and permeability. One important characteristics of two-phase graft copolymers are in their facility for blending well with their respective homopolymers. The $A_nB_n$ and $(AB)_n$ type star block copolymers were synthesized by means of organized-polymerization of block macromonomers. The phase separation is easier in star block copolymers than in analogous block copolymers.

The core-shell type polymer microspheres were synthesized by crosslinking reaction of block and graft copolymer films with microphase-separated spherical cores with crosslinking reagents. The micelle of core-shell microspheres moves like a pseudolatex. These core-shell microspheres have the possibility to form a crystalline array in near the overlap concentration. Thus, macromonomers can be utilized as a starting material to prepare some high-performance materials.

## REFERENCES

1. R. Milkovich and M.T. Chiang, U.S. Pat. 3842050(1974) and subsequent patents.
2. G. O. Schulz and R. Milkovich, *J. Appl. Polym. Sci.*, 27, 4773 (1982).
3. G .O. Schulz and R. Milkovich, *J. Polym. Sci., Polym. Chem. Ed.*, 22, 1633 (1984).

4. R. Asami, M. Takaki, and H. Hanahata, *Macromolecules*, 16, 628 (1983).
5. P. Masson, E. Franta, and P. Rempp, *Makromol. Chem., Rapid Commun.*, 3, 499 (1982).
6. S. Yamashita, R. Shibatani, K. Takakura, and K. Imai, *Kobunshi Ronbunshu*, 39, 187 (1982).
7. K. Ito, Y. Masuda, T. Shintani, T. Kitano, and Y. Yamashita, *Polym. J.*, 6, 443 (1983).
8. K. Ishizu, S. Minematsu, and T. Fukutomi, *J. Appl. Polym. Sci.*, 43, 2107 (1991).
9. R. Asami and M. Takaki, *Makromol. Chem. Suppl.*, 12, 163 (1985).
10. P. R. Rao, P. Masson, P. Lutz, and G. Beinert, *Polym. Bull.*, 11, 115 (1984).
11. P. Rempp, P. Lutz, P. Masson, and E. Franta, *Makromol. Chem. Suppl.*, 8, 3 (1984).
12. M. Takaki, R. Asami, S. Tanaka, H. Hayashi, and T. E. Hogen-Esch, *Macromolecules*, 19, 2900 (1986).
13. F. Severini, M. Pegerare, and L. Saija, *Angew. Makromol. Chem.*, 133, 111 (1985).
14. P. Lutz, P. Masson, G. Beinert, and P. Rempp, *Polym. Bull.*, 12, 79 (1984).
15. B.C. Anderson, G.D. Andrews, P. Arthur, Jr., H.W. Jacobson, L.R. Melby, and W.H. Sharkey, *Macromolecules,* 14, 1601 (1981).
16. K. Ishizu and T. Fukutomi, *J. Polym. Sci., Polym. Chem. Ed.*, 27, 1259 (1989).
17. K. Hatada, H. Nakanishi, K. Ute, and T. Kitayama, *Polym. J.*, 18, 581 (1986).
18. T. Kitayama, S. Kishiro, E. Masuda, and K. Hatada, *Polym. Bull.*, 25, 205 (1991).
19. K. Ishizu and K. Mitsutani, *J. Polym. Sci., Polym. Lett. Ed.*, 26, 511 (1988).
20. P. Masson, G. Beinert, E. Franta, and P. Rempp, *Polym. Bull.*, 7, 17 (1982).
21. W.W. Schulz, J. Kaladas, and D.W. Schulz, *J. Polym. Sci., Polym. Chem. Ed.*, 22, 3795 (1984).
22. K. Ito, H. Tsuchida, A.Hayashi, T. Kitano, E. Yamada, and T. Matsumoto, *Polym. J.*, 17, 827 (1985).
23. M. Miyamoto, S. Sasagawa, R. Terada, S. Nagaoka, and Y. Mori, *Kobunshi Ronbunshu,* 42, 367 (1985).
24. S. Nagaoka, H. Takiuchi, K. Yokota, Y. Mori, H. Tanzawa, and T. Kikuchi, *Kobunshi Ronbunshu*, 39, 165, 173 (1982).
25. T. Hamaide, N. Mariaggi, J.L. Foureys, P. LePerchec, and A. Guyot, *J. Polym. Sci., Polym. Chem. Ed.*, 22, 3091 (1984).
26. L.J. Mathias, J.B. Canterberry, and M. South, *J. Polym. Sci., Polym. Lett. Ed.*, 20, 473 (1982).
27. T. Suzuki and T. Tomono, *J. Polym. Sci., Polym. Chem. Ed.*, 22, 2829 (1984).
28. Y. Kawakami, Y. Miki, T. Tsuda, R.A.N. Murthy, and Y. Yamashita, *Polym. J.*, 14, 913 (1982).
29. Y. Kawakami, R.A.N. Murthy, and Y. Yamashita, *Makromol. Chem.*, 185, 9 (1984).
30. Y. Kawakami, T. Aoki, and Y. Yamashita, *Macromolecules,* 18, 580 (1985).
31. G. C. Cameron and M. S. Chisholm, *Polymer*, 26, 437 (1985).
32. Y. Tezuka, A. Fukushima, and K. Imai, *Makromol. Chem.*, 186, 685 (1985).
33. G.D. Andrews and L.R. Melby, in New Monomers and Polymers, B.M. Culbertson and C.U. Pittrnan, Jr., Eds., Plenum, New York, 1984, p. 357.

34. O.W. Webster, W.R. Hertler, D.Y. Sogah, W.B. Farnham, and T.V. RajanBabu, *J. Am. Chem. Soc.*, 105, 5706 (1983).
35. R. Asami, Y. Kondo, and M. Takaki, in Recent Advances in Anionic Polymerization, T.E. Hogen-Esch and J. Smid, Eds., Elsevier, Amsterdam, 1986, p. 381.
36. T. Aida and S. Inoue, *Macromolecules,* 14, 1162 (1981).
37. T. Yasuda, T. Aida, and S. Inoue, *Makromol. Chem., Rapid Commun.,* 3, 585 (1982).
38. S. Inoue, *J. Macromol. Sci.-Chem.*, A25, 571 (1988).
39. T. Yasuda, T. Aida, and S. Inoue, *J. Macromol. Sci.-Chem.*, A21, 1035 (1984).
40. J.P. Kennedy and M. Hiza, *J. Polym. Sci., Polym. Chem. Ed.*, 21, 1033 (1983).
41. J.P. Kennedy and C.Y. Lo, *Polym. Bull.,* 13, 441 (1985).
42. M.F. Farona and J.P. Kennedy, *Polym. Bull.*, 11, 359 (1984).
43. R. Asami, M. Takaki, K. Kita, and E. Asakura, *Polym. Bull.*, 2, 713 (1980).
44. R. Asami, M. Takaki, K. Kyuda, and E. Asakura, *Polym. J.*, 15, 139 (1983).
45. J. Sierva-Vergas, J.G. Zilliox, P. Rempp, and E. Franta, *Polym. Bull.*, 3, 83 (1980).
46. J. Sierva-Vergas, P. Masson, G. Beinert, P. Rempp, and E. Franta, *Polym. Bull.*, 7, 277 (1982)
47. M. Takaki, R. Asami, and T. Kuwabara, *Polym. Bull.*, 7, 521 (1982).
48. R. Asami and M. Takaki, *Makromol. Chem., Suppl.*, 12, 163 (1985).
49. E.J. Goethals and M. Vlegels, *Polym. Bull.*, 4, 521 (1981).
50. S. Kobayashi, M. Kaku, S. Sawada, and T. Saegusa, *Polym. Bull.*, 13, 447 (1985).
51. S. Aoshima, K. Ebara, and T. Higashimura, *Polym. Bull.*, 14, 425 (1985).
52. M. Sawamoto, S. Aoshima, and T. Higashimura, *Makromol. Chem., Macromol. Symp.*, 13/14, 513 (1988).
53. M. Sawamoto, T. Enoki, and T. Higashimura, *Polym. Bull.,* 16, 117 (1986).
54. F.A. Waite, *J. Oil Col., Chem. Assoc.*, 54, 342 (1971).
55. D.J. Walbridge, in Dispersion Polymerization in Organic Media, K.E.J. Barret, Ed., Wiley, London, 1975, Chapter 3.
56. K. Ito, N. Usami, and Y. Yamashita, *Macromolecules,* 13, 216 (1980).
57. Y. Yamashita, Y. Tsukahara, K. Ito, K. Okada, and Y. Tajima, *Polym. Bull.*, 5, 335 (1981).
58. C. Bonardi, B. Boutevin, Y. Pietrasanta, and M. Taha, *Makromol. Chem.*, 186, 261 (1985).
59. T. Fukutomi, A. Yokota, and K. Ishizu, *J. Polym. Sci., Polym. Chem. Ed.*, 22, 2983 (1984).
60. K. Ishizu, K. Mitsutani, and T. Fukutomi, *J. Polym. Sci., Polym. Lett. Ed.*, 25, 287 (1987).
61. Y. Yamashita, K. Ito, H. Mizuno, and K. Okada, *Polym. J.*, 14, 255 (1982).
62. K. Ishizu, T. Ono, T. Fukutomi, and K. Shiraki, *J. Polym. Sci., Polym. Lett. Ed.*, 25, 131 (1987).
63. K. Ishizu and T. Fukutomi, *J. Polym. Sci., Polym. Lett. Ed.*, 25, 168 (1987).
64. K. Ishizu, A. Murakami, and T. Fukutomi, *J. Polym. Sci., Polym. Chem. Ed.,* 24, 1441 (1986).

65. B. Boutevin, Y. Pietrasanta, H. Taha, and T. E. Sarraf, *Polym. Bull.*, 10, 157 (1983).
66. B. Boutevin, Y. Pietrasanta, H. Taha, and T. E. Sarraf, *Makromol. Chem.*, 184, 2401 (1983).
67. M. Akashi, I. Kirikihara, and N. Miyauchi, Angew. *Makromol. Chem.*, 132, 81 (1985).
68. T. Fukutomi and K. Ishizu, *J. Polym. Sci., Polym. Lett. Ed.*, 25, 175 (1987).
69. K. Ishizu and M. Amemiya, *J. Membr. Sci.*, 54, 75 (1990).
70. P. Ferruti, E. Martuscelli, L. Nicolais, M. Palma, and F. Riva, *Polymer,* 18, 387 (1977).
71. P. Ferruti, D. Arnoldi, M.A. Marchisio, E. Martsuscelli, M. Palma, F. Riva, and L. Provenzale, *J. Polym. Sci., Polym. Chem. Ed.*, 15, 2151 (1977).
72. V. Percec, P.L. Rinaldi, and B.C. Auman, *Polym. Bull.*, 10, 397 (1983).
73. Y. Tsukahara, K. Mizuno, A. Segawa, and Y. Yamashita, *Macromolecules*, 22, 1546 (1989).
74. Y. Tsukahara, K. Tsutsumi, Y. Yamashita, and S. Shimada, *Macromolecules*, 23, 5201 (1990).
75. D.J. Meier, *Polym. Prepr.* Am. Chem. Soc. Div. Polym. Chem., 18, 340 (1977).
76. E. Helfand, *Macromolecules*, 8, 552 (1975).
77. E. Helfand and Z.R. Wasserman, *Macromolecules*, 9, 879 (1976).
78. E. Helfand and Z.R. Wasserman, *Macromolecules*, 11, 960 (1978).
79. E. Helfand and Z.R. Wasserman, in Developments in Block Copolymers, Vol. 1, I. Goodman Ed., Applied Science Publishers, Barking, Essex, England, 1982, p. 99.
80. G.E. Molau, in Block Polymers, S.L. Aggarwal Ed., Plenum Press, New York, 1970, p. 79.
81. T. Ono, H. Minamiguchi, T. Soen, and H. Kawai, *Kolloid Z. Z. Polym.*, 250, 394 (1972).
82. K. Ishizu, K. Inagaki, K. Bessho, and T. Fukutomi, *Makromol. Chem.*, 185, 1169 (1984).
83. T. Hashimoto, H. Tanaka, and H. Hasegawa, *Macromolecules*, 18, 1864 (1985).
84. R.J. Ceresa Ed., in Block and Graft Copolymerization, Wiley, New York, 1973.
85. M.J. Folkes Ed., in Processing, Structure and Properties of Block Copolymers, Elsevier Applied Science, London, 1985.
86. A. Noshay and J.E. McGrath Eds., in Block Copolymers, Academic Press, New York, 1977.
87. J.J. Burke and V. Weiss Eds., in Block and Graft Copolymers, Syracuse University Press, New York, 1973.
88. T. Ohta and K. Kawasaki, *Macromolecules*, 19, 2621 (1986).
89. P.G. de Gennes, in Scaling Concepts in Polymer Physics, Cornell University Press, Ithaca, New York, 1979, Chapter 9.
90. L. Leibler, *Macromolecules,* 13, 1602 (1980).
91. M.O. de la Cruz and I.C. Sanchez, *Macromolecules,* 19, 2501 (1986).
92. I.V. Berlinova and I.M. Panayotov, *Makromol. Chem.*, 190, 1515 (1989).
93. C. Tsitsilianis, P. Chaumont, and P. Rempp, *Makromol. Chem.*, 191, 2319 (1990).

94. C. Tsitsilianis, S. Graff, and P. Rempp, *Eur. Polym. J.*, 27, 243 (1991).
95. R.W. Penisi and L.J. Fetters, *Macromolecules*, 31, 1094 (1988).
96. R.P. Quirk, B. Lee, and L. Schock, *Makromol. Chem., Macromol. Symp.*, 53, 201 (1992).
97. R.P. Quirk and F. Ingatz-Hoover, in Recent Advances in Anionic Polymerization, T.E. Hogen-Esch and J. Smid, Eds., Elsevier, New York, 1987, p. 393.
98. G. Einert and J. Herz, *Makromol. Chem.*, 181, 59 (1980).
99. H. Eschwey and W. Burchard, *Polymer*, 16, 180 (1975).
100. K. Ishizu, S. Yukimasa, and R. Saito, *Polymer*, 33, 1982 (1992).
101. K. Ishizu, S. Yukimasa, and R. Saito, *Polymer Commun.*, 32, 386 (1991).
102. T. Hashimoto, M. Shibayama, and H. Kawai, *Macromolecules*, 13, 1237 (1980).
103. M. Shibayama and T. Hashimoto, *Macromolecules*, 19, 740 (1986).
104. H. Tanaka and T. Nishi, *J. Chem. Phys.*, 82, 4326 (1986).
105. K. Ishizu and K. Kuwahara, *J. Polym. Sci., Polym. Chem. Ed.*, in press.
106. M. Morton, T.E. Helminiak, S.D. Gadkary, and F. Bueche, *J. Polym. Sci.*, 57, 471 (1962).
107. J.E.L. Roovers and S.B. Bywater, *Macromolecules*, 7, 443 (1974).
108. N. Hadjichristidis and J.E.L. Roovers, *J. Polym. Sci., Polym. Phys. Ed.*, 12, 2521 (1974).
109. L.A. Utacki and J.E.L. Roovers, *Macromolecules*, 6, 366 (1973).
110. L.J. Fetters and M. Morton, *Macromolecules*, 7, 552 (1974).
111. L.K. Bi and L.J. Fetters, *Macromolecules*, 9, 732 (1976).
112. T. Hashimoto, Y. Ijichi, and L.J. Fetters, *J. Chem. Phys.*, 89, 2463 (1988).
113. Y. Ijichi, T. Hashimoto, and L.J. Fetters, *Macromolecules*, 22, 2817 (1989).
114. D.J. Worsfold, J.G. Zilliox, and P. Rempp, *Can. J. Chem.*, 47, 3379 (1969).
115. P. Rempp and E. Franta, *Pure Appl. Chem.*, 30, 229 (1972).
116. A. Kohler, J.G. Zilliox, P. Rempp, J. Polacek, and T. Koessler, *Eur. Polym. J.*, 8, 627 (1972).
117. J.G. Zilliox, P. Rempp, and J. Parrod, *J. Polym. Sci.*, C22, 145 (1968).
118. L.K. Bi and L.J. Fetters, *Macromolecules*, 8, 90 (1975).
119. D.B. Alward, D.J. Kinning, E.L. Thomas, and L.J. Fetters, *Macromolecules*, 19, 215 (1986).
120. K. Ishizu, K. Shimomura, and T. Fukutomi, *J. Polym. Sci., Polym. Chem. Ed.*, 29, 923 (1991).
121. K. Ishizu, K. Shimomura, R. Saito, and T. Fukutomi, *J. Polym. Sci., Polym. Chem. Ed.*, 29, 607 (1991).
122. K. Ishizu, S. Yukimasa, and R. Saito, *J. Polym. Sci., Polym. Chem. Ed.*, to be submitted.
123. A. Prochazka, M.K. Baloch, and Z. Tuzar, *Makromol. Chem.*, 180, 2521 (1979).
124. Z. Tuzar, B. Bednar, C. Konak, M. Kubin, S. Svobodova, and K. Prochazka, *Makromol. Chem.*, 183, 399 (1982).
125. B. Bednar, J. Devaty, B. Koupalova, and J. Kralicek, *Polymer*, 25, 1178 (1984).
126. M.H. Park, R. Saito, K. Ishizu, and T. Fukutomi, *Polymer Commun.*, 29, 230 (1988).
127. M.H. Park, K. Ishizu, and T. Fukutomi, *Polymer*, 30, 202 (1989).

128. R. Saito, K. Ishizu, T. Nose, and T. Fukutomi, *J. Polym. Sci., Polym. Chem. Ed.*, 28, 1793 (1990).
129. R. Saito, K. Ishizu, and T. Fukutomi, *Polymer,* 31, 679 (1990).
130. R. Saito, K. Ishizu, and T. Fukutomi, *Polymer,* 32, 531 (1991).
131. R. Saito, K. Ishizu, and T. Fukutomi, *Polymer,* 32, 2258 (1991).
132. R. Saito, K. Ishizu, and T. Fukutomi, *J. Appl. Polym. Sci.,* 43, 1103 (1991).
133. K. Ishizu and T. Fukutomi, *J. Polym. Sci., Polym. Lett. Ed.*, 26, 281 (1988).
134. K. Ishizu, *Polymer*, 30, 793 (1989).
135. K. Ishizu and A. Onen, *J. Polym. Sci., Polym. Chem. Ed.*, 27, 3721 (1989).
136. F. Buhler and W. Gronski, *Makromol. Chem.*, 187, 2019 (1986).
137. J. Glazer, *J. Polym. Sci.*, 14, 225 (1954).
138. M. Daoud and J.P. Cotton, *J. Phys.* (Les Ulis, Fr.) 43, 531 (1982).
139. T.A. Witten, P.A. Pincus, and M.E. Cates, *Europhys. Lett.*, 2, 137 (1986).
140. T.A. Witten and P.A. Dincus, *Macromolecules,* 19, 2509 (1986).
141. A. Halperin, *Macromolecules*, 20, 2943 (1987).
142. K. Ishizu, *J. Colloid. Interface Sci.*, to be submitted.
143. J.B. Berkowitz, M. Yamin, and R.M. Fuoss, *J. Polym. Sci.*, 28, 69 (1958).
144. M. Akashi, I. Kirikihira, and N. Miyauchi, *Angrew. Macromol. Chem.*, 132, 81 (1985).
145. S. Kobayashi, H. Uyama, I. Yamamoto, and Y. Matsumoto, *Polym. J.*, 22, 759 (1990).
146. D. Chao and K. Ito, *Polym. Prepr. Jpn.*, 40, 2588 (1991).
147. D.A. Tomalia, H. Baker, J. Dewald, M. Hall, S. Martin, J. Roeck, J. Ryder, and P. Smith, *Polym. J.*, 17, 117 (1985).
148. D.A. Tomalia, H. Baker, J. Dewald, M. Hall, G. Kallos, S. Martin, J. Roeck, J. Ryder, and P. Smith, *Macromolecules*, 19, 2466 (1986).
149. G.R. Newkome, Z. Yao, G.R. Baker, and V.K. Gupta, *J. Org. Chem.,* 50, 2004 (1985).
150. C.J. Hawker and J.M.J. Frechet, *J. Chem. Soc., Chem. Commun.*, 15, 1010 (1990).
151. C.J. Hawker and J.M.J. Frechet, *J. Am. Chem. Soc.*, 112, 7638 (1990).
152. K.L. Wooley, C.J. Hawker, and J.M. Frechet, *J. Am. Chem. Soc.*, 113, 4252 (1991).
153. C.J. Hawker, R. Lee, and J.M. Frechet, *J. Am. Chem. Soc.*, 113, 4583 (1991).
154. C.J. Hawker and J.M.J. Frechet, *Polymer,* 33, 1507 (1992).
155. T.H. Mourey, S.R. Turner, M. Rubinstein, J.M.J. Frechet, C.J. Hawker, and K.L. Wooley, *Macromolecules*, 25, 2401 (1992).
156. R.L. Lescanec and M. Muthukumar, *Macromolecules,* 23, 2280 (1990).

# Macromonomers Chapter 3

## STEREOREGULAR MACROMONOMERS WITH RESPECT TO MAIN CHAIN AND/OR SIDE CHAIN

**Koichi Hatada and Tatsuki Kitayama**

Department of Chemistry
Faculty of Engineering Science
Osaka University
Toyonaka, Osaka 560, JAPAN





Macromolecular Design: Concept and Practice (Macromonomers, Macroinitiators, Macroiniferters, Macroinimers, Macroiniters, and Macroinifers) Pages 85 - 127

## 1. INTRODUCTION

Many papers have been published on the syntheses and applications of macromonomers[1-5]. Especially, macromonomers of high purity and narrow molecular weight distribution (MWD) have been desired for the syntheses of graft or comb-like polymers which have well-defined structures. Characteristic features of the radical polymerization of macromonomers have been studied by kinetic experiments[6-14]. However, almost no attention had been paid to the tacticity of the macromonomers. Recently we prepared highly stereoregular poly(methyl methacrylate) (PMMA) macromonomers by utilizing vinylbenzylmagnesium chloride initiated polymerizations[15-17] and by stereospecific living polymerizations[15,18-24], and studied the radical and anionic polymerizations and copolymerizations in detail[15-24].

PMMA is one of the typical polymers whose tacticity can be widely varied by changing the reaction conditions and whose properties in bulk and in solution depend strongly on the tacticity[25,26]. Glass transition temperature (Tg) of PMMA is much lower for the isotactic (*it*-) PMMA than for the syndiotactic (*st*-) PMMA [26,27]. The $^{13}$C NMR spin-lattice relaxation times of the carbons in *it*-PMMA are larger than those of the corresponding carbons in *st*-PMMA, indicating that the *it*-polymer chains have greater segmental mobility than the syndiotactic ones[28]. Therefore, control of stereoregularity of PMMA macromonomer will provide another means for controlling the properties of graft and comb-like polymers.

In this paper, tacticity dependence on polymerization behavior of these stereoregular macromonomers and properties of the resulting stereoregular polymacromonomers are described. The preparation of polymacromonomers and graft copolymers stereoregular with respect to main chain and/or side chain are also described. These are the new type of stereoregular polymers.

# 2. STEREOREGULAR POLY(METHYL METHACRYLATE) MACROMONOMER HAVING STYRENE GROUP —— CONTROL OF STEREOSTRUCTURE OF BRANCHES IN GRAFT OR COMB-LIKE POLYMERS

## 2.1 Preparation of Macromonomer

The preparation of stereoregular PMMA is usually carried out by anionic polymerization. There are two methods for introducing the end functionality to the anionically prepared PMMA; one is the polymerization of MMA with initiator containing a vinyl group (*Initiator method*) and the other is reaction of the PMMA living end with a vinyl compound containing an appropriate functional group (*Terminator method*). The preparations of the macromonomers by these two methods are described below.

### 2.1.1 Initiator Method

Many papers have been published on the polymerizations of vinyl monomers with Grignard reagents. However, reports on the polymerizations with the Grignard reagent containing a vinyl group, which appear to be well suited to the synthesis of macromonomers containing exactly one polymerizable double bond at the chain end, are very few. Tomoi *et al.*[29] synthesized a highly branched polystyrene using a mixture of *m*- and *p*-isomers of vinylbenzylmagnesium chloride (VBzMgCl) as an initiator in toluene in the presence of hexamethylphosphoramide (HMPA). Asami and his coworkers[30] attempted to prepare a macromonomer by polymerizing MMA with *p*-VBzMgCl, but they found that the resulting polymer contained much more than one vinylbenzyl group per polymer chain.

We polymerized MMA with *o*-VBzMgCl to obtain PMMA macromonomers with a variety of tacticities. The polymers prepared at -78°C were found to containone vinyl benzyl group at the chain end[17].

$$\text{(o-}CH_2{=}CH\text{-}C_6H_4\text{)-}CH_2MgCl + n\ CH_2{=}C(CH_3)(C{=}O)OCH_3 \longrightarrow \text{(o-}CH_2{=}CH\text{-}C_6H_4\text{)-}CH_2\text{-}\left(CH_2\text{-}C(CH_3)(C{=}O)OCH_3\right)_n\text{-}H \quad (1)$$

The results of the polymerizations in toluene and tetrahydrofuran (THF) are summarized in Table 1. Most of the reaction products were composed of methanol-insoluble polymer and methanol-soluble oligomer. The tacticities and molecular weights of the oligomers distinctively lower than those of the polymers, particularly in the polymerization in toluene, indicating the coexistence of two different propagating species in the polymerization. The polymers prepared in toluene at -78 and -89°C were highly isotactic and isotacticity decreased at 0°C. PMMAs prepared in THF at -78 and -98°C, mostly soluble in methanol, were highly syndiotactic and had fairly narrow MWD.

Table 1 Polymerization of MMA with *o*-Vinylbenzylmagnesium Chloride[a]

| Solvent | Temp. (°C) | Time (hr) | Fraction | Yield (%) | $\bar{M}_n$[b] | $\bar{M}_w/\bar{M}_n$[c] | Tacticity(%) mm | mr | rr | $f$[d] |
|---|---|---|---|---|---|---|---|---|---|---|
| Toluene | 0 | 3 | MeOH-insol. | 24.1 | 11200 | 3.05 | 61 | 17 | 23 | 0.92 |
| | | | MeOH-sol. | 32.8 | 2400 | 1.34 | 49 | 28 | 23 | 1.04 |
| Toluene | -78 | 24 | MeOH-insol. | 35.1 | 14500 | 2.10 | 94 | 4 | 2 | 0.95 |
| | | | MeOH-sol. | 37.1 | 2280 | 1.38 | 64 | 22 | 14 | 1.00 |
| Toluene | -89 | 90 | MeOH-insol. | 41.3 | 16400 | 1.77 | 95 | 5 | 2 | 1.19 |
| | | | MeOH-sol. | 21.1 | 2540 | 1.33 | 55 | 24 | 21 | 1.10 |
| THF | 0 | 3 | MeOH-insol. | 10.2 | 8800 | 1.28 | 3 | 18 | 79 | 0.94 |
| | | | MeOH-sol. | 70.5 | 2840 | 1.60 | 3 | 20 | 77 | 1.08 |
| THF | -78 | 24 | MeOH-insol. | 0.9 | 9780 | 3.20 | - | - | - | - |
| | | | MeOH-sol. | 86.0 | 2440 | 1.51 | 0 | 12 | 88 | 1.09 |
| THF | -98 | 48 | MeOH-insol. | 6.0 | 9930 | 1.12 | 1 | 4 | 95 | 0.93 |
| | | | MeOH-sol. | 90.6 | 4360 | 1.37 | 0 | 5 | 95 | 1.15 |

[a] MMA 40 mmol, initiator 4.0 mmol, solvent 40 mL.
[b] Determined by VPO. [c] Determined by GPC.
[d] Number of vinylbenzyl group per chain determined by $^1$H NMR spectroscopy.

In the polymerization of MMA with *n*-$C_4H_9MgCl$, the initiator reacts with the carbonyl group of MMA and the resulting butyl isopropenyl ketone copolymerizes with MMA monomer[31]. If this type of reaction occurs in the polymerization with VBzMgCl, the resulting PMMA is not suitable for a macromonomer. The numbers of vinyl group in the PMMAs prepared with *o*-VBzMgCl shown in Table 1 were mostly in the range from 0.9 to 1.1. Whether or not the positive deviation from unity means the existence of the vinylbenzyl isopropenyl ketone unit was checked by close inspection of vinyl proton region in the $^1$H NMR spectrum of the polymer of totally deuterated MMA

$$C_4H_9MgCl + CH_2{=}C(CH_3){-}C({=}O){-}OCH_3 \longrightarrow CH_2{=}C(CH_3){-}C({=}O){-}C_4H_9 + CH_3OMgCl \quad (2)$$

prepared with *o*-VBzMgCl in toluene at -78°C. In the spectrum the strong signals due to MMA units were suppressed and the vinylproton signals could be easily inspected. The results indicate that the PMMAs prepared with *o*-VBzMgCl contain one vinylbenzyl group at the $\alpha$-end and no other vinylbenzyl group in the molecules. Consequently these PMMAs are usable as a macromonomer. The attack of the initiator on the carbonyl group of MMA scarcely occurred in the polymerization with *o*-VBzMgCl, probably owing to steric hindrance around the benzylic carbanion in *o*-VBzMgCl. It is worth noting that a single initiator, *o*-VBzMgCl, can afford both highly isotactic (*mm*=94%) and syndiotactic (*rr*=96%) PMMAs by changing the polymerization medium. So *o*-VBzMgCl is the most suitable initiator for preparing PMMA macromonomers with a variety of tacticities.

PMMAs with a variety of tacticities were obtained with *p*- and *m*-VBzMgCl's in toluene and in THF. The PMMAs prepared at -78°C or lower temperatures were found to contain one vinylbenzyl group at the chain end and could also be used as macromonomers. The PMMA macromonomer prepared with *m*-VBzMgCl in THF at -110°C was highly syndiotactic (*rr*=97%) and had narrow MWD ($\overline{M}_w/\overline{M}_n$=1.19) [16].

### 2.1.2 Terminator Method

In the polymerization of MMA with VBzMgCl's mentioned in 2.1.1,methanol-soluble oligomer formed concomitantly with methanol-insoluble polymer and the initiator efficiency is usually less than unity; as a result, the molecular weight and its distribution can not be fully controlled.

The stereoregular macromonomers with finely controlled molecular weight and narrow MWD can be prepared by the end-capping reactions of highly stereoregular living PMMAs.

Polymerization of MMA with *t*-$C_4H_9$MgBr in toluene at -78°C proceeds in a living manner without side reactions to give highly isotactic and living PMMA with narrow MWD[32,33]. End-capping reaction of this highly *it*-PMMA anion with *p*-bromomethylstyrene (*p*-BMST) in the presence of HMPA or 1,8-diazabicyclo[5.4.0]undec-7-ene (DBU) gave an *it*-PMMA macromonomer containing one styrene group at the chain end[18,19].

Figure 1 shows the $^1$H NMR spectrum of the resultant macromonomer. The signals due to vinyl methylene and methine protons at 5.02-5.56 and 6.52-6.57 ppm, respectively, as well as those for phenyl protons at 6.99-7.22ppm are clearly observed. The benzyl methylene protons resonate at 2.51-2.94ppm. The peak at 0.81 ppm was assigned to *t*-$C_4H_9$- group originated from the initiator. It was ascertained, by quantitative NMR analysis in combination with deuterated monomer method, that one *t*-$C_4H_9$- group per PMMA chain, attached to the initiating end of the chain[33]. Therefore, the number of styrene group introduced at the other end was determined from the intensity measurement of the $^1$H NMR signals due to the *t*-$C_4H_9$- group and the vinyl group. The measurements indicated that most of the macromonomer molecules contained one styrene group at the end.

$$CH_2{=}CH{-}C_6H_4{-}CH_2Br \;+\; t\text{-}C_4H_9{-}\!\left(CH_2{-}\underset{\underset{OCH_3}{|}}{\underset{C=O}{\underset{|}{\overset{CH_3}{\overset{|}{C}}}}}\right)_{\!n}\!{-}MgBr \xrightarrow[\text{HMPA or DBU}]{\text{Toluene}}$$

$$t\text{-}C_4H_9{-}\!\left(CH_2{-}\underset{\underset{OCH_3}{|}}{\underset{C=O}{\underset{|}{\overset{CH_3}{\overset{|}{C}}}}}\right)_{\!n}\!{-}CH_2{-}C_6H_4{-}CH{=}CH_2 \;+\; MgBr_2 \qquad (3)$$

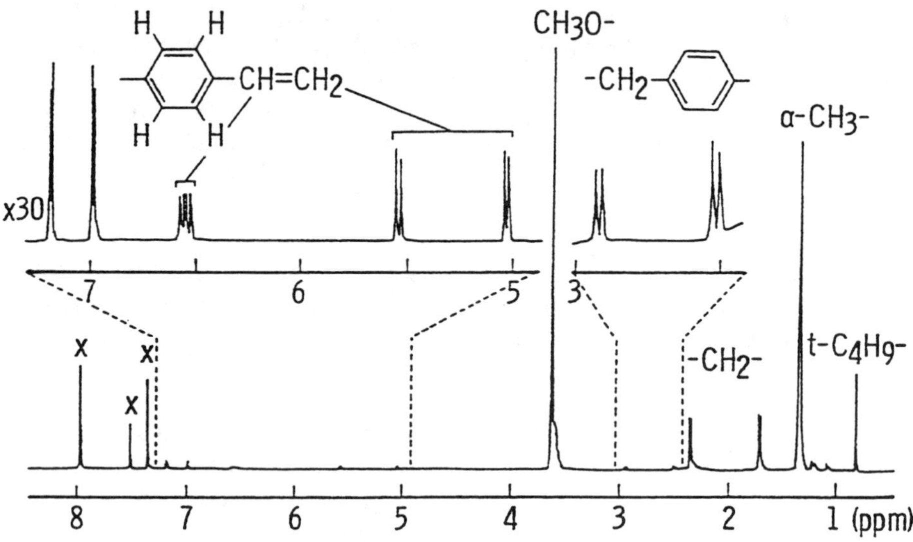


*Figure 1.* *[1]H NMR spectrum of it-PMMA macromonomer prepared through the end-capping reaction of living PMMA with p-BMST (500 MHz, $C_6D_5NO_2$, 110°C, HMDS) [18]. Signals labeled x are due to the solvent. Preparation of living PMMA: MMA 150 mmol, toluene 70 mL, t-$C_4H_9$MgBr 7.5 mmol, temp. -78°C, time 24 hr. Coupling reaction: p-BMST 38 mmol, HMPA 23 mmol, reaction time 12 hr at -78°C and further 12 hr at 0°C.*

Highly syndiotactic and living PMMA with narrow MWD can be prepared with *t*-$C_4H_9Li/R_3Al$ system (Al/Li$\geq$2) in toluene at low temperatures. The polymer molecules were found to contain one *t*-$C_4H_9$- group at one end and one methine proton at the other end and no alkyl group originated from $R_3Al$ in the chain[34]. The coupling reaction of the syndiotactic living PMMA with *p*-BMST in the presence of N,N,N',N'-tetramethylethylenediamine (TMEDA) proceeded quantitatively and a highly syndiotactic PMMA macromonomer was obtained. The [1]H NMR spectrum indicated that the macromonomer contained one styrene group at the chain end[18,19].

It should be noted that both the *it*- and the *st*-PMMA macromonomers thus obtained have the same chemical structure and are suitable macromonomers for testing tacticity dependence of the reactivities.

## 2.2 Radical Polymerization and the Mechanism [17-19]

Polymerization of *it*- or *st*-PMMA macromonomer with AIBN in toluene-$d_8$ was carried out at several temperatures and the reaction was followed by $^1$H NMR spectroscopy[18,19]. Characteristics of the macromonomers used are summarized in Table 2. As the polymerization proceeded, the signal intensity of the vinyl group decreased, and the signals of methoxy and $t$-$C_4H_9$- group became broader. The amount of macromonomer consumed was determined from the relative intensity of vinyl methylene proton signals (5.03-5.05 ppm and 5.52-5.56ppm) to $t$-$C_4H_9$- signal (0.90 ppm). Figure 2 shows the first order plots of monomer consumption of the *it*- and *st*-macromonomers at 60°C. These plots clearly indicate that the initial rate of polymerization ($R_p$) of the *it*-PMMA macromonomer is higher than that of the syndiotactic one if the molecular weights of the macromonomers are similar to each other (Table 3). The *it*-macromonomer gave the polymacromonomer with higher average degree of polymerization ($\overline{DP}$) than the *st*-macromonomer with similar molecular weight. *it*-PMMA macromonomer with higher molecular weight ($\overline{M}_n$ = 3690) showed higher $R_p$ ($1.24 \times 10^{-6}$mol/L·s) than that with the lower molecular weight ($\overline{M}_n$ = 2900) (*cf.* Tables 3 and 7). Molecular weight dependence of $R_p$ was also observed for the *st*-PMMA macromonomers to a smaller extent but the tendency was opposite to that of the *it*-macromonomer (Table 3).

In the radical polymerization kinetics, $R_p$ and $\overline{DP}$ are proportional to the rate constant of propagation ($k_p$) and inversely proportional to $k_t^{0.5}$ ($k_t$; the rate constant of termination) (*cf.* eqs. (4) and (5)), and the larger $R_p$ or $\overline{DP}$ does not necessarily mean larger rate constant in each reaction process and the interpretation is not straightfoward. However, the results strongly suggest the effect of tacticity of macromonomer on their reactivities. As mentioned previously, the mobility of PMMA chain strongly depends on its tacticity in bulk[25-27] and in solution[25,26,28], and it-PMMA chain has higher mobility than st-PMMA chain. The mobility of the macromonomer should affect both the propagation and termination processes in the polymerization. Thus the difference of Rp and $\overline{DP}$ in the polymerizations of it- and st-macromonomers should be caused by the difference of their mobilities.

ESR spectroscopy is the most promising way for the study on the nature of propagating radicals[39] and was recently applied to the radical polymerization of macromonomer[20,36,37]. We reported that the propagating radicals in the polymerization of st-PMMA macromonomer ($\overline{M}_n$ = 5380) could be observed by using ESR spectroscopy[20].

The *st*-PMMA macromonomer and AIBN in benzene solution frozen at -20°C was irradiated by UV light for 1 min and the ESR spectra were measured after the light

Then, the $k_t$ was calculated to be 54 $L \cdot mol^{-1} \cdot s^{-1}$. Similarly the $k_t$ values in the temperature range from 30 to 60°C were determined[20].

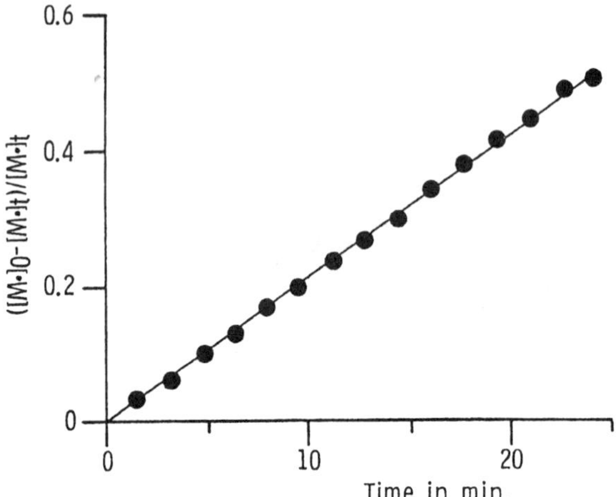


***Figure 4.*** *Second-order plot of peak height versus time at 30°C [20]. $[M\cdot]_0$, $[M\cdot]_t$: concentration of PMMA macromonomer at time zero and t, respectively.*

ESR signals of propagating radicals of *it*-PMMA macromonomer ($M_n$= 2900) could also be observed successfully at and below 30°C, and the $k_t$ values were determined in the temperature range from 10 to 30°C[36]. Since an Arrhenius plot of the $k_t$ gave a linear correlation, $k_t$ value at 60°C was estimated by extrapolation of the $k_t$'s at lower temperatures. The values of $k_t$ for *st*-macromonomer with $M_n$ of 2720 could be determined similarly in the temperature range from 20 to 50°C[36], and the $k_t$ at 60°C was estimated by extrapolation as in the case of *it*-macromonomer. The values are listed in Table 3.

*it*-Macromonomer showed larger $k_t$ value than *st*-macromonomer, which may be due to the higher segmental mobility[28] of *it*-PMMA chain. The larger $k_t$ value for the lower-molecular-weight macromonomer is explained by the lower segmental density around the propagating radicals than that of the higher molecular weight macromonomer. These $k_t$ values are very much smaller than $k_t$ for styrene polymerization, indicating that the termination reaction is extremely restricted in the polymerization of macromonomers.

As mentioned above, ESR spectroscopic study[20] indicated that the termination reaction of the PMMA macromonomers is bimolecular process as in the case of ordinary radical polymerization of low molecular monomers. The results suggest that the following equations (4) and (5) for $R_p$ and kinetic chain length ($\upsilon$), respectively, for the ordinary radical polymerization are applicable to our present case;

$$R_p = (k_p/k_t^{0.5})(f \cdot k_d)^{0.5}[I]^{0.5}[M] \quad (4)$$
$$\upsilon = (k_p/k_t^{0.5})(1/2)(f \cdot k_d)^{-0.5}[I]^{-0.5}[M] \quad (5)$$

where $f$, $k_d$ and $k_p$ are initiator efficiency and the rate constants for decomposition of initiator and for propagation, respectively. [I] and [M] represent concentrations of initiator and monomer, respectively.

The kinetic orders with respect to macromonomer and initiator have been discussed from kinetic measurement by several research groups. The reported values for macromonomers scatter largely depending on the structure of macromonomers (1.36-1.49[6,7] and 2.5-3.1[8]), while the values for initiators were only slightly lower than 0.5. $R_p$ was determined at different $[M]_0$ for the polymerizations of *it*- and *st*-macromonomers by $^1$H NMR spectroscopy. The results showed that the order with respect to macromonomer is almost unity[19].

The PMMA macromonomers were polymerized in toluene at 60°C using AIBN-$d_{12}$ and the initiator fragments in the resultant polymacromonomers were analyzed by $^2$H NMR spectroscopy to determine the $f$ values[35]. The $f$ values thus obtained are listed in Table 3, all of which are lower than that for bulk or solution polymerization of styrene (0.5-0.7) [40]. The results suggest that a large amount of the primary radicals fail to initiate the polymerization of macromonomer as compared with the case of the polymerization of ordinary monomer owing to high viscosity of the reaction media from the initial stage of polymerization.

From eq. (4) and the value of $k_d$ available from the literature[41], the $k_p$'s for the polymerization of the PMMA macromonomers were estimated from the values of $R_p$, $f$, $k_t$, and $k_d$. The results are summarized in Table 3.

Although both $k_p$'s and $k_t$'s for the PMMA macromonomers are smaller than the corresponding values of styrene, the $R_p$ for the macromonomers were larger than that of styrene (Table 3). This means that, even though $k_p$'s for the macromonomers are smaller than that of styrene, extremely smaller $k_t$ values for the macromonomers than that for styrene bring about the larger $R_p$'s for the macromonomers.

The $R_p$ for *it*-PMMA macromonomer is higher than that of *st*-PMMA macromonomer. The $k_t$ and $k_p$ values for the *it*-macromonomer was about 100 times and 10 times larger respectively than that of the *st*-macromonomer of a similar $M_n$. The effect of tacticity on the macromonomer polymerization with respect to termination reaction is much greater than the propagation reaction. This is probably because the termination process involves the reaction between two propagating radical centers both of which should be surrounded by many PMMA segments while the propagation process involves the reaction of one propagating radical with a macromonomer which consists of a single PMMA chain. The difference between $R_p$'s for *it*- and *st*-macromonomers is not so remarkable, since $R_p$ is proportional to $(k_p/k_t)^{0.5}$ and thus the increases of $k_p$ and $k_t$ affect $R_p$ in opposite ways. The difference in $R_p$ is mostly accounted for the difference in $f$ values as seen in Table 3; $(f_{it-}/f_{st-})^{0.5}$ =1.13. However, the larger $k_p$ and $k_t$ values for the *it*-PMMA macromonomer than those for the syndiotactic one indicate that the higher segmental mobility of *it*-PMMA chain contributes to increase these kinetic parameters.

The kinetic chain length ($\upsilon$) for the polymerization of PMMA macromonomers at 60°C was calculated using the kinetic parameters obtained in this work and eq. (5), and listed in Table 3 together with the observed $\overline{DP}$'s of the polymacromonomer. In all the cases, the $\overline{DP}$'s are apparently smaller than the calculated ones, indicating the chain transfer reaction to occur. The number of initiator fragment in a polymacromonomer molecule, $N$, was found to be less than unity (0.50-0.72) from the end group analysis by $^2$H NMR spectroscopy for the polymacromonomer prepared with totally deuterated AIBN[35]. The result also confirms the occurrence of the chain transfer reaction. Thus, in the radical polymerization of macromonomers, high segmental density around the propagating radicals not only decreases the rates of elementary processes ($k_t$ and $k_p$) but also changes the reaction pathway. The $N$ values for *st*-macromonomers (0.50-0.65) were smaller than that for *it*-macromonomer (0.72), indicating the higher frequency of the chain transfer reaction for the *st*-macromonomer, probably because the termination reaction between polymacromonomer radicals is more strongly restricted for the *st*-macromonomer than for the isotactic one. Therefore, the stereoregularity of PMMA macromonomer affects not only the rates of elementary processes but also the frequency of the transfer reaction.

The polymerizations of *st*-macromonomer ($\overline{M}_n$ = 5380) with AIBN in toluene-$d_8$ were followed by $^1$H NMR spectroscopy to obtain $R_p$ values at various temperatures. The Arrhenius plot of the $R_p$ values gave a linear relationship as shown in Figure 5 and the activation energy for the overall process of polymerization ($E_{overall}$) was determined to be 68 kJ·mol$^{-1}$. $E_{overall}$ is expressed as follows;

$$E_{overall}=E_p - E_t/2 + E_d/2 \tag{6}$$

where $E_p$, $E_t$, and $E_d$ represent the activation energies for the propagation, termination, and decomposition of the initiator, respectively.

The value of $E_t$ for this macromonomer was determined to be 66 kJ·mol$^{-1}$ by ESR spectroscopy[19,20], which is much larger than that for styrene polymerization (10 kJ·mol$^{-1}$)[38], indicating that the termination reaction in the polymerization of macromonomer was highly restricted. From the $E_t$ and $E_d$ for AIBN (128.9 kJ·mol$^{-1}$) [41], $E_p$ of *st*-PMMA macromonomer ($\overline{M}_n$ = 5380) was determined as 36 kJ·mol$^{-1}$, which is slightly larger than that for styrene polymerization (33 kJ·mol$^{-1}$)[38]. The result indicates that propagation reaction of macromonomer is also restricted as compared with that of styrene, although the extent of the restriction in propagation reaction is less remarkable than that in termination reaction. The situation is very similar to the above mentioned fact that the tacticity of macromonomer affects much greater on the termination reaction than on the propagation reaction.

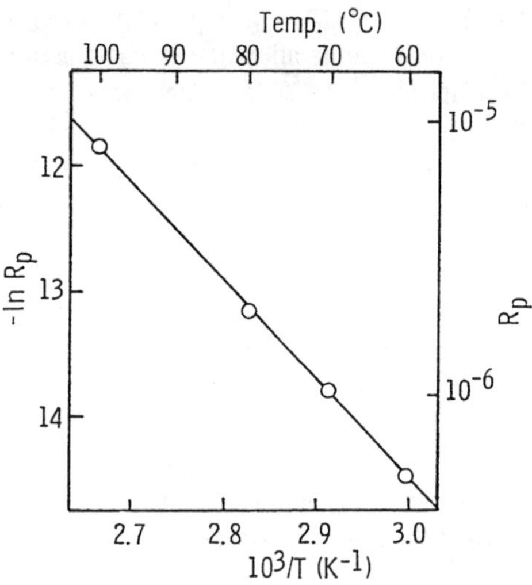


***Figure 5.*** *Arrhenius plot of $R_p$ in the polymerization of st-PMMA macromonomer ($M_n$=5380) with AIBN in toluene-$d_8$ [19]. $[M]_0$=0.05 mol/L, $[M]_0/[I]_0$=20 mol/mol.*

In the radical polymerization of the PMMA macromonomers prepared with *o*-VBzMgCl, the *it*-macromonomer showed higher reactivity than syndiotactic one[17].

## 2.3 Radical Copolymerization[18,19,22]

Copolymerization of macromonomer ($M_1$) with low molecular weight monomer ($M_2$) is a simple method to examine the reactivity of macromonomers. For the analysis of copolymerization of macromonomer, Jaacks' single point experiment method[42] has been widely applied[4,9-14]. In this method copolymerization is carried out under the condition, $[M_1] << [M_2]$,and only one of the reactivity ratios can be determined as follows:

$$r_2 = \frac{\log([M_2]_t/[M_2]_0)}{\log([M_1]_t/[M_1]_0)} \qquad ([M_1] << [M_2]) \tag{7}$$

Then, the reactivity of $M_1$ (macromonomer) can be estimated by $1/r_2 = k_{21}/k_{22}$, which represents the relative addition rate of $M_1$ and $M_2$ to $M_2$ radical. The reason why the Jaacks' method has been widely used instead of the standard procedure, which gives both $r_1$ and $r_2$, is that the analysis of the molar copolymer composition is very difficult for the copolymers formed under the conditions, $[M_1] \geq [M_2]$, because of the large difference between molecular weights of macromonomer and low molecular weight comonomer.

Our stereoregular PMMA macromonomers have one $t$-$C_4H_9$- group per molecule at the chain end, the $^1$H NMR signal of which is observed separately from those of the monomeric units. For copolymers with styrene, from the intensity measurements of the $^1$H NMR signals due to the $t$-$C_4H_9$- group in the macromonomer units and the phenyl groups of both monomer units allow the molar fraction in the copolymer to be accurately and easily determined over a wide range of copolymer composition. Therefore, both $r_1$ and $r_2$ for the copolymerization of these macromonomers with styrene could be determined based on the Mayo-Lewis' ordinary copolymerization theory. The data should give us more profound knowledge on the copolymerization of macromonomer.

The copolymerization of *it*- ($\overline{M}_n$=2900) and *st*- ($\overline{M}_n$=2720) PMMA macromonomers ($M_1$) with styrene ($M_2$) was carried out in toluene at 60°C with AIBN[18,19]. The reactions were terminated before the conversions of macromonomer and styrene reached 10%. Both the starting macromonomers were soluble in methanol and could be easily removed from the reaction mixtures by reprecipitation from toluene solution into methanol. The resultant copolymers were analyzed for the copolymer composition by $^1$H NMR spectroscopy as mentioned above.

The monomer reactivity ratios, $r_1$ and $r_2$, and the values of $1/r_2$ are summarized in Table 4. The $r_1$ and $r_2$ values for both the copolymerizations of *it*- and *st*-macromonomers with styrene indicate that the difference between the apparent reactivities of each macromonomer and styrene in the copolymerization is rather small but styrene has higher reactivity than the macromonomers toward both the macromonomer radicals ($M_1\cdot$) and styrene radical ($M_2\cdot$). Then the macromonomer units in the copolymers tend to be isolated in styrene unit sequences. Since the values of $1/r_2$ ($=k_{21}/k_{22}$) correspond to the relative reactivity of the macromonomers to styrene toward styrene radical, the larger value of $1/r_2$ for the *it*-PMMA macromonomer than that for the syndiotactic one indicates the higher reactivity of *it*-macromonomer; $k_{21}$ for the *it*-macromonomer is 1.26 times as large as $k_{21}$ for the *st*-macromonomer. In the above discussion, only the reactivity in the propagation process should be concerned, and thus the higher mobility of *it*-PMMA chain than that of *st*-PMMA chain[28] may be related to the higher reactivity of the *it*-macromonomer.

Table 4 Reactivity Ratio in the Radical Copolymerization of Stereoregular PMMA Macromonomer ($M_1$) and Styrene ($M_2$) [19]

| $M_1$ | $r_1$ [a] | $r_2$ [a] | $1/r_2=k_{21}/k_{22}$ |
|---|---|---|---|
| *it*-Macromonomer [b] | 0.79 | 1.29 | 0.775 |
| *st*-Macromonomer [c] | 0.40 | 1.62 | 0.617 |

[a] Cited from ref.[18].
[b] $\overline{M}_n$=2900. [c] $\overline{M}_n$=2720.

Comparison of $r_1$ values for *it*- and *st*-macromonomers may not lead to simple conclusion on the reactivities of the macromonomer radicals. However, if one assumes that the effect of tacticity of the macromonomer on the rate constants for cross propagations, $k_{12}$ and $k_{21}$, are the same, that is,

$$\frac{(k_{12})_{it-}}{(k_{12})_{st-}} = \frac{(k_{21})_{it-}}{(k_{21})_{st-}} \tag{8}$$

then the ratio $(r_1/r_2)_{it-}/(r_1/r_2)_{st-}$ is described as follows:

$$\frac{(r_1/r_2)_{it-}}{(r_1/r_2)_{st-}} = \frac{(k_{11}\cdot k_{21}/k_{12}\cdot k_{22})_{it-}}{(k_{11}\cdot k_{21}/k_{12}\cdot k_{22})_{st-}}$$

$$= \frac{(k_{11})_{it-}}{(k_{11})_{st-}}$$

$$= 2.5 \tag{9}$$

The assumption should be reasonable since the macromonomer units in the copolymer tend to be isolated as mentioned above. The value of 2.5 means that the effect of tacticity of PMMA macromonomer is more evident in the homopropagation (the addition process of macromonomer to macromonomer radical) than in the cross propagation (the addition to styrene radical). This may be due to the fact that in the homopropagation process both the PMMA chain in the propagating chain end and that in the incoming monomer mutually interact to affect the reaction process.

The ratio of $k_p$ for the *it*-macromonomer to that for *st*-macromonomer ($\sim 10$) (*cf.* Table 3) is larger than the ratio of $(k_{11})_{it-}/(k_{11})_{st-}$ in the copolymerization with styrene ($\sim 2.5$) (cf eq. (9)). The smaller tacticity effect in the copolymerization may be ascribed to the lower density of PMMA segments around the propagating radicals due to the incorporation of styrene units. Thus, the monomer sequence distribution around the propagating radical may be an important factor to determine the rate of propagation, that is, the penultimate effect should be taken into consideration in the strict sense. The ratio of reactivities of *it*- and *st*-PMMA macromonomers toward styrene radical could be estimated as 1.26 from $r_2$ values as mentioned above. Therefore, the tacticity effect increased in the following order.

| | $k_{21}$ | $k_{11}$ | $k_p$ | $k_t$ |
|---|---|---|---|---|
| *it-/st-* | ~1.26 | ~2.5 | ~10 | ~100 |

The order is consistent with the increasing number of PMMA chains involved in each process, indicating that the mutual interaction between the PMMA chains plays an important role in determining the rate constant for propagation process.

Copolymerization of macromonomers is another straightforward method to estimate their relative reactivities. In order to confirm the effect of tacticity of PMMA macromonomers on their reactivities, radical copolymerizations of *it*- and *st*-PMMA

each other. The solubility parameter of PMMA chain (18.4-19.5)[50-55]is closer to the value of chloroform (19.0)[56] than the value of toluene (18.2)[56] or DMSO (24.6)[56], although the literature values[50-55] of PMMA scatter a little. The interaction between macromonomer and solvent molecules is expected to decrease in the order of chloroform > toluene > DMSO. The order should correspond to the increasing order of macromonomer-macromonomer interaction in these solvents, and the order is consistent with the increasing order of $R_p$ in the homopolymerization of *it*- and *st*-PMMA macromonomers. The results, therefore, suggest the importance of solubility of macromonomers or macromonomer-solvent interaction in the polymerization of macromonomers as suggested by Asami and his coworkers[57].

Table 7 Rate of Polymerization and Copolymerization for *it*- ($M_1$) and *st*- ($M_2$) PMMA Macromonomers at 60°C in Several Solvents[a][22]

| Solvent | $\delta$[b] | $R_p$X10$^6$ (mol/L·s) | | |
|---|---|---|---|---|
| | | *it*- | *st*- | *it*-/*st*- |
| Chloroform | 19.0 | 0.33 | 0.21 | 0.24 |
| Toluene | 18.2 | 1.24 | 0.74 | 0.33 |
| DMSO | 24.6 | 2.92 | 2.45 | 0.41 |

[a] $[M]_0$=0.05 mol/L, $[M]_0/[I]_0$=20 mol/mol.
[b] Solubility parameter ($MPa^{1/2}$) cited from ref.[56].

## 2.4 Radiation Polymerization[23]

Radiation-induced polymerizations of *it*- and *st*-PMMA macromonomers were carried out in bulk state with $^{60}$Co radiation at dose rate of 0.84 Mrad/hr[23]. The $M_n$'s of *it*- and *st*-macromonomers used were 2900 and 2720, respectively (*cf.* Table 2). The *it*-PMMA macromonomer was irradiated with 20 Mrad of $\gamma$-ray at room temperature. The GPC curve for the reaction product after irradiation is shown in Figure 6B together with that for the starting macromonomer (Figure 6A). A broad peak was observed at higher molecular weight side than that of the starting macromonomer. Since PMMA is a typical polymer which decomposes by irradiation of $\gamma$-ray or electron beam[58], the higher molecular weight fraction in the GPC curve corresponds to the polymacromonomer but not to crosslinking products. Thus, the bulk polymerization by $\gamma$-ray irradiation is possible for *it*-PMMA macromonomer at room temperature.

The product after irradiation was heated at 80°C for 24hr and the reaction mixture was analyzed by GPC. A shoulder at the lower molecular weight side of the polymacromonomer peak (Figure 6C) should be due to the polymer formed during the post-polymerization. The lower molecular weight of the polymacromonomer from post-polymerization may be due to higher polymerization temperature and lower macromonomer concentration.

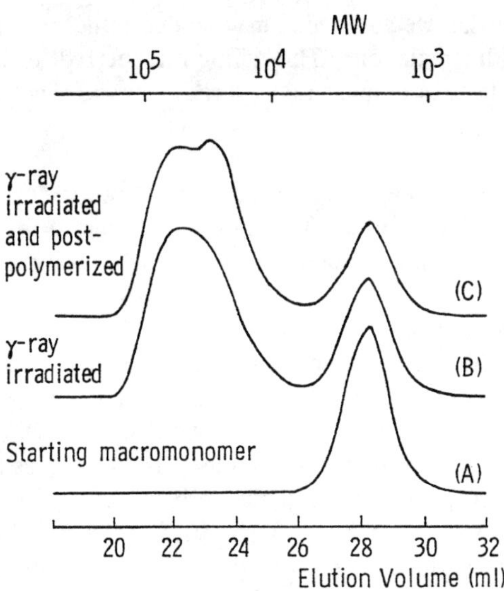


*Figure 6. GPC curves of the starting it-PMMA macromonomer (A), reaction product of it-PMMA macromonomer irradiated by γ-ray (20 Mrad) (B), and of the reaction product of the post-polymerization at 80°C for 24 hr (C) [23].*

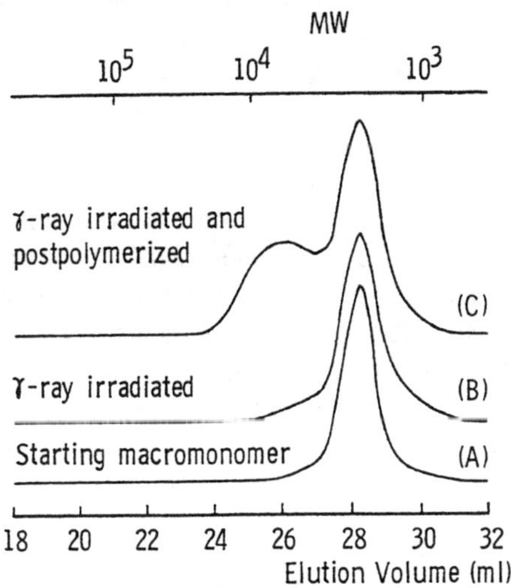


*Figure 7. GPC curves of the starting st-PMMA macromonomer (A), reaction product of st-PMMA macromonomer irradiated by γ-ray (20 Mrad) (B), and of the reaction product of the post-polymerization at 80°C for 24 hr (C) [23].*

A similar experiment was carried out for the *st*-PMMA macromonomer. GPC curves shown in Figure 7 indicate that the *st*-PMMA macromonomer is hardly polymerized by γ-irradiation but post-polymerized after heating at 80°C. The GPC curve for the irradiated macromonomer (Figure 7B) shows tailing to the side of lower molecular weight, which may be due to decomposition or chain scission of PMMA unit with irradiation. The tailing disappeared in the curve in Figure 7C, indicating that the decomposition product was polymerized during the post-polymerization.

The Tg's of the *it*- and *st*-macromonomers were 21 and 95°C, respectively. The great difference in reactivities between these macromonomers should arise from the large difference in their Tg's, that is, their segmental mobilities in bulk.

The *it*-macromonomer was polymerized at various doses of $\gamma$-ray and the products were analyzed by GPC. The conversion reached to 75% after 5 Mrad irradiation while the conversion reached to only 84% even with 80 Mrad irradiation. The UV detector used in these experiments was operated at wavelength of 254 nm. Sensitivity at 254 nm of styrene end-group of the macromonomer is 24 times as large as that for phenyl group in the polymacromonomer and the carbonyl group in PMMA chain is hardly detected. For the reaction mixture after 80 Mrad irradiation, the content of the unreacted starting macromonomer was estimated to be 2% from the relative intensities of GPC curves detected by RI and UV detectors, while the relative peak intensities of the lower molecular weight fraction in the whole GPC trace obtained with RI detector was 16%. The results indicated that 14% of macromonomer was transformed to unimer by chain-transfer reaction or termination reaction. The UV/RI intensity ratio of the peak due to polymacromonomer was almost the same as that for the polymer obtained by usual radical polymerization, indicating that no vinyl group residue was introduced or remained in the poly(macromonomer). Thus, the polymerization proceeded in a conventional manner of vinyl addition polymerization.

The *st*-macromonomer was also polymerized at various doses at room temperature. Though the polymerization reaction proceeded more at higher doses, the main product formed was dimer. Since the active site in the dimer radical should locate in the middle of PMMA chain, the segmental mobility around the propagating radical was more strongly restricted than the unimer radical to prevent further propagation. Figure 8 shows GPC curves of the reaction mixtures post-polymerized at 80°C for 24hr after various doses of $\gamma$-ray irradiation. As $\gamma$-ray dose increased, the conversion increased and the molecular weight of the polymacromonomer decreased, indicating increase of the number of active species in the polymerization system.

In conclusion, the mobility of macromonomer in bulk state strongly affected bulk polymerization of stereoregular PMMA macromonomer, and the polymerization easily proceeded for *it*-macromonomer while the polymerization of *st*-macromonomer required for heating at 80°C after irradiation. The radiation induced polymerization in bulk would give a novel route for preparation of comb-like graft copolymer.

## 2.5 Ziegler Polymerization [59]

The *it*- and *st*-PMMA macromonomers were polymerized with a typical Ziegler type initiator,$(C_2H_5)_3Al$-$TiCl_4$ at -78, 0and 60°C. The results are shown in Table 8

Polymerization occurred at 0 and 60°C. The yields of the polymacromonomers increased and the molecular weights decreased with increasing polymerization temperature. The temperature dependence of the molecular weight indicates that the extent of chain transfer reaction is enhanced at higher temperature.

Fig 8

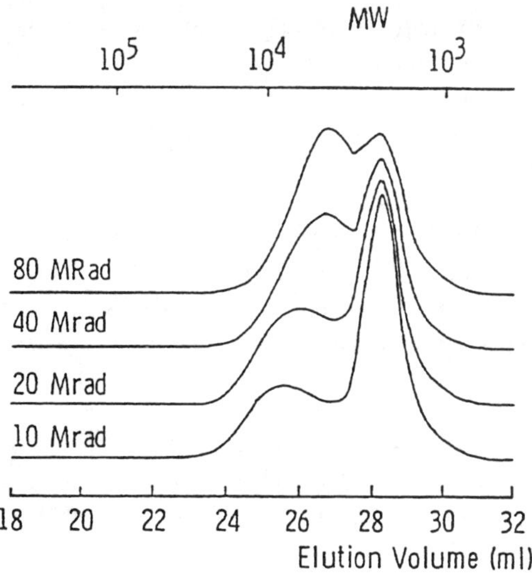


*Figure 8.* *GPC curves of post-polymerized st-PMMA macromonomer at 80°C for 24 hr after various doses of γ-ray irradiation [23].*

Table 8 Polymerization of *it*- and *st*-PMMA Macromonomers with $(C_2H_5)_3Al$-$TiCl_4$ in Toluene for 24 hr at Various Temperatures[a] [59]

| Macromonomer[b] | | Temp. | Polymacromonomer | | |
|---|---|---|---|---|---|
| Tacticity | $\bar{M}_n$ | (°C) | Yield(%) | $\bar{M}_n$[c] | $\bar{M}_w/\bar{M}_n$[d] |
| | | 60 | 47 | 44100 | 1.64 |
| *it*- | 2900 | 0 | 21 | 85400 | 1.77 |
| | | -78 | 0 | - | - |
| | | 60 | 30 | 28000 | 1.55 |
| *st*- | 2720 | 0 | 5 | 48200 | 1.81 |
| | | -78 | 0 | - | - |

[a] Macromonomer 0.20 mmol, toluene 2 mL, $TiCl_4$ 0.02 mmol, $(C_2H_5)_3Al$ 0.2 mmol
[b] See Table 2. [c] Determined by VPO.
[d] Determined by GPC.

The yields and the molecular weights of polymacromonomers were larger in the case of *it*-macromonomer than in the case of *st*-macromonomer. This clearly indicates that the *it*-macromonomer shows higher reactivity than the *st*-macromonomer even in Ziegler polymerization.

Polymerization of styrene with $(C_2H_5)_3Al\text{-}TiCl_4$ initiator in toluene gives highly isotactic polymer. Then, the polymacromonomers mentioned above might be isotactic in main chain. If it be so, the polymacromonomers would be new type of stereoregular polymers which is stereoregular in side chain as well as in main chain. The structural studies is now under way.

## 3. STEREOSPECIFIC POLYMERIZATION OF POLYISOBUTYLENE MACROMONOMER HAVING METHACRYLOYL GROUP—— CONTROL OF STEREOSTRUCTURE OF THE MAIN CHAIN OF COMB-LIKE POLYMERS [60]

Anionic polymerization of polystyrene macromonomers having methacryloyl function was carried out by Masson *et al.*[61] and Asami *et al.*[62]. However, the main objective of their works was to control the molecular weight and its MWD of the polymacromonomers, and the main-chain tacticity was not examined. In this chapter stereospecific polymerization of a polyisobutylene (PIB) macromonomer having methacryloyl moiety is described[60]. The methacrylate unit was selected as polymerizable functionality by reason that stereospecific polymerization of methacrylate has been well surveyed[26]. Anionic polymerization of this macromonomer led to the control of main chain stereoregularity of branched polymers.

### 3.1 Preparation of Macromonomer

The PIB macromonomer having methacryloyl function was prepared from a hydroxy terminated PIB, α-phenyl-ω-(2-hydroxy-1-methylethyl)-poly(1,1-dimethylethylene), (PIB-OH) ($\overline{M}_n$ = 1990, $\overline{M}_w/\overline{M}_n$=1.15, $\overline{DP}$=33.1) and methacryloyl chloride in the presence of triethylamine in a mixture of $CH_2Cl_2$ and $CHCl_3$ (9:1) at room temperature according to the procedure by Kennedy *et al.* (eq. (11))[63]. The hydroxy terminated PIB was obtained from Akron Cationic Development Corporation.

$$\text{C}_6\text{H}_5\text{-}\!\left(\!\begin{array}{c}\text{CH}_3\\|\\\text{C}\\|\\\text{CH}_3\end{array}\!\text{-CH}_2\right)_{\!n}\!\text{-}\begin{array}{c}\text{CH}\\|\\\text{CH}_3\end{array}\!\text{-CH}_2\text{-OH} + \text{CH}_2\text{=}\begin{array}{c}\text{CH}_3\\|\\\text{C}\\|\\\text{COCl}\end{array}$$

[PIB-OH]

$$\xrightarrow{(C_2H_5)_3N} \text{C}_6\text{H}_5\text{-}\!\left(\!\begin{array}{c}\text{CH}_3\\|\\\text{C}\\|\\\text{CH}_3\end{array}\!\text{-CH}_2\right)_{\!n}\!\text{-}\begin{array}{c}\text{CH}\\|\\\text{CH}_3\end{array}\!\text{-CH}_2\text{-O-}\begin{array}{c}\text{C}\\\|\\\text{O}\end{array}\!\text{-}\begin{array}{c}\text{CH}_3\\|\\\text{C}\end{array}\!\text{=CH}_2 \qquad (11)$$

[PIB macromonomer] (yield=95.5%, $\overline{Mn}$=2250)

The [1]H NMR spectra of the starting alcohol, PIB-OH, and the resulting PIB macromonomer are shown in Figure 9. The PIB-OH shows the signals of -$\underline{CH_2}$-OH and phenyl protons at 3.4 ppm and 7.0-7.5 ppm, respectively (Figure 9B). The functionality of hydroxy group of PIB-OH was determined from the intensity ratio of these two signals as 0.918. In the spectrum of the PIB macromonomer (Figure 9A), the signals of -$\underline{CH_2}$-OH disappeared completely and new signals ascribable to vinylidene protons of the methacryloyl group appeared at 5.54 and 6.11 ppm. The functionality of methacryloyl group of the macromonomer was determined from the intensity ratio of the signals due to vinylidene protons and phenyl protons as 0.918. The values of the functionalities for PIB-OH and PIB macromonomer clearly indicate the complete conversion of -$CH_2OH$ group into -$CH_2$-O-CO-C($CH_3$)=$CH_2$ group.

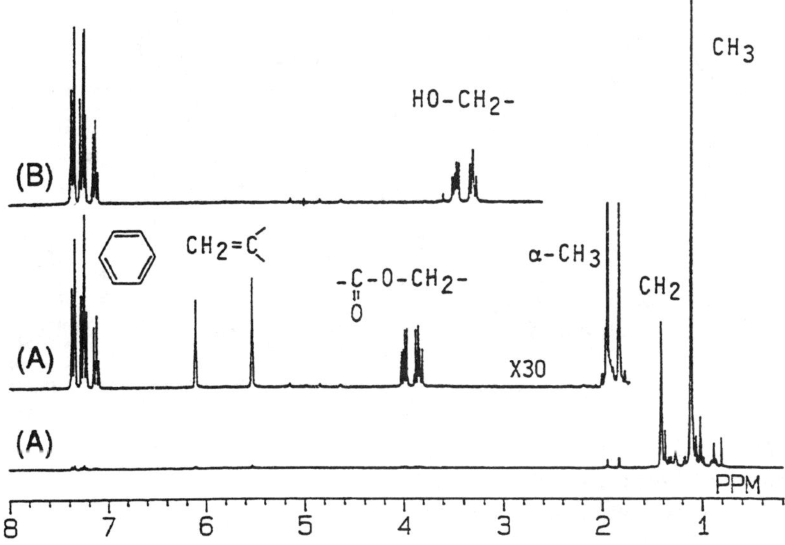


*Figure 9. 270 MHz [1]H NMR spectra of PIB macromonomer (A) and the starting hydroxy-terminated PIB (B) [60].*

## 3.2 Stereospecific Polymerization of PIB Macromonomer

The PIB macromonomer was polymerized by various anionic initiators including $t$-$C_4H_9MgBr$[32,33], $t$-$C_4H_9Li/(C_2H_5)_3Al$[34] and 1,1-diphenylhexyllithium (DPHLi) [64-67], which can afford living polymers of MMA under appropriate conditions. The results are shown in Table 9. Polymerizations with $t$-$C_4H_9MgBr$ and DPHLi gave polymacromonomers in high yields. Polymerizations with $n$-$C_4H_9Li$ or $t$-$C_4H_9Li/(C_2H_5)_3Al$ gave rather lower yields than those with DPHLi or $t$-$C_4H_9MgBr$.

Figure 10 shows the [13]C NMR spectra of the polymethacrylate backbone signal regions for the polymacromonomers prepared with $t$-$C_4H_9MgBr$ in toluene at -60°C and DPHLi in THF at -78°C. The signals of C=O, quaternary and $\alpha$-$CH_3$ carbons show the peak splittings due to triad tacticity and the signal of $CH_2$ carbon the splittings due to diad tacticity. The peak assignments shown in the Figure 10 were made by comparing the peak splittings with those observed in the spectrum of PMMA. Figure 10A clearly indicates that the polymacromonomer prepared with $t$-$C_4H_9MgBr$ is highly isotactic and the *mm* triad fraction amounts to 91% (Table 9). As far as we are aware, this is the first example of stereospecific polymerization of macromonomer. The polymacromonomer prepared with DPHLi in THF is syndiotactic-rich as in the case of PMMA formed under the same conditions.

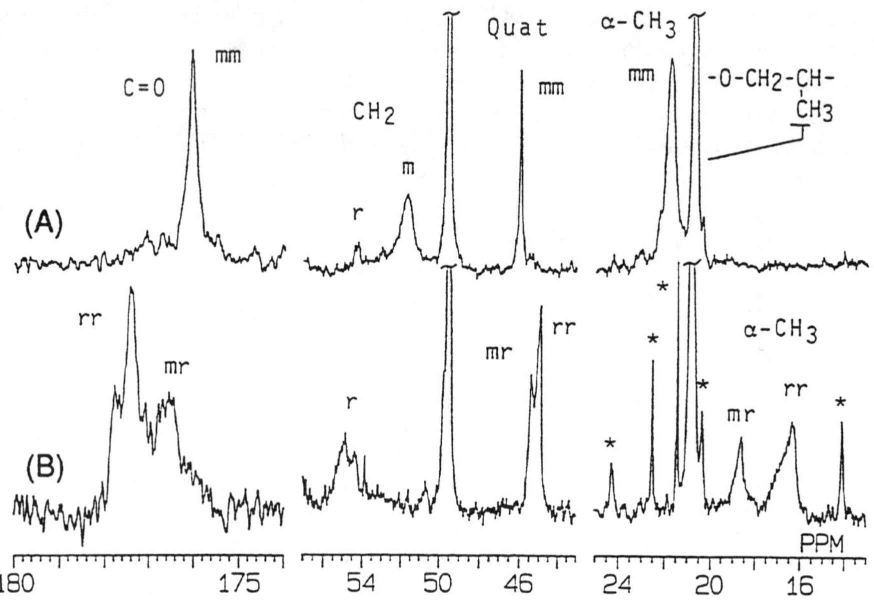


***Figure 10.*** *125 MHz $^{13}C$ NMR spectra of polymacromonomers prepared with t-$C_4H_9MgBr$ in toluene at -60°C (A) and DPHLi in THF at -78°C (B). Asterisked peaks are due to the initiator fragment.*

Table 9 Anionic Polymerization of PIB Macromonomer[a] [60]

| Initiator | $[M]_0/[I]_0$ | Solv.[b] | Temp. (°C) | Time (hr) | Conv.[c] (%) | $\bar{M}_n$ VPO | $\bar{M}_n$ GPC | $\bar{M}_n$ Calcd | $\bar{M}_w/\bar{M}_n$ | Tacticity(%)[d] *mm* | *mr* | *rr* |
|---|---|---|---|---|---|---|---|---|---|---|---|---|
| $t$-$C_4H_9MgBr$ | 8.3 | Tol. | -60 | 168 | 91 | 26600 | 24800 | 17000 | 1.52 | 91 | 9 | 0 |
| $n$-$C_4H_9Li$ | 4.3 | Tol. | -78 | 72 | 56 | 30200 | 25600 | 5500 | 2.55 | 83 | 17 | 0 |
| $t$-$C_4H_9Li$/ $(C_2H_5)_3Al$ | 8.8 | Tol. | -60 | 168 | 34 | 17300 | 14100 | 6700 | 1.32 | 5 | 43 | 52 |
| DPHLi[e] | 8.9 | THF | -78 | 48 | 95 | 22500 | 21300 | 19000 | 1.26 | 3 | 39 | 58 |
| DPHLi[e] | 8.0 | THF | -100 | 120 | 94 | 23200 | 19300 | 16900 | 1.20 | 2 | 36 | 62 |

[a] Macromonomer 0.20-0.22 mmol, solvent 5 mL.
[b] Tol.: toluene, THF: tetrahydrofuran. [c] Determined by GPC.
[d] Determined from $^{13}C$ NMR signals of α-$CH_3$ group.
[e] 1,1-Diphenylhexyllithium.

The fractions of triad tacticity for the polymacromonomers prepared with various initiators were determined from the a-$CH_3$ carbon signals and are shown in Table 9. The syndiotacticity of the polymacromonomer prepared with DPHLi in THF increased with decreasing polymerization temperature and the polymer formed at -100°C had 62% of triad syndiotacticity and $\overline{M}_w/\overline{M}_n$ value of 1.20. Syndiotactic-rich poly(macromonomer) was also formed by $t$-$C_4H_9Li/(C_2H_5)_3Al$, which gives highly *st*-PMMA in toluene at low temperatures[34]. These syndiotactic-rich polymacromonomers have narrow MWD as shown in Table 9.

The $\overline{DP}$ of the polymacromonomer prepared with DPHLi in THF at -78°C was determined as 8.9 from the intensity ratio of the signals due to $CH_3$ group of the initiator fragment (14.1 ppm) and $\alpha$-$CH_3$ group of methacryloyl unit. The value was in good agreement with the value expected from the amounts of the initiator charged and the macromonomer consumed. The MWD observed by GPC was fairly narrow ($\overline{M}_w/\overline{M}_n$=1.26). These results suggest the living character of this polymerization. The MWD of the polymacromonomer prepared with $t$-$C_4H_9MgBr$ was fairly narrow, though the $\overline{M}_n$ was a little larger than the calculated value. Further investigation is under way to examine the living character of this polymerization.

The polymerization with $n$-$C_4H_9Li$ also gave *it*-polymacromonomer but the MWD was much broader than those of the other polymacromonomers and the $\overline{M}_n$ was much larger than the calculated value. These results in the $\overline{M}_n$ and MWD are due to the side reaction of $n$-$C_4H_9Li$ with carbonyl group of methacryloyl unit of the macromonomer. The $^1$H NMR spectrum of the low molecular weight fraction of the polymerization product formed with $n$-$C_4H_9Li$ showed the signals due to the $CH_2$ group of PIB-OH observed at 3.4 ppm, together with those due to the unreacted PIB macromonomer. The PIB-OH should form along with butyl isopropenyl ketone through the reaction of $n$-$C_4H_9Li$ with the carbonyl group, as shown in eqs. (12) ~ (15).

$$n\text{-}C_4H_9Li + CH_2{=}C(CH_3)\text{-}C({=}O)\text{-}O\text{-}CH_2\text{-}CH(CH_3)\text{-}\!\left(CH_2\text{-}C(CH_3)_2\right)_n\!\text{-}C_6H_5 \longrightarrow n\text{-}C_4H_9\text{-}CH_2\text{-}C^-(CH_3)\text{-}C({=}O)\text{-}O\text{-}CH_2\text{-}CH(CH_3)\text{-}\!\left(CH_2\text{-}C(CH_3)_2\right)_n\!\text{-}C_6H_5\ Li^+ \ [I] \qquad (12)$$

$$CH_2{=}C(CH_3)\text{-}C(n\text{-}C_4H_9)(OLi)\text{-}O\text{-}CH_2\text{-}CH(CH_3)\text{-}\!\left(CH_2\text{-}C(CH_3)_2\right)_n\!\text{-}C_6H_5 \longrightarrow CH_2{=}C(CH_3)\text{-}C({=}O)\text{-}C_4H_9 + LiO\text{-}CH_2\text{-}CH(CH_3)\text{-}\!\left(CH_2\text{-}C(CH_3)_2\right)_n\!\text{-}C_6H_5 \qquad (13)$$

$$[I] \xrightarrow{\text{macromonomer}} n\text{-}C_4H_9\!\left(CH_2\text{-}C(CH_3)(C{=}O\text{-}O\text{-}CH_2\sim)\right)_m\!CH_2\text{-}C^-(CH_3)(C{=}O\text{-}O\text{-}CH_2\sim)\ Li^+ \ [II] \qquad (14)$$

$$[II] + CH_2{=}\underset{\underset{C_4H_9}{|}}{\underset{C=O}{\overset{CH_3}{C}}} \longrightarrow n\text{-}C_4H_9\!\left(CH_2\text{-}\underset{\underset{\underset{\underset{\wr}{CH_2}}{O}}{C=O}}{\overset{CH_3}{C}}\right)_{m+1}\!CH_2\text{-}\underset{\underset{C_4H_9}{C=O}}{\overset{CH_3}{C^-}}\ Li^+ \quad [III] \qquad (15)$$

The butyl isopropenyl ketone is much more reactive than methacrylate monomer[31,68-70] so that it readily reacts with polymacromonomer anion [I] to form the less reactive anion [II]. Therefore, in the polymerization with $n$-$C_4H_9Li$, there exist at least two types of propagating anions which are very much different in reactivity. The lithium alkoxide of PIB-OH formed during the polymerization may coordinate with the propagating anions to change their reactivity and the stereoregularity. These are the reasons why the polymacromonomer prepared with $n$-$C_4H_9Li$ shows broader MWD and lower isotacticity. DPHLi and $t$-$C_4H_9MgBr$ are bulky enough not to be involved in the side reaction and gave the polymacromonomers with much narrower MWD. Suppression of such a side reaction is of primary importance for the control of the anionic polymerization of the macromonomer.

## 4. STEREOSPECIFIC COPOLYMERIZATION OF STEREOREGULAR POLY(METHYL METHACRYLATE) MACROMONOMER HAVING METHACRYLOYL END GROUP [24].

In this chapter the attempt of the synthesis of graft polymer* that is stereoregular with respect to main chain as well as side chain is described. For the first step toward this objective, stereospecific polymerizations of stereoregular PMMA macromonomers were performed.

### 4.1 Preparation of Macromonomer

Highly isotactic and highly syndiotactic PMMA macromonomers having methacryloyl group as a polymerizable function were prepared from the corresponding living PMMA anions formed with $t$-$C_4H_9MgBr$ [32,33] and $t$-$C_4H_9Li/(n\text{-}C_4H_9)_3Al$ (1/3)[34], respectively, in toluene at -78°C in the following manner (Equation 16).

---

*(*In the IUPAC's definitions [Basic Definitions of Terms Relating to Polymer (1974), Pure Appl. Chem.,* ***40,*** *477 (1974)], graft polymer and graft copolymer are defined in the different categories, structure-based and source-based, respectively. In the structure-based definition, graft polymer is defined as "a polymer whose molecules have one or more species of block connected to the main chain as side chains, these side chains having constitutional or configurational features different from the constitutional units comprising the main chain, exclusive of junction points". The polymers dealt in this section are mostly graft polymers in the structure-based definition. The exception is the polymer comprised of st-PMMA main chain and st-PMMA side chains, which should be called branched polymer. For avoiding the confusion, however, it is also called "graft polymer" in this section).*

$$t\text{-}C_4H_9\text{-}\!\left(CH_2\text{-}\underset{\substack{|\\ C=O\\ |\\ OCH_3}}{\overset{\substack{CH_3\\ |}}{C}}\right)_{\!n}\!\text{-}X \xrightarrow[\text{Additive/Toluene, }-78^{\circ}C]{CH_2=CH\text{-}CH_2I} t\text{-}C_4H_9\text{-}\!\left(CH_2\text{-}\underset{\substack{|\\ C=O\\ |\\ OCH_3}}{\overset{\substack{CH_3\\ |}}{C}}\right)_{\!n}\!\text{-}CH_2\text{-}CH=CH_2$$

X = MgBr or Li

$$\xrightarrow[2)\ H_2O_2/NaOH/H_2O]{1)\ 9\text{-}BBN/THF} t\text{-}C_4H_9\text{-}\!\left(CH_2\text{-}\underset{\substack{|\\ C=O\\ |\\ OCH_3}}{\overset{\substack{CH_3\\ |}}{C}}\right)_{\!n}\!\text{-}CH_2CH_2CH_2OH$$

$$\xrightarrow[(C_2H_5)_3N/THF]{CH_2=\overset{CH_3}{C}H\text{-}COCl} t\text{-}C_4H_9\text{-}\!\left(CH_2\text{-}\underset{\substack{|\\ C=O\\ |\\ OCH_3}}{\overset{\substack{CH_3\\ |}}{C}}\right)_{\!n}\!\text{-}CH_2CH_2CH_2O\text{-}\underset{\substack{\|\\ O}}{C}\text{-}\overset{\substack{CH_3\\ |}}{C}=CH_2 \qquad (16)$$

The functionality of allyl-ended *st*-PMMA was determined to be 0.99 from the $^1$H NMR intensity ratio of the signals due to *t*-$C_4H_9$- group at 0.82 ppm and olefinic protons in allyl group at 4.92-4.98 and 5.54-5.67ppm. In the $^1$H NMR spectrum of the hydroxy-ended *st*-PMMA, the allyl signals disappeared completely but the functionality of hydroxy group could not be estimated from the spectrum. In order to determine the functionality trichloroacetyl isocyanate was reacted with the hydroxy group in an NMR sample tube and the $^1$H NMR spectrum was again measured for the intensity measurements of the signals due to methylene proton of -$CH_2OCONHCOCCl_3$ (4.17 ppm) and *t*-$C_4H_9$- protons. From the measurement the functionality of hydroxy group was determined to be 0.96. The use of trichloroacetyl isocyanate has already been applied successfully to hydroxy-ended poly(ethylene oxide)[71] and poly(propylene oxide)[72].

The final product, *st*-PMMA macromonomer containing methacryloyl group, was analyzed for the functionality by the measurement of $^1$H NMR signals due to vinylidene protons of methacryloyl end group and *t*-$C_4H_9$- protons. The functionality was found to be 0.89. The characteristics of the macromonomer is shown in Table 10.

Table 10 Characteristics of *st*-PMMA Macromonomer Having Methacryloyl Function [24]

| DP[a] | | $\bar{M}_w$[b] / $\bar{M}_n$ | Tacticity(%)[c] | | | F[c] (%) |
|---|---|---|---|---|---|---|
| $^1$H NMR | $^{13}$C NMR | | *mm* | *mr* | *rr* | |
| 32.1 | 32.2 | 1.07 | 1 | 10 | 89 | 89 |

[a] See text for the determination.
[b] Determined by GPC.
[c] Determined by $^1$H NMR spectroscopy.

## 4.2 Anionic Copolymerization

Anionic copolymerization of *st*-PMMA macromonomer with MMA or ethyl methacrylate (EMA) was carried out in THF with DPHLi at -78°C. The reaction products were recovered by precipitation into hexane and thus should contain the macromonomer if it exists. However, the $^1$H NMR spectra of the products did not

is predominantly syndiotactic. The triad tacticities of the graft polymers were determined by the method described in 4.3 using $^{13}C$ NMR spectroscopy. The results showed that the main chains of the graft polymers were highly syndiotactic, particularly in the case of the graft polymer with EMA. This is the first example of preparation of graft polymer that is stereoregular with respect to both main chain and side chain. It should be noted that the chemical structures of these graft polymers are also completely analyzed.

Anionic copolymerization of the macromonomer with MMA or EMA was also carried out in toluene at -78°C using $t$-$C_4H_9MgBr$ as an initiator. Polymerization of MMA under the same conditions proceeds in a living manner to give highly isotactic PMMA[32,33]. Thus the graft polymer prepared with this initiator should contain one $t$-$C_4H_9$- group at the chain end.

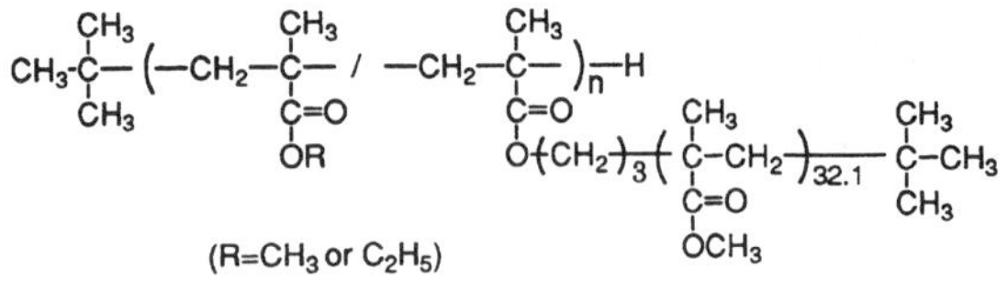

$$CH_3\text{-}C(CH_3)_2\text{-}(\text{-}CH_2\text{-}C(CH_3)(C{=}O\text{-}OR)\text{-} / \text{-}CH_2\text{-}C(CH_3)(C{=}O\text{-}O\text{-}(CH_2)_3\text{-}(C(CH_3)(C{=}O\text{-}OCH_3)\text{-}CH_2)_{32.1}\text{-}C(CH_3)_3)\text{-})_n\text{-}H \quad (19)$$

(R=$CH_3$ or $C_2H_5$)

The copolymer with MMA separated from the reaction product by repeated precipitation in chloroform-hexane mixtures was bimodal and was further fractionated into higher and lower molecular weight fractions. The separated fractions were characterized by GPC and NMR as shown in Table 12.

Table 12 Characteristics of Graft Copolymer of *st*-PMMA Macromonomer (Mac) and Comonomer (RMA) Prepared with $t$-$C_4H_9MgBr$ in toluene at -78°C [24]

| RMA | RMA[a]/Mac | Mn GPC | Mn NMR | Mw[b]/Mn | Tacticity(%)[c] mm | mr | rr | Number of branch |
|---|---|---|---|---|---|---|---|---|
| MMA[d] | 22.2 | 77300[f] | 100200 | 2.63 | 99 | 1 | 0 | 18.1[h] |
| MMA[e] | 23.9 | 15300[f] | 14000 | 1.27 | 96 | 4 | 0 | 2.45[h] |
| EMA | 4.5 | 22000[g] | - | 2.10 | 84 | 11 | 5 | 5.74[i] |
| EMA | 9.1 | 41400[g] | - | 3.38 | 85 | 7 | 8 | 9.51[i] |

[a] Monomeric unit ratio of RMA and *st*-PMMA macromonomer in the copolymer.
[b] Determined by GPC based on the calibration against standard polystyrene samples.
[c] Tacticity of the main chain determined by $^{13}C$ NMR spectroscopy.
[d] The higher molecular weight fraction of the copolymer.
[e] The lower molecular weight fraction of the copolymer.
[f] Measured in chloroform. [g] Measured in THF.
[h] Determined by $^{1}H$ NMR spectroscopy.
[i] Estimated from composition and Mn determined by GPC.

The compositions of these two fractions were fairly close to each other, and tacticities of the main chains were highly isotactic. The results suggest that there exist at least two types of active species which are different in reactivity but very similar in stereospecificity and monomer selectivity. A similar phenomenon has been observed for the copolymerization of MMA with EMA[73] or butyl methacrylate [74] by $t$-$C_4H_9MgBr$ in toluene at low temperature. Anyway, it should be noted that the graft polymer, which is highly isotactic in main chain and highly syndiotactic in side chain, could be obtained. The stereoregular branched PMMAs showed peculiar properties in solution and in bulk as described in chapter 5.

The copolymer with EMA were also found to be highly isotactic in main chain as shown in Table 12, that is, the copolymers were stereoregular graft polymers with respect to main chain and side chain. Thus the anionic copolymerization of *st*-PMMA macromonomer and methacrylate monomers could produce the graft polymers with controlled main-chain stereoregularities with highly stereoregular side chains.

## 4.3 Determination of Main Chain Tacticity of Graft Polymers

The main chain stereoregularity** of the graft polymers prepared with DPHLi and $t$-$C_4H_9MgBr$ were found to be syndiotactic and isotactic, respectively, from the methylene proton signals in the $^1H$ NMR spectra. For the quantitative determination, $^{13}C$ NMR spectra of the graft polymers measured in chloroform-*d* at 55°C were used.

Figure 12 illustrates the carbonyl carbon spectra of the graft polymers prepared from *st*-PMMA macromonomer and MMA or EMA with DPHLi, together with those of the macromonomer and poly(EMA) prepared with DPHLi in THF at -78°C. Commonly observed signals in these spectra, except for that of poly(EMA), are mostly due to MMA units in the macromonomer chains. The spectrum of the graft polymer with EMA main chain (Figure 12C) showed new peaks ascribed to heterotactic (176.7 ppm) and syndiotactic (177.4ppm) sequence of EMA units. Comparing the spectrum of the graft polymer (Figure 12C) with that of poly(EMA) (Figure 12D), the signals due to EMA-centered syndiotactic and heterotactic sequences are expected to be observed in the shift ranges shown in Figure 12C. The signals due to isotactic sequence were not observed in the spectrum, but are expected to be observed at 175.9-176.5ppm, where the graft polymer prepared with $t$-$C_4H_9MgBr$ shows the characteristic signals. The signals due to the carbonyl carbons of the methacrylate units derived from the methacryloyl function at the end of the macromonomer are also expected to exhibit tactic splittings similar to those of MMA and EMA units, although they were not distinguishable in the spectrum. In the determination of the main-chain tacticity, however, it is assumed that carbonyl carbon signals due to the main-chain methacrylate units of the macromonomer are observed in the same chemical shift ranges as those of MMA and EMA units. The chemical shift ranges for *mr* triad

---

***The main chain tacticity should be defined as cotacticity in a strict sense. However, the term "tacticity" is used in the following discussion for the simplicity.*

(*mr*(EMA)+ *mr*(MMA)) partly overlaps with that for *mm* triad of MMA units (*mm*(MMA)). Since the signal due to *mm* triad of MMA units was not observed in the spectrum of the macromonomer (Figure 12A), the estimation of signal intensity of *mr* triad was not interfered by the overlapping.

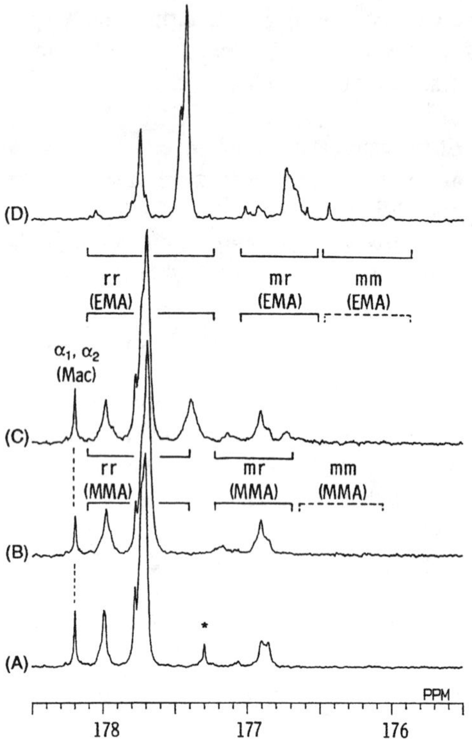


***Figure 12.** 67.8MHz $^{13}$C NMR spectra of carbonyl carbons of st-PMMA macromonomer (A), the graft polymers with MMA (B) and with EMA (C) prepared with DPHLi in THF at -78°C, and poly(EMA) prepared under the same conditions (D). The spectra were measured in chloroform-d at 55°C.*

** The signal was not observed in the spectrum of st-PMMA [75] and is probably due to the ω end unit,* $—O\text{-}(CH_2)_3\text{-}\underset{COOCH_3}{\overset{CH_3}{C}}\text{-}CH_2—$ , *of the macromonomer.*

An average tacticity of the main chain was determined by subtraction of the signals due to MMA units in the macromonomer chains as follows. The sharp resonance at 178.2 ppm in Figure 12A was assigned to the carbonyl carbons in the first and second MMA units at the initiating chain end ($\alpha_1$ and $\alpha_2$ units)[75], and the same signal was also observed in the spectra of the graft polymers. The DP (32.2) of the macromonomer estimated from the intensities of this signal and those of all the carbonyl carbons agreed well with that determined from the $^1$H NMR spectrum (32.1). The possibility that the signal due to EMA units overlaps with the $\alpha_1$ and $\alpha_2$ signal was ruled out by the fact that no peak was observed at 178.2ppm in the spectrum of poly(EMA) prepared with DPHLi in THF (Figure 12D). Thus, the signal due to $\alpha_1$ and $\alpha_2$ units was used as an intensity standard for the estimation of the contributions of the macromonomer unit signals in the spectra of the graft polymers. In the spectrum of the graft polymer with MMA, signals from the main-chain MMA units should also overlap with those from the macromonomer unit. The contributions from the macromonomer units were subtracted using the signal intensity of the $\alpha_1$ and $\alpha_2$ units as a standard. Tacticities thus determined are shown in Table 11. The main-chain tacticities of all the graft polymers are predominantly syndiotactic. Due to the limit of S/N ratio, the isotactic triad fractions were assumed to be negligible.

The tacticities of the graft polymers prepared with $t$-$C_4H_9MgBr$ in toluene were estimated similarly and are shown in Table 12.

## 4.4 Uniform Polymacromonomer

Synthetic polymers usually have distributions of molecular weight, stereoregularity and copolymer composition, and thus properties of a polymer sample are usually observed as averaged values. It would be a great help for the fundamental studies on the structures and properties of polymers if polymers or oligomers without distributions of molecular weight, stereoregularity and copolymer composition could be obtained. We recently succeeded in isolating *it*- and *st*-PMMAs that are uniform with respect to molecular weight and almost uniform with respect to tacticity up to 29-mer[76-78]. For the experiments *it*- and *st*-PMMAs prepared by stereospecific living polymerizations[32-34] were used and the separation was done by using supercritical fluid chromatography (SFC). The SFC technique allows us to isolate the uniform PMMAs up to 100-mer[79]. However, higher oligomers would be difficult to separate by SFC at present.

One of the ways of preparing uniform polymer with higher molecular weight is stepwise reaction of uniform oligomer or polymerization of uniform macromonomer. Thus, the uniform syndiotactic 28-mer of MMA was isolated by SFC from the *st*-PMMA macromonomer prepared with $t$-$C_4H_9Li/(n$-$C_4H_9)_3Al$ in toluene at -78°C (*cf.* 4.1). The isolated uniform *st*-PMMA macromonomer was radically polymerized with AIBN in benzene at 60°C, and the reaction mixtures were subjected to GPC analysis[80]. Since the difference in the molecular weight between n- and (n+1)-mers is large, each homologue of the oligo-macromonomer can be distinguished up to at least pentamer even by GPC (Figure 13). It is clear from the GPC curves that average molecular weight of the product increased with increasing polymerization time. This may be due to the long-lived character of the propagating radical of the macromonomer. Polymerization of uniform macromonomer followed by GPC separation would allow us to prepare uniform polymers with high molecular weight. The extensive studies are now under way.

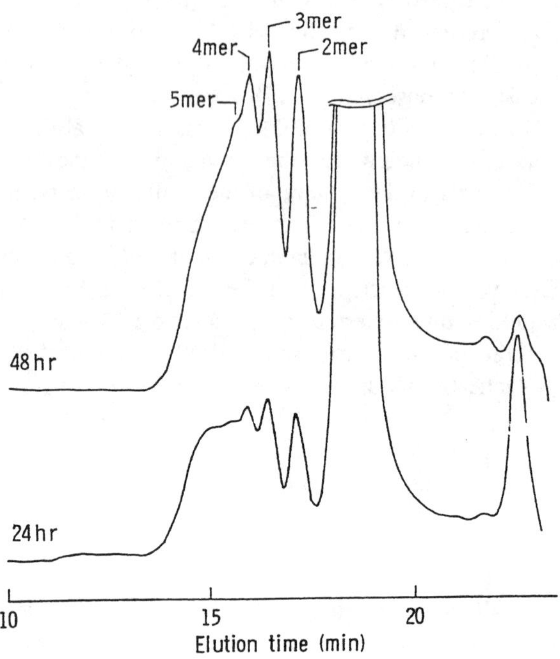


***Figure 13.*** *GPC curves for the reaction mixtures of radical polymerization of uniform st-PMMA macromonomer (DP=28) with AIBN in benzene at 60°C for 24 and 48 hr [80].*

# 5. PROPERTIES OF STEREOREGULAR POLYMACROMONOMERS

## 5.1 Glass Transition Temperatures

Copolymerizations of *it*-PMMA macromonomer having styrene end-functionality with styrene were carried out with AIBN at 60°C in toluene at various initial monomer ratios and the resultant copolymers were analyzed by DSC (Table 13). All the copolymers showed a single glass transition and the Tg increased continuously with an increase in the styrene content from the Tg of the polymacromonomer to that of polystyrene. The result is reasonable since the copolymer is considered not to be phase-separated. It should be noted that incorporation of only 7% of the macromonomer unit in the polystyrene chain decreased the Tg by 39°C[81].

Tg's of the radically prepared polymers of *it*- and *st*-PMMA macromonomers lay between those of the linear PMMAs with similar $\overline{M}_n$ and the starting macromonomers. Tg of the polymer of the *st*-macromonomer was higher than that of the *it*-macromonomer as in the case of linear PMMAs (See Table 14 in 5.2)[18,21].

Table 13 Radical Copolymerization of *it*-PMMA Macromonomer[a] ($M_1$) with Styrene ($M_2$) Prepared with AIBN in Toluene at 60°C for 24hr[b][81]

| $M_1$ in feed | Yield (%) | $\overline{M}_n$[c] | $\overline{M}_w/\overline{M}_n$[d] | $\overline{DP}$ | $M_1$ in polymer | Tg (°C) |
|---|---|---|---|---|---|---|
| 1.00[e] | 70 | 75000 | 1.45 | 29.8 | 1.00 | 44 |
| 0.22 | 32 | 20300 | 1.43 | 31.9 | 0.22 | 50 |
| 0.14 | 21 | 21000 | 1.38 | 50.2 | 0.13 | 52 |
| 0.08 | 18 | 12600 | 1.52 | 46.1 | 0.07 | 55 |
| 0[f] | 28 | 6200 | 1.72 | 59.6 | 0 | 94 |

[a] $\overline{M}_n$ = 2580(NMR), 2520(VPO), $\overline{M}_w/\overline{M}_n$=1.11, *mm:mr:rr* = 95:4:1, *f*=0.97
[b] $M_1$ 0.2mmol, toluene 2mL, $[M_1+M_2]_0/[I]_0$=100 mol/mol.
[c] Determined by VPO. [d] Determined by GPC.
[e] $M_1$ 0.2 mmol, toluene 0.75mL, $[M_1]_0/[I]_0$=40 mol/mol.
[f] $M_2$ 20 mmol, toluene 20 mL, $[M_2]_0/[I]_0$=100 mol/mol.

## 5.2 Solution Properties

Table 14 shows the intrinsic viscosities ($[\eta]$'s) and Huggins' coefficients ($k_H$'s) of the *it*- and *st*-PMMA macromonomers, the poly(PMMA macromonomer)s derived therefrom, and *it*- and *st*-PMMAs with similar molecular weights to the corresponding poly(PMMA macromonomer)s. The $[\eta]$'s of the polymacromonomers were lower than those of the corresponding linear PMMAs with the similar molecular weight, and the $k_H$'s were larger than those of the linear PMMAs. Although the conformity does not exist in the literature as regards the influence of branching on the viscosity, most results indicate that the increased value of $k_H$ is obtained with a higher degree of branching[82]. So the results indicate that the obtained poly(macromonomer)s are actually the comb-like polymer and the molecules have smaller hydrodynamic volume in solution than the linear PMMA.

Stereoregularity of the PMMA branches of polymacromonomer is another important factor affecting the solution properties. Figure 14 illustrates GPC curves of polymers derived from *it*- and *st*-PMMA macromonomers together with those of linear PMMAs, $\overline{M}_n$'s of which were very similar to each other (about 6 x $10^4$). The polymacromonomers exhibited larger elution volumes at peak maxima than the linear PMMAs. Effect of tacticity on the elution volume is evident not only for the linear PMMAs but also for the polymacromonomers; *it*-polymers have smaller elution volumes than the corresponding *st*-polymers.

Table 16 [24] lists the [$\eta$]'s and the $k_H$'s of the branched PMMAs in chloroform and in acetone. The branched PMMAs with *it*- and *st*-main chains were prepared by copolymerizing MMA and *st*-PMMA macromonomer with *t*-$C_4H_9MgBr$ in toluene and with DPHLi in THF, respectively (see 4.2). The *st*-graft polymer, which is syndiotactic in main chain, showed lower viscosity than *st*-PMMA with a similar $M_n$, as commonly observed for branched polymers[82]. The $k_{HS}$' for these *st*-polymers did not differ significantly. We reported that homopolymers of *it*- and *st*-PMMA macromonomers exhibited larger $k_H$ values than the corresponding linear PMMAs[18,21]. The smaller difference of $k_H$ values in the present case may be due to the smaller extent of branching in the branched polymer than in the polymacromonomers.

Table 16 Intrinsic Viscosity ([$\eta$]) and Huggins' Coefficient ($k_H$) of the Graft Polymers of MMA and PMMA Measured in Chloroform or Acetone at 35°C

| Polymer | Number of branch | $\overline{DP}$ of main chain | Chloroform [$\eta$]/dL·g$^{-1}$ | Chloroform $k_H$ | Acetone [$\eta$]/dL·g$^{-1}$ | Acetone $k_H$ |
|---|---|---|---|---|---|---|
| *st*-Graft polymer[a] | 6.62 | 103.9 | 0.139 | 0.86 | 0.086 | 1.57 |
| *it*-Graft polymer[b] | 2.45 | 61.0 | 0.111 | 1.23 | 0.072 | 4.99 |
| *st*-PMMA[c] | | 226.1 | 0.220 | 0.88 | 0.118 | 1.35 |
| *it*-PMMA[d]+ *st*-Macromonomer[e] (42:58) | | | 0.070 | 1.23 | 0.057 | 1.82 |
| *st*-Macromonomer[e] | | 32.1 | 0.055 | 1.25 | 0.040 | 1.48 |

a The graft polymer prepared with DPHLi in THF (see Table 11).
b The lower molecular weight fraction of the graft polymer prepared with *t*-$C_4H_9MgBr$ in toluene (see Table 12).
c *st*-PMMA ($\overline{M}_n$=22700, $\overline{M}_w/\overline{M}_n$=1.26, *mm*:*mr*:*rr*=0:11:89).
d *it*-PMMA ($\overline{M}_n$=5370, $\overline{M}_w/\overline{M}_n$=1.10, *mm*:*mr*:*rr*=97:2:1).
e *st*-PMMA macromonomer ($\overline{M}_n$=3400, $\overline{M}_w/\overline{M}_n$=1.07, *mm*:*mr*:*rr*=1:10:89).

Direct comparison of [$\eta$]'s of the *it*- and *st*-graft polymers is not meaningful because of the different number of branches per chain and the $M_n$. However, $k_H$ values for these graft polymers are apparently different, especially in acetone. As already mentioned in 2.3, *it*- and *st*-PMMAs form crystalline stereocomplex in the solid state and in solution. The large $k_H$ of *it*-graft polymer (4.99), or strong dependence of reduced viscosity on concentration, suggests the possibility of association of the polymer in acetone. Absolute peak intensities in the $^1$H NMR spectrum of the *it*-graft polymer in acetone-$d_6$ (1 wt/v %) at 35°C was 61% of the expected value and thus indicated that 39% of the polymer did not show the signals due to association, that is, stereocomplex formation. Viscosity and $^1$H NMR measurements of a mixture of *it*-PMMA with $\overline{DP}$=53.1 and *st*-PMMA macromonomer shown in Table 16 as a reference sample for the *it*-graft polymer were carried out under the same conditions. The mixture also exhibited large $k_H$ (1.82) and diminished $^1$H NMR signal intensity in acetone-$d_6$ (68% of the theoretical value). Although the degree of association in acetone as observed from $^1$H NMR intensity is not different between the mixture and the *it*-graft polymer, the $k_H$ value for the latter is almost three times as large as that for the former. The *it*-graft polymer may undergo, at least in part, intramolecular association between *it*-main chain and *st*-PMMA branches and thus exist as shrinked particles. This might be one of the reasons for the large $k_H$ value, since spherical particles in solution usually exhibit large $k_H$ values[82]. The intermolecular association among the *it*-graft polymer molecules may be weakened by the occurrence of the intramolecular association as compared with that in the mixture. Such a weak association is expected to be sensitive to concentration and thus cause further enhancement of concentration dependence of $\eta_{sp}/c$, i.e. larger $k_H$.

## 5.3 Stereocomplex Formation

Isotactic and syndiotactic PMMAs form crystalline stereocomplex in the solid state and in solution as described in 2.3 and 5.2. Figure 15 illustrates DSC curves of the copolymer of *it*- and *st*-PMMA macromonomers (cf. Table 6[22]) prepared in DMSO before and after annealing at 140°C for 24 hr. Single glass transition was observed at 76°C for the unannealed copolymacromonomer, the temperature of which is the intermediate value for *it*- (21°C) and *st*-PMMA macromonomers (95°C). The copolymacromonomer annealed showed the peak due to melting of stereocomplex at 165°C (ΔH=0.52 cal/g), while a mixture of *it*- and *st*-macromonomers or that of *it*- and *st*-polymacromonomers did not show the endothermic peak, indicating that the copolymacromonomer forms stereocomplex much easier than the mixtures. Thus, the copolymerization of stereoregular PMMA macromonomers provides a new type of stereoregular graft polymer, having the branches of different tacticities[22].

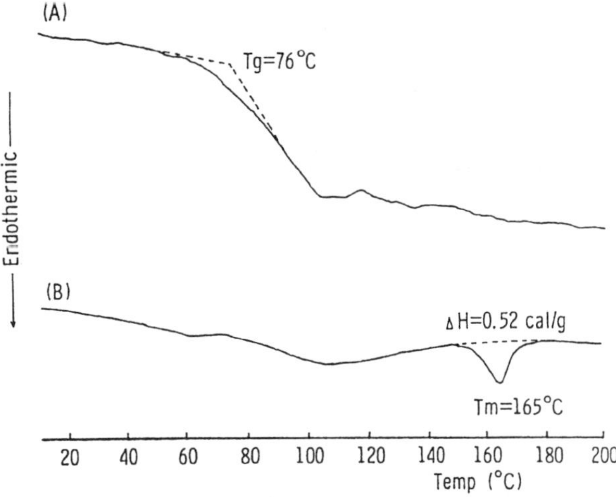


*Figure 15. DSC curves for poly(it-PMMA macromonomer-co-st-PMMA macromonomer) prepared in DMSO at 60°C (A) and that annealed at 140°C for 24 hr (B) [22].*

The complex formation between PMMA and polymacromonomer or graft copolymer with PMMA branches was also examined by annealing the mixture at 140-150°C. The branched polymers with *st*-PMMA branches did not form stereocomplex with linear *it*-PMMA under these conditions. However, the complex formation took place between linear *st*-PMMA and poly(*it*-PMMA macromonomer) or copolymer of styrene (Table 17). The melting points and heat of fusion (ΔH) of the complexes from the branched polymers were lower than those of the complex of linear PMMAs, indicating restricted formation of stereocomplex from branched polymers. The complex with the graft copolymer showed larger ΔH than that with polymacromonomer, probably because the lower segmental density of *it*-PMMA branches in the graft copolymer allows more *it*-PMMA segments to be involved in the complex formation. Anyway, the observations are apparently consistent with the double helix model proposed by Challa and his coworkers[47]; the helix model implies the difficulty of *st*-PMMA branches in the comb-like polymer or graft polymer to be involved in the complex formation with linear *it*-PMMA chains. Thus, the results shown in Table 17 seem to support the helix model.

Table 17 Stereocomplex Formation between Linear *st*-PMMA[a] and *it*-PMMA, Poly(*it*-PMMA macromonomer) or Poly(*it*-PMMA macromonoer-*co*-styrene)[b]

| Polymer | $\overline{DP}$ | Anneal. Conditions | | mp (°C) | ΔH (cal/g) |
|---|---|---|---|---|---|
| | | Temp.(°C) | Time(hr) | | |
| Linear | 26 | 140 | 168 | 186 | 0.06 |
| Linear | 290 | 150 | 24 | 195 | 1.9 |
| Polymacromnomer | 27[c)] | 140 | 24 | 186 | 0.21 |
| Graft Copolymer[d)] | 27[c)] | 140 | 24 | 183 | 0.54 |

[a] $\overline{DP}$=603, *rr*=96.
[b] The ratio of *it*-PMMA and *st*-PMMA was 1:2.
[c] $\overline{DP}$ of *it*-PMMA branch.
[d] Poly(*it*-PMMA macromonomer-*co*-styrene) (0.22:0.78).

## REFERENCES

1. R. Milkovich and M. T. Chiang, U.S.Patent, 3,786,116(1974); R. Milkovich, *Polym. Prepr.,Polym. Chem. Div., Am. Chem. Soc.*, **21**, 40 (1980)
2. P. F. Rempp and E. Franta, *Adv. Polym. Sci.*, **58**, 1 (1984)
3. Y. Yamashita, *J. Appl. Polym. Sci., Appl. Polym. Symp.*, **36**, 193 (1981)
4. R. Asami and M. Takaki, *Makromol. Chem., Rapid Commun.*, **12**, 163 (1985)
5. K. Ito, *Kobunshi Kako (Polymer Applications)*, **35**, 262 (1986)

6. Y. Tsukahara, K. Mizuno, S. Segawa, and Y. Yamashita, *Polym. Prepr. Jpn.*, **37**, 336 (1988); *Macromolecules*, **22**, 1546 (1989)
7. K. Tanaka, K. Ito, and E. Yamada, *Polym. Prepr. Jpn.*, **37**, 337 (1989)
8. M. Takaki and R. Asami, *Polym. Prepr. Jpn.*, **37**, 335 (1988)
9. G. Schulz and R. Milkovich, *J. Polym. Sci.*, **22**, 1633 (1984)
10. P. Rempp, P. Lutz, P. Masson, and E. Franta, *Makromol. Chem., Suppl.*, **13**, 471 (1985)
11. J. P. Kennedy and M. Hiza, *J. Polym. Sci., Polym. chem. Ed.*, **21**, 1033 (1983); J. P. Kennedy and C. Y. Lo, *Polym. Bull.*, **13**, 343 (1985)
12. K. Ito, H. Tsuchida, A. Hayashi, T. Kitano, W. Yamada and T. Matsumoto, *Polym. J.*, **17**, 827 (1985); K. Ito, H. Tsuchida, and T. Kitano, *Polym. Bull.*, **15**, 425 (1986)
13. G. G. Cameron and M. S. Chisholm, *Polymer*, **26**, 437 (1985)
14. K. Muehlbach and V. Percec, *J. Polym. Sci., Polym. Chem.*, **25**, 2605 (1987)
15. K. Ute and K. Hatada, *Kobunshi Kako (Polymer Applications)*, **36**, 366 (1986)
16. K. Hatada, H. Nakanishi, K. Ute, and T. Kitayama, *Polym. J.*, **18**, 581 (1986)
17. K. Hatada, T. Shinozaki, K. Ute, and T. Kitayama, *Polym. Bull.*, **19**, 231 (1988)
18. K. Hatada, T. Kitayama, K. Ute, E. Masuda, T. Shinozaki, and M. Yamamoto, *Polym. Bull.*, **21**, 165 (1989)
19. E. Masuda, S. Kishiro, T. Kitayama, and K. Hatada, *Polym. J.*, **23**, 847 (1991)
20. K. Hatada, T. Kitayama, and E. Masuda, *Makromol. Chem., Rapid Commun.*, **11**, 101 (1990)
21. K. Hatada, T. Kitayama, and K. Ute, "*Progressin Pacific Polymer Science*", B. C. Anderson, Y. Imanishi Eds, Springer-Verlag (1991) pp.131.
22. T. Kitayama, E. Masuda, and K. Hatada, *Polym. J.*, to be submitted
23. T. Kitayama, E. Masuda, K. Hatada, and M. Hatada, *Polym. Bull.*, to be submitted
24. T. Kitayama, O. Nakagawa, S. Kishiro, T. Nishiura, and K. Hatada, *Polym. J.*, **25**, *** (1993)
25. H. Yuki and K. Hatada, *Adv. Polym. Sci.*, **31**, 1 (1979)
26. K. Hatada, T. Kitayama, and K. Ute, *Prog. Polym. Sci.*, **13**, 189 (1988)
27. E. V. Thompson, *J. Polym. Sci., A-2*, **4**, 199 (1966)
28. K. Hatada, T. Kitayama, Y. Okamoto, K. Ohta, Y. Umemura, and H. Yuki, *Makromol. Chem.*, **179**, 485 (1978)
29. M. Tomoi, T. Onozawa, K. Seki, and H. Kakiuchi, *Polym. J.*, **7**, 372 (1975)
30. R. Asami, Y. Kondo, M. Takagi, and T. Ito, The 12th Chubu Regional Meeting of Chemical Societies, Japan, Nagoya, October 1981, 1G20
31. K. Ute, T. Kitayama, and K. Hatada, *Polym. J.*, **18**, 249 (1986)
32. K. Hatada, K. Ute, K. Tanaka, T. Kitayama, and Y. Okamoto, *Polym. J.*, **17**, 977 (1985)
33. K. Hatada, K. Ute, K. Tanaka, Y. Okamoto, and T. Kitayama, *Polym. J.*, **18**, 1037 (1986)
34. T. Kitayama, T. Shinozaki, E. Masuda, M. Yamamoto, and K. Hatada, *Polym. Bull.*, **20**, 505 (1988); T. Kitayama, T. Shinozaki, T. Sakamoto, M. Yamamoto, and K. Hatada, *Makromol. Chem. Suppl.*, **15**, 167 (1989)
35. T. Kitayama, S. Kishiro, E. Masuda, and K. Hatada, *Polym. Bull.*, **25**, 205 (1991)

36. E. Masuda, T. Kitayama, K. Hatada, and M. Kamachi, *Makromol. Chem..*, to be submitted
37. Y. Tsukahara, K. Tsutsumi, Y. Yamashita, and S. Shimada, *Macromolecules*, **22**, 2869 (1989)
38. M. S. Matheson, E. E. Auer, E. B. Bevilacqua, and E. J. Hart, *J. Am. Chem. Soc.*, **73**, 1700 (1951)
39. M. Kamachi, *Adv. Polym. Sci.*, **82**, 207 (1987)
40. J. C. Bevington, *Trans. Faraday Soc.*, **51**, 1392 (1955)
41. J. P. Van Hook and A. V. Tobolsky, *J. Am. Chem. Soc.*, **80**, 779 (1958)
42. V. Jaacks, *Makromol. Chem.*, **161**, 161 (1972)
43. W. H. Watanabe, C. F. Ryan, P. C. Fleischer, Jr., and B. S. Garrett, *J. Phys. Chem.*, **65**, 896 (1961)
44. A. M. Liquori, G. Anzuino, M. D'Alagni, P. De Santis, and M. Savino, *Nature*, **206**, 358 (1965)
45. A. M. Liquori, P. De Santis, M. Savino, and M. D'Alagni, *J. Polym. Sci., Part B*, **4**, 943 (1966)
46. E. J. Vorenkamp, F. Bosscher, and G. Challa, *Polymer*, **20**, 59 (1979)
47. F. Bosscher, G. T. Brinke, and G. Challa, *Macromolecules*, **15**, 1442 (1982)
48. H. Kusanagi, H. Tadokoro, and Y. Chatani, *Macromolecules*, **9**, 531 (1976)
49. K. Hatada, K. Ute, and Y. Terawaki, *Polym. Prepr. Jpn.*, **34**, 821 (1985)
50. J. Alfrey, A. O. Goldberg, and J. A. Price, *J. Colloid Sci.*, **5**, 251 (1950)
51. G. M. Bristow and W. F. Watson, *Trans. Faraday Soc.*, **54**, 1731 (1958)
52. A. V. Tobolsky, "*Propertiesand Structure of Polymers*", Wiley, New York, 1960, p.64, 66
53. P. A. Small, *J. Appl. Chem.*, **3**, 71 (1953)
54. D. Mangaraj, S. Parta, and S. Rashid, *Makromol. Chem.*, **65**, 39 (1963)
55. Union Carbide Corp., *Private communication to H. Burrell, Polymer Hand book, 1st Ed.*
56. E. A. Grulke, "Solubility Parameter Values" in "Polymer Handbook", ed. by J. Brandrup and E. H. Immergut, Wiley, New York, 1989, p.519
57. R. Asami and M. Takaki, *27th IUPAC Symposium on Macromolecules, Strasbourg 1981, Preprint I*, p.240; M. Takaki, R. Asami, H. Hanahata, N. Sukenaga, and K. Sogawa, *Polym. Prepr. Jpn.*, **30**, 860 (1981)
58. M. Dole, "*Polymers that primary degrade on irradiation*," in "*The radiation chemistry of macromolecules*", ed. by M. Dole, Academic Press, New York, 1973. chap. 6.
59. K. Hatada, T. Kitayama, and E. Masuda, unpublished results.
60. T. Kitayama, S. Kishiro, and K. Hatada, *Polym. Bull.*, **25**, 161 (1991)
61. P. Masson, E. Franta, and P. Rempp, *Makromol. Chem. Rapid Commun.*, **3**, 499 (1982)
62. R. Asami, K. Kyuda, M. Takaki, and Y. Ogawa, *Polym. Prepr. Jpn.*, **31**, 277 (1982)
63. J. P. Kennedy, and M. Hiza, *Polym. Bull.*, **10**, 146 (1983)
64. B. C. Anderson, G. D. Amdrews, P. Arthur, Jr., H. W. Jacobson, A. J. Playtis, and W. H. Sharkey, *Macromolecules*, **14**, 1599 (1981)

65. R. K. Das, R. D. Allen, T. C. Ward, J. E. McGrath, and R. J. Dodson, *Polym. Prep. Am. Chem. Soc., Div. Polym. Chem.*, **25(1)**, 185 (1984)
66. P. Lutz, P. Masson, G. Beinert, and P. Rempp, *Polym. Bull.*, **12**, 79 (1984)
67. Z.-K. Cao, Y. Okamoto, and K. Hatada, *Kobunshi Ronbunshu*, **43**, 857 (1986)
68. K. Hatada, T. Kitayama, K. Fujikawa, K. Ohta, and H. Yuki, *Anionic Polymerization; Kinetics, Mechanism, and Synthesis*, J. E. McGrath ed., *ACS Symposium Series*, **166**, 327 (1981)
69. K. Hatada, T. Kitayama, S. Okahata, and H. Yuki, *Polym. J.*, **13**, 1045 (1981)
70. K. Hatada, T. Kitayama, S. Okahata, and H. Yuki, *Polym. J.*, **14**, 971 (1982)
71. R. De Vos and E. J. Goethals, *Polym. Bull.*, **15**, 547 (1986)
72. K. Miura, T. Kitayama, K. Hatada, and T. Nakata, *Polym. J.*, **22**, 671 (1990)
73. T. Kitayama, K. Ute, M. Yamamoto, N. Fujimoto, and K. Hatada, *Polym. J.*, **22**, 386 (1990)
74. K. Hatada, K. Ute, T. Kitayama, M. Yamamoto, T. Nishimura, and M. Kashiyama, *Polym. Bull.*, **21**, 489 (1989)
75. K. Ute, T. Nishimura, and K. Hatada, *Polym. J.*, **21**, 1027 (1989)
76. K. Hatada, K. Ute, T. Nishimura, M. Kashiyama, T. Saito, and M. Takeuchi, *Polym. Bull.*, **23**, 157 (1990)
77. K. Ute, N. Miyatake, T. Asada, and K. Hatada, *Polym. Bull.*, **28**, 561 (1992)
78. K. Hatada, T. Kitayama, and K. Ute, *Makromol. Chem., Symp.*, xx,xx (1993)
79. K. Ute, N. Miyatake, Y. Ohsugi, and K. Hatada, unpublished results.
80. K. Hatada, T. Nishiura, K. Ute, and N. Miyatake, *Polym. Bull.*, to be submitted
81. E. Masuda, T. Kitayama and K. Hatada, unpublished results.
82. M. Stickler and N. Sutterlin, "*Polymer Handbook*", 3rd Ed., ed. by J. Brandrup and E. H. Immergut, John Wiley & Sons, Inc., VII/183 (1989).
83. W. R. Moore and R. J. Hort, *J. Polym. Sci.*, A1, 929 (1963)
84. E. Masuda, T. Kitayama and K. Hatada, unpublished results.

# Macromonomers Chapter 4

## POLY(ETHYLENEOXIDE) MACROMONOMERS

**Koichi Ito**

Toyohashi University of Technology
Department of Materials Science
Tempaku-cho, Toyohashi 441
JAPAN





Macromolecular Design: Concepts and Practice (Macromonomers, Macroinitiators, Macroiniferters, Macroinimers, Macroiniters, and Macroinifers) Pages 129 - 160

# 1. INTRODUCTION

Poly(ethylene oxide), PEO is one of the well-known, water-soluble, nonionic polymers with a variety of practical applications [1,2]. Its macromonomers have also been a subject of many recent publications dealing with the synthesis of well-defined graft copolymers by copolymerization with any conventional monomers [3-40], and regular comb polymers by homopolymerization [4,41-48]. Their potential applications may include various fields such as paints, cosmetics, medicals [11,39,40], electronics as solid ionconductors [35-37], polymeric catalysts and reagents [8,19,38,71] etc.

One of the most important and interesting features of PEO macromonomers may be the amphiphilic nature by themselves, which are represented by the structure

$$RO{-}\!\left[CH_2CH_2O\right]_n\!{-}V$$

where V is the $\alpha$ end group, which can either polymerize or copolymerize and R is the $\omega$ end group. The macromonomers are soluble in a very wide range of solvents including water, alcohols, benzene, and even petroleum, depending on the nature of the end groups, R and V, and the PEO chain length, n. Such an amphiphilicity can never be realized with conventional monomers, and this makes the polymerization chemistry very fascinating.

This chapter is intended to review our recent works on PEO macromonomers, including synthesis, polymerization, and copolymerization, with particular stress on the amphiphilic aspect involved.

# 2. SYNTHESES OF PEO MACROMONOMERS

The PEO macromonomers we used are *p*-vinylbenzyl ether or methacrylate of poly(ethylene glycol) monoalkyl ethers:

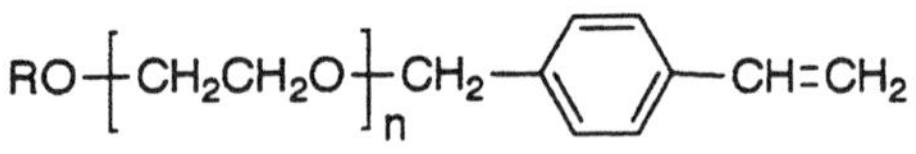

$$RO{-}\!\left[CH_2CH_2O\right]_n\!{-}CH_2{-}C_6H_4{-}CH{=}CH_2$$

(R–PEO–VB–n)

$$RO{-}\!\left[CH_2CH_2O\right]_n\!{-}\underset{\underset{O}{\|}}{C}{-}\overset{\overset{CH_3}{|}}{C}{=}CH_2$$

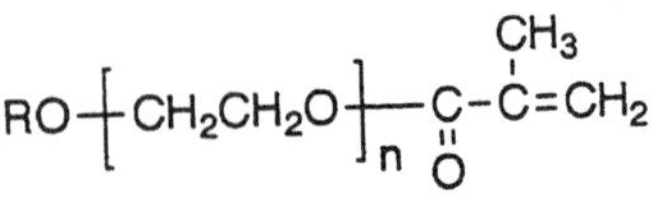

(R–PEO–MA–n)

They are abbreviated by R-PEO-VB-n and R-PEO-MA-n, to specify the $\omega$-alkyl group by R; $C_1$ for methyl, $tC_4$ for *tert*-butyl, $C_4$ for *n*-butyl, $C_8$ for *n*-octyl, $C_{12}$ for *n*-dodecyl, and $C_{18}$ for *n*-octadecyl. The number-average degree of polymerization of the PEO chain is specified by the number n, while VB and MA stand for the $\alpha$-end groups, *p*-vinylbenzyl and methacryloyl, respectively.

The macromonomers with $R=C_1$, $C_4$, $tC_4$, and $C_8$ were prepared under vacuum by living anionic polymerization of ethylene oxide (EO) with corresponding potassium alkoxides in tetrahydrofuran (THF) at 40°C, followed by end capping with *p*-vinylbenzyl chloride (VBC) or methacryloyl chloride (MAC) [15,16,44]. EO conversions were almost quantitative after polymerization for about 16 hours (overnight) to afford macromonomers with essentially pure $\alpha$-and $\omega$-end groups after end-capping. The molecular weight distribution was also usually sharp with Mw/Mn less than 1.1.

$$CH_3OCH_2CH_2OK \text{ disp.} \xrightarrow{EO} CH_3O\text{-}[CH_2CH_2O]_n\text{-}K$$

$$\xrightarrow{\text{VBC or MAC}} C_1\text{-PEO-VB-n or } C_1\text{-PEO-MA-n}$$

$$\underset{(R = C_4,\ tC_4,\ C_8)}{ROK} \xrightarrow{EO} RO\text{-}[CH_2CH_2O]_n\text{-}K$$

$$\xrightarrow{\text{VBC or MAC}} \text{R-PEO-VB-n or R-PEO-MA-n}$$

$\omega$-Hydroxy macromonomers, with R=H, were also prepared by polymerization of EO with a *tert*-butyldimethylsilylprotected alkoxide, followed by end-capping with VBC or MAC and desilylation with tetra-*n*-butylammonium fluoride (TBAF) [45];

$$tBuMe_2SiOCH_2CH_2OK \xrightarrow{EO} \xrightarrow{\text{VBC or MAC}} \xrightarrow{TBAF}$$

$$HO\text{-}[CH_2CH_2O]_n\text{-}CH_2\text{-}C_6H_4\text{-}CH{=}CH_2 \quad \text{(H-PEO-VB-n), or}$$

$$HO\text{-}[CH_2CH_2O]_n\text{-}C(=O)\text{-}C(CH_3){=}CH_2 \quad \text{(H-PEO-MA-n)}$$

Some macromonomers were also prepared under argon or nitrogen from commercially available poly(ethylene glycol) monoalkyl ethers by alkoxidation with sodium or sodium hydride followed by reaction with VBC or MAC;

## 3.1 General Polymerization Behavior and Kinetics [44]

In general, the PEO macromonomers, with the chain length $n>20$,polymerized more rapidly than their lower molecular weight homologs with $n=1$-4 both in benzene (Fig.1) and in water (Fig.2). This can be reasonably ascribed to a lower rate of diffusion-controlled termination of the macromonomer. Most important to be noted, however, is the fact that the polymerization in water is surprisingly higher, in spite of the lower initiator concentration, than that in benzene. The former is almost quantitative after about 5h, while the latter is very slow and levels off below 50% conversion after 4 days or so. Thus in benzene, the employed condition of very low concentration of both the macromonomer and the initiator on the molar basis may well result in the "dead-end" phenomenon [51] as observed. In water, however, these amphiphilic macromonomers will aggregate to form micelles with their hydrophobic groups in the core and the hydrophilic PEO chains in the surrounding shell. This aggregation will locally concentrate and orientate the hydrophobic polymerizing groups to enhance their polymerization. The difference in the initiators used, 4,4'-azobis(4-cyanovaleric acid) (AVA) in water and AIBN in benzene, cannot be a factor because the former decomposed a little more slowly than the latter, contrary to the observed polymerization rates.

The rate of polymerization, $R_p$, of $tC_4$-PEO-VB and -MA was estimated from initial slope of the conversion vs time plots with the results given in Fig.3. It is clear in these log-log plots that $R_p$ in water is about two orders of magnitude higher than that in benzene at the same concentration level. $R_p$ can be expressed experimentally as follows;

$$Rp \propto [M]^{1.5} [I]^{1/2} \qquad \text{in benzene} \qquad (1)$$

and

$$Rp \propto [M][I]^{1/2} \qquad \text{in water} \qquad (2)$$

Eq. (2) is in accordance with the conventional relationship in radical polymerization [51] as

$$R_p = k_p\left(\frac{2f k_d}{k_t}\right)^{1/2} [M][I]^{1/2} \qquad (3)$$

where $k_p$ and $k_t$ are the rate constants of propagation and termination, respectively, while $k_d$ and f are decomposition rate constant and the initiator efficiency of the initiator, respectively.

The square-root dependence of [I], as observed both in benzene and in water, clearly supports the conventional bimolecular termination of the polymer radicals to prevail also for the polymerization of the macromonomers, at least under the conditions employed. Since these radical termination reactions are now well-known to be

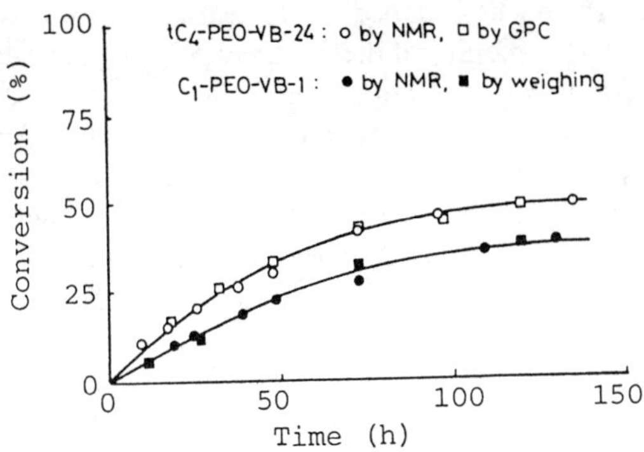


Figure 1. Conversion vs time plots for polymerization in benzene. [M] = 45 mmol/L and [AIBN] = 2.25 mmol/L at 60°C.

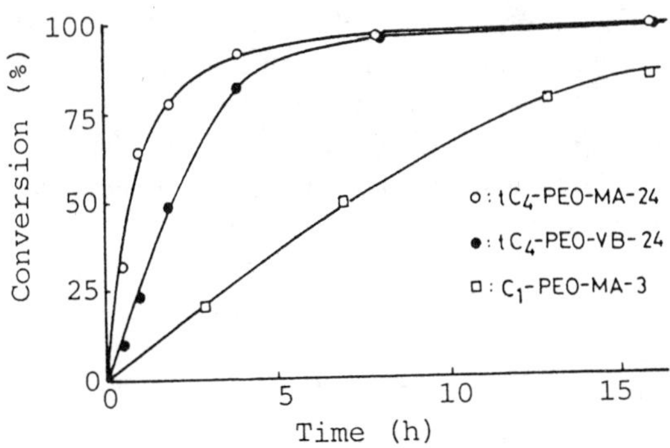


Figure 2. Conversion vs time plots for polymerization in water. [M] = 45 mmol/L and [AVA] = 0.45 mmol/L at 60°C.

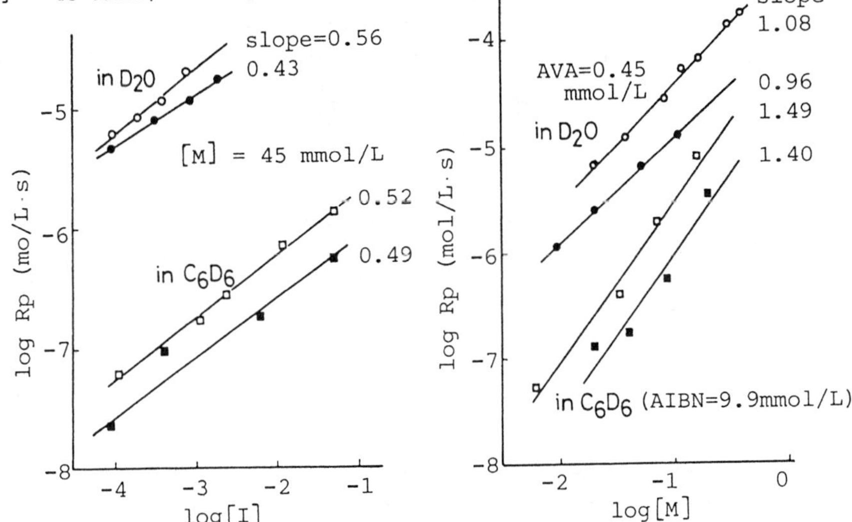


Figure 3. Log-log plots of Rp vs [I] and [M] for polymerization of $tC_4$-PEO-VB-24 (●,■) and $tC_4$-PEO-MA (○,□).

diffusion- controlled, they should be retarded by increasing viscosity of the medium, thus by restricting the segmental motion of the involved polymer radicals. This can be a reason for the higher rate of polymerization of the macromonomer as compared to the low molecular weight model monomer, as found in Fig.1 and 2. The dependence on [M] may become higher than the first order as observed in benzene, as a result of the inverse dependence of $k_t$ on the medium viscosity, which increases with the monomer concentration, because the monomer here is itself a polymer of a reasonable molecular weight. The result in eq. (1) is thus reasonable in reflecting this situation.

In water, however, an apparently normal kinetic behavior, eq.(2), was observed with an unusually high polymerization rate. This can be expected in the context of the micellar polymerization by considering that $k_t$ may be extremely low and remains constant in the organized micellar environment, independent of the overall monomer concentration, which is much higher than the critical micellar concentration (cmc) as will be shown later. The linear dependence of $R_p$ on [M] may simply reflect a linear dependence of the concentration of the micelles, which are the sites of polymerization, on the overall concentration. This situation may appear to be somewhat similar to that of the conventional emulsion polymerization but it is different that the monomers here are themselves the micelle-forming surfactants and that the system remains apparently clear throughout the polymerization, as in benzene. Therefore, some chain transfers reported for PEO-VB macromonomers with relatively low PEO chain length [7] and the self-gellation reported for PEO-MA macromonomers [42] appear to be excluded or negligible at least in our polymerization condition. The same normal kinetics as in eq. (2) has been reported by Yeoh and co-workers [52] for the polymerization of a surfactant monomer such as sodium 11-[N-(methacryl)amido] undecanoate.

The degree of polymerization, DP, was also estimated by means of light scattering measurements in THF [47] for the polymacromonomers from $C_1$-PEO-VB-n and plotted according to the conventional kinetics as follows:

$$DP = \frac{2k_p}{(2k_t k_d f)^{1/2}} \frac{[M]}{[I]^{1/2}} \tag{4}$$

under the assumption of a recombination termination and no chain transfer. The result in Fig.4 clearly shows that the polymerizations of the PEO macromonomers with n=22, 44, and 103 in water attained much higher DPs than those of either the monomer with n=3 or styrene in benzene, supporting again the micellar polymerization mechanism of the former. As a reference, $R_p$ and DP for polymerization of styrene were calculated by eqs. (3) and (4) using the literature data for $k_p$, $k_d$, and $k_t$, to be as small as $8 \times 10^{-8}$ mol/L·s and 22, respectively, at [M]=45 mmol/L and [I]=0.45mmol/L.

In view of the amphiphilic properties of the PEO macromonomers, it remains another challenging problem to examine the polymerization in a nonpolar hydrocarbon medium. The PEO macromonomers, $C_{12}$-PEO-MA-n or $C_{12}$-PEO-VB-n, wiht n=10 to 23, were found to be soluble throughout polymerization in water, benzene, and cyclohexane. Fig.5 typically shows the solvent effect on the time-conversion plots for

$C_{12}$-PEO-MA-14. Very clearly, the polymerization in cyclohexane is faster than that in water, but by far slower than that in water. The effects of the ω-alkyl groups and the PEO chain lengths were also significant in cyclohexane but just reversed from those in water. Thus a macromonomer with a lower ω-alkyl group and/or with a higher PEO chain length was found to polymerize more rapidly. These results are just as expected by assuming inverse micelle formation in cyclohexane, since either longer PEO chain or lower ω-alkyl group will enhance the hydrophilicity and so favor the aggregation to form the inverse micelles [48].

The polymerization in cyclohexane, however, was not so markedly enhanced as compared to that in water. This appears to be related to the fact that the inverse micelle formation of the nonionic surfactants are in general of open association type without any definite cmc or aggregation number. Therefore in cyclohexane we assume somewhat loosely organized, inverse micelles which will favor the polymerization to some significant extent as observed. The aggregation, even if it may be transient, will probably depress a diffusion-controlled termination between the highly branched polymer radicals.

## 3.2 Effects of α- and ω-Terminal Groups and PEO Chain Lengths on the Polymerizability [44]

Fig.6 shows that the ω-alkyl groups of the macromonomer, R-PEO-VB, have little effect on their polymerizability in benzene and that they polymerize more rapidly than the lower molecular weight homologue, $C_1$-PEO-VB-4. This results is in line with the previous discussions. Fig.7 shows that the chain lengths of PEO have also little effect on the polymerization in benzene. These results indicate that the chemical reactivity of the terminal double-bond determining $k_p$ and the mutual segmental diffusion of the polymer radicals determining $k_t$ do not so significantly depend on the ω-alkyl groups or on the PEO chain lengths, at least within the range studied, i.e., $C_1$ to $C_{18}$ for R and 17 to 60 for n.

In water, however, significant effects are observed. From the results with R-PEO-VB-n in Figs. 7, 8, and 9, we can summarize the effects of R and n on $R_p$ in the decreasing order as follows : $R=C_{18} > H \sim C_8 > C_1 > C_4 > tC_4$, and $n=17\text{-}25 > 45 > 60$. As to the effect of n lower than 17, the model monomer, $C_1$-PEO-VB-4 is insoluble in water, but its polymerization appears to be much slower than $C_1$-PEO-VB-17, considering the results in benzene in Fig.6together with those in water for R-PEO-MA-n in Fig. 2. Therefore, the polymerizability of $C_1$-PEO-VB-n has its maximum around n=10-25. A similar trend also appears to be true with the DPs of the polymacromonomers in Fig.4. These results are expected because the micelle formation will be unfavored with either too high or too low chain lengths of PEO. In this manner, the results observed in water reflect the effects of R and n on the formation of the micelle, the structure of which will in turn affect kp and/or kt, as will be discussed later.

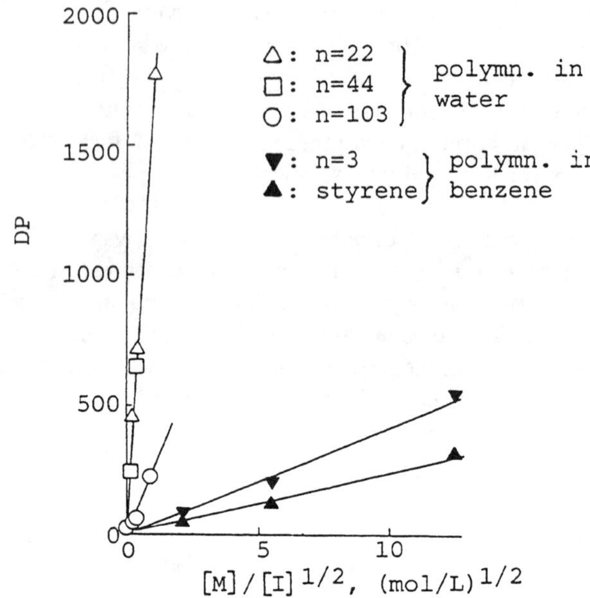


Figure 4. DP of poly(macromonomer)s as a function of the feed composition. $C_1$-PEO-VB-n with n = 22, 44, and 103 were polymerized in water with AVA, while the monomer of n = 3 and styrene were polymerized in benzene with AIBN

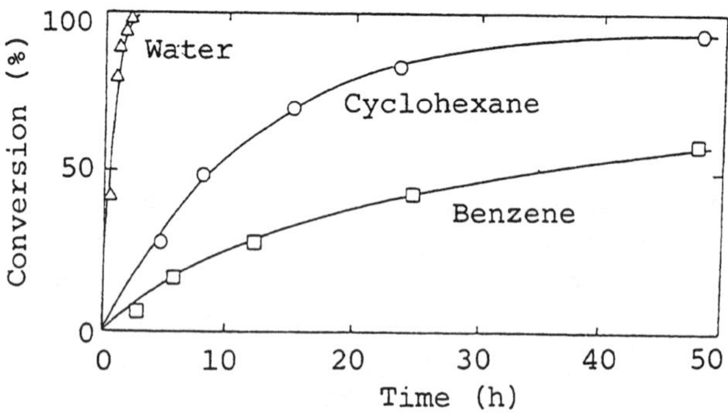


Figure 5. Conversion vs time plots for polymerization of $C_{12}$-PEO-MA-14, [M] = 45 mmol/L in water with AVA (0.45 mmol/L) (△), in cyclohexane (○), and in benzene (□) with AIBN (2.25 mmol/L).

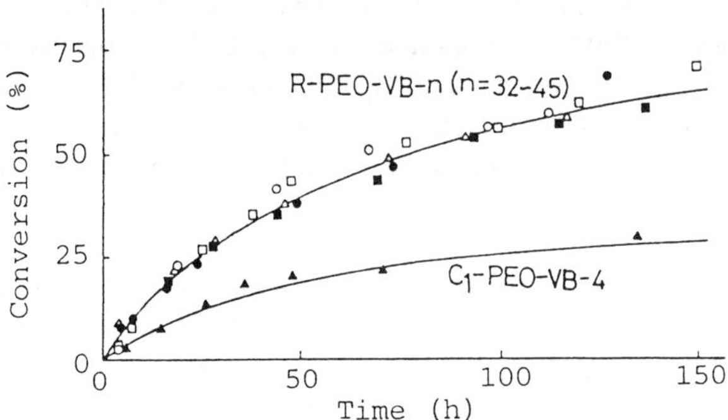


Figure 6. Polymerization of $C_1$-PEO-VB-45 (●), $C_4$-PEO-VB-39 (■), $tC_4$-PEO-VB-32 (△), $C_8$-PEO-VB-43 (○), $C_{18}$-PEO-VB-35 (□), and $C_1$-PEO-VB-4 (▲) in benzene with [M] = 45 mmol/L and [AIBN] = 2.25 mmol/L.

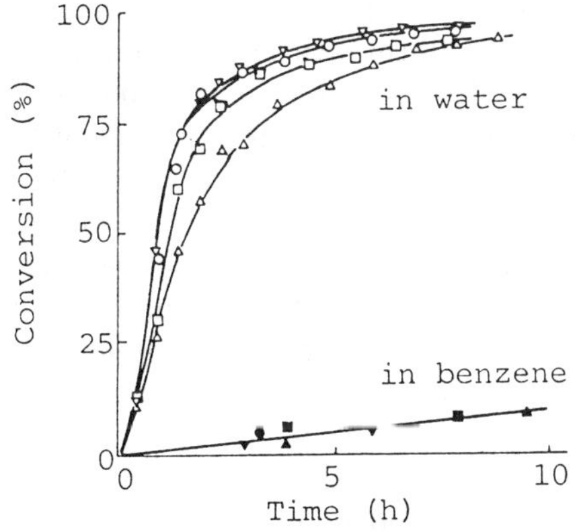


Figure 7. Polymerization of $C_1$-PEO-VB-17 (▽), -25 (○), -45 (□), and -60 (△) in water with [M] = 45 mmol/L and [AVA] = 0.45mmol/L. Solid symbols refer to polymn. in benzene with AIBN =2.25mmol/L.

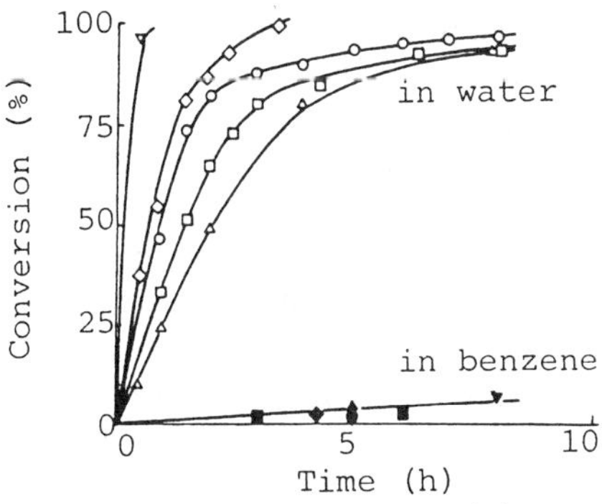


Figure 8. Polymerization of $C_1$-PEO-VB-25 (○), $C_4$-PEO-VB-39 (□), $tC_4$-PEO-VB-32 (△), $C_8$-PEO-VB-43 (◇), and $C_{18}$-PEO-VB-35 (▽) in water with [M] = 45 mmol/L and [AVA] = 0.45 mmol/L. Solid symbols refer to polymn. in benzene with AIBN = 2.25 mmol/L.

Table 2
Micelle Formation of Macromonomer Models in Water at 25°C*

| macromonomer model | cmc, $10^5$ mol/L | $M_m$ | m | $\langle s^2 \rangle^{1/2}$, m | $V_m$, $m^3$ | $m/V_m$, $m^{-3}$ |
|---|---|---|---|---|---|---|
| H-PEO-Bz-53 | $1._{1}$ | $8._{7}x10^{4}$ | $4._{5}x10$ | $5._{1}x10^{-8}$ | $1._{2}x10^{-21}$ | $3._{8}x10^{22}$ |
| $C_1$-PEO-Bz-25 | $3._{3}$ | $2._{1}x10^{6}$ | $1._{8}x10^{3}$ | $1._{7}x10^{-7}$ | $4._{5}x10^{-20}$ | $4._{0}x10^{22}$ |
| $C_1$-PEO-Bz-45 | $3._{0}$ | $1._{1}x10^{6}$ | $5._{2}x10^{2}$ | $1._{7}x10^{-7}$ | $4._{5}x10^{-20}$ | $1._{2}x10^{22}$ |
| $C_4$-PEO-Bz-39 | $5._{8}$ | $3._{4}x10^{4}$ | $1._{9}x10$ | $7._{6}x10^{-8}$ | $3._{9}x10^{-21}$ | $4._{9}x10^{21}$ |
| $tC_4$-PEO-Bz-32 | $7._{0}$ | $3._{0}x10^{4}$ | $1._{9}x10$ | $1._{3}x10^{-7}$ | $2._{1}x10^{-20}$ | $9._{0}x10^{20}$ |
| $C_8$-PEO-Bz-43 | $1._{8}$ | $1._{0}x10^{5}$ | $4._{7}x10$ | $6._{1}x10^{-8}$ | $2._{0}x10^{-20}$ | $2._{3}x10^{22}$ |
| $C_{18}$-PEO-Bz-35 | $1._{8}$ | $1._{3}x10^{5}$ | $6._{8}x10$ | $1._{2}x10^{-8}$ | $1._{4}x10^{-23}$ | $4._{9}x10^{24}$ |
| $C_1$-PEO-IB-25 | $7._{1}$ | $2._{3}x10^{4}$ | $1._{9}x10$ | $2._{5}x10^{-8}$ | $1._{4}x10^{-22}$ | $1._{4}x10^{23}$ |

* Subscript figures in the first decimal place indicate uncertainty of ±0.1 to ±0.3 in determining cmc, $M_m$, and $\langle s^2 \rangle^{1/2}$.

be forced more or less to assume looplike conformations. $R=C_8$ may stand intermediate between (b) and (c). Model (c) was presented as such because of the highly hydrophobic nature of the long alkyl groups ($C_{18}$), which will align themselves very compactly in the core with the other, less hydrophobic *p*-vinylbenzyl groups residing at the interphase to the surrounding hydrophilic shell composed of PEO chains.

These models, even if only schematic, appear reasonable to account for the observed parallel trend in the polymerizability ($R_p$) and the molecular packing density in the micelle ($m/V_m$). Among the macromonomers investigated, $C_{18}$-PEO-VB polymerized most rapidly, probably as a result of the most closed packing or orientation of the double bonds as in the model (c). H- and $C_1$-PEO-VB polymerized moderately rapidly with the double bonds concentrated in the micelle core as in the model (a). It is reasonable that the moderately short PEO chain is more favorable for organization into the micelle and therefore also for polymerization as observed, because too long PEO will make the macromonomer too hydrophilic for organization of the $\alpha$-terminal double bonds. The higher polymerizability of H-PEO- as compared to $C_1$-PEO macromonomers is also reasonable because of the more compact, denser micelle formation of the former. The macromonomers with $\omega$-alkyl groups of $C_4$, and ${}^tC_4$ polymerized most slowly probably because of loose packing of their double bonds together with the $\omega$-alkyl groups as in the model (b).

In any case, the high polymerizability of these amphiphilic PEO macromonomers in water appears to be closely related to their organization into the micelle. This organization will clearly favor the propagation reaction (increase in $k_p$) and hinder the bimolecular termination (decrease in $k_t$). It is to be noted here that a simple accumulation of the double bonds or an increased local monomer concentration will only explain the enhanced $R_p$ but cannot explain by itself the enhanced rate of <u>conversion</u> as observed. The linear dependence of $R_p$ on [M], as in eq. (2) or (3), means the independence of the conversion on [M], because the conversion, which corresponds to -d[M]/[M] or its integrated form, already includes normalization in term of the monomer concentration. Therefore, by virtue of eq. (3), the enhanced <u>conversion</u> requires an increase in the term $k_p/k_t^{1/2}$, i.e., either an increase in $k_p$ or a decrease in $k_t$, under a reasonable assumption of constant $k_d$ and f.

In conclusion, amphiphilic PEO macromonomers were found to polymerize in water with an unusually high rate that has not been observed for a solution polymerization of a conventional monomer or macromonomer. The organization of the macromonomers into the micelles as in Fig.12 was considered to be the most important factor for the results.

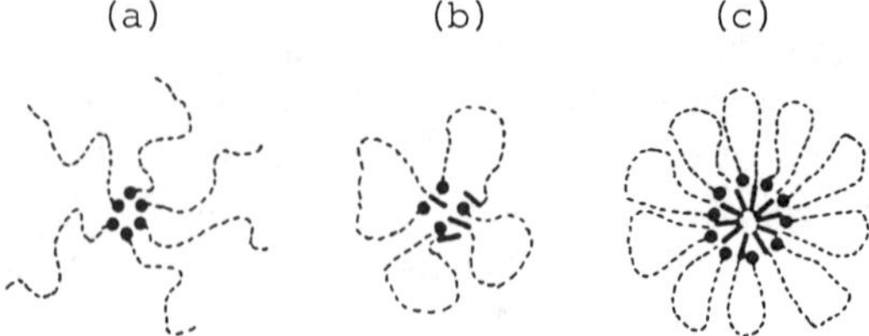


Figure 12. Schematic models of micelles of macromonomers in water for (a) R = H or $C_1$, (b) R = $C_4$ or $tC_4$, (c) R = $C_{18}$.

## 3.4 Solution Properties of Regular Comb Polymers - Polymacromonomers [47]

Polymerization of macromonomers provides a series of model branched polymers. In particular, homopolymerization affords regular comb polymers of a well-defined structure. Thus, a PEO macromonomer, $C_1$-PEO-VB-n readily polymerizes to give a polystyrene with PEO side chains, which are regularly and densely spaced, each on every benzene ring of the backbone, as in Fig.13.

The length or the degree of polymerization, n, and its distribution of the side chains (arms or teeth) can be either definitely known by independent characterization of the macromonomer used or easily controlled in preparation of the macromonomer by virtue of the living anionic polymerization of EO. The size of the main chain or the average number of side chains, f, may be, at least qualitatively, controlled by changing the feed compositions such as the monomer and the initiator concentrations in the context of conventional polymerization kinetics as in eq. (4).

The comb polymers are expected to exhibit a variety of characteristics depending on the value of n and f, since a polymer with a very small n relative to f will represent a linear polymer on the one hand, while that with a very large n and very small f will behave like a starlike polymer on the other hand. Both large n and f will most probably represent a typical comb or brush polymer. Thus it will be interesting to examine how n and f affect the solution characteristics of the comb polymers.

Table 3 summarized all the comb polymers investigated with the results of the characterization of the molecular weight, Mw, measured by light scattering and that estimated by GPC calibrated with the linear PEO standard samples, together with the intrinsic viscosity, all measured in THF. A universal calibration was constructed in Fig.14 by plotting log(Mw[$\eta$]), with Mw measured by light scattering, as a function of the elution volume, Ve, in GPC. All the plots, including the comb polymers as well as the corresponding linear homopolymers, can be clearly represented by a single curve, supporting not only the usefulness of the universal calibration but also the validity of the Mw thus determined by LS. In contrast, the apparent molecular weight based on the GPC calibration with linear standard polymers are severely

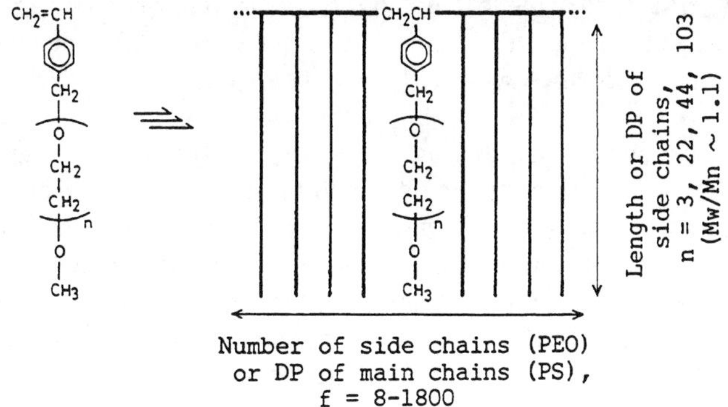


Figure 13. The comb polymers derived from $C_1$-PEO-VB-n.

Table 3
Characterization of Comb Polymers, Poly($C_1$-PEO-VB-n)

| n | $10^{-4}M_W$(LS) | $10^{-4}M_W$(GPC) | [η] dL/g | f |
|---|---|---|---|---|
| 3 | 16 | 10 | 0.21 | 554 |
| 3 | 5.3 | 2.6 | 0.11 | 190 |
| 3 | 2.2 | 1.0 | 0.067 | 80 |
| 22 | 229 | 25 | 0.81 | 1760 |
| 22 | 92 | 21 | 0.46 | 711 |
| 22 | 58 | 17 | 0.28 | 446 |
| 22 | 8.3 | 1.8 | 0.08 | 64 |
| 22 | 5.4 | 0.6 | 0.055 | 41 |
| 44 | 149 | 13 | 0.16 | 648 |
| 44 | 56 | 12 | 0.12 | 244 |
| 44 | 3.0 | 3.4 | 0.09 | 13 |
| 44 | 1.8 | 1.8 | 0.08 | 8 |
| 103 | 116 | 18 | 0.22 | 215 |
| 103 | 33 | 14 | 0.21 | 61 |
| 103 | 29 | 12 | 0.22 | 51 |
| 103 | 13 | 8.1 | 0.26 | 24 |
| 103 | 9.1 | 5.1 | 0.24 | 17 |

underestimated, particularly at higher molecular weights, i.e., for the comb polymers with a longer and/or higher number of side chains, as is clear in Fig. 15. The polymers with the shortest side chains, n=3, as well as polystyrene (PS) samples fall nearly on the theoretical line with Mw(LS)=Mw(GPC), as expected.

As to the molecular weight distribution of the comb polymers, GPC also yielded poor estimation based on the usual calibration with the linear standard polymers. For an example, a comb polymer of n=103 and f=215 in Table 3, the usual calibration gave Mw/Mn=1.13 while recalibration using the Mw data measured by LS estimated Mw/Mn as high as 3.0-4.9. Similar trends were also true for the other comb polymers. In this manner, the usual GPC calibration based on the linear standard polymers appears to be almost meaningless for branched polymers in that too low values are estimated for Mn, Mw, and Mw/Mn.

From the above discussions, it is clear that the present comb polymers assume very compact dimensions in solution as compared to those of the linear polymers of the corresponding molecular weights. Now it will be interesting to examine what conformations they should have in solution depending on the length of the side chains. Fig. 16. shows the double-logarithmic plots of the intrinsic viscosity and the Mw by LS, both measured in THF at 25°C, according to the Mark-Houwink-Sakurada equation:

$$[\eta] = K \cdot Mw^{a} \tag{5}$$

The value of the exponent a around 0.6-0.7 was observed for the comb polymers with relatively short side chains (n=3 and 22) as well as for linear PEO and PS, clearly supporting their expanded coil conformations, as expected. The each isolated chain in the solutions of these comb polymers apparently behaves like an expanded-coil linear polymer chain, as a whole, although with a much more contracted dimension, as compared to a real linear polymer with the same molecular weight, because of a geometric requirement for the comb structure.

Most striking in Fig. 16 is the observation that the comb copolymers with longer chains exhibited very low or even nearly zero values for the exponent at least within the Mw range investigated; a=0.15 for n=44, and a=-0.06 for n=103. Since the zero exponent is required for a rigid sphere, the results suggest a very densely filled, nondraining, rigid spherelike conformation, as a whole, for each of these comb polymer chains. Although further extensive studies should remain to confirm more explicit conformations, the nearly zero exponent appears to be the first experimental observation for synthetic branched polymers to out knowledge. At any rate, the little dependence of the viscosity on the molecular weight, as observed for the present regular comb polymers, is very interesting in nature and appears to find applications particularly in the field of paint technology.

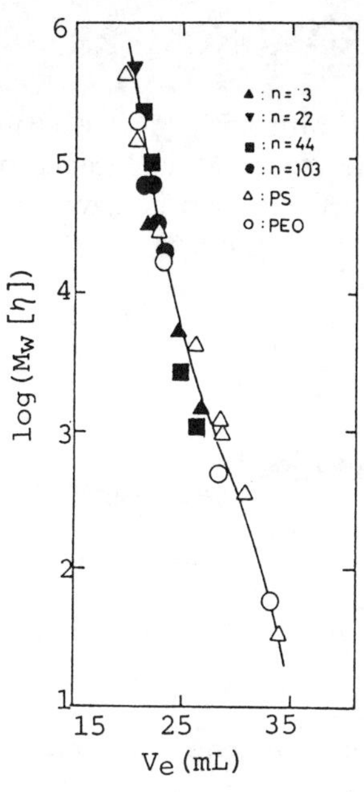


Figure 14. Universal calibration of GPC data for comb polymers and corresponding homopolymers.

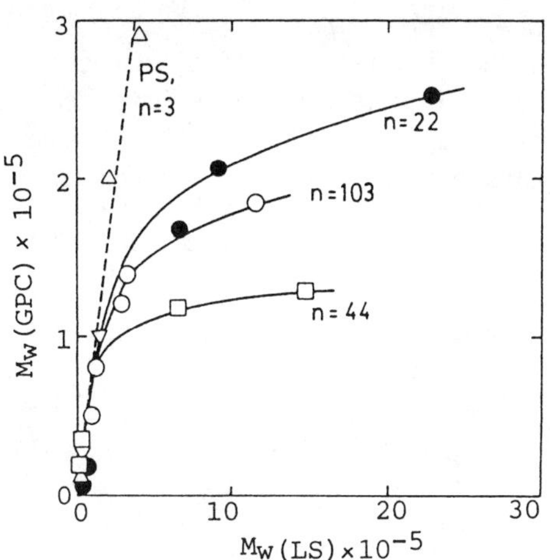


Figure 15. Comparison of $M_w$'s determined by GPC and LS. A broken line is theoretical for $M_w(GPC) = M_w(LS)$.

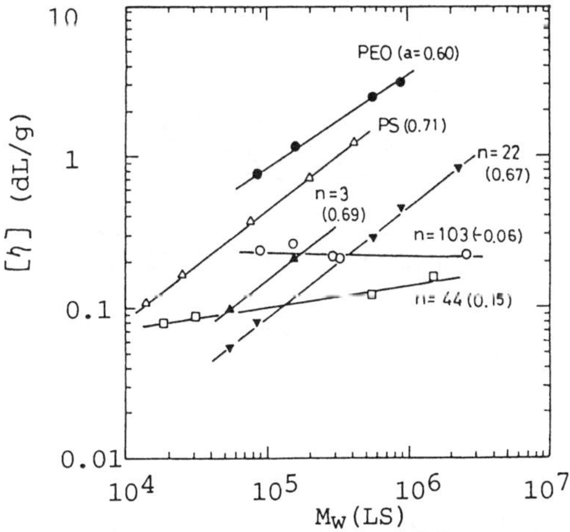


Figure 16. Double-logarithmic plots of [η] vs $M_w(LS)$ for PEO (●, a=0.60), PS (△,0.71), and comb polymers of n=3 (▲, 0.69), n=22 (▼,0.67), n=44 (□,0.15), and n=103 (○,-0.06). Values in parentheses denote the exponent, a, in the Mark-Houwink-Sakurada equation; $[\eta] = KM^a$.

## 4. COPOLYMERIZATION OF PEO MACROMONOMERS

Since the first publications by ICI group [53] and Milkovich [54], syntheses of well-defined graft copolymers have been the most important application of the technique, as illustrated in Fig.17. A number of PEO macromonomers have also been prepared and copolymerized with conventional monomers [3-40].

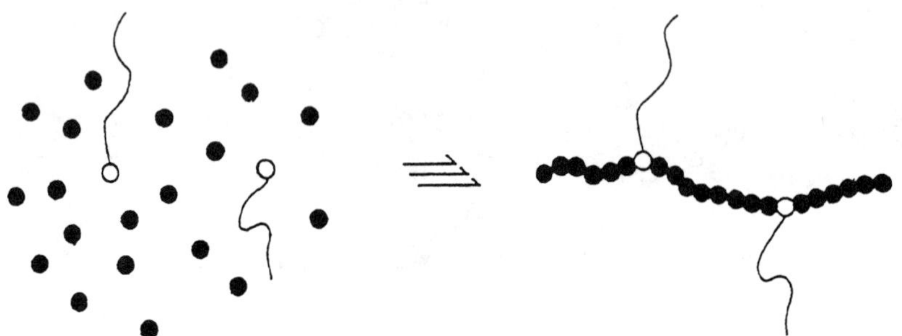

Figure 17. Synthesis of graft copolymers by the macromonomer technique.

It will be appropriate to start our discussion with some fundamental points [15,56] involved in the macromonomer technique which consists of the copolymerization between a conventional monomer ($M_1$) with a relatively low molecular weight and a macromonomer ($M_2$) with a high molecular weight, therefore the molar ratio $[M_1]/[M_2]$ being generally very higher than unity. In such a case, the conventional Mayo-Lewis composition equation

$$\frac{d[M_1]}{d[M_2]} = \frac{1 + r_1 [M_1]/[M_2]}{1 + r_2 [M_2]/[M_1]} \tag{6}$$

reduces to, in the limiting condition of $[M_1]/[M_2] >> 1$ [54,55]

$$\frac{d[M_1]}{d[M_2]} = r_1 \frac{[M_1]}{[M_2]} \tag{7}$$

This is equivalent to saying that we are effectively dealing with only two of four conventional propagation reactions:

$$-M_1\bullet + M_1 \xrightarrow{k_{11}} -M_1\bullet \tag{8}$$

$$-M_1\bullet + M_2 \xrightarrow{k_{12}} -M_2\bullet \tag{9}$$

with

$$r_1 = k_{11}/k_{12} \tag{10}$$

together with the implicit condition of the stationary state that the other cross-propagation

$$-M_2\bullet + M_1 \xrightarrow{k_{21}} -M_1\bullet \tag{11}$$

should occur with a rate equal to that of the reaction (9), so that

$$k_{12}\,[-M_1\bullet][M_2] = k_{21}\,[-M_2\bullet][M_1] \tag{12}$$

Hence, with the condition $[M_1]/[M_2]>>1$,we may neglect the contribution of polymer radical -$M_2\bullet$ on the composition equation.

Therefore, the $r_1$ thus determined should approximate most simply the relative reactivities of the monomers $M_1$ and $M_2$ in competition reactions with the $M_1$-ended radical which should approximates a homopoly-$M_1$ ($M_1$-consecutive) radical. We thus need not consider in general complication due to penultimate effects or depropagation, which might possibly be important in propagation reactions involving a more crowded radical as suspected in the case of increased composition of the macromonomer ($M_2$). It should also be noted that the $r_1$ values obtained by copolymerization of a common monomer ($M_1$) with various monomers ($M_2$) can be a measure of the relative reactivities of the monomers ($M_2$) by taking their reciprocals, $1/r_1=k_{12}/k_{11}$.

Now we turn to the results of the application of equation (6) and (7) to the composition data of copolymerization of styrene ($M_1$) with $tC_4$-PEO-VB-29, as an example. As shown in Table 4, the copolymer compositions were necessarily very rich in styrene on a molar basis but could be satisfactorily determined by $^1$H-NMR and GPC, although the latter method was somewhat less sensitive to the change in composition at a low conversion. Application of the method of Kelen and Tüdös [57], which is based on eq (6), gave $r_1=1.36\pm0.08$and $r_2=1.47\pm1.19$,with error of 95% confidence limit. The corresponding copolymer composition curves are shown in Fig.18. Evidently, under the present condition of $[M_1]/[M_2]>>1$,as is usual for the macromonomer technique, the value of $r_2$ involves too much error to be significant, while the evaluation of $r_1$ by use of the approximate equation, (7), gave a very reasonable result, indicating that the copolymer composition is essentially controlled by the $r_1$ parameter alone. Similar conclusions have been drawn in many other systems including macromonomers such as polystyrene [54] and poly(dimetylsiloxane) [58].

Table 4

Copolymerization of Styrene ( $M_1$) and $tC_4$-PEO-VB-29 ($M_2$)
(AIBN ca 20mg, THF ca 4.7g, heptadecane ca 1.0g, 60°C)

| | Feed | | | | | $M_1$ in polymer, mol% | |
|---|---|---|---|---|---|---|---|
| Run | $M_1$, g | $M_2$, g | $M_1$, mol% | Time, h | Yield, wt% | NMR | GPC |
| 1 | 0.803 | 1.202 | 90.3 | 3.0 | 6.1 | 92.3 | 92.4 |
| 2 | 1.001 | 1.000 | 93.3 | 3.0 | 7.7 | 94.7 | 94.4 |
| 3 | 1.201 | 0.801 | 95.4 | 3.0 | 5.8 | 96.4 | 96.5 |
| 4 | 1.396 | 0.604 | 97.0 | 3.0 | 6.4 | 97.8 | - |
| 5 | 1.605 | 0.407 | 98.2 | 3.2 | 8.1 | 98.7 | 98.6 |

## 4.1 Solution Copolymerization of PEO Macromonomers

Along with extensive studies on preparation and practical applications of macromonomers, there has been considerable progress in understanding the macromonomer's reactivity in copolymerization with a conventional trunk-forming comonomer, which is of fundamental importance in determining the primary structure of the resulting graft copolymers [59,60]. Among others, Tsukahara [61] and Gnanou [62] and their coworkers have shown with polystyrene macromonomers that their reactivities are essentially determined by the chemical nature of their terminal copolymerizing groups, so long as they are thermodynamically compatible with, or at least able to interpenetrate, without difficulty, the trunk polymer chains, i.e., the polymer radicals from the comonomers employed. The so-called kinetic excluded-volume effect, if any, appears to be negligibly small. Very recently, however, Radke and Müller [63] have pointed out a considerable decrease in the reactivity of the PMMA macromonomer toward the highly multi-branched poly-MMA macroradical. This can be expected from the depressed $k_p$ for homopolymerization of the macromonomers [50], as described, suggesting a complicated mechanism involving the penultimate effects in the copolymerization with higher macromonomer compositions.

An apparent decrease in the macromonomers reactivity compared to that of the low molecular weight analogue, as often observed in such cases including polyisobutylene [64] and poly(dimethylsiloxane) [58,65,66], could be reasonably ascribed to thermodynamic repulsion or incompatibility between the macromonomer and the trunk polymer from the comonomer, resulting in phase separation in extreme cases. Monomer partition at the polymerization sites, which may differ in the composition from the bulk, can also be a factor for changing the apparent reactivity in the cases including a selective solvent [67] or in a heterogeneous copolymerization system [18,53].

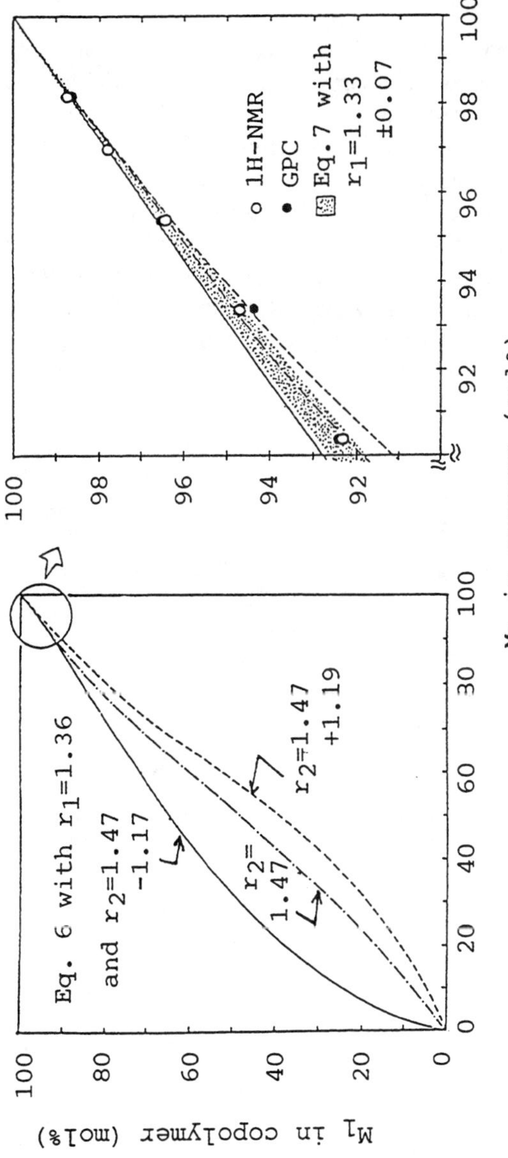


Figure 18. Copolymer composition curves for copolymerization of styrene ($M_1$) and PEO macromonomer, $tC_4$-PEO-VB-29 ($M_2$).

We have studied systematically the solution copolymerization of the PEO macromonomers, with the results summarized in Table 5 for copolymerization with styrene as $M_1$ [15-17,20,45,68]. The $r_1$ values were determined by the approximate composition equation, (7), at the condition $[M_1]/[M_2] >> 1$.

A consistent value of $r_1$, independent of the composition, was confirmed in each case, so that $r_1$ can be simply taken as a measure of the relative reactivity of the comonomer ($M_1$) and the macromonomer ($M_2$) toward a poly-$M_1$ or $M_1$-consecutive radical. Thus, in Table 5, $1/r_1$ measures the relative reactivity of the macromonomer toward polystyryl radical. On referring to the $r_1$ values of styrene in copolymerization with the low molecular weight homologues [15,69] such as *p*-methylstyrene ($r_1=0.84$) and methyl methacrylate ($r_1=0.52$), the $r_1$ values in Table 5 are apparently higher, particularly with increasing PEO chain length, n. Thus the relative reactivity of the PEO macromonomers with styryl or methacryloyl end groups decrease with increasing PEO chain length, as is clear in the plots in Fig.19 [15]. Rempp *et al.* [4] reported a similar reduction in $1/r_1$ but with somewhat complicated dependence on the polymerization condition.

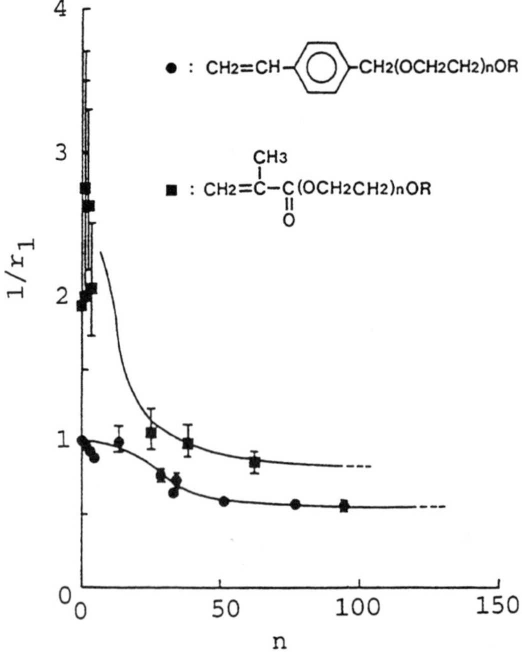


Figure 19. Relative reactivities of PEO macromonomers as a function of the PEO chain length, n, in copolymerization with styrene as $M_1$.

Table 5
Copolymerization of Styrene ($M_1$) with PEO Macromonomers ($M_2$)*

| $M_2$ | $r_1$ | Solvent | Ref. |
|---|---|---|---|
| H-PEO-VB-46 | 1.45±0.06 | benzene | 45 |
| " 65 | 1.68±0.06 | " | " |
| $C_1$-PEO-VB-25 | 1.31±0.09 | benzene | 45 |
| " 45 | 1.48±0.03 | " | " |
| " 60 | 1.71±0.11 | " | " |
| $tC_4$-PEO-VB-13 | 1.01±0.10 | THF | 15 |
| " 29 | 1.33±0.07 | " | " |
| " 33 | 1.53±0.05 | " | " |
| " 34 | 1.36±0.10 | " | " |
| " 51 | 1.71±0.05 | " | " |
| " 77 | 1.77±0.07 | " | " |
| " 94 | 1.79±0.12 | " | " |
| " 23 | 1.66±0.03 | benzene | 16 |
| " 23 | 1.42±0.05 | MIBK | " |
| " 93 | 1.86±0.26 | benzene | 16 |
| " 93 | 1.46±0.09 | " | " |
| $C_{18}$-PEO-VB-35 | 1.5 ± 0.1 | benzene | 20 |
| $C_1$-PEO-$C_4$-S-29 | 1.47±0.02 | benzene | 68 |
| $tC_4$-PEO-MA-3 | 0.49±0.09 | THF | 15 |
| " 25 | 0.94±0.13 | " | " |
| " 35 | 1.02±0.12 | " | " |
| " 62 | 1.18±0.11 | " | " |
| H-PEO-MA-53 | 0.83±0.12 | benzene | 45 |
| " 67 | 1.12±0.08 | " | " |
| $C_{12}$-PEO-MA-4 | 0.74±0.04 | benzene | 17 |
| " 9 | 0.79±0.05 | " | " |
| " 20 | 0.82±0.02 | " | " |
| " 25 | 1.04±0.04 | " | " |

* Initiator = AIBN, at 60°C.

Since PEO and polystyrene are incompatible, we conclude as before that the thermodynamic repulsion between these different polymer chains is a main reason for the apparently reduced reactivity of the terminal double bond of the macromonomer. This explanation appears reasonable by the further facts that the reduction in the macromonomer's reactivity is more pronounced in a better solvent, benzene vs MIBK [16], and that no such reduction was observed in the copolymerization involving the compatible polymer pairs such as benzyl methacrylate ($M_1$) with $C_1$-PEO-VB-25 or $C_{12}$-PEO-MA-9 [17].

Generally speaking, however, more detailed experiment as well as interpretation of the copolymerization involving the macromonomers certainly remain to be an important subject in future to establish the macromonomer technique for the syntheses of truly well-defined graft copolymers.

## 4.2 Heterogeneous Copolymerization of PEO Macromonomers

Just as the oil-soluble macromonomers have been the basis of nonaqueous dispersions [53], water-soluble macromonomers should be very useful emulsifiers or dispersants in aqueous system, an important field to be developed in view of the current problems involving the environmental pollution. Particularly important, the macromonomers, copolymerized with conventional monomers, are bound to the polymer particles so that the resulting soap-free emulsions or dispersions should be almost permanently stable to flocculation.

H- or $C_1$-PEO-MA macromonomers have been applied to aqueous emulsion or aq.-alcoholic dispersion polymerization of styrene or methyl methacrylate [21,33,34]. We have also used $C_{12}$-PEO-MA [18] and $C_1$-PEO-$C_4$-S [68] for soap-free emulsion polymerization of styrene or benzyl methacrylate. The copolymerization parameter, $r_1$, was evaluated by the same method as in the solution copolymerization, i.e., by use of the approximate eq. (7) at the condition $[M_1]/[M_2] >> 1$, where $[M_1]$ and $[M_2]$ are the monomer molar concentrations in feed. In general, however, the observed change in the "apparent" monomer reactivity ratio in these heterogeneous system appears only to reflect the corresponding change in the monomer composition near the polymerization site ("effective" monomer composition), $([M_1]/[M_2])_{eff}$, since the approximate composition equation should be then [18,54,67]

$$\frac{d[M_1]}{d[M_2]} = r_1 \frac{[M_1]}{[M_2]} = r_{1,\,true} \left( \frac{[M_1]}{[M_2]} \right)_{eff} = r_{1,\,true}\; k \frac{[M_1]}{[M_2]} \tag{13}$$

where $r_{1,true}$ is the true monomer reactivity ratio and k is a factor due to the monomer partition;

$$k = \left( \frac{[M_1]}{[M_2]} \right)_{eff} \Big/ \frac{[M_1]}{[M_2]} = \frac{r_1}{r_{1,\,true}} \tag{14}$$

Therefore, under a reasonable assumption that $r_{1,true}$ will not change so much with the polymerization medium, the change in the "apparent" reactivity ratio, $r_1$, should merely correspond to the change in the effective monomer composition from that in the feed.

Table 6 summarized the apparent $r_1$ values obtained. In emulsion system with $C_{12}$-PEO-MA-9 and benzyl methacrylate (BzMA), $r_1$ is independent of the nature of the initiator and also very close to that obtained in the benzene solution. This fact suggests that not only monomer composition at the polymerization site (active center) but also the reactivity ratio are the same as those of the solution (homogeneous) polymerization. This may be expected because the macromonomer used, $C_{12}$-PEO-MA-9, is compatible with poly-BzMA [17] and the reactivity ratio in the radical copolymerization will not change so much with the medium. In contrast, however, the reversed "apparent" reactivity was observed in heptane in which the monomer clear solution turned into a polymer suspension upon polymerization. Since BzMA is soluble in the medium, the result suggests that the polymerization occurred preferentially near on the inversed micelle surface where the macromonomers should be enriched.

For styrene as $M_1$, $C_{12}$-PEO-MA-20 was relatively suitable to get a stable emulsion at least to low conversions. The $r_1$ value, however, was significantly less than unity, indicating the higher incorporation of the macromonomer from the monomer mixture, as compared to the solution system. This "apparent" higher reactivity of the macromonomer appears to indicate that the polymerization proceeded preferentially near on the surface of the micelles or the particles which had solubilized or emulsified styrene monomers in the core, surrounded by the PEO macromonomers in the shell. The system flocculated, however, at higher conversions.

On the other hand, a PEO macromonomer with hydrophobically enhanced, styrylalkyl end group, $C_1$-PEO-$C_4$-S-29, was found to be very effective in producing a stable, monosized polystyrene emulsion to a complete conversion [68], as Fig.20. The $r_1$ is ideally unity, indicating the azeotropic conversion of the macromonomer and styrene, Fig.21, to afford the graft copolymers with polystyrene trunks and PEO branches, which act as effective anchors and steric stabilizers, respectively. The value of $r_1$ appears to indicate that all the PEO macromonomers are organized into the micelles or particles with their hydrophobic *p*-styrylalkyl end group headed toward the core phase of styrene monomers, while the hydrophilic PEO chains extended into the aqueous phase. This situation will effectively remove the repulsive interaction which might operate between polystyryl radicals and the PEO chains to retard their reaction as observed in the solution system. A model in Fig.22 will explain well such an effective latex formation stabilized by steric graft PEO chains.

## 5. CONCLUSION

Amphiphilic PEO macromonomers exhibited a variety of polymerization and copolymerization behaviors, depending on the medium, the $\alpha$- and $\omega$-terminal groups,

Table 6
Comparison of Copolymerization Parameters in Solution and Emulsion Systems

| $M_1$ | $M_2$ | initiator | medium | Appearance monomer | Appearance polymer | $r_1$ | ref |
|---|---|---|---|---|---|---|---|
| BzMA | $C_{12}$-PEO-MA-9 | AIBN | benzene | solution | solution | 1.23±0.09 | 18 |
| " | " | KPS | water | emulsion | emulsion | 1.41±0.07 | " |
| " | " | AVA | water | emulsion | emulsion | 1.38±0.07 | " |
| " | " | AIBN | water | emulsion | emulsion | 1.37±0.03 | " |
| " | " | AIBN | heptane | solution | suspension | 0.40±0.02 | " |
| styrene | $C_{12}$-PEO-MA-20 | AIBN | benzene | solution | solution | 0.82±0.02 | 18 |
| " | " | AVA | water | emulsion | emulsion | 0.43±0.03 | " |
| styrene | $C_1$-PEO-$C_4$-S-29 | AIBN | benzene | solution | solution | 1.47±0.02 | 59 |
| " | " | AIBN | water | emulsion | emulsion | 1.00±0.04 | " |

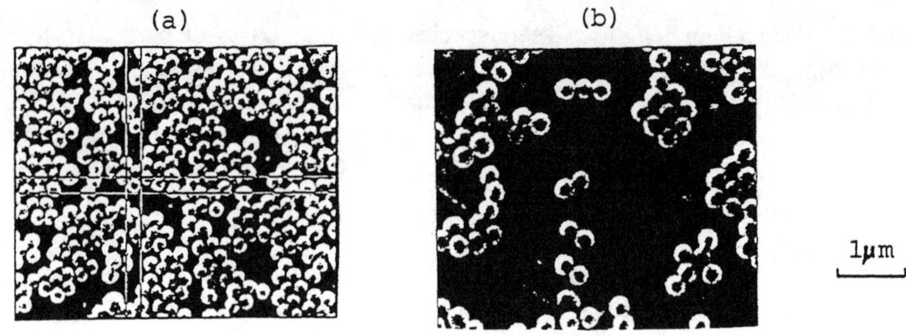


Figure 20. SEM of polystyrene particles from soap-free emulsion copolymerization of styrene ($M_1$) and $C_1$-PEO-$C_4$-S-29 ($M_2$). (a) $[M_1]/[M_2]$=146 mol/mol, (b) $[M_1]/[M_2]$=472 mol/mol.

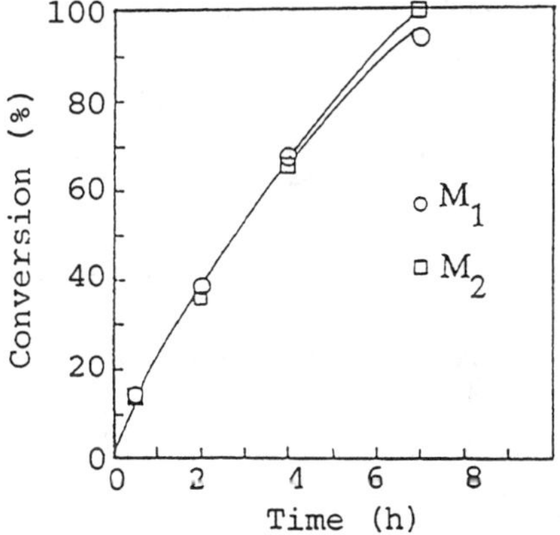


Figure 21. Copolymerization of styrene ($M_1$) and $C_1$-PEO-$C_4$-S-29 ($M_2$). $M_1$=1g, $[M_1]/[M_2]$ = 75.4mol/mol, AIBN=1mol%, water=10mL.

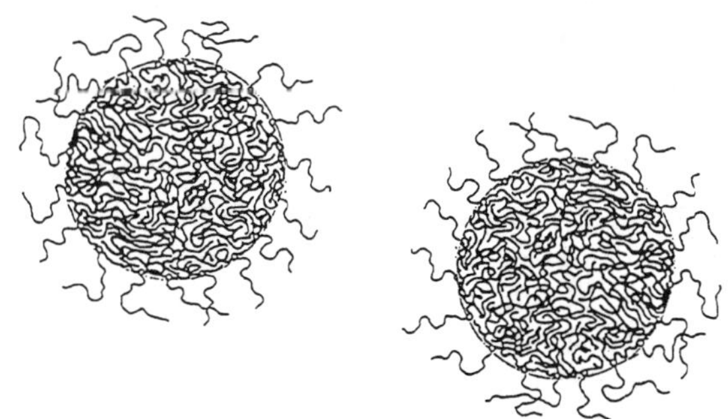

Figure 22. A schematic model of sterically stabilized polymer particles obtained by soap-free emulsion or dispersion copolymerization.

and the PEO chain lengths. Their specific organization is of particular interest for controlling the polymerization and the resulting polymer structure. Although the branched polymers, in general, remain to be characterized in more strict manners, the amphiphilic polymer systems, relatively easily accessed via the macromonomer technique, appear to provide a promising field to be explored, in view of a variety of potential applications.

The present review included only the monofunctional PEO macromonomers, but telechelic or multifunctional macromonomers will also be useful particularly in designing the crosslinked polymer systems [70]. We have prepared bis(*p*-vinylbenzyl ether) of poly(ethylene glycol), VB-PEO-VB [71,72], and found an interesting polymerization behavior in water. A very rapid gelation occurred via the polymerization of the double bonds which appear to be rigidly associated or organized in the hydrophobic region, as suggested by $^1$H-NMR. Copolymerization of styrene and/or *p*-chloromethylstyrene with a small amount of VB-PEO-VB afforded polystyrene beads which effectively act as the solid catalysts or supports to encourage nucleophilic reactions in aqueous environment [71].

## REFERENCES


1. M. J. Schick, ed., *"Nonionic Surfactants: Physical Chemistry"* Marcel Dekker, New York, N.Y., 1987.
2. F. E. Bailey, Jr., J. V. Koleke, *"Alkylene Oxide and Their Polymers"*, Marcel Dekker, New York, N.Y., 1990.
3. P. Masson, G. Beinert, E. Franta, P. Rempp, *Polym. Bull.*, 7, 17 (1982).
4. P. Rempp, P. Lutz, P. Masson, P. Chaumont, E. Franta, *Makromol. Chem., Suppl.*, 13, 47 (1985).
5. Y. Gnanou, P. Rempp, *Makromol. Chem.*, 188, 2111 (1987).
6. T. Hamaide, A. Revillon, A Guyot, *Eur. Polym. J.*, 20, 855 (1984).
7. T. Hamaide, A.Revillon, A. Guyot, *Eur. Polym. J.*, 23, 27, 787 (1987).
8. T. Hamaide, N. Mariaggi, J. L. Foureys, P. LePercec, *J. Polym. Sci., Polym. Chem. Ed.*, 22, 3091 (1984).
9. L. J. Mathias, J. B. Canterberry, M. South, *J. Polym. Sci., Polym. Let. Ed.*, 20, 473, (1982).
10. T. Suzuki, T. Tomono, *J. Polym. Sci., Polym. Chem. Ed.*, 22, 2829 (1984).
11. I. V. Berlinova, I. M. Panayatov, *Makromol. Chem.*, 188, 2141 (1987).
12. I. V. Berlinova, I. M. Panayatov, *Makromol. Chem.*, 190, 1515 (1989); I. V. Berlinova, N. G. Vladimirov, I. M. Panayatov, *Makromol. Chem., Rapid Commun.*, 10, 163 (1989).
13. Y. Tanizaki, *Yukagaku (Oil Chem.)*, 31, 253 (1982).
14. S. Kobayashi, T. Mizutani, M. Kaku, T. Saegusa, *Polym. Bull.*, 8, 169 (1983).
15. K. Ito, H. Tsuchida, A. Hayashi, T. Kitano, E. Yamada, T. Matsumoto, *Polym. J.*, 17, 827 (1985).
16. K. Ito, H.Tsuchida, T. Kitano, *Polym. Bull.*, 15, 425 (1986).

17. K. Ito, S. Yokoyama, F. Arakawa, Y. Yukawa, N. Iwashita, Y. Yamasaki, *Polym. Bull.*, 16, 337 (1986).
18. K. Ito, S. Yokoyama, F. Arakawa, *Polym. Bull.*, 16, 345 (1986).
19. S. Itsuno, K. Yamazaki, F. Arakawa, T. Kitano, K. Ito, E. Yamada, T. Matsumoto, *Kobunshi Ronbunshu*, 43, 91 (1986).
20. K. Ito, F. Arakawa, H. Tanaka, K. Hashimura, S. Itsuno, *Rep. Asahi Glass Found. Ind. Technol.*, 54 105 (1989).
21. F. Hoshino, M. Sasaki, H. Kawaguchi, *Polym. J.*, 19, 383 (1987).
22. G. Lecesque, V. Moitié, B. Bacle, P. Depraéter, *Polymer*, 29, 2271 (1988).
23. E.Wu, I. M. Khan, J. Smid, *Polym. Bull.*, 20, 455 (1988).
24. K. S. Kazanskii, N. V. Ptitsyna, *Macromol. Chem.*, 190, 225 (1989).
25. E. Tsuchida, H. Ohno, N. Kobatashi, H. Ishizaka, *Macromolecules*, 22, 1771 (1989).
26. I. M. Khan, D. Fish, Y. Dalaviz, J. Smid, *Makromol. Chem.*, 190, 1069 (1989).
27. R. K. Drummond, J. Klier, J. A. Alameda, N. A. Pappas, *Macromolecules*, 22, 3816 (1989).
28. H.-Q. Xie, J. Liu, H. Li, *J. Macromol, Sci.Chem.*, A27, 725 (1990).
29. J. Feck, V.Möhring, *Makromol. Chem.*, 191, 3045 (1990).
30. Ph. Trijasson, Y. Frere, Ph. Gramain, *Makromol. Chem., Rapid Commun.*, 11, 239 (1990).
31. Y. Qiu, X. Yu, L. Feng, S. Yang, *Macromol. Chem.*, 193, 1377 (1992).
32. G. Bo, B. Wesslén, K. B. Wesslén, *J. Polym. Sci.: Polym. Chem.* A30, 1799 (1992).
33. M. Akashi, D. Chao, E. Yashima, N. Miyauchi, *J. Appl. Polym. Sci.*, 39, 2027 (1990); I. Capek, M. Risa, M. Akashi, *Polym. J.*, 24, 959 (1992).
34. M. J. Westby, *Coll. Polym. Sci.*, 266, 46 (1988), C. Prestidge, Th. F. Tadros, *J. Coll. Interf. Sci.*, 124, 660 (1988).
35. D. W. Xia, D. Soltz, J. Smid, *Solid State Ionics*, 14, 211 (1984); D. Fish, D. W. Xia, J. Smid, *Makromol. Chem., Rapid Commun.*, 6, 761 (1985).
36. E. Tsuchida, N. Kobayashi, H. Ohno, *Macromolecules*, 21, 96 (1988); N. Kobayashi, T. Hamada, H. Ohno, E. Tsuchida, *Polym. J.*, 18, 661 (1986); N. Kobayashi, M. Uchiyama, K. Shigehara, E. Tsuchida, *J. Phys. Chem.*, 89, 987 (1985).
37. D. J. Bannister, G. R. Davies, I. M. Ward, J. E. McIntyre, *Polymer*, 25, 1600 (1984); J. M. Cowie, A. C. S. Martin, A. M. Firth, *Br. Polym. J.*, 20, 247 (1988).
38. T. Wakui, W. Y. Xu, C.-S. Chen, J.Smid, *Makromol. Chem.*, 187, 533 (1986).
39. S. Nagaoka, H. Takiuchi, K. Yokota, Y. Mori, T. Kikuchi, *Kobunshi Ronbunshu*, *39, 165, 173 (1982); M. Miyamoto, S. Sasagawa, R. Terada, S. Nagaoka, Y. Mori,* Kobunshi Ronbunshu, 42, 367 (1985).
40. H. Miyama, A. Kuwano, N. Fujii, S. Nagaoka, Y. Mori, M. Noisshiki, *Kobunshi Ronbunshu*, 42, 623 (1985).
41. I. Nwanko, D. M. Xia, J. Smid, *J. Polym. Sci., Polym. Phys.*, B, 26, 581 (1988).
42. P. Gramain, Y. Frere, *Polym. Commun.*, 27, 16 (1966); *Makromol. Chem.*, 188, 593 (1987).

43. F. Yang, Ph. Dejardin, Y. Frere, Ph. Gramain, *Makromol. Chem.*, 191, 1197 (1990); *Makromol. Chem.*, 191, 1209 (1990).
44. K. Ito, K. Tanaka, H. Tanaka, G. Imai, S. Kawaguchi, S. Itsuno, *Macromolecules*, 24, 2348 (1991).
45. K. Ito, K. Hashimura, S. Itsuno, E. Yamada, *Macromolecules*, 24, 3977 (1991).
46. D. Chao, S. Itsuno, K. Ito, *Polym. J.*, 23, 1045 (1991).
47. K. Ito, Y. Tomi, S. Kawaguchi, *Macromolecules*, 25, 1543 (1992).
48. K. Ito, H. Kobayashi, *Polym. J.*, 24, 199 (1992).
49. Y. Tsukahara, K. Tsutsumi, Y. Yamashita, S. Shimada, *Macromolecules*, 23, 5201 (1990).
50. E. Masuda, S. Kishiro, T. Kitayama, H. Hatada, *Polym. J.*, 23, 847 (1991).
51. G. Odian, *"Priciplesof Polymerization"*, 3rd ed., Wiley, New York, 1991.
52. K. W. Yeoh, C. H. Chew, L. M. Gan, L. L. Koh, *Polym. Bull.*, 22, 123 (1989); K. W. Yeoh, C. H. Chew, L. M. Gan, L. L. Koh, H. H. Teo, J. *Macromol. Sci., Chem.*, A26, 663 (1989).
53. K. E. J. Barrett, ed., *"Dispersion Polymerization in Organic Media"*, Wiley, 1975.
54. R. Milkovich, M. T. Chiang, *USP*, 3786116 (1974); G. O. Schultz, R. Milkovich, *J. Appl. Polym. Sci.*, 27, 4773 (1982); *J. Polym. Sci., Polym. Chem. Ed.*, 22, 1633 (1984).
55. J. Sierra-Vargas, E. Franta, P. Rempp, *Makromol. Chem.*, 182, 2603 (1981).
56. K. Ito, in Y. Yamashita, ed., *"Chemistry and Industry of Macromonomers"*, Ch. 3, IPC, Tokyo, 1989.
57. T. Kelen, F. Tüdös, *J. Makromol. Sci., Chem.*, A9, 1 (1975).
58. G. G. Cameron, M. S. Chisholm, *Polymer*, 26, 437 (1985).
59. G. F. Meijs, E. Rizzardo, *J. Macromol. Sci.-Rev.*, C30, 305 (1990).
60. Y. Tsukahara, *J. Soc. Rubber Ind., Jpn.*, 65, 562, (1992).
61. Y. Tsukahara, M. Tanaka, Y. Yamashita, *Polym. J.*, 19, 1121 (1987).
62. Y. Gnanou, P. Lutz, *Makromol. Chem.*, 190, 557 (1989).
63. W. Radke, A. H. E. Müller, *Makromol. Chem., Symp.*, 54/55, 583, (1992).
64. J. P. Kennedy, M. Hiza, *J. Polym. Sci., Polym. Chem. Ed.*, 21, 1033 (1983); J. P. Kennedy, C. Y. Lo, *Polym. Bull.*, 13, 343 (1985).
65. G. G. Cameron, M. S. Chisholm, *Polymer*, 27, 437, 1420 (1986).
66. Y. Tsukahara, N. Hayashi, X.-L. Jiang, Y. Yamashita, *Polym. J.*, 21, 377 (1989).
67. V. Percec, J. H. Wang, *J. Polym. Sci., Part A*: *Polym. Chem.*, 28, 1059 (1990); V. Percec, U. Eppl, J. H. Wang, H. A. Schneider, *Polym. Bull.*, 23, 19 (1990); V. Percec, J. H. Wang, *Makromol. Chem., Symp.*, 54/55, 561 (1992).
68. D. Chao, S. Itsuno, K. Ito., *Polym. Prep. Jpn.*, 40, 2588 (1991); D. Chao, K. Ito, *34th IUPAC Macro. Symp., Prague*, 1-P19 (1992).
69. J. Brandrup, E. H. Immergut, ed., *"Polymer Handbook"*, 3rd ed., Wiley, New York, 1989.
70. K. Ito, *J. Soc. Rubber Ind., Jpn.*, 65, 553 (1992).
71. S. Itsuno, I. Moue, K. Ito, *Polym. Bull.*, 21, 365 (1989).; *J. Chem. Soc., Chem. Commun.*, 1599 (1991).
72. Y. Shibata, S. Kawaguchi, K. Ito, *Polym. Prep. Jpn.*, 41, 2020 (1992).

# MACROMONOMERS: CHARACTERIZATION, POLYMERIZATION REACTIVITY AND APPLICATIONS

**Yasuhisa Tsukahara**

Department of Materials Science
Kyoto Institute of Technology
Kyoto 606
JAPAN





Macromolecular Design: Concept and Practice (Macromonomers, Macroinitiators, Macroiniferters, Macroinimers, Macroiniters, and Macroinifers) Pages 161 - 227

## 1. INTRODUCTION

A macromonomer is a reactive oligomer or polymer in which a polymerizable functional group is connected to a chain end. Macromonomers consist of two parts; one part is a polymer chain having some kinds of physical properties and the other part is the end functional group having some extent of chemical reactivity for polymerization reactions. **Figure 1** shows a polystyrene macromonomer having a vinylbenzyl end group; i.e.,styrene end group, as an example. Change of the chemical structure of the polymer chain, part-(a) in **Figure 1**, provides macromonomers having various kinds of properties, while macromonomers having different chemical reactivity can be obtained by changing the polymerizable end group corresponding to part-(b).

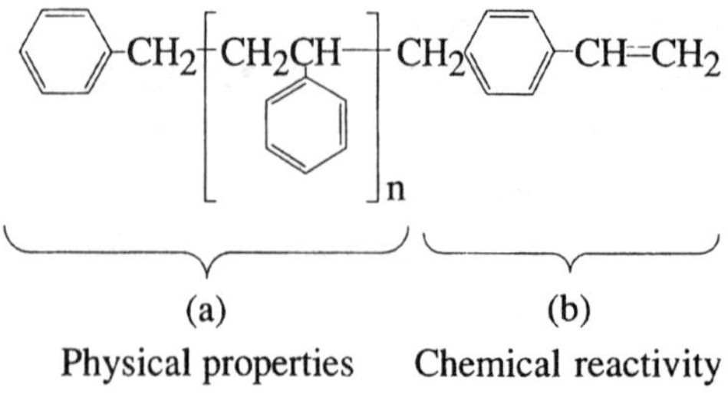

$$C_6H_5-CH_2-\left[CH_2CH(C_6H_5)\right]_n-CH_2-C_6H_4-CH{=}CH_2$$

(a) Physical properties    (b) Chemical reactivity

*Figure 1 An example of the chemical structure of a macromonomer [7].*

These variations make macromonomers very useful building blocks in developing new polymeric materials. Some variation of polystyrene macromonomers having a different end group are shown in **Table 1**. Examples of methacryloyl-ended macromonomers having a different polymer chain are shown in **Table 2**.

There have been many reports on the preparation of macromonomers as precursors for functional graft copolymers [1-8] since the patent of macromonomers by Milkovich [2] appeared. In these cases, macromonomers are mono-functional i.e., there is one polymerizable end group per molecule. However, more recently, telechelic or bi-functional and multi-functional macromonomers as shown in **Figure 2** are being used for the formation of functional polymer networks or gels [8,9]. Macromonomers are also members of the end functional reactive oligomers. One can utilize the unique properties of oligomers by using macromonomers different from those of small monomers and polymers like non-volatility, high solubility, and so on. These properties can also provide wide applications for macromonomers.

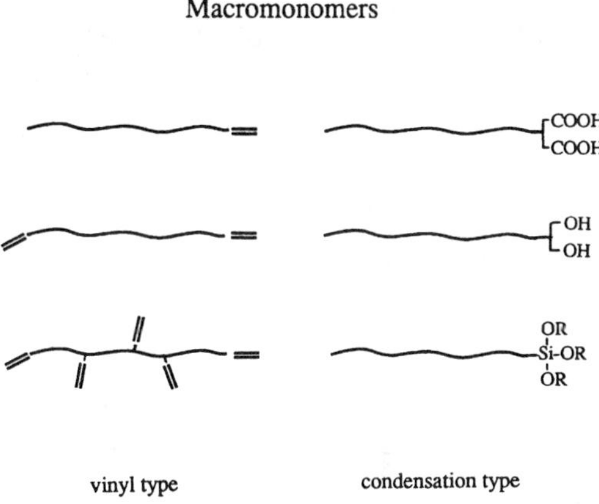


**Figure 2** *Variations of macromonomers.*

In addition to the preparation and applications of macromonomers, the studies on characterization and polymerization reactivity of macromonomers are also very important, but our understanding of these subjects is not sufficient enough at present. The most of the characterization methods of the end functional group of macromonomers and the information about their reactivity are very common in more wide field of the end functional oligomers or telechelic polymers [10]. The obtained results can be shared in both fields.

In this chapter, characterization and polymerization behavior of macromonomers are reviewed on the basis of our study. Some characteristic features of polymerization products and applications are also described.

# Table 1

| Structure | No. | Structure | No. |
|---|---|---|---|
| $PSt-CH_2CH=CH_2$ | 1 [2)] | $PSt-CH_2CH(OH)-CH_2(OH)$ | 8 [2)] |
| $PSt-CH_2CH_2OCH=CH_2$ | 2 [2)] | $PSt-CH_2-C_6H_4-CH=CH_2$ | 9 [13)] |
| $PSt-CH_2C(Ph)_2CH_2C(=O)OCH=CH_2$ | 3 [2)] | $PSt-C(=O)OCH_2CH(OH)CH_2OC(=O)C(CH_3)=CH_2$ | 10 [14),15)] |
| $PSt-CH_2CH_2OC(=O)C(CH_3)=CH_2$ | 4 [11),12)] | $PSt-CH_2CH_2OCH_2-C_6H_4-C(CH_3)=CH_2$ | 11 [12)] |
| $PSt-CH_2CH_2OCH_2-C_6H_4-CH=CH_2$ | 5 [11),12)] | $PSt-OC(=O)-$(norbornenyl) | 12 [16)] |
| $PSt-CH_2CH_2OC(=O)CH=CHC(=O)OH$ | 6 [11),12)] | $CH_2=CH-C_6H_4-CH_2-PSt$ | 13 [17)] |
| $PSt-CH_2CH(-O-)CH_2$ | 7 [2)] | $CH_2=CH-C_6H_4-CH_2-PSt-SC(=S)N(Et)_2$ | 14 [18)] |

Table 2

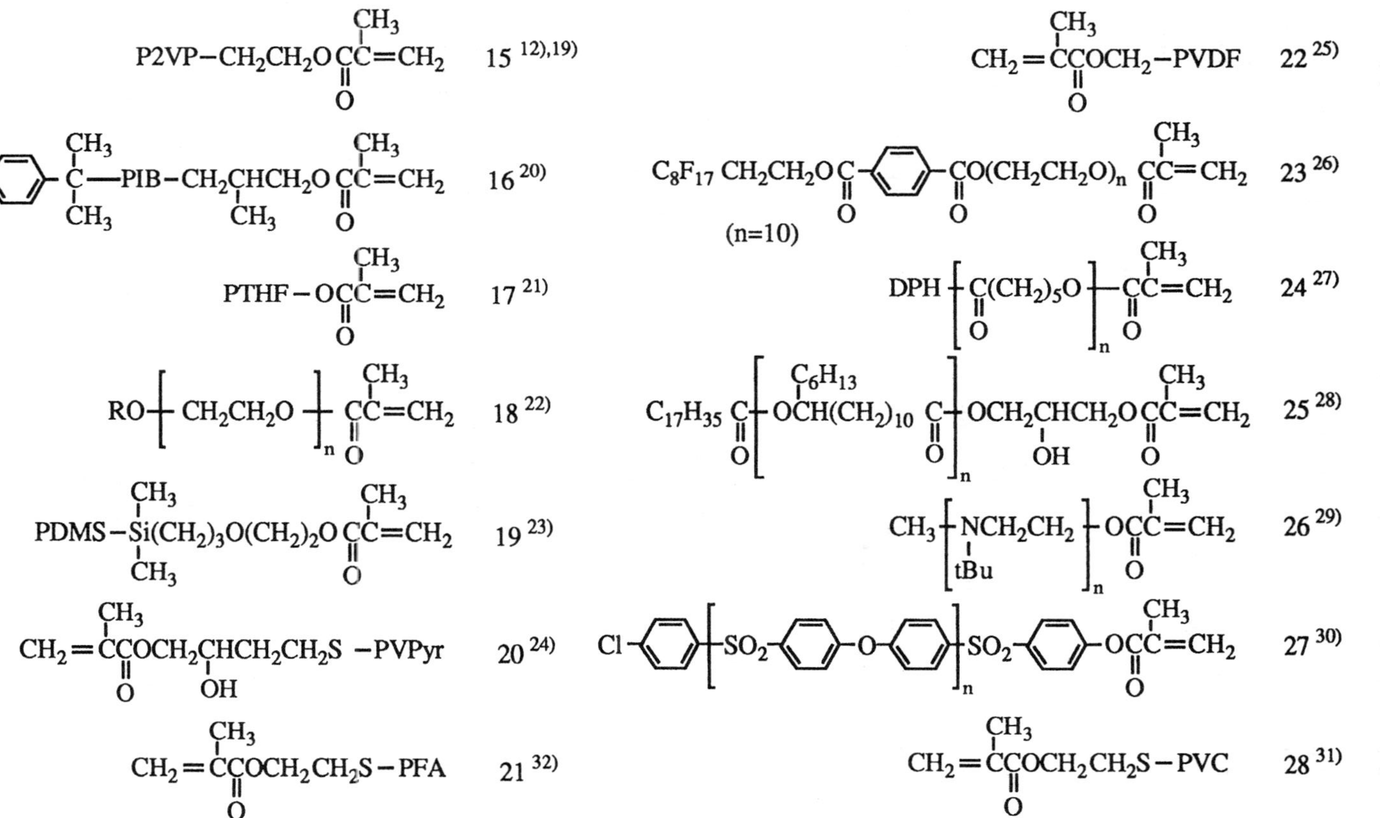
15 [12),19)]
16 [20)]
17 [21)]
18 [22)]
19 [23)]
20 [24)]
21 [32)]
22 [25)]
23 [26)]
24 [27)]
25 [28)]
26 [29)]
27 [30)]
28 [31)]

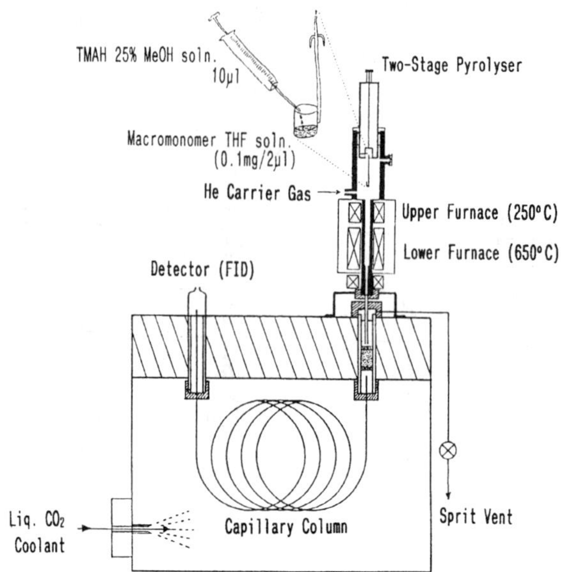


**Figure 3** *Schematic representation of the flow diagram of a PyGC system using a two stage pyrolyser.*

comonomer under appropriate conditions. The end group functionality is at least considered to be greater than this value. The values from PyGC are well consistent with those from $^1$H-NMR and the maximum conversion, indicating that PyGC is a new technique to investigate the end group of macromonomers quantitatively.

## 2.2 End Groups of Macromonomers from Radical Chain Transfer Method

Preparation of macromonomers by radical chain transfer reaction is another important method which has been widely employed to produce the end functional polymers [42]. **Scheme 2** shows an example of the preparation of PMMA macromonomer by the radical chain transfer reaction, in which methyl methacrylate (MMA) is polymerized with $\alpha,\alpha'$-azobis(isobutyronitrile) (AIBN) in the presence of thioglycolic acid (TGA) to give poly(methyl methacrylate) (PMMA) prepolymer in the first step followed by the reaction of the terminal carboxylic acid with glycidyl methacrylate (GMA) to convert the prepolymer to macromonomer in the second step [28, 43-45].

Some examples of polymerization conditions of the first step of **Scheme 2** are shown in **Table 5** together with DP of the formed prepolymers which were determined by several different methods. **Figure 4** shows the GPC curves of the formed PMMA prepolymers in which the peak shifts toward high $M_w$ direction with increase of [S]/[M]. This indicates that TGA plays a role as both a molecular weight modifier and a carrier for the terminal carboxyl group.

**Table 3.** Polystyrene Macromonomers Having a Methacryloyl End Group and Corresponding Polystyrene for Pyrolysis Gas Chromatograph

| Code | Mn (by GPC) | DP[a)] | Chemical Structure |
|---|---|---|---|
| $\alpha$-secBu-$\omega$-methacryloyloxyethyl polystyrene | | | |
| MA-PSt1500 (A-1) | 1500 | 13 | |
| MA-PSt3200 (A-2) | 3200 | 29 | |
| MA-PSt4400 (A-3) | 4400 | 41 | Sec-Bu-$(CH_2\text{-}CH(C_6H_5))_n$-$CH_2CH_2OC(=O)\text{-}C(CH_3)=CH_2$ |
| MA-PSt8100 (A-4) | 8100 | 76 | |
| MA-PSt8400 (A-5) | 8400 | 79 | |
| MA-PSt12400(A-6) | 12400 | 118 | |
| $\alpha$-benzyl-$\omega$-methacryloyloxyethyl polystyrene | | | |
| MA-PStB3000(B-1) | 3000 | 27 | $C_6H_5\text{-}CH_2\text{-}(CH_2\text{-}CH(C_6H_5))_n\text{-}CH_2CH_2OC(=O)\text{-}C(CH_3)=CH_2$ |
| MA-PStB5800(B-2) | 5800 | 54 | |
| $\alpha$-secBu polystyrene | | | |
| PSt1400 | 1400 | 13 | Sec-Bu-$(CH_2\text{-}CH(C_6H_5))_n$-H |
| $\alpha$-benzyl polystyrene | | | |
| PStB4600 | 4600 | 43 | $C_6H_5\text{-}CH_2\text{-}(CH_2\text{-}CH(C_6H_5))_n$-H |

a) With consideration of the end groups.

DP of prepolymers determined by titration and that of macromonomer by $^1$H-NMR in **Table 5** correspond to the number of MMA monomer units per residue of the chain-transfer agent at the chain end, while DP obtained by GPC corresponds to the number of monomer units per all types of end groups. **Figure 5** shows the pyrograms for the prepolymer (run-1 in **Table 5**) simultaneously taken with (a) a sulfur-specific flame photometric detector (FPD) and (b) an ordinary flame inonization detector (FID) [45].

It can be seen in **Figure 5-(b)** that there are many minor peaks together with two strong peaks corresponding to MMA monomer and benzothiophene added as an internal standard in the pyrogram. PMMA main chains are almost quantitatively depolymerized to monomers under the given PyGC conditions. On the other hand, the pyrolysis products derived from the chain transfer agent residue can be selectively observed as five main peaks corresponds to $H_2S$, $CH_3SH$, SCO, $CS_2$, and $CH_3SCH_3$ with sulfur specific FPD in **Figure 5-(a)** in addition to the isolated single peak of benzothiophene. Therefore, the relative amounts of $HOOCCH_2S$- at the chain end to those of MMA units ($RA_S$) in the prepolymer samples can be estimated by relating the peak intensities of the five main peaks in FPD mode and that of MMA in FID and nitrogen-phosphorous detector (NPD) mode after normalization with reference to benzothiophene. The relative amounts of the functional end groups in the case of corresponding macromonomers can be estimated by above method.

**Table 4**
**End Group Functionality of PSt Macromonomers with a MA End Group Determined by Py-GC, $^1$H-NMR, and Maximum Conversion**

| Sample Code | Mn of GPC | mole% of MA[a)] | end group functionality (%) by PyGC | by $^1$H-NMR | from max conv.[b)] |
|---|---|---|---|---|---|
| | [ sec-Bu-$(St)_n$-$CH_2CH_2OCOC(CH_3)=CH_2$] | | | | |
| A-1 | 1500 | 7.8 | 115 | 95 | 96 |
| A-2 | 3200 | 3.4 | 87 | 70 | 86 |
| A-3 | 4400 | 2.5 | 70 | 88 | 82 |
| [c)]A-5 | 8400 | 1.3 | 55 | 36 | 65 |
| A-6 | 12400 | 0.9 | 113 | 86 | 86 |
| | [Ph-$CH_2$-$(St)_n$-$CH_2CH_2OCOC(CH_3)=CH_2$] | | | | |
| B-1 | 3000 | 3.7 | 97 | 97 | 94 |
| B-2 | 5800 | 1.9 | 93 | 100 | 91 |

a) The relative mole% of the methacryloyl (MA) end group to St monomer units assuming that every molecule has one MA end group, i.e., 100% end group functionality. b) Obtained in the radical copolymerization of PSt macromonomers with MMA in benzene at 60°C for 48 hrs with AIBN. c) Low end group functionality sample.

The information about the residue of the initiator at the chain end is also obtained from the pyrograms of the prepolymers taken with FID and NPD. Among the peaks of various products, the peaks reflecting the cyanoisopropyl end groups derived from AIBN were assigned by NPD and GC/MS [46]. One can determine the relative amounts of the cyanoisopropyl end groups to MMA units ($RA_N$) by taking into consideration the relative peak intensities. The degree of polymerization of the prepolymers and the corresponding macromonomers are calculated by taking into consideration of both $HOOCCH_2S$- group and cyanoisopropyl group at the chain end as follows:

$$DP = 1/(RA_S + RA_N) \qquad (1)$$

**Table 5 Polymerization of MMA with AIBN in the Presence of Thioglycolic Acid as a Radical Chain-Transfer Agent[a)]**

| | feed[b)] (mol/L) | | | [S]/[M] | X[c)] | GPC | | tit.[d)] | PyGC[d)] | | $^1$H-NMR[d)] |
|---|---|---|---|---|---|---|---|---|---|---|---|
| run | [M] | [S] | [I] | x$10^2$ | % | Mn | DP | DP | DP | DP[e)] | DP[e)] |
| 1 | 4.498 | 0.350 | 0.045 | 7.781 | 15.1 | 2700 | 26 | 26(25) | 21(20) | 30(28) | 29(28) |
| 2 | 4.507 | 0.100 | 0.045 | 2.219 | 12.3 | 7400 | 73 | 79(71) | 73(62) | 91(73) | 113(101) |
| 3 | 4.498 | 0.030 | 0.045 | 0.667 | 14.7 | 19000 | 189 | 306(224) | 198(144) | 359(201) | 292(214) |
| | | | | | | Cs[f)] | 0.47 | 0.49[g)] | 0.61[g)] | | |

a) Polymerizations were carriedout with AIBN in benzene at 60°C. Polymerization time was 30 min in each run.
b) [M], [S], and [I] are the concentration of MMA, TGA, and AIBN, respectively. c) X=conversion.
d) The value in parentheses are DP values corrected by the fraction of the cyanoisopropyl end group using $DP/DP_0$ values in Table-7. e) DP determined after the conversion to the corresponding macromonomer.
f) Chain-transfer constant for TGA determined from DP. g) Calculated from DP values in parentheses.

with high end group functionality is obtained as seen in **Table 6**. This indicates that the prepolymers, and the corresponding macromonomers, having high terminal functionality are produced by the radical chain transfer method with constant amount of sufficient chain transfer agent.

**Table 6 Conversion of Prepolymers to Macromonomers[a)]**

| | feed, mg | | conversion[b)] |
|---|---|---|---|
| run | prepolymer | GMA[c)] | $DP_{tit}$(Prep)/$DP_{NMR}$(Mac) |
| 1 | 400.3 | 43.6 | 0.90 |
| 2 | 400.0 | 16.4 | 0.70 |
| 3 | 399.6 | 6.1 | 1.00 |

a) Reactions of prepolymers with GMA were carried out by adding 6.0 mg of N,N-dimethyllaurylamine and 2 mol% of hydroquinone to prepolymers in 1.6 g of xylene (reaction time, 6 h; reaction temperature, 140°C). b) Estimated with the degree of polymerization of prepolymer to that of macromonomer using uncorrected DP with the fraction of the cyanoisopropyl end group. Prp=prepolymer, Mac=macromonomer. c) Glycidyl methacrylate.

**Table 7**

**Estimation of the Fraction of the Cyanoisopropyl End Group at the PMMA Chain End Polymerized with AIBN in the presence of Thioglycolic Acid[a)]**

| | [M] | | $Rp\times10^4$ | $ktRp/kp^2[M]^2$ | fraction, % | | |
|---|---|---|---|---|---|---|---|
| run | mol/L | 1/DP[b)] | mol/Lsec | $\times10^3$ | $DP/DP_0$[c)] | kinetic data[d)] | PyGC[e)] |
| 1 | 4.498 | 0.0385 | 3.77 | 1.79 | 3.7 | 4.7 | 5.9 |
| 2 | 4.507 | 0.0137 | 3.08 | 1.46 | 10.4 | 10.7 | 15.6 |
| 3 | 4.498 | 0.00529 | 3.67 | 1.74 | 26.8 | 32.9 | 27.2 |

a) Polymerization conditions are shown in Table-5. b) Determined by GPC using a PSt standard calibration curve. c) $DP/DP_0\times100$, using $DP_0^{-1}=1.42\times10^{-3}$ determined by GPC for PMMA polymerized with [S]=0. d) Calculated from DP x ($ktRp/kp^2[M]^2$) with Rp in the Table and $k_p=5.15\times102$ and $k_t=2.55\times10^7$ (L/mol sec)[73]. e) Determined by using $RA_S$ and $RA_N$ in the text.

# 3. POLYMERIZATION REACTION OF MACROMONOMERS

Polymerization of macromonomers have the following characteristic features in comparison with the polymerization of the conventional small monomers: 1) distribution of reaction species in the polymerization media becomes often heterogeneous due to the interaction between polymer chains or polymer and solvent, 2) high viscosity of the polymerization media from the beginning of the polymerization reaction, 3) low monomer concentration 4) propagating steps are repeats of a polymer-polymer reaction, and 5) high segment density around the propagating active site due to the specific multibranched structure [48-51]. These features suggest that the polymerization reaction of macromonomers might be unusually sensitive to the diffusion controlled step, thus, making it very interesting to study. Investigation of the polymerization behavior of macromonomers is just to know how the polymer chain (part (a) in **Figure 1**) influences the polymerization reactivity of the end functional group (part (b)).

## 3.1 Homopolymerization

### 3.1.1 Degree of Polymerization of Polymacromonomers

The attainable degree of polymerization (DP) of polymacromonomers was at first considered to be very low due to the steric effect associated with the specific multibranched structure of the propagating site. High segment density around the propagating site was considered to prevent the propagating active site from approaching polymerizable end group of macromonomers via the steric effect or the excluded volume effect. The effect of this hindrance might increase as the number of branches around the propagating active site i.e., the degree of polymerization of the growing chain increases [49].

In fact, DP of the radical polymerization products of ω-vinylbenzoxy and ω-methacryloyl poly(tetrahydrofuran) macromonomers (VBO-PTHF, MA-PTHF) reported by Asami et al. [21,52,53], those of ω-vinylbenzyl and methacryloyl polystyrene (VB-PSt, MA-PSt) by Remmp et al. [54,55], and those of ω-vinylbenzyl poly(ethylene oxide) (VB-PEO) by Ito et al. [56,57] were considerably small. **Table 8** shows some examples of the molecular weights and the corresponding DP of polymacromonomers in literature. It is seen in **Table 8** that DP of polymacromonomers is small irrespective of the difference of macromonomers. Anionic polymerization with diphenylmethyl potassium or sec-BuLi of MA-PSt give almost the same DP as those obtained by radical polymerization products [55]. In cationic polymerization of VB-PSt with $CF_3SO_3H$, only oligomers with DP around 2 - 6 were obtained [58]. Group transfer polymerization of MA-PSt was recently reported by Asami and Takaki [59] using alkylsilylacetal (1-methoxy-1-(trimethylsiloxy)-2-methyl-1-propene) as an initiator and tris(dimethyl-amino)sulfonium difluoro-trimethylsilicate as a catalyst. The polymerizations at 0°C gave the bimodal products, whereas polymerizations at -78°C gave the unimodal products with DP consistent with the feed mole ratio of macromonomer

Table 8. Degrees of polymerization of poly(macromonomer)s and polymerization conditions of macromonomers.

| macromonomer | | polymerization condition | | | | polymerization products | | | |
|---|---|---|---|---|---|---|---|---|---|
| | | $C(10^{-2}mol/L)$[a)] | | | | | | | |
| Code[c)] | Mw | [M] | [I] | Solvent | Temp (°C) | yield (%) | Mw[b)] ($x10^4$) | $D_p$ | ref. |
| | | - Radical Polymerization[d)] - | | | | | | | |
| VBO-PTHF | 9300 | - | - | Bz | 60 | 52 | 6.5(G) | 7 | 53 |
| | | - | - | | | 78 | 5.6(G) | 6 | |
| | 9600 | 4.3 | 0.8 | | | 46 | 10.6(G) | 11 | |
| | | 4.3 | 4.3 | | | 65 | 4.8(G) | 5 | |
| MA-PTHF | 6900 | 7.4 | 0.8 | Bz | 60 | 96 | 23.2(G) | 58 | |
| | | 7.6 | 4.0 | | | 92 | 17.6(G) | 22 | |
| | | 5.5 | 5.7 | | | 40 | 7.2(G) | 2 | |
| | 7000 | - | - | | | 36 | 4.9(G) | 7 | |
| | | - | - | | | 76 | 4.2(G) | 6 | |
| MA-PSt | 1100 | - | - | Bz | 60 | 45 | 2.4(L) | 22 | 55 |
| | 2100 | - | - | | | 15 | 13.0(L) | 62 | |
| | 5400 | 1.99 | 1.64 | Bz | 60 | 8 | 3.0(L) | 7 | 51 |
| | | 7.91 | 1.64 | | | 71 | 17.9(L) | 41 | |
| | | 11.8 | 1.64 | | | 80 | 57.5(L) | 131 | |
| | | 14.0 | 1.64 | | | 64 | 99.7(L) | 227 | |
| | 13100 | 2.81 | 1.50 | Bz | 60 | - | 23.8(L) | 18 | 49 |
| | | 3.89 | 1.50 | | | - | 94.7(L) | 72 | |
| | | 6.00 | 1.50 | | | - | 543.0(L) | 415 | |
| VB-PSt | 13600 | 2.66 | 1.64 | Bz | 60 | 42 | 16.7(L) | 13 | 51 |
| | | 3.98 | 1.64 | | | 57 | 48.9(L) | 37 | |
| | | 5.33 | 1.64 | | | 71 | 224 (L) | 170 | |

| | | | | | | | | | |
|---|---|---|---|---|---|---|---|---|---|
| VB-iPMMA | 2300 | 10.0 | 0.25 | Bz | 60 | 12 | 14.1(VPO) | 61 | 60 |
| VB-sPMMA | 2400 | 10.0 | 0.25 | | | 10 | 14.0(VPO) | 58 | |
| - Anionic Polymerization - | | | | | | | | | |
| MA-PSt | 1100 | 10[e] | DPM-K | THF | -70 | - | 2.1(L) | 19 | 55 |
| | 1100 | 40[e] | | | | - | 7.3(L) | 66 | |
| IB-PSt | 4000 | 10[e] | $CH_3CH(\ )K$ | | -30 | - | 0.5(L) | 1 | |
| VB-PSt | 3000 | 70[e] | s-BuLi | | 40 | - | 9.7(L) | 32 | |
| - Cationic Polymerization - | | | | | | | | | |
| VB-PSt | 6000 | 2.0 | 1.4[f] | $CH_2Cl_2$ | 0 | 93 | 1.4(G) | 2 | 58 |
| | | 2.0 | 1.4[f] | | -78 | 92 | 3.4(G) | 6 | |
| VO-PEO | 234-336 | - | $BF_3Et_2O$ | | - | - | low(?) | - | 76 |

a) [M]: macromonomer concentration, [I]: initiator concentration.
b) Molecular weight determined from (G): GPC, (L): light scattering, and (VPO): vapor pressure osmometry.
c) VB-PTHF: ω-vinylbenzyl polytetrahydrofuran, MA-PTHF: ω-methacryloyl polytetrahydrofuran, MA-PSt: ω-methacryloyl polystyrene, VB-PSt: ω-vinylbenzyl polystyrene, VB-iPMMA: ω-vinylbenzyl isotactic poly(methyl methacrylate), VB-sPMMA: ω-vinylbenzyl syndiotactic poly(methyl methacrylate), IB-PSt: ω-isopropenylbenzyl polystyrene, VO-PEO: ω-vinyloxy poly(ethylene oxide).
d) Azobisisobutyronitrile(AIBN) is used as an initiator in radical polymerization.
e) Ratio of [M]/[I].
f) Initiator is $CF_3SO_3H$.

to initiator. However, the polymerization products are rather oligomers. Recently, Hatada et al. [60] reported the synthesis of stereoregular ω-ortho-vinylbenzyl-poly(methyl methacrylate) (VB-PMMA). The product, isotactic and syndiotactic VB-PMMA were polymerized with AIBN and gave the polymerization product with DP of ca.60. Hashimoto and Sumitomo [61] have reported the synthesis of ω-vinylbenzylpolyamide macromonomers (VB-PAMD, MW=3900-8400)from bicyclic oxalactam where DP of polymerization products were 5-8. So far, radical polymerizations of macromonomers have been primarily conducted. Reports on the anionic, cationic and other types of polymerizations are few in comparison with the radical polymerization.

We [51] have investigated the radical polymerization of polystyrene macromonomer having a methacryloyl (MA) end group or a vinylbenzyl (VB) end group. In the case of polymerization of macromonomers, large variation of DP or conversion is observed because the viscosity of polymerization media vary with the concentration and the molecular weight of the macromonomer which changes the diffusion control effect.

**Figure 6** shows DP of polymacromonomers versus macromonomer concentration in feed, [M], plots in the radical polymerization of MA-PSt4400, MA-PSt12400, VB-PSt4980, and VB-PSt13200 with AIBN in benzene at 60°C. It is seen that DP of polymacromonomers determined by a low angle light scattering detector [49] are strongly dependent on [M] in an unique manner i.e, during the polymerization of macromonomers of MA-PSt4400 and VB-PSt4980, the molecular weights of poly(MA-PSt4400)s and poly(VB-PSt4980) are very small at the [M] $< 0.08$ mol/L where DP are around 10. DP increases rapidly as [M] increases in the concentration region of 0.08-0.14mol/L and then suddenly decreases to the value corresponding to DP=1 around 0.15 mol/L. Above this [M], polymerization of the macromonomer is strongly suppressed. This tendency is much more drastic for the high molecular weight macromonomers of MA-PSt12400 and VB-PSt13200. However, DP (or $M_w$) did not depend much on the initiator concentration [I], in the concentration range of [I] $= 4.2 \times 10^{-3}$ -$2.56 \times 10^{-2}$ mol/L, and has a tendency to decrease gradually with [I] due to the decrease of [M] with conversion [49]. When the $M_w$ of a macromonomer is around 1000, DP reaches more than one thousand at high [M].

The rapid increase of DP observed in **Figure 6** was related to the increase in viscosity of polymerization media with [M] due to the macromolecular character of the macromonomer. The excluded volume effect can also participate in the rapid increase of DP especially for the large molecular weight macromonomers. The critical concentration of the overlapping ($C^*$) for MA-PSt12400 can be estimated as $1.1 \times 10^{-2}$ mol/L in benzene assuming a random coil. Onset of rapid increase of DP with [M] in **Figure 6** occurs at [M] around $3C^*$. Below $C^*$, the macromonomers might encounter the excluded volume effect or the osmotic effect to interpenetrate the coil of the growing chain to bring the polymerizable end group close to the propagating radical site. On the other hand, the sudden decrease of DP at high [M] might be related to the vitrification effect due to the glass transition phenomenon [62,63] where the segmental diffusion of the end group and the translational diffusion of the whole chain is strongly restricted. This concentration is related to both polymerization temperature and glass transition temperature of the polymerization media. More recently, Ito et al. reported the radical polymerization of poly(ethylene oxide) macromonomers. They reported that large molecular weight polymacromonomers were obtained in water due to the polymerization in micelles. These results indicate that the high segment density around the propagating radical does not necessarily limit the growing to large DP.

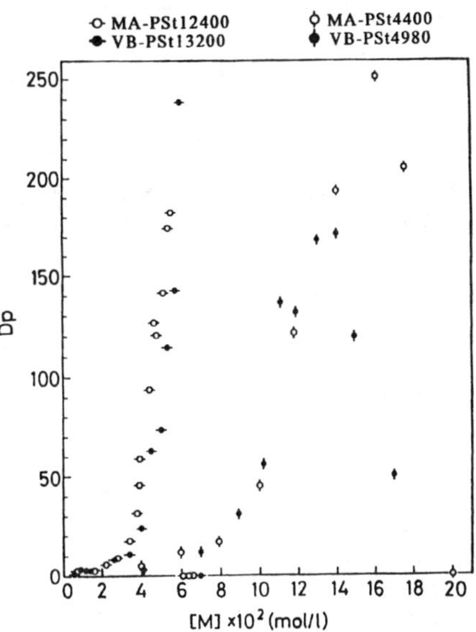


**Figure 6** *Comparison of DP of polymacromonomers versus [M] plots in the radical polymerization of MA-PSt4400, MA-PSt12400, VB-PSt4980, and VB-PSt13200.*

### 3.1.2 Time-Conversion Curve and Polymerization Rate

Radical polymerization rate ($R_p$) of macromonomers and its dependence on initiator and macromonomer concentrations have been investigated by several authors [49,51,52,67,68,75].With the polymerization of macromonomers, it is very difficult to measure $R_p$ by dilatometry which is often used to determine the polymerization kinetics of the conventional small monomers. However, the $R_p$ of macromonomers is determined by measuring the change in the peak area of the unreacted macromonomer in GPC during the process of polymerization. Also, the decrease of NMR and UV peak intensities of the polymerizable end group is used to determine $R_p$.

**Figure 7** shows the typical time-conversion curves of the radical polymerization of macromonomers [51], MA-PSt4400 and VB-PSt4980 in benzene at 60°C with $[M]=11x10^{-2}$mol/L and $[I]=1.64x10^{-1}$mol/L. In this case, the polymerization solutions were viscous from the initial stage with this [M]. It is seen in **Figure 7** that $R_p$ of

MA-PSt4400 having a methacryloyl end group is greater than that of VB-PSt4980 having a vinylbenzyl end group where the molecular weight of these macromonomers are nearly identical. It is also seen in **Figure 7** that the degree of conversion (X) increases or $R_p$=dX/dt decreases monotonically as the polymerization proceeds. The same feature in the time-conversion curves is also seen at different [M] and for macromonomers with different $M_w$ indicating that this feature is very common and characteristic of the polymerization of macromonomers. However, this feature of the time- conversion curves is quite different from that in the polymerization of small monomers in the presence of the gel effect where a remarkable autoacceleration of the polymerization rate is normally observed as polymerization proceeds under isothermal condition. Absence of acceleration of $R_p$ with X is also seen in the $M_w$ versus X relationship where $M_w$ almost constant during polymerization reaction proceeds [49]. These features are reasonably explained on the basis that the gel effect appears from the beginning of the polymerization reaction due to the presence of macromolecular monomers.

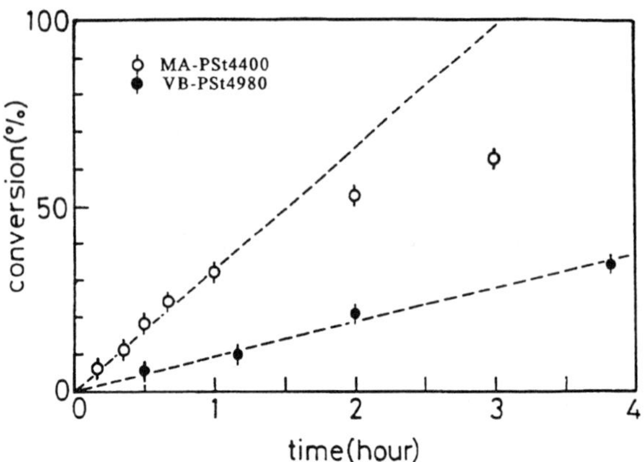


**Figure 7** *Typical time-conversion curves in the radical polymerization of MA-PSt4400 (open symbol) and VB-PSt4980 (closed symbol). Polymerizations were carried out in benzene at 60°C at [M]=0.11 mol/L and [I]=0.164 mol/L.*

In the radical polymerization of small monomers, $R_p$ is known to be expressed by the following in the absence of the gel effect [46,62-66];

$$R_p = k_p(k_d f/k_t)^{1/2} \; [M] \; [I]^{1/2} \tag{2}$$

where f is the initiator efficiency, $k_p$, $k_t$ and $k_d$ are the rate constants for propagation, termination and initiator decomposition, respectively. The dependence of $R_p$ on [M] is first order and on [I] is one-half order in the absence of the diffusion controlled effect. The deviation from eq (2) is observed under the gel effect and other diffusion control effects. The mode of termination affects the kinetic order with respect to [I].

In **Figure 8**, $R_p$ values of MA-PSt4400 and VB-PSt4980 determined from the initial slope of time-conversion curves at different monomer concentrations [51]. It is seen from the figure that $R_p$ of MA-PSt4400 is substantially greater than that of VB-PSt4980. $R_p$ of MA-PSt macromonomers of other molecular weights also greater than those of the corresponding molecular weights of VB-PSt macromonomers. This means that the polymerization reactivity of the end group bearing a long PSt chain can still influence $R_p$. The larger value of $R_p$ of the MMA monomer than that of the St monomer is preserved in the presence of a strong diffusion control effect. In addtion, $R_p$ increases with $M_w$ of macromonomers at the same [M]. This is because the gel effect becomes greater as $M_w$ increases. The slopes of the straight lines fitted to the data on the log$R_p$ versus log[M] graph are 2.6 for MA-PSt4400 and 2.8 for VB-PSt4980 in **Figure 8**. Obviously, these higher orders of $R_p$ than unity with respect to [M] can be attributed to [M]-dependent diffusion control effect; i.e., variations of f and $(k_p/k_t^{1/2})$ with [M] in equation (2). The increase in viscosity of polymerization media influences on f and it becomes low at high [M] due to the reduction of mobility of the primary radicals (the cage effect) [47,63].

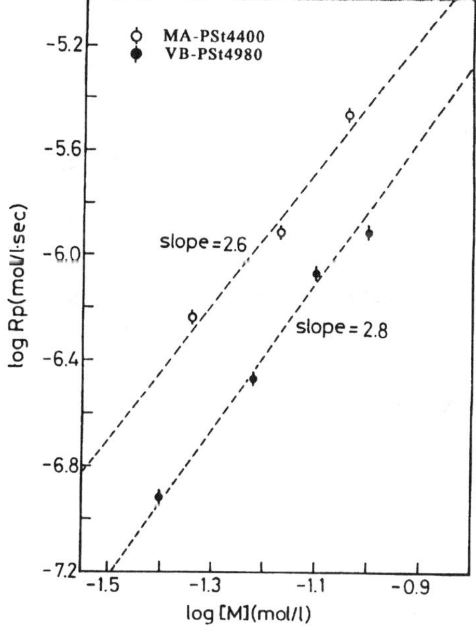


**Figure 8** *Double-logarithmic plots of polymerization rate ($R_p$) versus feed concentration of macromonomer [M] in the radical polymerization of MA-PSt4400 and VB-PSt4980. Polymerizations were carried out in benzene at 60°C at [I] = $1.64 \times 10^{-2}$ mol/L.*

It was observed [49] that variation of $R_p$ upon [I] depended on [M]. The observed kinetic order with respect to [I] is almost 1/2 at low [M], whereas at high [M], the order is less than 1/2. For the propagating radical of small DP, that is formed at low

[M], the bimolecular termination of polymacromonomer radicals can takes place with little difficulty. However, at high [M], the propagating radicals of large DP is formed. Thus, the bimolecular termination becomes more difficult due to the high segment density around the radical site as schematically shown in **Figure 9**. The peripheral segment density of the propagating radical increases as DP increases (discussed in section 4). This might lead to the primary termination to be more important. When [M] is very high, the unimolecular termination also takes place. Eventually, the observed kinetic order with respect to [I] are given by the balance of these factors [49,51].

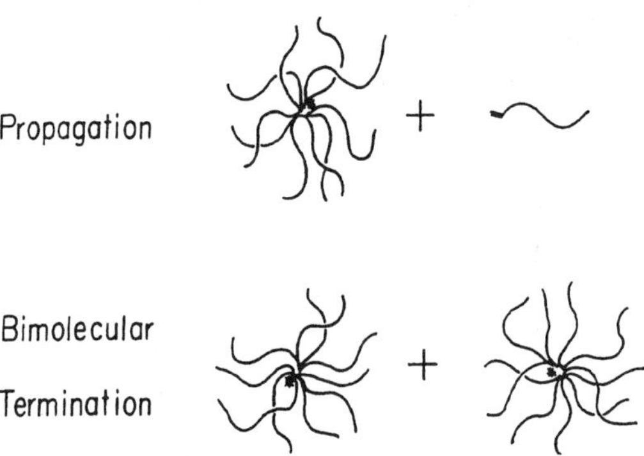


**Figure 9** *Schematic representation of the propagation reaction and the bimolecular termination reaction in the polymerization of macromonomers.*

## 3.1.3Specific Features of the Propagating Radical of Macromonomers and Rate Constants

Since the radical site is located near the center of the growing chain of the multi-branched structure, the local segment density around the radical site is very high. This causes the propagating radical to be stable and detectable by ESR spectroscopy during polymerization. An example of ESR spectra of the polymerization of MA-PSt12400 is shown in **Figure 10**[50)]. Polymerizations were carried out with AIBN in benzene at 60°C during which ESR spectra were taken. In **Figure 10**, spectrum (a) is for the polymerization of MA-PSt12400 at low [M], (b) is at high [M], and (c) is for the polymerization of MMA in the presence of equimolar concentration of inert PSt, $M_w$ of which is nearly identical to that of MA-PSt12400. The polymerization conditions are shown in **Table 9**. **Figure 10-(b)** represents a nine line ESR spectrum which consists of a broad five-line spectrum superimposed by a weak four line spectrum. This spectrum becomes sharper at -196°C and almost identical with that of the spectrum of the MMA propagating radical [69], which is characterized by nine lines consisting of five strong lines having a hyperfine splitting constant of 23.4G (intenisty distribution is 1:4:6:4:1) and of four weak lines, appearing between the five lines.

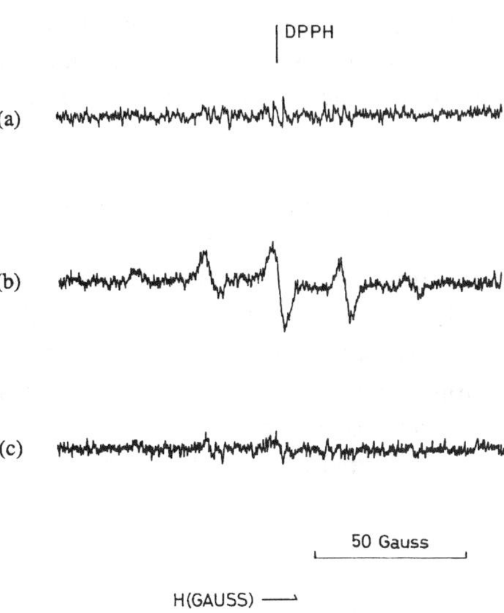


**Figure 10** *ESR spectra measured during polymerization of MA-PSt12400 and MMA + inert PSt. Polymerizations were carried out with AIBN in benzene at 60°C(See Table 9). The vertical line in the figure is DPPH, g=2.0036.*

**Table 9**
**Polymerization Conditions of Methacryloyl-Ended PSt Macromonomers and MMA/PSt Mixture for ESR Measurements[a)]**

| run | monomer | feed concentration [M] $10^{-2}$(mol/L) | feed concentration [I] $10^{-1}$(mol/L) |
|---|---|---|---|
| a | MA-PSt12400 | 1.1 | 1.64 |
| b | MA-PSt12400 | 4.0 | 1.64 |
| c | MMA+PSt | 4.0 | 1.64 |

a) ESR measurements were carried out during the polymerization.

Table 11 Comparison of Polymerization Rate of Macromonomers with that of Small monomers

| monomer | rate constant (L/mol·sec) $k_p$ [b] | $k_t$ [b] $(\times 10^{-3})$ | $(k_p/k_t^{1/2})$ | f | $R_p$ [a] $(10^{-6}$ mol/L·sec) cal | obs | DP cal | DP obs |
|---|---|---|---|---|---|---|---|---|
| MMA | 515 | 26000 | 0.101 | 0.5 | 10.1 | - | 8.6 | - |
| | 187 | 72100 | 0.0220 | " | 2.20 | | 1.9 | |
| | 714 | 720000 | 0.0266 | " | 2.66 | | 2.3 | |
| St | 165 | 60000 | 0.0213 | 0.8 | 2.70 | - | 1.4 | - |
| | 176 | 72000 | 0.0207 | " | 2.62 | | 1.4 | |
| | 80 | 64000 | 0.0100 | " | 1.27 | | 0.7 | |
| MA-PSt4400 | 18 [c] | 4.2 [c] | 0.278 | 0.05~0.1 [c] | 8.82 ~12.4 | 10.4 | 53.2 ~75.1 | 50~80 |
| VB-PSt4980 | 4 [c] | 0.63 [c] | 0.159 | 0.03~0.05 [c] | 3.90 ~5.05 | 2.81 | 43.0 ~55.5 | 40~70 |

a) Calculated by eq (2) with [M]=0.11 (mol/L), [I]=0.164 (mol/L), and $k_d = 1.01 \times 10^{-5}$ $(sec^{-1})$ for AIBN at 60°C.
b) from ref.47 and ref.73.
c) from ref.51.
d) DP is assumed to equal to the kinetic chain length $\nu_p$ defined by eq (6).

It is seen in the table that $(k_p/k_t^{1/2})$ values of the macromonomers in **Table 11** are 10-20 times larger than those of the corresponding small monomers. The initiator functionality (f) obtained for MA-PSt4400 is around 0.05-0.1[51]. f increases with the monomer concentration but decreases with the viscosity of polymerization media, thus f depends on [M] in a complex fashion. Reduction of $k_t$ with [M] is much greater than that of $k_p$ resulting in rapid increase of $(k_p/k_t^{1/2})$. This might explain the rapid increase of DP and $R_p$ with [M] in **Figures 6 and 8** through Equations (2) and (6). **Table 11** indicates that the radical polymerization reactivity of macromonomers is not lower but equal or even higher than those of the corresponding small monomers when these are compared at the same monomer and initiator concentrations. It is of great interest to know the values of $k_p$ and $k_t$ of macromonomers as functions of the molecular weight and concentration of the macromonomer.

## 3.2 Copolymerization Reactivity of Macromonomers

Macromonomers can be used as a building block of polymeric materials most effectively when they are copolymerized with the various kind of small monomers to produce the wide variety of well-defined graft copolymers. This method is called as "Macromonomer method" and has been widely employed during this decade [1]. There are some excellent reviews on this subject [2,3,77,78]. In preparation of graft copolymers by the macromonomer method, it is of great importance to get the information on the copolymerization reactivity of macromonomers. Because the specific features in the polymerization system of macromonomers mentioned before influence the copolymerization reactivity, thus, it cannot be predicted by the intrinsic reactivity of the polymerizable end group alone.

In some cases, the copolymerization reactivity of a macromonomer is nearly identical to that of the small monomer corresponding to the polymerizable end group. The copolymerization can be predicted by intrinsic reactivity of the polymerizable end group which is related to the induced, resonance, and steric character of the substituents on the double bond of the end group. On the other hand, it is also observed that the long polymer chain attached on the polymerizable end group influences the copolymerization reactivity in some cases.

### 3.2.1 Copolymerization Reactivity Estimated by Monomer Reactivity Ratio

The copolymerization reactivity of macromonomers is discussed in terms of the monomer reactivity ratios on the basis of the copolymerization Equation (7) with data at low degrees of conversion under the terminal model assumption [47,79];

$$\frac{d[M_1]}{d[M_2]} = \frac{[M_1](r_1[M_1] + [M_2])}{[M_2](r_2[M_2] + [M_1])} \tag{7}$$

where, $[M_1]$ and $[M_2]$ are the molar concentrations of monomer-1 ($M_1$) and monomer-2 ($M_2$); $r_1$ and $r_2$ are the respective monomer reactivity ratios defined as the ratio of the rate constant for the self-propagation to the cross-propagation i.e.,$(k_{11}/k_{12})$ and $(k_{22}/k_{21})$.

$r_1$ and $r_2$ for the copolymerization of macromonomers are normally evaluated by measuring the copolymer composition as a function of the feed composition and fitting the calculated curves to the experimental data using a particular pair of $r_1$ and $r_2$ as in the case of small monomers. Fineman-Ross [80] and Kelen-Tudos [81] methods are used to determine $r_1$ and $r_2$. However, the molar concentration of macromonomers is much smaller than the small comonomers in most cases which makes it difficult to determine the exact values of the monomer reactivity ratio for macromonomers. In fact, when $M_1$ is the small monomer and $M_2$ is the macromonomer, estimated $r_2$ values for the macromonomers are often scattered and sometimes negative. In this case, the ratio of the molar composition of macromonomer $[M_2]$ to that of comonomer $[M_1]$ is much smaller than unity. Then, Equation (7) becomes the following simple one parameter equation by neglecting the contribution from the propagating radical $\sim M^*_2$ (Equation 8):

$$d[M_1]/d[M_2] = r_1[M_1]/[M_2] \tag{8}$$

From Equation (8), the copolymerization reactivity of the macromonomer is evaluated by $1/r_1$; the ratio of the rate constant for the propagating comonomer radical adding macromonomer to the rate constant for its own comonomer. This $1/r_1$ is often used to discuss the copolymerization reactivity of macromonomers. $1/r_1$ increase when the relative copolymerization reactivity of the macromonomer increases.

On the other hand, variation of the copolymer composition with conversion is also used to evaluate $r_1$ and $r_2$ by the Skeist [82-85] Equation (9) which gives the relation between conversion and instantaneous feed compositions of $M_1$, $M_2$, $f_1$ and $f_2$:

$$X = 1 - \frac{[M]}{[M]_o} = 1 - \left(\frac{f_1}{(f_1)_o}\right)^{\alpha} \left(\frac{f_2}{(f_2)_o}\right)^{\beta} \left(\frac{(f_1)_o - \delta}{f_1 - \delta}\right)^{\gamma} \qquad (9)$$

here $f_1$ and $f_2$ are the instantaneous feed composition (mole fraction) of $M_1$ and $M_2$, $(f_1)_o$ and $(f_2)_o$ are their initial values, respectively. [M] is the total monomer concentration ($[M]=[M_1]+[M_2]$) and $[M]_o$ is its initial value. $\alpha$, $\beta$, $\gamma$, and $\delta$ are given by

$$\alpha = \frac{r_2}{(1 - r_2)}, \quad \beta = \frac{r_1}{(1 - r_1)}$$
$$\gamma = \frac{(1 - r_1 r_2)}{(1 - r_1)(1 - r_2)}, \quad \delta = \frac{(1 - r_2)}{(2 - r_1 - r_2)} \qquad (10)$$

When the $[M_2]/[M_1]$ is small, the time conversion curve calculated from Equation (9) becomes insensitive to the value of $r_2$.

Copolymerizations of macromonomers with conventional small monomers have been investigated by many authors. Initially, the attentions have been paid for the confirmation of the introduction of the polymerizable end group on macromonomers. However, the studies involving the evaluation of the monomer reactivity ratios or the copolymer composition distribution are recently increasing.

In **Table 12**, examples of the reported monomer reactivity ratios for the radical copolymerization of macromonomers with small comonomers are shown. When the copolymerization reactivity of the macromonomer is almost the same to that of the small monomer corresponding to the polymerizable end group, the copolymerization data accumulated for the small monomer pairs [73] can be used to predict the copolymerization behavior of macromonomers.

On the other hand, as shown in **Table 12**, there are several reports in which the effect of the polymer chain attached on the polymerizable end group (part (a) in Figure-1) has been pointed out. You will see some examples discussing the effect of the macromolecular characters on the basis of the monomer reactivity ratio as followings.

Schulz and Milkovich [85] investigated the radical copolymerization behavior of ω-methacryloyl polystyrene (MA-PSt, MW=11,000) with MMA, EA, and BMA in benzene(Bz) and dimethylsulfoxide(DMSO). It is shown that the monomer reactivity ratios are almost the same to those of the copolymerization of small monomers corresponding to the polymerizable end group. However, the lower copolymerization reactivity of the macromonomer in suspension polymerization in aqueous media is also shown due to the phase separation during the copolymerization.

Kennedy et al. [86,87] investigated the radical copolymerization of polyisobutylene macromonomers having a vinylphenyl end group (VP-PIB, MW=4,200,9,600,48,000) with MMA and St. It was shown that the copolymerization reactivity of the VP-PIBs was identical to that of St and was not affected by solvent, molecular weight of the VP-PIB, and conversion in the absence of phase separation. On the other hand, in the radical copolymerization of PIB having a methacryloyl end group with MMA (MA-PIB), it was observed that the copolymerization reactivity was lower than that of MMA and the reactivity lowered with conversion due to the incompatibility effect [20].

Kennedy et al. [88] prepared polyisobutenyldicyclopentadiene **31** (DCp-PIB, MW=27,000)by cationic polymerization of isobutylene with dicyclopentadienyl chloride and diethylaluminumchloride as a catalyst. This macromonomer was copolymerized with olefin monomers like ethylene and propylene to produce the three component graft copolymer as follows:

$$\text{DCp}-(CH_2-C(CH_3)_2-)_n CH_3 \ (\mathbf{31}) + CH_2=CH_2 + CH_2=CH(CH_3) \longrightarrow$$

$$\sim -CH_2-CH(CH_3) \sim CH_2-CH_2-[\text{DCp}]-CH_2CH_2 \sim$$
$$[\text{DCp}]-\sim(CH_2-C(CH_3)_2\sim)_n CH_3$$

It was shown that the copolymerization reactivity of DCp-PIB toward propylene was lowered by the steric hindrance of the PIB chain.

Ito et al. [22] investigated the radical copolymerization of VB-PEO and MA-PEO with St in THF to study the effect of the molecular weight of the macromonomers on the copolymerization reactivity. They found that the copolymerization reactivities of the macromonomers evaluated from $1/r_1$ inversely proportional to the molecular weight as shown in **Figure 12**. The reduction of the relative reactivity is discussed in terms of the thermodynamic repulsive interaction between unlike polymers or the aggregation of PEO chains in solution. Ito et al. [89] also investigated the radical copolymerization of VB-PEO and MA-PEO having a lauryl and stearyl group at the $\alpha$-chain end with St and benzyl methacrylate (BzMA) in benzene, in which the reduction of the relative reactivities was again observed.

Moreover, $\alpha$-dodecyloxy MA-PEO macromonomer was copolymerized with BzMA and St in the dispersion polymerization. In the copolymerization of the macromonomer with BzMA in aqueous dispersion, the apparent reactivities of the macromonomers were the same as those observed in benzene. However, the observed reactivities increased in the aqueous dispersion with St. It was suggested that a preferential polymerization occurred near the surface of formed micellar particles [90].

Table 12. Typical values of monomer reactivity ratio for macromonomers in radical copolymerization with small comonomers.

| Macromonomer | | Comonomer | monomer reactivity ratio | | | | remarks[d] | ref |
|---|---|---|---|---|---|---|---|---|
| | | | MMS[b] | | SMS[c] | | | |
| type[a] | MW | $M_1$ | $r_1$ | $r_2$ | $r_1$ | $r_2$ | | |
| MA-PSt | 11,000 | MMA | 1.0 | - | 1.0 | 1.0 (MMA) | no | 85 |
| | | EA | 0.73 | - | 0.28 | 2.0 ( " ) | " | " |
| | | BA | 0.82 | - | 0.37 | 1.8 ( " ) | " | " |
| | | St | 0.61 | - | 0.52 | 0.49( " ) | " | " |
| | 12,400 | MMA | 0.96-1.05 | - | 1.0 | 1.0 ( " ) | " | 93 |
| | | St | 0.47-0.49 | - | 0.52 | 0.49( " ) | " | " |
| | 14,000 | HEMA | 2.0-2.3 | - | 1.05 | 0.29( " ) | no? | 103 |
| | 23,000 | | | | | | | |
| | 3,180 | HEMA | 1.7 | - | 1.05 | 0.29( " ) | yes | 104 |
| VB-PSt | 3,000 | MMA | 0.47 | 0.24 | 0.41 | 0.44(pMeSt) | no | 53 |
| | 6,200 | " | 0.89 | 0.22 | " | " ( " ) | " | " |
| MA-PMMA | 1,650 | SMA | 1.0 | 1.0 | 1.0 | 1.0 (MMA) | no | 43 |
| | 4,080 | " | 1.0 | 1.0 | 1.0 | 1.0 ( " ) | " | " |
| MA-PIB | 3,200 | MMA | >1.0 | - | 1.0 | 1.0 (MMA) | yes? | 20 |
| | 4,800 | " | >1.0 | - | 1.0 | 1.0 ( " ) | " | " |
| VP-PIB | 4,200 | MMA | 0.5 | - | 0.44 | 0.39(ipSt) | no | 87 |
| | 9,600 | " | 0.5 | - | " | " ( " ) | " | " |
| | 48,000 | " | 1.1 | - | " | " ( " ) | " | " |
| | 4,200 | St | 1.2 | - | 1.22 | 0.89( " ) | | 87 |
| | 9,600 | " | 2.1 | - | " | " ( " ) | | " |
| | 48,000 | " | 6.6 | - | " | " ( " ) | | " |
| MA-PTHF | 6,500 | -VN | 1.28 | 0.21 | 1.00 | 0.40(MMA) | no | 53 |
| A-PTHF | 10,000 | " | 2.69 | 0.40 | 2.13 | 0.13(BA) | " | " |

| | | | | | | | | |
|---|---|---|---|---|---|---|---|---|
| MA-PDMS | 500 | St | 1.07 | - | 0.52 | 0.46(MMA) | yes | 91 |
| | 1,160 | " | 1.06 | - | " | " ( " ) | " | " |
| | 9,750 | " | 1.55 | - | " | " ( " ) | " | " |
| | 500 | AN | 0.41 | 1.93 | 0.15 | 1.20(MMA) | yes | 92 |
| | 1,110 | " | 0.89 | 5.4 | " | " ( " ) | " | " |
| | 1,020 | MMA | 1.0 | - | 1.0 | 1.0 (MMA) | no | 94 |
| | 8,670 | " | 1.4 | - | " | " ( " ) | yes | " |
| $MA-(EO)_n$ | n = 1[e] | St | 0.37 | 0.43 | 0.52 | 0.46(MMA) | yes | 22 |
| | 25 | " | 0.94 | - | " | " ( " ) | " | " |
| | 35 | " | 1.02 | - | " | " ( " ) | " | " |
| | 62 | " | 1.18 | - | " | " ( " ) | " | " |
| $MA-(EO)_n-R$ | 0 | BzMA | 1.20 | 0.94 | 1.14 | 0.85(MMA) | yes | 89 |
| | 5 | " | 1.24 | - | " | " ( " ) | " | " |
| | 10 | " | 1.23 | - | " | " ( " ) | " | " |
| | 20 | " | 1.49 | - | " | " ( " ) | " | " |
| $VB-(EO)_n$ | 1 | St | 0.86 | 0.76 | 1.0 | 1.9 (St ) | yes | 22 |
| | 13 | " | 1.01 | - | " | " ( " ) | " | " |
| | 33 | " | 1.53 | - | " | " ( " ) | " | " |
| | 51 | " | 1.71 | - | " | " ( " ) | " | " |
| | 94 | " | 1.79 | - | " | " ( " ) | " | " |
| VB-PEO | 2,100 | MMA | 2.0 | 0.06 | 0.06 | 0.46(St ) | yes | 105 |
| | 6,000 | St | 1.2 | 0.86 | 0.86 | 1.0 ( " ) | " | " |
| VBE-PPO | 2,380 | MMA | 0.555 | - | 0.48 | 0.298(VBE) | yes | 96 |
| | 26,600 | | ~0.778 | | | | | |
| | " | BMA | 0.475 | - | 0.62 | 0.32 ( " ) | " | " |
| | | | ~0.899 | | | | | |

a) Abbriviations are MA-PIB: methacryloyl polyisobutylene ,VP-PIB: vinylphenyl polyisobutylene, see text for others.

b) Copolymerization data for macromonomer with small comonomer.

c) Copolymerization data for the corresponding small monomer with small comonomer.

d) Redcution of copolymerization reactivity for the macromonomer.

e) Number of EO monomer units.

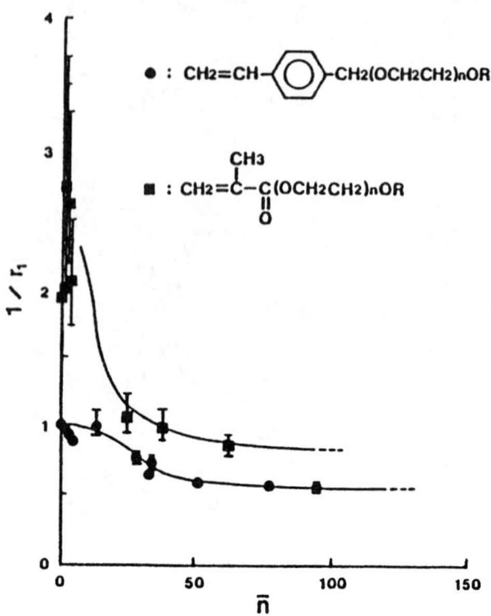


**Figure 12** *Relative reactivities of PEO macromonomers as a function of their degree of polymerizations.*

Cameron and Chisholm [91,92] prepared poly(dimethylsiloxane) terminated with 3-methacryloyloxypropyl-(dimethylchlorosilane) (MA-PDMS) and investigated the radical copolymerization of the MA-PDMS with St and AN. $r_1$(St) increased as the molecular weight of MA-PDMS increased as shown in **Figure 13**. This indicates that the copolymerization reactivity of MA-PDMS derived from $1/r_1$ decreases as the molecular weight of the macromonomer increases.

$$n\text{-BuLi} + n\left[\text{-Si}(CH_3)_2\text{-O-}\right]_3 \xrightarrow{\text{THF}} n\text{-Bu}\left[\text{-Si}(CH_3)_2\text{-O-}\right]_{3n-1}\text{-Si}(CH_3)_2\text{-O}^-Li^+$$

$$\xrightarrow{ClSi(CH_3)(CH_3)_2CH_2CH_2CH_2OC(=O)C(CH_3)=CH_2} n\text{-Bu}\left[\text{-Si}(CH_3)_2\text{-O-}\right]_{3n}\text{-Si}(CH_3)_2CH_2CH_2CH_2OC(=O)C(CH_3)=CH_2 \qquad 32$$

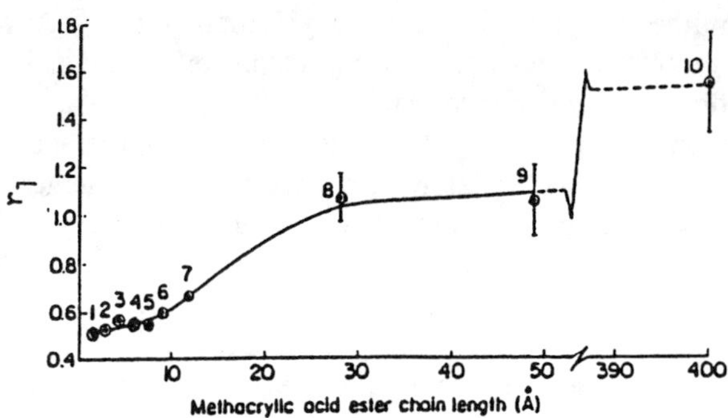


__Figure 13__ $r_1$ *for the copolymerization of n-alkyl methacrylate ester with styrene versus length of the extended ester side chain. For points 1-7, numbers correspond to numbers of carbon atoms in alkyl groups. Points 8-10, from MA-PDMS macromonomers.*

Radical copolymerizations of the same kind of MA-PDMS (MW=1,020, 8,640) with MMA in benzene and phenetol were carried out to examine the incompatibility effect between unlike polymers. $r_1$(MMA) for MA-PDMS8640 deviated from the azeotropic behavior and $r_1$ increased as conversion increased while $r_1$(MMA) for MA-PDMS1020 was almost unity and unchanged during the copolymerization as shown later [94].

Asami and Takaki et al. [53] prepared PTHF macromonomers having a vinylbenzyloxy (VBO-PTHF), vinylphenoxy (VPO-PTHF), methacryloyl (MA-PTHF), and acryloyl (A-PTHF) end group by cationic polymerization of THF with triethyloxonium tetrafluoroborate ($Et_3OBF_4$) followed by the coupling reaction of living PTHF cation with sodium p-vinylbenzylmethoxide, sodium p-vinylphenylmethoxide, sodium methacrylate, and sodium (or triethylammonium) acrylate, respectively. It is shown that radical copolymerizations of MA-PTHF (MW=6,500) and A-PTHF (MW=10,000) with **b**-vinylnaphthalene (**b**-VN) give $r_1$(**b**-VN)=1.28, $r_2$(MA-PTHF)=0.21 and $r_1$(**b**-VN)=2.69, $r_2$(A-PTHF) from the determination of the copolymer composition by UV spectra. Comparison of these values with $r_1$(**b**-VN)=1.00, $r_2$(MMA)=0.40 and $r_1$(**b**-VN)=2.06, $r_2$(MA)=0.11 of the corresponding small monomers indicates that copolymerization reactivity of these macromonomers are lower than those of the corresponding small monomers.

Severini et al. [95] reported the preparation of methacryloyl capped poly(4-vinylpyridine) macromonomers (MA-P4VP, viscosity-average molecular weight = 12000-54000) by anionic polymerization according to Spiegelman and Parravano, the molecular weight distributions of which were rather broad. The macromonomers were copolymerized with butylacrylate (BA) using AIBN in DMF. They obtained the monomer reactivity ratio $r_2$= 8.3 for the copolymerization of MA-P4VP ($M_1$) with BA ($M_2$) and $r_2$= 0.46~0.9 for 4-vinylpyridine ($M_1$) with BA ($M_2$) indicating the large reduction of the relative reactivities of the macromonomers [95].

Akashi et al. [24] prepared styryl terminated oligovinylpyrrolidone macromonomers ($M_n$=720-4400) by the esterification of carboxyl group terminated vinylpyrrolidone oligomers with chloromethylstyrene. The macromonomers were copolymerized with St in ethanol using BPO as an initiator to obtain a graft copolymer having a hydrophobic backbone and hydrophilic branches. The content of the macromonomer in the graft copolymer decreased as the molecular weight increased. They suggested the effect of the incompatibility between the macromonomers and the growing chains.

Muhlbach and Percec [96] synthesized a series of poly(2,6-dimethyl-1,4-phenylene oxide) macromonomers having a vinylbenzyl ether **33** (VBO-PPO, $M_n$=1000-27000) and investigated the influence of the molecular weight of the macromonomers on the copolymerization reactivity in details.

$$Br-C_6H_2(CH_3)_2-O-[C_6H_2(CH_3)_2-O]_n-C_6H_2(CH_3)_2-O-CH_2-C_6H_4-CH=CH_2 \qquad 33$$

The macromonomers ($M_2$) were copolymerized with butylmethacrylate (BMA) and methyl methacrylate (MMA) ($M_1$) in toluene using AIBN. The macromonomer reactivity ($1/r_1$) is observed to increase with its molecular weight, $M_n$, up to about 5000-7000, but above these values, the reactivity decreases as shown in **Figure 14**. When graft copolymers are prepared using the macromonomer of molecular weight less than 5000 (in the case of $M_1$ was MMA) or 7000 (in the case of BMA), the graft copolymer shows a single glass transition temperature, while they shows two glass transition temperatures corresponding to the respective component when $M_n$ of the macromonomer of greater than 5000-7000. The results are discussed in terms of the effects of the phase separation and the kinetic excluded volume[92].

Tsuruta et al. [97] obtained linear polyamides having a vinyl end group **34** (VB-PAM) in the course of the study on lithium alkylamide catalyzed addition reaction of amines to 1,4-divinylbenzene (DVB). They carried out radical copolymerization of macromonomers **34** with St to produce polyamine-polystyrene graft copolymers. Nabeshima and Tsuruta [98] evaluated the monomer reactivity ratios in the radical copolymerizations of VB-PAM ($M_n$=3050, 3750) with HEMA and N,N'-diethyl-N-(4-vinylphenetyl)ethylenediamine (EDAS) with HEMA in benzene. EDAS corresponds to VB-PAM of $D_p$=1. The monomer reactivity ratios estimated are $r_1$(HEMA)=0.47, $r_2$(VB-PAM)=0.16 and $r_1$(HEMA)=0.56, $r_2$(EDAS)=0.48 indicating the considerable reduction of the copolymerization reactivity of the macromonomers.

$$CH_2=CH-C_6H_4-CH_2CH_2N(Et)CH_2CH_2N(Et)-[CH_2CH_2-C_6H_4-CH_2CH_2N(Et)CH_2CH_2N(Et)]_{n-1}-CH_2CH_2-C_6H_4-CH_2CH_2N(Et)CH_2CH_2NH(Et) \qquad 34$$

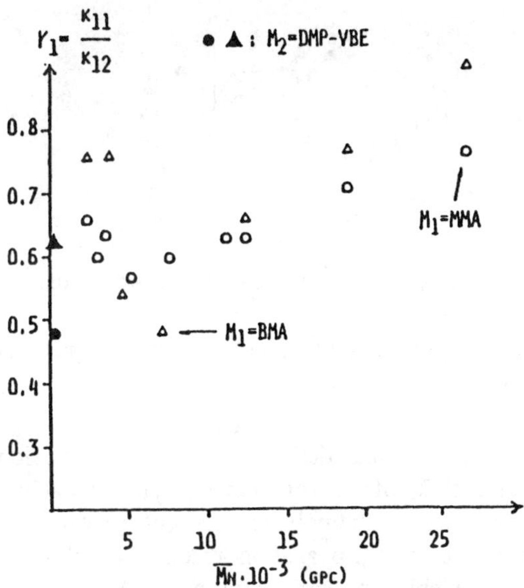


__Figure 14__ *Plots of $r_1$ versus the number average molecular weight of VB-PPO in the copolymerization of $M_1$= MMA (○), BMA (Δ ) with $M_2$=VB-PPO.*

Recently, Niwa et al. [26] synthesized an amphiphilic macromonomer **35** having fluoroalkyl end group (n=10) and copolymerized **35** with St in chloroform using AIBN. They reported that the relative copolymerization reactivity of **35** was higher than that of MMA due to the effect of aggregation of the macromonomers in the copolymerization mixture.

$$CF_3(CF_2)_7CH_2CH_2O\underset{\underset{O}{\|}}{C}-C_6H_4-\underset{\underset{O}{\|}}{C}O(CH_2CH_2O)_n-\underset{\underset{O}{\|}}{C}-\overset{\overset{CH_3}{|}}{C}=CH_2 \qquad 35$$

There are some reports on the effect of the concentration and the conversion of macromonomers on the observed copolymerization reactivity. Arnold et al. [99] investigated the anionic copolymerization of butadiene(M1) with VB-PSt (M2) using BuLi/t-BuONa as an initiator. It is observed that the copolymerization reactivity of the macromonomer estimated by $1/r_1$ varies with the molecular weight, feed composition, and the conversion of the macromonomer as shown in **Figure 15**. Percec et al. [100,101] reported the increase of $1/r_1$ with increase of [M] in the radical copolymerization of MMA(M1) or BMA(M1) with VBO-PPO (**33**). They are discussing the observed copolymerization reactivity in terms of the micellar formation in the reaction media.

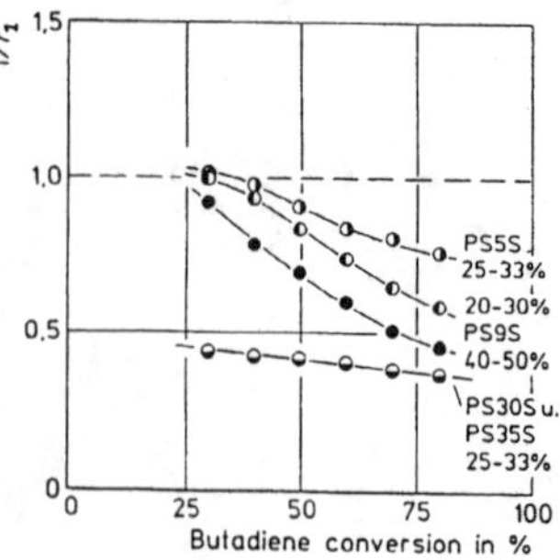


Figure 15 $1/r_1$ *versus butadiene conversion plots*

### 3.2.2 Factors Influencing on Copolymerization Reactivity of Macromonomers

It is seen from the previous section that there is no general conclusion on whether or not the copolymerization reactivity of macromonomer is molecular weight dependent. Even whether or not the large size of the polymer chain of macromonomers influences the rate ratio for the self to the cross-propagation reaction is still not clear at present. Several reasons can be considered for various discussions on the copolymerization reactivity.

First, it is difficult to evaluate the exact monomer reactivity ratios for both macromonomer and comonomer, because of the very low feed concentration of macromonomer. Second, it is often ambiguous whether the reaction mixture is truly homogeneous or not in each report. Third, it is uncertain what chemical structure at the polymer chain end should be considered as the polymerizable end group. The intrinsic reactivity of the polymerizable end group (part (b) in Figure 1) and the corresponding small monomer depend on the choice of the chemical structure. These factors are always involved in estimation of the copolymerization reactivity irrespective of the nature of the propagating active site, such as whether it is radical or ionic.

The observed reactivity involves the effects of various kind of interactions between polymer segments or between polymer segment and solvent. These interactions cause the aggregation (micell formation) or seggregation (phase separation) of the particular reaction species. Orientation of the reactive end group is also possible in some cases. In these cases, the local environments around the reaction site in macromonomer systems might be very different from that of copolymerizations between small monomers under homogeneous reaction condition.

The results of Percec et al. [96] on the radical copolymerization of the macromonomer **33** with MMA or BMA, Ito et al. [90] on poly(oxyethylene) macromonomers with St, and of Niwa et al. [26] on the fluoro-containing macromonomer **35** with styrene are interpreted in terms of the aggregation or orientation of the perticular component. In these cases, the local composition are greater than the feed composition. Thus, the observed reactivity of the macromonomer is higher than that of the small monomer corresponding to the polymerizable end group.

When the polymerization solvent is a good solvent for the macromonomer and the comonomer chain, homogeneous distribution of the macromolecular species is expected. However, a small difference between solvent quality for the macromonomer and the comonomer chain can cause the distribution of the reaction speices inhomogeneous in reaction mixture compared with the case of the corresponding small monomer pairs.

This situation becomes more prominent when the copolymerization is carried out in a preferential or selective solvent, usually a good solvent for the macromonomer, in which the formed polymer from the comonomer seggregate into its own phase and the distribution of reaction species becomes heterogeneous. Tsukahara et al. [93,94] investigated the radical copolymerization of ω-methacryloyl polystyrene macromonomer (MA-PSt) and polydimethylsiloxane (MA-PDMS) with MMA in cyclohexane (preferential solvent for MA-PSt at 60°C) or in phenetol (poor solvent for MA-PDMS, θ-temperature is 83-89°C) as well as in benzene (mutual good solvent). In cyclohexane, the polymerization mixture became opaque soon after the polymerization began. Fine particles of PMMA with diameter ca. 0.1μm are formed during the early stage of the copolymerization as shown in **Figure 16**. These particles are stabilized by the simultaneously formed graft copolymers. In this case, the large reduction was observed in the copolymerization reactivity of the macromonomers, because the macromonomer molecules can not penetrate into the seggregated particles.

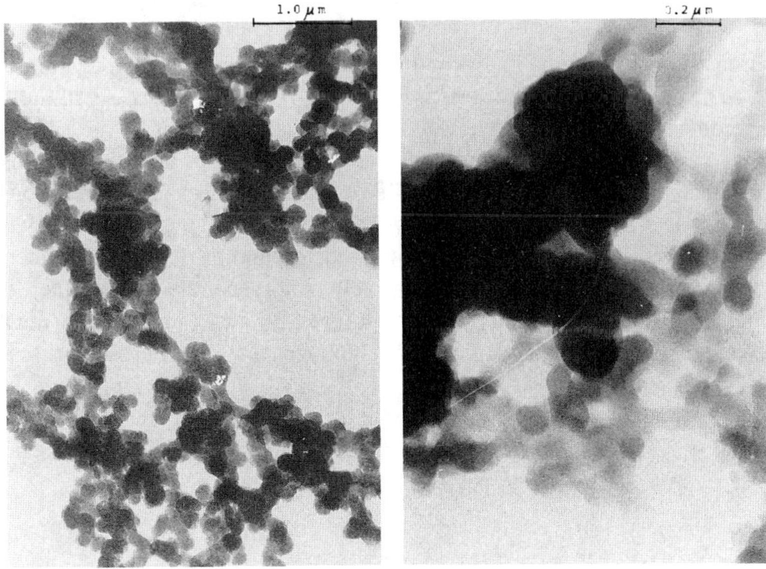


**Figure 16** *Electron micrographs of the products in the copolymerization of MA-PSt macromonomer ($M_n$=12400) with MMA in cyclohexane.*

However, even under homogeneous and transparent reaction conditions, the reduction of the copolymerization reactivity of macromonomers is still observed in the same macromonomer-comonomer pair. One possible explanation for this is the dynamic exluded volume effect [62,136]. In addition to this, the restriction of the degree of interpenetration of unlike polymer species in the reaction media might be another reason, which is related to the repulsive interaction between the macromonomers and the propagating comonomer chains or the incompatibility effect as shematically shown in **Figure 17** [93]. Let us consider this to some extent.

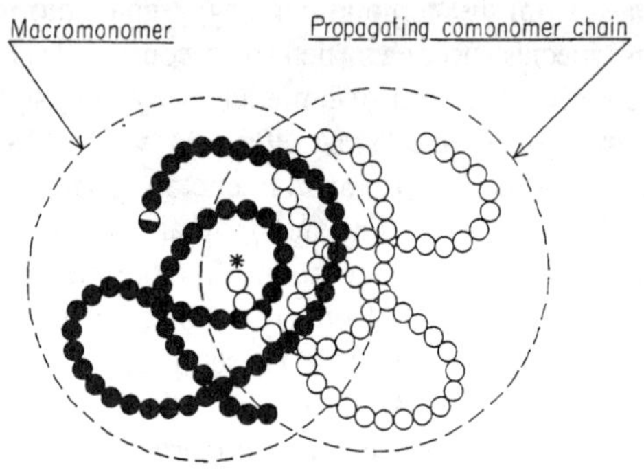


**Figure 17** *Schematic representation of the interpenetration of the macromonomer and the propagating comonomer chain in the reaction media.*

If macromonomers cannot interpenetrate sufficiently into the propagating comonomer chains to bring their end groups near the radical sites, the decrease in the reactivity of the macromonomer can be observed in comparison with that of the small monomers. In this case, the distribution of the reaction species is nonhomogeneous (microscopically) where Flory's principle for the size effect on the reaction rate might not hold [102]. In other words, when the copolymerization reactivity of a macromonomer is equal to that of the corresponding small monomer, then the macromonomer chain should extensively interpenetrate into the propagating comonomer chain in copolymerization solutions. Because, in that case, the distribution of the polymerizable end group must be homogeneous in microscopic scale as in the case of small monomers.

Here, let $\chi_{AB}^{eff}$ be the effective thermodynamic interaction parameter between polymer-A and polymer-B in solution. This is defined by

$$\chi_{AB}^{eff} = \chi_{AB}(1 - \Phi_s) \qquad (11)$$

where, $\chi_{AB}$ is Flory-Huggins interaction parameter between polymer-A and polymer-B and $\Phi_s$ is the volume fraction of solvent. Then, the partial segment of each polymer having degree of polymerization, $N_p$, which can be considered compatible and interpenetrate each other even when the entities of the polymer chains are incompatible with each other, is defined by $N_p < (1/\chi_{AB}^{eff})$ [93].

In **Table-13**,$N_p$ calculated for the copolymerization mixture of MA-PSt with MMA and MA-PDMS with MMA in benzene is shown together with the total monomer concentration, $C^t$, and $\chi_{AB}^{eff}$ [94]. By comparing the calculated $N_p$ values and DP of macromonomer, it can be seen in **Table 13** that almost the entire macromonomer chain can be compatible with the propagating comonomer chain in the MSt-MMA system. This means that, the macromonomer chain can extensively interpenetrate into the coil of the propagating

Table 13. Effective Polymer-Polymer Interaction Parameter, $\chi_{AB}^{eff}$, and Degree of Partial Interpenetration, $N_p$

| Copolymerization Component (M1) | Macromonomer (M2) Code | DP | total[a)] monomer conc. $C^t$ (g/ml) | local composition $\phi$ | $\chi_{AB}^{eff}$ | $N_p$ |
|---|---|---|---|---|---|---|
| MMA | MA-PSt | 122 | $C^t$=0.310>$C^*$ | 0.237 | 0.0063-0.0142 | 70-159 |
| MMA | MA-PDMS | 11 | $C^t$=0.241<$C^*$ | 0.734 | 0.380-0.510 | 2-3 |
| MMA | MA-PDMS | 143 | $C^t$=0.251<$C^*$ | 0.253 | 0.131-0.176 | 6-8 |

a) At this $C^t$, the reaction mixtures are honmogeneous through the copolymerization reaction.
$C^* = (1/2)\Sigma(3M_w/4\pi<R_g>_i^{3/2}N_A)$

comonomer chain from the thermodynamic point of view. On the other hand, in the MA-PDMS copolymerization system, only the small part of the macromonomer chain can be interpenetrate into the propagating comonomer chain. In this case, the concentration of the polymerizable end group of macromonomers around the propagating radical sites becomes lower than that in the absence of the incompatibility effect.

It should be mentioned that the degree of interpenetration ($N_p$) between unlike polymer chains depends on the quality of solvent -C toward each polymer chain, i.e. difference between $\chi_{AC}$ and $\chi_{AC}$ asymmetric nature of the solvent, in addition to $\chi_{AB}$. Especially, when the total monomer concentration is low or copolymerization is stopped at low conversion, the interaction between polymer species and solvent, i.e., the asymmetric nature of polymerization solvent, becomes prominent. On the other hand, the interaction between unlike polymer species is important at high total monomer concentration or in high conversion copolymerizations. **Table 14** shows the relation between the value of the interaction parameters between polymer components and the observed reduction in copolymerization reactivity of macromonomers. It is seen in the **Table 14** that $\chi_{AB}$ is large when the reduction in the copolymerization reactivity is observed. For understanding of the restriction of interpenetration between unlike polymer species on the copolymerization reactivity of macromonomers, a study of the copolymerization between unlike macromonomers might be useful and interesting.

Table 14. Interaction parameter and reported copolymerization reactivity

| Polymer Components A/B | $\chi_{AB}$(30°C) | Observed Reactivity Reduction | Research Group |
|---|---|---|---|
| PSt/PMMA | 0.036 | negligible | Schulz-Milkovich |
| | | negligible | Tsukahara-Yamashita |
| PSt/PEO | 0.055 | remarkable | Ito et al. |
| PSt/PIB | 0.39 | remarkable | Kennedy et al. |
| PSt/PTHF | 0.027 | negligible | Asami-Takaki |
| PSt/PDMS | 0.43 | remarkable | Cameron et al. |
| PMMA/PDMS | 0.66 | remarkable | Tsukahara-Yamashita |

Furthermore, evaluation of the chemical composition distribution (CCD) in addtion to the average composition is also interesting and desirable to understand the copolymerization behavior of macromonomers in details [106-109]. **Figure 18** shows examples of the CCDs of the radical copolymerization products of the macromonomer MA-PSt12400 with MMA at different feed compositions. The copolymerization conditions and the average copolymer compositions are shown in **Table 15**. CCDs were evaluated from a high performance liquid chromatography (HPLC). The HPLC was run in the reverse-phase adsorption mode with combination of an octadecyl modified silica gel column and a linear gradient of tetrahydrofuran and acetonitrile. It is seen in **Figure 18** that CCDs of the graft copolymers are broad irrespective of the copolymer composition compared with those of the corresponding random copolymers of MMA and St. These CCD are well described by the calculation of Stejskal and Kratochvil [109].

Table 15
**Synthesis of Graft Copolymer Samples**

| sample code | feed: macromonomer[b] (g) | feed: macromonomer[b] (wt %)[c] | feed: MMA (g) | feed: AIBN (g) | feed: benzene (mL) | reactn time (h) | conversion (wt %): total monomer | conversion (wt %): macro-monomer | copolymer[a]: S content (wt %) | copolymer[a]: $10^{-4}M_n$ |
|---|---|---|---|---|---|---|---|---|---|---|
| A | 1.7400 | 29.9 | 3.5019 | 0.0346 | 16.7 | 2.5 | 16.5 | 11.5 | 27.0 | 5.8 |
| B | 2.3304 | 50.0 | 2.0032 | 0.0199 | 13.3 | 2.5 | 10.8 | 8.9 | 46.3 | 10.7 |
| C | 4.0003 | 77.4 | 1.0054 | 0.0205 | 16.7 | 2.5 | 9.8 | 8.8 | 74.4 | 29.2 |
| D | 1.1707 | 50.2 | 1.0004 | 0.0099 | 6.7 | 12.0 | 50.0 | 43.0 | 46.3 | 8.6 |
| E | 2.3300 | 50.0 | 2.0016 | 0.0199 | 13.3 | 48.0 | 83.0 | 74.5 | 48.3 | 6.9 |
| F | 2.3306 | 50.0 | 2.0015 | 0.0200 | 13.5 | 1.5 | 6.1 | 4.9 | 44.4 | 10.5 |

[a] The values measured after removal of the unreacted macromonomer. [b] $M_n$ of the macromonomer is 12 400; $M_w/M_n$ = 1.06 by GPC. [c] Feed compositions corrected by the end functionality of the macromonomer ($f$ = 0.86).

It is interesting to see the influence of conversion on the CCD in **Figure 19**. With the increase of the conversion, the CCD curves seems to shift toward the macromonomer-rich composition direction. This might be addressed by the reduction in the relative copolymerization reactivity of the macromonomer with MMA. This was not detected by the estimation of the monomer reactivity ratio alone. However, the reduction is small under homogeneous copolymerization condition. It should be noted here that the obtained CCD is sensitive to the purification or isolation procedure of the copolymers before the mesurement. Thus, one must be careful about the isolation procedure. Further study is neccesary to draw the general picture on the copolymerization reactivity of macromonomers. Furthermore, the study of the copolymerization behavior of rigid macromonomers or those having a mesogenic group are of great interest.

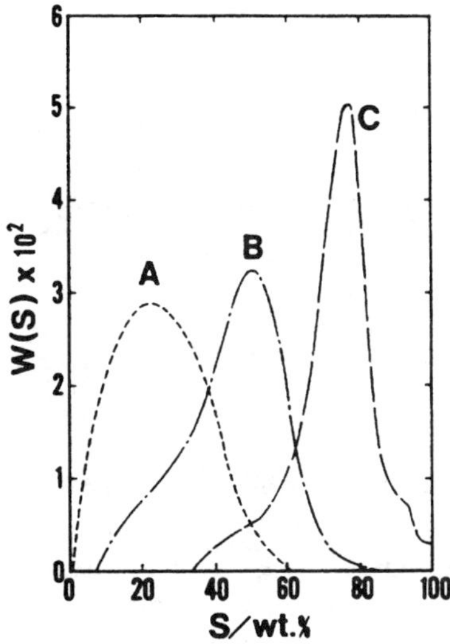


**Figure 18** *Chemical composition distribution (CCD) curves of graft copolymers prepared by the radical copolymerizations of MMA with MA-PSt12400. Copolymerization conditions are shown in Table 15.*

# 4. PROPERTIES AND APPLICATIONS OF POLYMERIZATION PRODUCTS OF MACROMONOMERS

## 4.1 Polymacromonomers

Polymerizations of macromonomer provide regular branched polymers with very high branch density as shown in **Figure 9**. Such polymacromonomers have 1) regular branch length, 2) regular branch number, and 3) regular branching period. The properties of such polymacromonomers are expected to be very diffrent from the corresponding linear polymers of the same molecular weight [112-114]. The knowledge about the polymacromonomers provide the understanding of the growing chain of macromonomers in polymerization reactions, too.

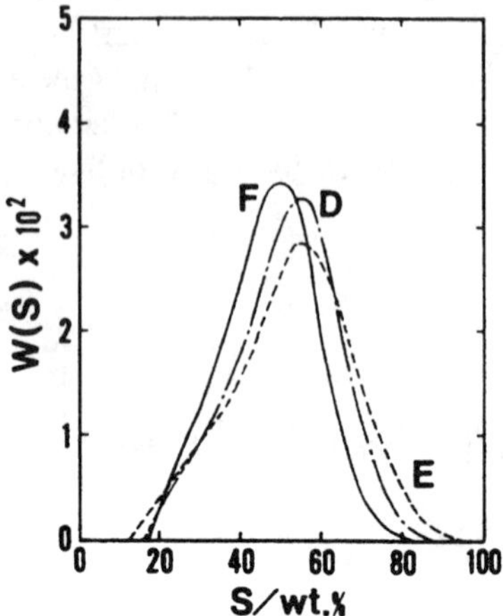


Figure 19 *Effect of conversion on the CCD curves of graft copolymer prepared using MA-PSt12400 (See Table 15).*

## 4.1.1 Molecular Size and Intrinsic Viscosity

**Figure 20** shows the intrinsic viscosity, $[\eta]$, of the polymacromonomers of polystyrene macromonomer in toluene as a function of the weight average molecular weight, $M_w$, together with those of the linear polystyrene standards [112]. The polymacromonomers were prepared from the radical polymerization of polystyrene macromonomers (MA-PSt) of various molecular weights as shown in the previous section. It is seen that the $[\eta]$ values of the polymacromonmers at low $M_w$ region are almost contant with respect to $M_w$ of polymacromonomers in contrast to those of the linear PSt. The exponent **a** in the Mark-Houwink-Sakurada equation (12) is almost zero for the polymacromonomers.

$$[\eta] = KM^a \tag{12}$$

On the other hand, at high $M_w$ region, $[\eta]$ goes up and increases with $M_w$. Furthermore, the $M_w$ for the onset of the rising of $[\eta]$ seems to shift toward higher $M_w$ region with increase of the length of the branch segments. The rising of $[\eta]$ with $M_w$ is not observed for the polymacromonomers having long branches in this molecular weight range in **Figure 20**. This $[\eta]$ versus $M_w$ relationship presumably indicates the change in the molecular shape of polymacromonomers from the star like shape to the comb like one. **Figure 20** also indicates that the segment density of the polymacromonomers in solution increases with increase of the number of branch while that of the linear PSt rapidly decreases with $M_w$. This might influence the polymerization behavior of the macromonomers via the multibranched structure of the propagating chain.

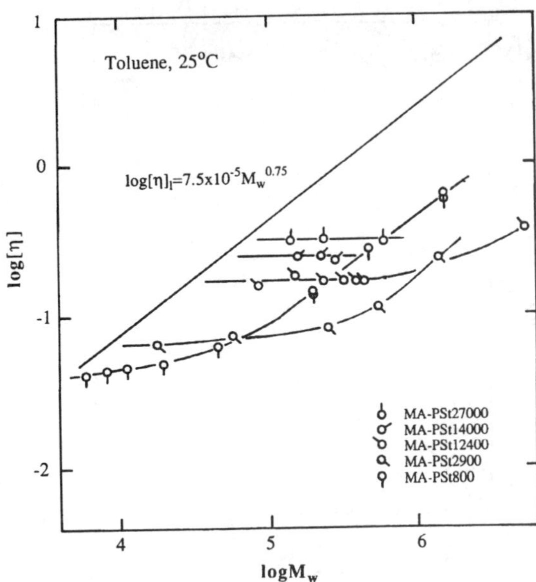


**Figure 20** *Log[η]-logM$_w$ plots of polymacromonomers having different branch length and numbers in toluene at 25°C. Streightline in the figurerepresentsLog[η]-logM$_w$ relationship for linear PSt.*

Increase in the compactness with the number of branch is also observed in the plots $M_w$ of polymacromonomers determined by a gel permeation chromatography equipped with a laser light scattering detector (LS) in addition to the conventional RI or UV detectors versus the apparent $M_w$ determined by the elution volume correlating with the calibration curve of linear polystyrene standards as shown in **Figure 21** [49]. The polymacromonomers in this Figure were prepared from MA-PSt12400 ($M_n$=12400, $M_w/M_n$=1.06) with different [M]. It is seen in **Figure 21** that when the apparent $M_w$ is about 2 x $10^5$, the $M_w$ reaches to 1 x $10^6$.

The product of $M_w$ and $[\eta]$ represents the hydrodynamic volume of the polymacromonomers. The hydrodynamic radius, Rh, of polymacromonomers estimated by $\{(3/4\pi)(M_w[\eta]/N_A)\}^{1/3}$, i.e., the radius of the equivalent sphere, is 2.2 to 32.7 nm for poly(MA-PSt800) with DP=6.9 to 1730 in toluene at 25°C, which are much smaller than those of linear PSts of the same molecular weights. The molecular dimension of polymacromonomers can also be estimated from GPC. For this purpose, the conventional $M_w$ versus $V_r$ calibration curve of linear polystyrene standards for GPC in toluene at 25°C was combined with the following empirical equation of the radius of gyration versus molecular weight relationship for the linear polystyrenes in toluene 25°C [115];

$$\langle S^2 \rangle_l = 1.38 \times 10^{-18} M_w^{1.19} \tag{13}$$

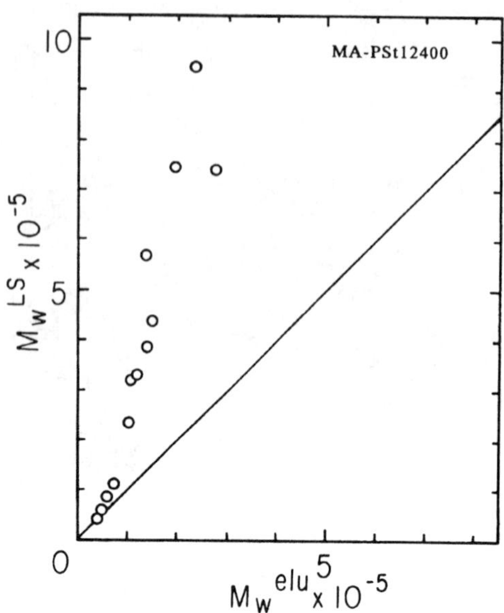


**Figure 21** *Plots of $M_w$ of polymacromonomers of MA-PSt12400 determined by LS responce curves versus the apparent one determined by the molecular weight-elution volume calibration curves of linear polystyrene standards.*

The size of polymacromonomers thus obtained are summarized in **Table 16** together with $M_w$ by LS [49]. It is seen in the Table that the molecular size of poly(macromonomer)s are around 10nm. These values are again much smaller than those of the linear polystyrenes of the same molecular weight. and these increases gradually with increase of the length and the number of the branch of polymacromonomers.

The ratio of the molecular size of polymacromonomer to that of the linear polymer is discussed by the g value, defined by the mean square radius of gyration of the polymacromonomer, $<S^2>_b$ to that of the linear polymer of the same molecular

$$g = <S^2>_b/<S^2>_l \quad (14)$$

weight; here the Flory's universal constant $\Phi$ $<S^2>_b$ is assumed to be equal for polymacromonomer and linear PSt. $<S^2>^{3/2}$ versus $V_r$ relationship obtained for the polymacromonomers from MA-PSt12400 are shown in **Figure 22** together with Mw versus $V_r$ plots [49 ]. Dependence of g upon DP is shown in **Figure 23** where DP equals to the ratio of Mw of polymacromonomer to that of macromonomer and is identical to the number of the branch. g value can be calculated theoretically. Under the theta condition, for the comb shaped polymer, g is given by [116,117]

$$g = (1 + f\gamma)^{-3}[1 + 2f\gamma + (2f + f^2)\gamma^2 + (3f^2 - 2f)\gamma^3] \quad (15)$$

where $\gamma$ is the ratio of the molecular weight of a branch to that of the backbone. g for the star shaped polymer is given by

$$g = (3f - 2)/f^2 \tag{16}$$

When the excluded volume effect exist, g value for the star shaped branched polymer near theta-temperature is given by [118,119]

$$g = [(3f - 2)/f^2][(1 + KbZ + \ldots)/(1 + 1.276Z + \ldots)] \tag{17}$$

where Z is the excluded volume parameter and Kb is given by

$$Kb = \frac{3}{[(f^{0.5}(3f-2)]}\left[\frac{67x2^{7/2}}{315}(f-1) - \frac{134}{315}(f-2) + \frac{4}{45}(11x2^{0.5}-138)(f-1)(f-2)\right] \tag{18}$$

For poly(MA-PSt12400)s, $\gamma$ in equation (15) equals to 126/Dp. The calculated curve of g versus f from equation (15) was almost identical with the curve from equation (16) at this DP range.

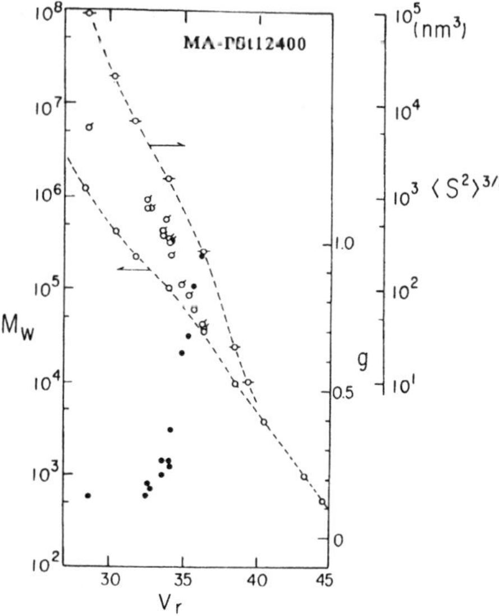


**Figure 22** *Plots of* $<S^2>^{3/2}$ *of poly(MA-PSt12400)s versus* $V_r$ *(-○-),* $M_w$ *of poly(MA-PSt12400)s* (ό) and linear polystyrene standards (○) *versus* $V_r$, *and g versus* $V_r$ *(●) in toluene at 25°C. Flow rate was 1.2 mL/min.*

Table 16 Characterization of poly(macromonomer)s of MA-PSt4400 and MA-PSt12400[a]

| run | [M] (mol/L) | $M_w$ x$10^{-4}$ LS | $M_w$ x$10^{-4}$ e.c.[b] | $D_p$ | Conv. (%) | $\langle S^2\rangle_b^{1/2}$ (nm) | $\langle S^2\rangle_l^{1/2}$ [c] (nm) | g |
|---|---|---|---|---|---|---|---|---|
| | | | | MA-PSt4400 | | | | |
| 1 | 1.99 | 3.0 | 1.8 | 6.8 | 8.4 | 3.98 | 5.4 | 0.539 |
| 2 | 3.99 | 3.7 | 2.4 | 8.4 | 32.2 | 4.77 | 6.12 | 0.607 |
| 3 | 5.50 | 7.5 | 3.4 | 17.1 | 57.5 | 5.81 | 9.35 | 0.386 |
| 4 | 6.00 | 6.7 | 2.9 | 15.3 | 64.0 | 4.93 | 8.8 | 0.317 |
| 5 | 6.89 | 6.8 | 3.8 | 15.5 | 51.7 | 6.21 | 8.84 | 0.493 |
| 6 | 7.91 | 17.9 | 4.0 | 40.7 | 71.3 | 6.21 | 15.7 | 0.156 |
| 7 | 10.0 | 26.2 | 6.2 | 59.5 | 77.8 | 7.77 | 19.7 | 0.156 |
| 8 | 10.4 | 18.5 | 6.5 | 42.0 | 75.5 | 8.54 | 16.0 | 0.285 |
| 9 | 10.9 | 44.7 | 11.0 | 102 | 80.0 | 11.7 | 27.0 | 0.188 |
| 10 | 11.6 | 50.1 | 12.6 | 114 | 81.8 | 12.7 | 28.9 | 0.193 |
| 11 | 11.8 | 57.5 | 10.4 | 131 | 80.0 | 9.36 | 31.4 | 0.0889 |
| 12 | 12.6 | 67.2 | 14.3 | 153 | 80.4 | 13.7 | 34.5 | 0.158 |
| 13 | 14.0 | 99.7 | 18.1 | 227 | 64.3 | 14.3 | 43.6 | 0.108 |
| 14 | 16.1 | 93.7 | 21.4 | 213 | 73.6 | 15.6 | 42.0 | 0.138 |
| 15 | 17.7 | 128 | 25.1 | 291 | 38.0 | 17.3 | 50.6 | 0.117 |
| | | | | MA-PSt12400[d] | | | | |
| 1 | 2.53 | 47.6 | 13.1 | 36.3 | 30.7 | 13.8 | 28.1 | 0.241 |
| 2 | 3.76 | 27.4 | 10.8 | 20.9 | 68.7 | 11.5 | 20.2 | 0.324 |
| 3 | 4.22 | 165 | 25.4 | 126 | 74.4 | 16.1 | 58.8 | 0.0750 |
| 4 | 4.68 | 154 | 43.3 | 118 | 69.9 | 17.5 | 56.4 | 0.0963 |
| 5 | 5.12 | 175 | 43.1 | 134 | 45.9 | 18.8 | 60.9 | 0.0953 |
| 6 | 5.37 | 195 | 52.3 | 149 | 42.4 | 17.3 | 64.9 | 0.0710 |
| 7 | 5.50 | 220 | 40.0 | 168 | 52.4 | 18.8 | 69.8 | 0.0725 |

a) Polymerizations were carried out in benzene at 60°C for 24 hrs with [I]= 1.64 x $10^{-2}$ (mol/L). [M] are as feed values (without correction by F).
b) Apparent molecular weight calculated with the calibration curve of the PSt standards. c) $\langle S^2\rangle^{1/2}$ of linear PSt of corresponding $M_w$ estimated from $\langle S^2\rangle=1.38\times10^{-18}M_w^{1.19}$. d) See also Tables I and III in ref.49.

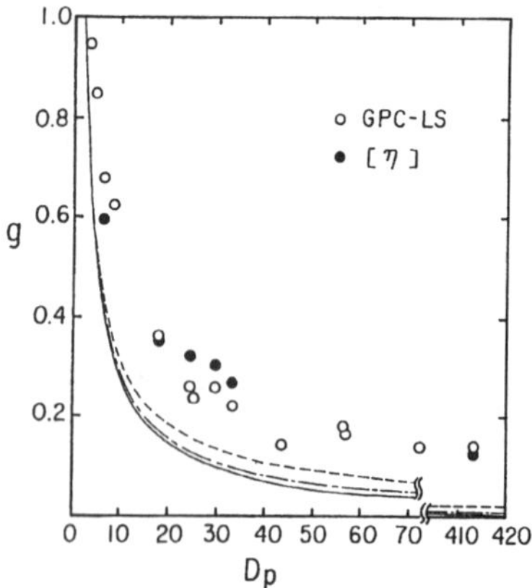


**Figure 23** *g values (open symbol) and intrinsic viscosity ratio [η] (closed symbol) versus degree of polymerization of poly(MA-PSt12400)s. Symbols are experimental values cal culated from eq 14 and from* $[\eta]_b/[\eta]_l$. *Solid curve, dotted curve, and broken curve are theoretical ones, each of which was calculated from eq 16 and 17 with z=0.1 and 0.5, respectively.*

The theoretical curve from equation (16) (solid curve) and from equation (17) with Z = 0.1 (dotted curve) and Z = 0.5 (broken curve) are shown in Figure 23 [49]. It is seen from the calculated curves in the figure that, from the theoretical point of view, the g value increases with increase in the excluded-volume effect. That is, the decrease in the molecular dimension due to branching is compensated in part by the excluded-volume effect. The experimental g value variation with Dp in the figure has almost identical tendency but rather higher than the calculated value. This is quite reasonable since the experimental g values were evaluated in a good solvent (toluene at 25oC), where a large excluded-volume effect can be expected. g variation with Dp in the figure provides the good evidence that there is no or negligible formation of irregular shaped polymacromonomers by the chain transfer reactions to polymer during polymerization.

It is seen in **Figure 23** that g values estimated from the elution volume are almost equal to those obtained from $[\eta]$ indicating that the molecular size thus obtained from GPC corresponds to the hydrodynamic volume of polymacromonomers. For further study, therefore, it is necessary to measure the molecular dimension by the light scattering or the small angle X-ray scattering technique to evaluate the size and shape of the polymacromonomers as functions of the branch length and number.

## 4.1.2 Bulk Properties

Because of the specific multibranched molecular structure, polymacromonomers show unique bulk properties compared with those of the corresponding linear polymers [113,114]. **Table 17** shows the characteristics of polymacromonomers pepared by the radical polymerization of polystyrene macromonomers of different molecular weights having a methacryloyl end group together with those of the linear polystyrene standards. In this table, the branch length of poly(macromonomer)s PM-1 to PM-3 and PM-4 to PM-6 is the same but the branch numbers are different. On the other hand, the molecular weights of PM-7 to PM-9 are ca. $1.4 \times 10^5$ and almost equal but those polymers have different branch lengths. The polydispersity index Mw/Mn of the poly(macromonomer)s are in the range of 1.5 to 2.8. These values depend on both the branch length and the number of branches.

Glass transition temperature, Tg, of the polymacromonomers is also shown in **Table 17** [114]. These polymacromonomers showed a clear single Tg in DSC curves, although they have a small fraction of the methacrylate component as the backbone. Tg of PM-7 to PM-9 having long branches is 99-100°C and almost the same to that of linear polystyrene of large molecular weight (LPS-2 to LPS-5) [73]. On the other hand, PM-1 and PM-3 having short branches, Mw of which is 890, showed much lower Tg. Tg of PM-4 to PM-6, Mw of which is 3200, is also 10-15°C lower than that of linear polystyrene of large molecular weight. Tg of PM-1 to PM-3 and PM-4 to PM-6 is almost the same irrespective of the number of branches. This may be because the molecular weight divided by the number of chain end in polymacromonomers is independent of their degree of polymerization.

## Table 17

Characteristics of Poly(macromonomer)s of PSt-macromonomers and Linear PSt Standards

| sample[a)] code | macromonomer Mwx10$^{-3}$ | Mw/Mn | poly(macromonomer) Mwx10$^{-3}$ [b)] | DP[c)] | Mw/Mn[b)] | Tg($^{o}$C)[d)] | film forming property |
|---|---|---|---|---|---|---|---|
| PM-1 | 0.89 | 1.11 | 230 | 258 | 2.22 | 56 | brittle |
| PM-2 | 0.89 | 1.11 | 489 | 549 | 2.13 | 56 | brittle |
| PM-3 | 0.89 | 1.11 | 1540 | 1730 | 1.99 | 68 | brittle |
| PM-4 | 3.1 | 1.07 | 62.7 | 19.6 | 1.86 | 89 | brittle |
| PM-5 | 3.1 | 1.07 | 529 | 165 | 2.52 | 84 | brittle |
| PM-6 | 3.1 | 1.07 | 1390 | 434 | 2.77 | 90 | brittle |
| PM-7 | 9.9 | 1.03 | 149 | 15.1 | 2.46 | 98 | brittle |
| PM-8 | 13.1 | 1.06 | 136 | 10.4 | 1.83 | 98 | brittle |
| PM-9 | 14.6 | 1.04 | 132 | 9.0 | 1.51 | 100 | brittle |
| LPS-1 | - | - | 5.4 | - | 1.03 | 83(81) | brittle |
| LPS-2 | - | - | 102 | - | 1.02 | 100(99) | good |
| LPS-3 | - | - | 190 | - | 1.04 | (99) | good |
| LPS-4 | - | - | 355 | - | 1.02 | (100) | good |
| LPS-5[e)] | - | - | 85.3 | - | 1.54 | (99) | good |

a) PM: poly(macromonomer), LPS: linear polystyrene

b) Determined by the intensity ratio of light scattering response to that of RI using LS-GPC.

c) Degree of polymerization calculated by
DP=(Mw of poly(macromonomer))/(Mw of macromonomer).
This value is equivalent to the number of branches per molecule.

d) From onset temperature in DSC curves. The values in the parentheses were calculated using Tg = 373 - (1.0 x 10$^{5}$/Mw).

e) Prepared by radical polymerization with 0.5 mole% of AIBN at 60 $^{o}$C for 6 hrs.

Since there are many chain ends per molecule in PM-1 to PM-6, the reduction in Tg of these polymacromonomers might be ascribed to the excess free volume around the chain ends. The chain end effect on the glass transition is described by the following relation:

$$Tg = Tg^{\infty} - (2Tg^{\infty} vM_0/v_m)/M \qquad (19)$$

here, $Tg^{\infty}$ is Tg for the polymer of infinite molecular weight, $M_0$ and M are the molecular weights of the monomeric unit and the polymer, $v_m$ is the free volume per monomeric unit, and v is the excess free volume at a chain end [122]. Tg of polymacromonomers calulated by eq (19 ) is shown in **Table 18**. Since the one chain end of the macromonomer is covalently connected to the backbone of polymacromonomers, twice the value of $M_w$ of the amcromonomer was used to calculate Tg. The value of $(2Tg^{\infty} vM_0/v_m)$ depends on the chemical structure of the end group. In **Table 18**, Tg is calculated with $(2Tg^{\infty} vM_0/v_m)=1.0x10^5$[73]. The end group of the branch of these polymacromonomers is sec-butyl group. Although the observed values are slightly higher than the calculated one, there are good agreement between them indicating the marked chain end effect. Observed Tg presumably involves the effect of branching points.

Table 18

Glass Transition Temperatures of Poly(macromonomer)s

| Sample | M.W. of branch chain | | Tg(°C) | |
|---|---|---|---|---|
| | Mn | Mw | obs | cal[a)] |
| Poly(MA-PSt 800) | 810 | 900 | 56-68 | 44 |
| Poly(MA-PSt 2900) | 2900 | 3100 | 84-90 | 84 |
| Poly(MA-PSt 12400) | 12400 | 13100 | 98 | 96 |
| Poly(MA-PSt 14000) | 14000 | 14600 | 100 | 97 |
| Poly(MA-PSt 27000) | 27000 | 28100 | - | 98 |

a) Calculated by the equation below using $Tg^{\infty}$=100 °C, $M_w$=2 Mw taking account for that only one chain end is free.

$$Tg = Tg^{\infty} - (2Tg^{\infty} vM_0/v_m)/M = Tg^{\infty} - 1.0 \times 10^5/2M_w$$

When PM-1 to PM-9 are cast onto Teflon or glass plates, many cracks are created by sample contraction during the solvent evaporation and the formed films are too brittle to handle, although no crack is formed in the film preparation of linear polystyrenes of LPS-2 to LPS-5 in the same molecular weight range under the same casting conditions.

The mechanical strength of the linear polymers like fracture strength and tensile strength is dependent on the molecular weight of the polymer molecules which decreases as the molecular weight of the polymer decreases [123-125].Turner reported that the tensile strength (TS) is decreases linearly with the inverse molecular weight,[125] and becomes zero below a certain critical molecular weight ($M_0$), where polymer materials become very brittle.

In **Table 19**, literature values of the critical molecular weight for the zero tensile strength ($M_0$) and for the chain entanglement ($M_c$) are listed for linear polystyrene

(PSt), poly(methyl methacrylate) (PMMA), and polyethylene (PE). $M_c$ is usually equal to $2M_e$ where $M_e$ is the molecular weight between entanglement points [126]. In **Table 19**, $M_0$ for linear PSt is in the range of $50x10^3$-$70x10^3$ and 1.5-2 times larger than $M_c$. For PMMA, $M_0$ is almost equal to $M_c$, and $M_0$ for PE is about $3M_c$. Both $M_0$ and $M_c$ depend slightly on temperature. Since the $M_0$ are close to $M_c$, molecular entanglements and the resulting entanglement network have been considered essential for the mechanical strength of polymeric materials [123-125,127].

Table 19

Critical Molecular Weight for the Zero Tensile Strength and the Chain Entanglement[a)]

| polymer | $M_0 x10^{-3}$ | $M_c x10^{-3}$ | ref |
|---|---|---|---|
| polystyrene | 50 (-196°C) | 30-35 (>$T_g$) | 123 |
| | 65 (21°C) | | 123 |
| | 72 (-196°C) | | 125 |
| | | 31 | 124 |
| | | 30-40 (160°C) | 128 |
| poly(methyl | 22 (-196°C) | 11 (>$T_g$) | 123 |
| methacrylate) | 21-23(-196°C) | | 125 |
| | | 28 | 124 |
| | | 31-32 (217°C) | 129 |
| polyethylene | 11 (-196°C) | 4 (>$T_g$) | 123 |
| | 12 (-196°C) | | 125 |
| | | 4 | 124 |

a) $M_0$ is the critical molecular weight for the zero tensile strength and $M_c$ is that for the chain entanglement.

In **Table 17**, the chain length of linear PSt (LPS-2 to 4) with $M_w=1x10^5$-$4x10^5$ is sufficient for chain entanglement, so that films of these polymers have sufficient mechanical strength and good film forming property, while the low molecular weight PSt (LPS-1) is brittle because its Mw is less than $M_0$. In the case of polymacromonomers, on the other hand, the molecular weight corresponding the maximum contour length between the most separated chain ends is the sum of the Mw of two branches and that of the chain backbone, which is, for example $21.3x10^3$- $30.1x10^3$ for PM-7 to PM-9 and $8.4x10^3$ to $49.8x10^3$ for PM-4 to PM-6. Thus, the maximum contour length of these macromonomers is insufficient for the formation of the chain entanglement networks (except for PM-3 and 5), although the molecular weight of these polymacromonomers is large enough for the entanglement. In another words, the branch length of the polymacromonomers in this work is not enough to interpenetrate into neighboring polymacromonomer molecules for entanglements.

The critical molecular weight for chain entanglements depends on the molecular cross sectional area; $M_c$ is an increasing function of the molecular cross sectional area [124]. This can be seen in **Table 19** by comparing $M_c$ values for PSt and PMMA with that for PE. In the case of polymacromonomers, the molecular cross sectional area, i.e.,the effective molecular cross sectional area for entanglement, may be very large due to the high segment density around the center backbone segment. Thus, $M_c$ is much greater than for linear PSt. This might be a partial reason for the brittleness of PM-3 and PM-6. The degree of polymerization of PM-3 is 1730, thus the molecular weight of the backbone is 173 x $10^3$ which is much larger than $M_0$ of PMMA in **Table 19**. Therefore, it seems that many branch segments of the polymacromonomers prevent them from the molecular entanglement because of the large molecular cross sectional area. In addition, as the molecular cross sectional area increases, the chain stiffness might increase which can affect $M_c$.

The molecular weight distribution also affects the mechanical strength. However, the film forming property of LPS-5 (Mw/Mn=1.54) is as good as those of LPS-2 to 4, indicating that the effect of the molecular weight distribution is minor. Furthermore, the fact Tg of these polymacromonomers is almost the same or lower than that of linear PSt of large molecular weight indicates that the brittleness of polymacromonomer is not associated to the slowness of the conformational relaxation due to the high branching density.

Therefore, the difference in the brittleness and the film forming property between linear PSts and the polymacromonomers might be ascribed to the absence of chain entanglement networks in polymacromonomers. A very high degree of polymerization or very short branches is probably necessary for the entanglement formation in polymacromonomers in bulk. The compact dimension and the absence of the molecular entanglement of polymacromonomers might be utilized for the viscosity regulation in polymer processing.

## 4.2 Graft Copolymers

By replacing one component of the monomer pairs by a macromonomer in the conventional copolymerization reactions, various kinds of graft copolymers can be prepared. There is almost no limitation in selection of both macromonomers and comonomers in this "Macromonomer method". Moreover, the molecular structure of the produced graft copolymer is well-defined by far compared with those obtained by conventional grafting methods such as the free radical or the radiation grafting methods, i.e., the branch length is controlled by the molecular weight of the macromonomer, and the branch number and the copolymer composition is controlled by the feed composition and other copolymerization conditions on the basis of the knowledge of the copolymerization reactivity of the macromonomers. Therefore, many applications are promising and actually there are many reports on this subjects. Here, some of our studies are shown as examples.

### 4.2.1 Hydrogels Utilizing Hydrophobic Microdomains

When a hydrophobic macromonomer is copolymerized with a hydrophilic comonomer, graft copolymers consisting of a hydrophilic backbone with hydrophobic branches are produced. The membranes or films of the graft copolymer behave as a hydrogel in swollen state with water utilizing hydrophobic microdomains as the crosslinking points. If one adopts a macromonomer having a rigid polymer chain of high Tg and the comonomer giving a soft segment, the produced graft copolymers show the properties of the thermoplastic elastomers (TPE) which are very important materials in polymer industry.

Synthesis and characterization of hydrogels from poly(2,3-dihydroxypropylmethacrylate-g-styrene) (GDS) was investigated [14,15]. The graft copolymers were synthesized by the macromonomer method i.e., by the radical copolymerization of the polystyrene macromonomer with isopropylideneglyceryl methacrylate (IPGMA) followed by the acidic hydrolysis of the acetal group of IPGMA monomeric units with 85% trifluoroacetic acid according to **Scheme-3** [14]. Poly(2,3-dihydroxypropylmethacrylate) (PDHPMA) is an interesting polymer as a component of hydrogels because of the much higher affinity for water than poly(hydroxyethyl methacrylate) (PHEMA) [103,130-132]. Since water is essential to life, hydrogels are expected to have unique functions and properties for biomedical and biological applications [133,134].

The characteristics of the graft copolymers are shown in Table 20 as examples. The number of PSt branches per graft copolymer (n) and the number average molecular weight of the backbone segment between the two adjoining branches ($M_n^s$) are calculated by $n = M_n \times W_s/M_n^b$ and $M_n^s = M_n \times (1-W_s)/n$ respectively, where $M_n$ is the number average total molecular weight of GDS graft copolymer, $M_n^b$ is the number average molecular weight of the PSt branch, and $W_s$ is the weight fraction of St component in the graft copolymer.

The films of the graft copolymers immediately swell when immersed into water as shown in **Figure 24**. The swelling degree approaches to the equilibrium value within 30 min. These hydrogels are soft and stable to show good mechanical strength enough to endure the handling for the measurements of swelling degree and oxygen permeability. The appearance of hydrogels, the equilibrium water content (H) and the equilibrium swelling degree (Q) of hydrogels are shown in **Table 20**.

In **Figure 25**, Q is replotted against $M_n^s$. It is seen that log Q versus log $M_n^s$ plots of these graft copolymers almost fall on a straight line although, there is some degree of scattering of data. The microdomain structure of graft copolymers is well-characterized by measuring of SAXS intensity distribution [14]. In **Figure 26** the relation between $(D_{wet}/D_{dry})^3$ and Q is shown. Where $D_{wet}$ is the nearest neighbouring distance between hydrophobic microdomains in swollen state and $D_{dry}$ is that in dry state. It is seen that $(D_{wet}/D_{dry})^3$, the swelling degree in colloidal scale, is well consistent with the swelling degree Q in macroscopic scale.

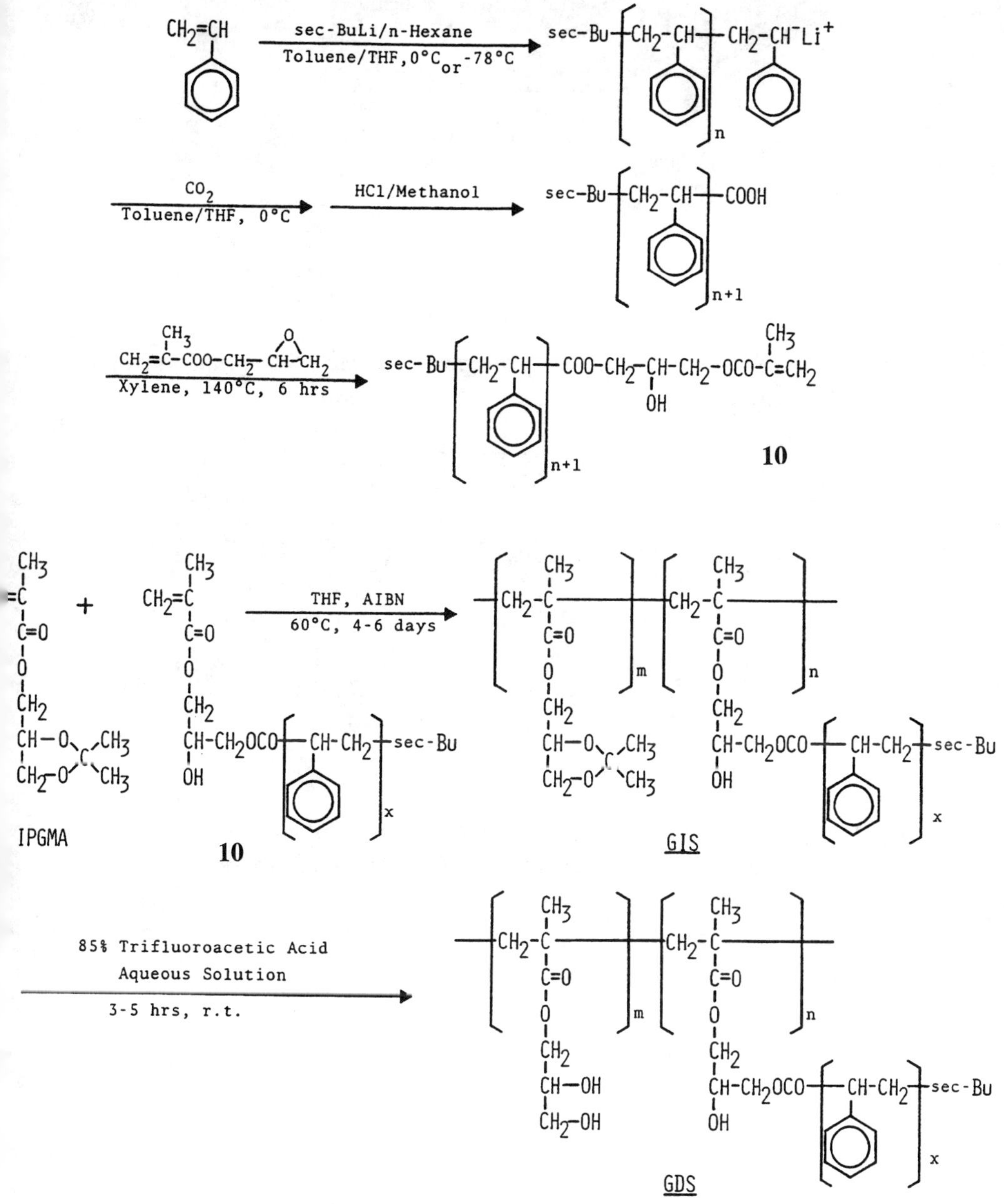

$CH_2=CH$ (styrene) $\xrightarrow[\text{Toluene/THF, 0°C or -78°C}]{\text{sec-BuLi/n-Hexane}}$ sec-Bu$-[CH_2-CH(C_6H_5)]_n-CH_2-CH^-(C_6H_5)\,Li^+$

$\xrightarrow[\text{Toluene/THF, 0°C}]{CO_2}$ $\xrightarrow{\text{HCl/Methanol}}$ sec-Bu$-[CH_2-CH(C_6H_5)]_{n+1}-COOH$

$\xrightarrow[\text{Xylene, 140°C, 6 hrs}]{CH_2=C(CH_3)-COO-CH_2-CH(O)CH_2}$ sec-Bu$-[CH_2-CH(C_6H_5)]_{n+1}-COO-CH_2-CH(OH)-CH_2-OCO-C(CH_3)=CH_2$

**10**

$CH_2=C(CH_3)-C(=O)-O-CH_2-CH(-O-)-CH_2-O-C(CH_3)_2$ (IPGMA) + $CH_2=C(CH_3)-C(=O)-O-CH_2-CH(OH)-CH_2OCO-[CH(C_6H_5)-CH_2]_x-$sec-Bu (**10**) $\xrightarrow[\text{60°C, 4-6 days}]{\text{THF, AIBN}}$

$-[CH_2-C(CH_3)(C(=O)-O-CH_2-CH(-O-)-CH_2-O-C(CH_3)_2)]_m-[CH_2-C(CH_3)(C(=O)-O-CH_2-CH(OH)-CH_2OCO-[CH(C_6H_5)-CH_2]_x-\text{sec-Bu})]_n-$

GIS

$\xrightarrow[\text{3-5 hrs, r.t.}]{\text{85\% Trifluoroacetic Acid Aqueous Solution}}$

$-[CH_2-C(CH_3)(C(=O)-O-CH_2-CH(OH)-CH_2-OH)]_m-[CH_2-C(CH_3)(C(=O)-O-CH_2-CH(OH)-CH_2OCO-[CH(C_6H_5)-CH_2]_x-\text{sec-Bu})]_n-$

GDS

**Scheme 3**

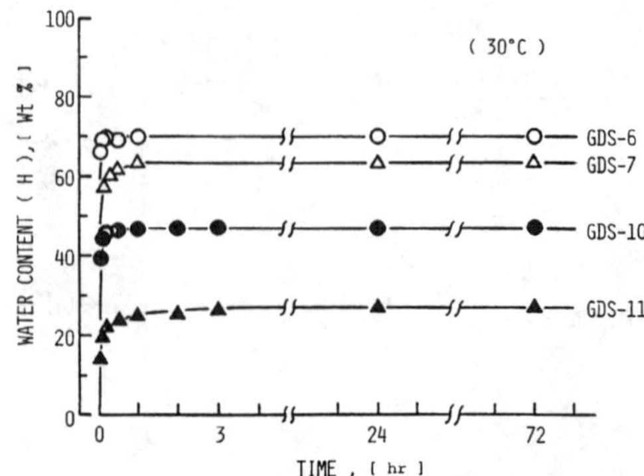


**Figure 24** *Change of water content (H) of the graftcopolymers GDS with incubation time at 30°C.*

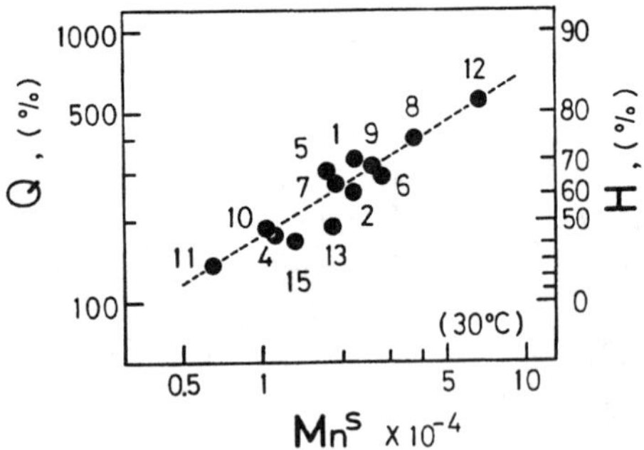


**Figure 25** *Swelling degree (Q) of the graftcopolymers GDS plotted against the molecular weight of the backbone segment between two adjoining PSt branches ($M_n^s$).*

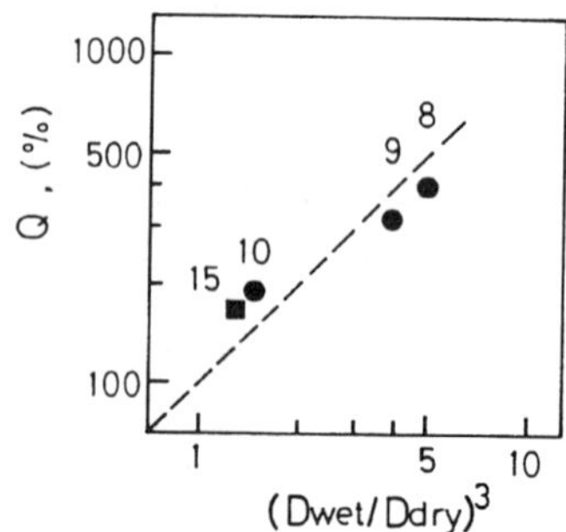


**Figure 26** *Relationship between the macroscopic swelling dgree (Q) and the microscopic one ($(D_{wet}/D_{dry})^3$).*

**Table 20** Characteristics of Graft Copolymers and Hydrogel membranes.

| Sample Code | $\bar{M}n \times 10^{-3}$ a) | | | number of branches | $Mn^s$ | Composition b) (wt%) | | Hydrogel Properties | | | |
|---|---|---|---|---|---|---|---|---|---|---|---|
| | total | GMA-backbone | PSt-branch | | $\times 10^{-3}$ | PSt | PGMA | H (%) | Q (%) | $P_m \times 10^9$ | Appearance c) |
| GDS-1 | 41.4 | 31.3 | 7.23 | 1.4 | 22.3 | 24.5 | 75.5 | 70.1 | 334 | - | t.l. |
| GDS-2 | 49.4 | 37.6 | 7.23 | 1.7 | 22.1 | 24.3 | 75.7 | 60.9 | 256 | - | t.l. |
| GDS-4 | 66.0 | 51.6 | 3.13 | 4.6 | 11.2 | 21.6 | 78.4 | 43.9 | 178 | 1.57 | t.l. |
| GDS-5 | 75.7 | 68.4 | 1.88 | 3.9 | 17.5 | 9.6 | 90.4 | 66.9 | 302 | 3.42 | t.p. |
| GDS-6 | 48.8 | 39.9 | 6.80 | 1.4 | 28.5 | 19.0 | 81.0 | 65.6 | 291 | 2.95 | t.l. |
| GDS-7 | 35.4 | 30.0 | 3.35 | 1.6 | 18.8 | 15.2 | 84.8 | 63.3 | 272 | 2.07 | t.l. |
| GDS-8 | 89.9 | 78.6 | 5.37 | 2.1 | 37.4 | 12.3 | 87.7 | 75.0 | 400 | 3.37 | t.p. |
| GDS-9 | 101.1 | 83.9 | 5.37 | 3.2 | 26.2 | 17.0 | 83.0 | 68.4 | 316 | 3.04 | t.p. |
| GDS-10 | 78.8 | 52.5 | 5.37 | 4.9 | 10.7 | 33.4 | 66.6 | 47.0 | 189 | 1.53 | t.l. |
| GDS-11 | 79.6 | 43.6 | 5.37 | 6.7 | 6.5 | 45.1 | 54.9 | 27.3 | 138 | 0.81 | t.p. |
| GDS-12 | 150.2 | 112.6 | 22.47 | 1.7 | 67.4 | 25.0 | 75.0 | 81.8 | 549 | 5.76 | t.l. |
| GDS-13 | 108.7 | 49.4 | 22.47 | 2.6 | 18.7 | 54.6 | 45.4 | 46.8 | 188 | 1.63 | t.l. |
| GDS-15 | 69.5 | 36.0 | 12.40 | 2.7 | 13.3 | 48.3 | 51.7 | 40.4 | 168 | 1.80 | t.p. |

a) determined from GPC equipped with light scattering detector.

b) determined from $^1$H-NMR spectra and elemental analysis.

c) t.p. = transparent , t.l. = translucent

The oxygen permeability coefficient ($P_m$) of hydrogels is an important factor to be measured for their biomedical applications. The $P_m$ values of the hydrogels are shown in **Table 20**. $P_m$ ranges from 0.81 x $10^{-9}$ to 5.76 x $10^{-9}$ [ml(STP)cm/cm$^2$sec. cmHg] depending on the water content H. These values are almost the same as the other poly(HEMA) hydrogels containing about 70% of water (1.5 x $10^{-9}$- 5.1 x $10^{-9}$) [103]. **Figure 27** shows the $P_m$ variation with $M_n^s$ where $P_m$ is well correlated to $M_n^s$ as in the case of the swelling degree in **Figure 25**. The relationship between $P_m$ and the water content H is shown in **Figure 28** [14]. It is seen in the figure that $P_m$ values are directly related to the water content of the hydrogel and log $P_m$ versus H plots fall on almost a straight line including the values of PSt and water. This result indicates that the oxygen permeability coefficient of the hydrogel is controlled by $M_n^s$ via H. The hydrogels utilizing macromonomers are useful in such applications like soft contact lenses and intraocular lenses [1, 133].

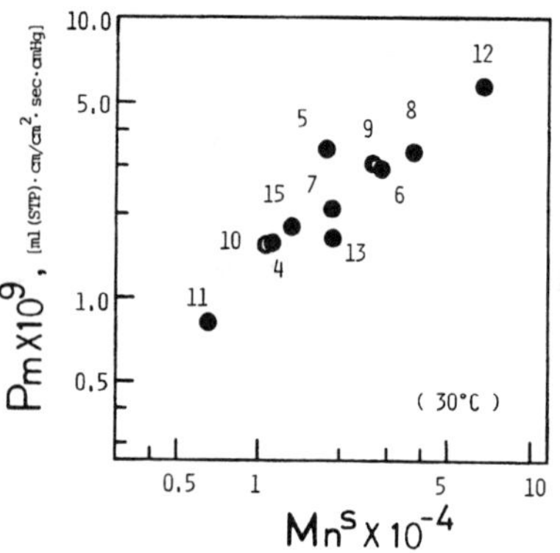


**Figure 27** *Oxygen permeability coefficient ($P_m$) of the hydrogels from the graft copolymers GDS plotted against $M_n^s$ in water.*

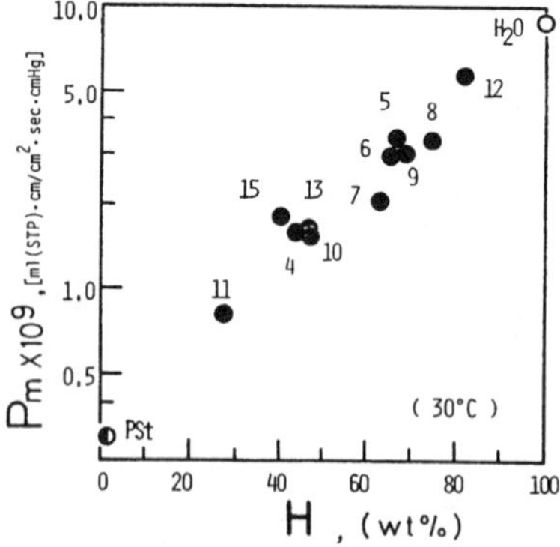


**Figure 28** *Relationship between $P_m$ and H of the hydrogels. The values of PSt and water are included for comparison.*

## 4.2.2 Surface Active Polymeric Additives

Another application of the graft copolymers is surface active polymeric additives for the surface modification of polymeric materials. Compatibilizers in polymer blends and alloys are a kind of interfacial modifiers. Surface modification of polymeric materials is very important in such fields as biomaterials, chromatography, coating, adhesives, antistatic properties, and many others. Surface modification utilizing the surface adsorption or surface accumulation of the additives having desirable properties, like antistatic and antioxidative properties, has been very popular and used for a long period [135].

**Figure 29** shows the schematic representations of the molecular architectures for the surface active polymeric additives, where A is the surface active component, B is the anchor component compatible with the matrix polymer, and C is the functional component to provide the desired properties to the surface. The permutation of ABC in the copolymers has no particular meaning and depends on the relative difficulty in preparation. The block copolymers prepared by the living anionic polymerization is well known to have the most well-defined structure, however, the use of monomers having polar groups and ionic groups are rather limited in the anionic polymerization. Graft copolymers prepared by the macromonomer method are more efficient in some cases.

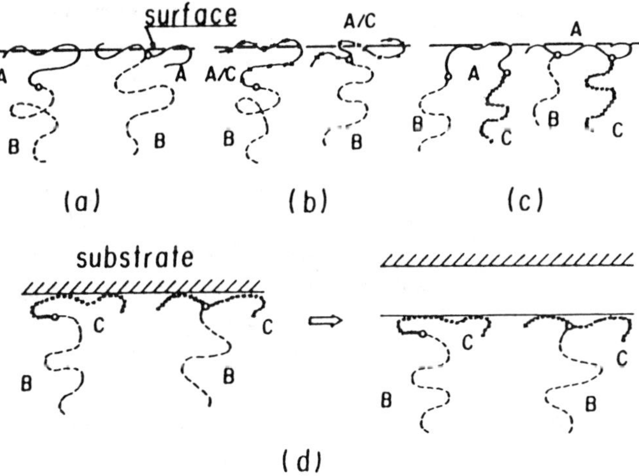


**Figure 29** *Schematic representation of the molecular architecture of block and graft copolymers as polymeric additives for the surface modification of polymeric materials. A: Surface active component, B: Anchor component, and C: Functional Component.*

Many applications can be expected utilizing the surface accumulation of the surface active graft copolymers. Effect of the surface layer on the gas permeability through the surface modified membrane is an example [138]. The membrane was prepared by casting of polystyrene (Mn = $93 \times 10^3$) containing 1 wt% of the graft copolymers poly(methacrylic acid-co-methyl methacrylate-g-styrene) onto glass slides from solution. This is an asymmetric membrane, one surface of which is covered by polystyrene segment thus hydrophobic and another surface is enriched by methacrylic acid component thus hydrophilic [137].The gas permeabilities of carbon dioxide, water and benzene vapors through the membrane were measured by using Rouse type permeability meter. Some examples are shown in **Table 21** and **Figure 30** [138].

Table 21 Permeability Coefficient P of Carbon Dioxide Through Surface Modified PSt Membrane.

| From side | P(ccstp · $cm^{-1}s^{-1}cmHg^{-1}$) |
|---|---|
| air | $5.58 \times 10^{-9}$ |
| substrate | $7.76 \times 10^{-9}$ |

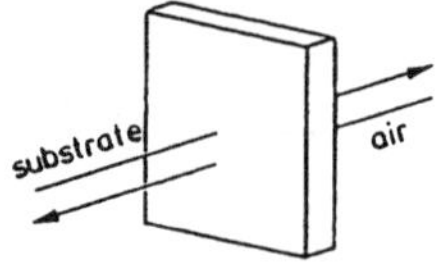


'Air' denotes the air side surface placed toward the high pressure side of the permeation cell and 'substrate' is the reverse of 'air'.

These results show that permeability of the gases depends on the permeation direction. That is, the permeabilities of water vapor and carbon dioxide are higher when the hydrophilic surface (glass side surface) is placed toward the high pressure side in permeation cell, while benzene vapor permeates in an opposite manner. In the case of the polystyrene membrane containing no graft copolymer, there was no difference in the gas permeability between the directions of which surface was placed toward the high pressure side. Further more, there was no difference in the sorption isotherm of benzene between the membrane containing the graft copolymer and that of polystyrene alone.

These results are readily understood in terms of the asymmetric nature of the membrane as mentioned above [138]. The asymmetric nature of the membrane is demonstrated in **Figure 31** in which the asymmetric membrane potential $\Delta(\Delta\Psi)$ for the membrane and polystyrene membrane is plotted against KCl concentration. $\Delta(\Delta\Psi)$ of the surface modified membrane is much greater than that of polystyrene membrane. The same surface effect is reported on the permeability of oxygen through PMMA membrane containing the polydimethylsiloxane graft copolymer [139], where the specific interaction of oxygen molecule and graft copolymer segments at surface might be important to enhance the oxygen permeability.

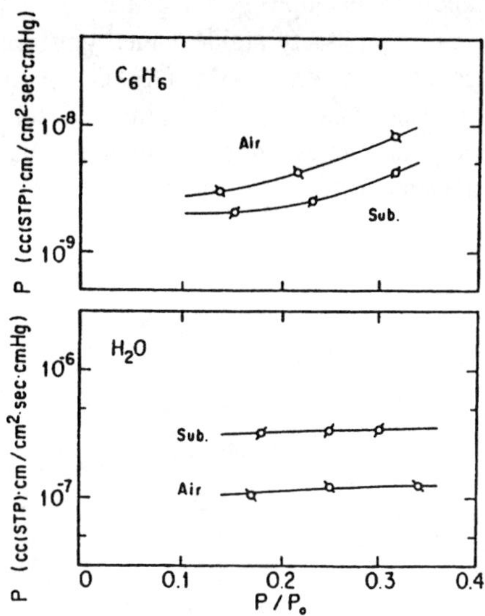


**Figure 30** *Gas permeation behavior through the polystyrene membrane containing 1wt% amount of the poly(methacrylic acid-co-(methyl methacrylate)-g- (MA-PSt12400)).*

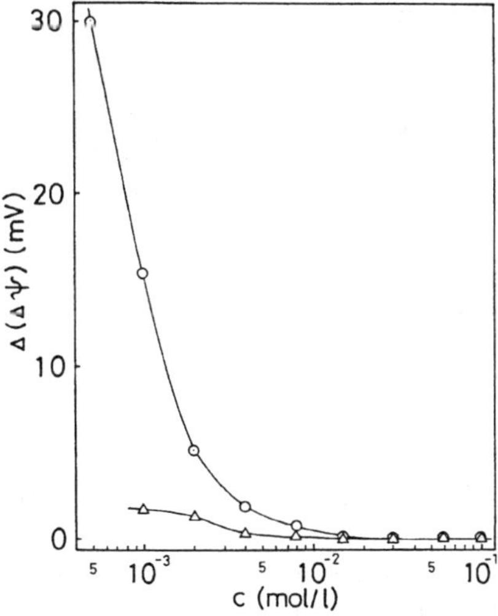


**Figure 31** *Asymmetric membrane potential* **Δ(ΔΨ)** *versus KCl concentration C for the membrane used in Figure 30 (O) and polystyrene membrane (Δ ).*

The effect of addition of the graft copolymer on the interaction between red cells and the polymer plate is another example [140, 141]. The polymer plates of PMMA containing 1 or 5 wt% of poly( 2-acrylamide-2-methylpropanesulfonic acid-g-MMA) (GMS) **36** and poly(diacetone acrylamide-g-MMA) (GMD) **37** cast from DMF solution are immersed into the aqueous solution containing red cells and blood plasma for a fixed time, then, the polymer plates adhered by red cells are obtained. Then, the polymer plate is placed in the rheoscope and the shear stress is loaded as shown in **Figure 32**.

$$-[CH_2-C(CH_3)(C(=O)-O-CH_3)]_x-[CH_2-CH(C(=O)-NH-C(CH_3)_2-CH_2-SO_3-H)]_y-[CH_2-C(CH_3)(C(=O)-O-CH_2-CH(OH)-CH_2-O-C(=O)-CH_2-S(CH_2-C(CH_3)(C(=O)-O-CH_3))_n-H)]_z-$$ **36**

$$-[CH_2-C(CH_3)(C(=O)-O-CH_3)]_x-[CH_2-CH(C(=O)-NH-C(CH_3)_2-CH_2-C(=O)-CH_3)]_y-[CH_2-C(CH_3)(C(=O)-O-CH_2-CH(OH)-CH_2-O-C(=O)-CH_2-S(CH_2-C(CH_3)(C(=O)-O-CH_3))_n-H)]_z-$$ **37**

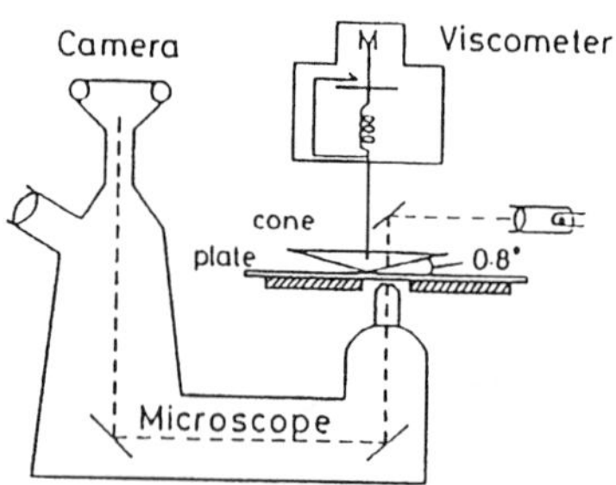


**Figure 32** *Schematic diagram of the rheoscope.*

The degree of deformation and the change of the number of red cells on the plate surface are recorded by taking photographs as a function of the shear stress. Informations about the interaction of red cells and the polymer surface can be obtained from the results. The typical plots of the number of remained red cells on the surface versus shear stress are shown in **Figure 33** [141]. It is seen that the number of the red cell on PMMA plates is clearly influenced by the addition of the graft copolymer. However, it should be noted that the incubation time and the concentration of albumin and fibrinogen in the plasma also influence on the interaction between the red cells and the polymer plates.

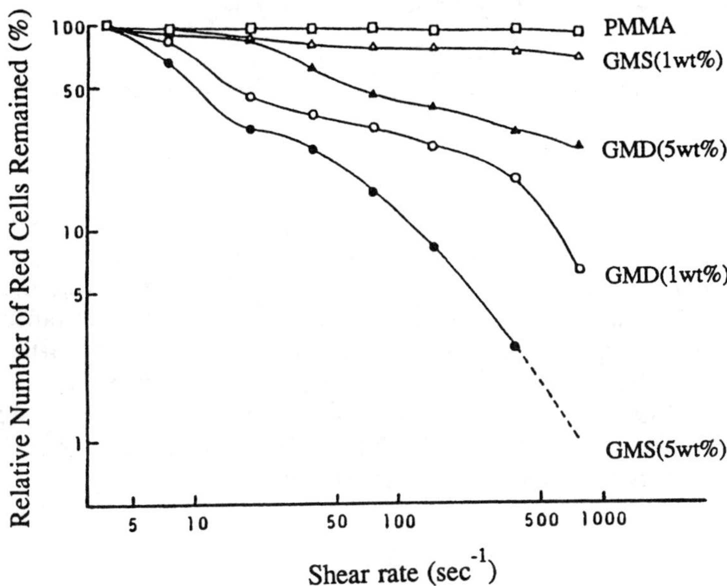


<u>**Figure 33**</u> *Dependence of the number of red cells remained on the sample surface on the shear rate measured by the rheoscope.*

The last example is the stabilization of polymer surface by the surface active polymeric UV-stabilizers. It is important to develop the UV-stabilized polymeric materials in the outer space where the strong UV light damages organic materials severely. The surface active polymeric UV-stabilizers **38** and **39** are prepared by radical copolymerization of dimethylsiloxane macromonomer and PSt macromonomer with 2-(2-hydroxy-2-vinylphenyl)-2H-benzotriazole and perfluoroalkylethyl acrylate (FA) [142]. The UV-stabilization of PSt films by the graft copolymers **38** and **39** is measured by the amount of the carbonyl groups produced by photo-oxidation. Addition of small amounts of surface active polymeric UV-stabilizer **38** and **39** provide more effective UV-stabilization to PSt than the polymeric UV-stabilizer having no surface active segments.

$-[CH-CH_2]_x-[CH-CH_2]_y-[CH-CH_2]_z-$ with side groups $C(=O)-O-CH_2CH_2(CF_2)_6CF_3$; OH, N, N, N (benzotriazole); $CH_2-[CH-CH_2]_n-$Sec-Bu **38**

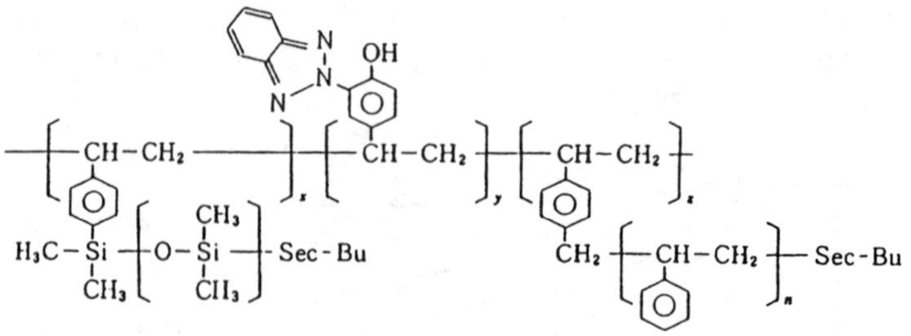

39

There are also many other applications of the surface active graft copolymers; such as, compatibilizers for polymer blends, dispersants for dispersion polymerization, improvers of paint, and so on wherever surface or interface is concerned. For applications like a membrane, homogeneous coverage of surface modified layer is required, while for applications to provide the biocompatibility or anti-static surfaces, homogeneous surface coverage is not necessarily required. The surface structure of surface accumulated graft copolymers might be related to the spreading coefficient of the graft copolymer on the surface. Further studies on the surface structure will be nessesary to understand more on these applications.

## 5. CONCLUSION

Macromonomers can be considered as a useful building block in developing new polymeric materials. One can use an oligomer or polymer as a reaction unit or a functional unit by using macromonomers in addition to the small monomer as the basic unit. This might provide wide variety in materials design of polymers. For this purpose, the understanding of the polymerization reactivity, charateristics, and properties of macromonomers is of fundamental and great importance.

**Acknowledgement**

The author is deeply indebted to Prof. Emeritus Y. Yamashita, Nagoya University, for his continuous suggestion and encouragement throughout this study. He also wishes to express his thanks to Mr. K. Tsutsumi, Dr. H. Ohtani, Prof. Y. Okamoto, and Prof. S. Tsuge, Nagoya University, and Prof. S. Teramachi, Kogakuin University, for their collaboration on the works in this chapter.

## REFERENCES

1. Y. Yamashita, *"Chemistry and Industry of Macromonomers"*, IPC, Tokyo (1989).
2. R. J. Milkovich and M. T. Chiang, *U. S. Patents*, 3,786,116 (1974); R. J. Milkovich, *ACS Polym. Prepr.*, 21, 40 (1980).
3. P. Rempp and E. Franta, *Adv. Polym. Sci.*, 58, 1 (1984).

4. Y. Chujo and Y. Yamashita, in *"Telechelic Polymers: Synthesis and Applications"*, Ed., by E. J. Goethals, CRC Press, Boca Raton (1989), Chap.8.
5. K. Ito, *Nippon Setchaku Kyoukai-Shi*, 24, 374 (1988); *Koubunshi Kakou . (Polymer Applications),* 35, 262 (1986).
6. R. Asami, *Koubunshi Kakou (Polymer Applications)*, 34, 439 (1986).
7. Y. Tsukahara, *Nippon Gomu Kyokaishi*, 66, 562 (1992).
8. J. P. Kennedy and B. Ivan, *"CarbocationicMacromolecular Engineering: Thoery and Practice"*,Hanser Publishers,Munich (1991).
9. S. Kohjiya and Y. Tsukahara, *Nippon Gomu Kyokaishi*, 66, 635 (1992).
10. E. J. Goethals, Ed., *"Telechelic Polymers: Synthesis and Applications"*, CRC Press, Boca Raton (1989).
11. P. Masson, E. Franta and P. Rempp, *Makromol. Chem., Rapid Commun.*, 3, 499 (1983).
12. P. Rempp, P. Lutz, P. Masson and E. Franta, *Makromol. Chem., Suppl.*, 8, 3 (1984).
13. R. Asami, M. Takaki and H. Hatanaka, *Macromolecules*, 16, 628 (1986).
14. Y. Tsukahara, C. H. Tsai, Y. Yamashita and Y. Muroga, *Polym. J.*, 19, 1033 (1987).
15. C. H. Tsai, Y. Tsukahara and Y. Yamashita, *Kobunshi Ronbunshu*, 45, 277 (1988).
16. R. L. Morton and T. J. McCarthy, *Am. Chem. Soc., Polym. Prepr.*, 28(1), 174 (1987).
17. Y. Nagasaki and T. Tsuruta, *Makromol. Chem.*, 187, 1583 (1986).
18. T. Otsu, K. Yamashita and K. Tsuda, *Macromolecules*, 19, 2219 (1986).
19. P. Ragunath Rao, P. Lutz, J. Lamps, P. Masson and P. Rempp, *Polym.Bull.*, 15, 69 (1986).
20. J. P. Kennedy and M. Hiza, *J. Polym. Sci., Polym. Chem. Ed.*, 21, 1033 (1983).
21. M. Takaki, R. Asami and T. Kuwabara, *Polym. Bull.*, 7, 521 (1982).
22. K. Ito, H. Tsuchida, A. Hayashi, T. Kitano, E. Yamada and T. Matsumoto, *Polym. J.*, 17, 827 (1985).
23. Y. Kawakami, Y. Miki, T. Tsuda, R. A. N. Murthy and Y. Yamashita, *Polym. J.*, 14, 913 (1982).
24. M. Akashi, I. Kirihara and N. Miyauchi, *Angew. Makromol. Chem.*, 132, 81 (1985).
25. J. Oku, R. J. H. Chan, H. K. Hall, Jr. and O. R. Hughes, *Polym. Bull.*, 16, 481 (1986).
26. M. Niwa and N. Higashi, *Macromolecules,* 31, 1191 (1988).
27. Y. Gnanou and P. Rempp, *Makromol. Chem.*, 188, 2267 (1987).
28. K. E. J. Barrett, Ed., *"Dispersion Polymerization in Organic Media"*, John Wiley, New York (1985).
29. E. J. Goethals and M. A. Viegels, *Polym. Bull.*, 4, 521 (1981).
30. V. Percec, P. L. Rinaldi and B. C. Auman, *Polym. Bull.*, 10, 397 (1983).
31. B. Boutevin, Y. Pietrasanta, M. Taha and T. E. Sarraf, *Makromol. Chem.*, 184, 2401 (1983).

32. S. Kawakami, K. Hayashi and H. Matsuo, *Fluoro Chemistry, Prepr.*, 9, 72 (1983).
33. G. Moad, D. H. Solomon, S. R. Johns and R. I. Willing, *Macromolecules*, 17, 1094 (1984).
34. D. A. Cywar and D. A. Tirrell, *Macromolecules*, 19, 2908 (1986).
35. K. Hatada, T. Kitayama and E. Masuda, *Polym. J.*, 18, 395 (1986).
36. H. Ohtani, S. Ishiguro, M. Tanaka and S. Tsuge, *Polym. J.*, 21, 41 (1989).
37. H. Ohtani, M. Tanaka and S. Tsuge, *J. Anal. Appl. Pyrolysis*, 15, 167 (1989).
38. C. Watanabe, K. Teraishi, S. Tsuge, H. Ohtani and K. Hashimoto, *J. High Res. Chromatogr.*, 14, 269 (1991).
39. H. Ohtani, S. Ueda, S. Tsuge and Y. Tsukahara, *J. Anal. Appl. Pyrolysis*, in press.
40. Y. Tsukahara, K. Tsutsumi, Y. Okamoto, H. Ohtani and S. Tsuge, *Polym. Prepr. Japan,* 40, 204 (1991).
41. Y. Tsukahara, K. Tsutsumi, Y. Okamoto, H. Ohtani and S. Tsuge, in preparation.
42. W. Heitz, in *"Telechelic Polymers: Synthesis and Applications"*, Ed., by E. J. Goethals, CRC Press, Boca Raton (1989), Chap.4.
43. K. Ito, N. Usami and Y. Yamashita, *Macromolecules*, 13, 216 (1980).
44. Y. Yamashita, Y. Tsukahara, K. Ito, K. Okada and Y. Tajima, *Polym. Bull.*, 5, 335 (1981).
45. Y. Tsukahara, Y. Nakanishi, Y. Yamashita, H. Ohtani, Y. Nakashima, Y. F. Luo, T. Ando and S. Tsuge, *Macromolecules*, 24, 2493 (1991).
46. H. Ohtani, Y. Nakashima, Y. F. Luo, S. Tsuge and Y. Tsukahara, in preparation.
47. G. Odian, *"Principles of Polymerization, 2nd ed."*, John Wiley, New York (1981).
48. Y. Tsukahara, K. Mizuno, A. Segawa and Y. Yamashita, *Polym. Prepr. Jpn.,* 37 1622 (1988).
49. Y. Tsukahara, K. Mizuno, A. Segawa and Y. Yamashita, *Macromolecules*, 22, 1564 (1989).
50. Y. Tsukahara, K. Tsutsumi, Y. Yamashita and S. Shimada, *Macromolecules*, 22, 2869 (1989).
51. Y. Tsukahara, K. Tsutsumi, Y. Yamashita and S. Shimada, *Macromolecules*. 23. 5201 (1990).
52. R. Asami, M. Takaki, K. Kyuda and E. Asakura, *Polym. J.*, 15, 139 (1983).
53. R. Asami and M. Takaki, *Macromol, Chem., Suppl.*, 12, 163 (1985).
54. P. Masson, E. Franta and P. Rempp, *Makromol. Chem., Rapid. Commun.*, 3, 499 (1982).
55. P. Rempp, P. Lutz, P. Masson, P. Chaumont and E. Franta, *Makromol. Chem., Suppl.*, 13, 47 (1985).
56. K. Tanaka, K. Ito, E. Yamada and T. Matsumoto, *Polym. Prepr. Jpn.*, 36, 1712 (1987).

57. H. Tanaka, K. Hashimura, K. Tanaka and K. Ito, *Polym. Prepr. Jpn.*, 37, 1424 (1988).
58. R. Asami, N. Sukenaga, H. Ogawa and M. Takaki, *Polym. Prepr. Jpn.*, 31, 203, 1153 (1982).
59. R. Asami, M. Takaki and Y. Moriyama, *Polym. Bull.*, 16, 125 (1986).
60. K. Hatada, T. Shinozaki, K. Ute and T. Kitayama, *Polym. Bull.*, 19, 231 (1988).
61. K. Hashimoto, H. Sumitomo and M. Kawasumi, *Polym. J.*, 17, 1045 (1985).
62. I. Mita and K. Horie, *J. Macromol. Sci., Rev.*, C27, 91 (1987).
63. G. T. Russell, D. H. Napper and R. G. Gilbert, *Macromolecules,* 21, 2133, 2141 (1988).
64. E. Trommsdorf, H. Kohle and P. Lagally, *Makromol. Chem.*, 1, 169 (1948).
65. M. Tirrell, *Rubber Chem. Tech.*, 57, 523 (1984).
66. T. Tulig and M. Tirrell, *Macromolecules*, 14, 1501 (1981).
67. K. Ito, K. Tanaka, H. Tanaka, G. Imai, S. Kawaguchi and S. Itsuno, *Macromolecules*, 24, 2348 (1991).
68. K. Ito, K. Hashimura, S. Itsuno, E. Yamada, *Macromolecules*, 24, 3977 (1991).
69. B. Ranby, J. F. Rabek, "*ESR Spectroscopy in Polymer Research*", Springer-Verlag, Berlin (1977).
70. M. J. Ballard, R. G. Gilbert, D. H. Napper, P. J. Pomery and J. H. O'Donnell, *Macromolecules*, 17, 504 (1984).
71. S. E. Bresler, E. N. Kazbekov and V. N. Shadrin, *Makromol. Chem.*, 175, 2875 (1974).
72. M. Kamachi, *Adv. Polym. Sci.*, 82, 207 (1987).
73. J. Brandrup and E. H. Immergut, ed., "*Polymer Handbook*", 3rd ed., John Wiley & Sons, New York (1989).
74. K. Hatada, T. Kitayama, E. Masuda and M. Kamachi, *Makromol. Chem., Rapid Commun.*, 11, 101 (1990).
75. I. Capek, M. Akashi, to be appeared in *J. Macromol. Sci. Rev.*
76. L. J. Mathias, J. B. Cantaerberry and M. South, *J. Polym. Sci., Polym. Lett., Ed.*, 20, 473 (1982).
77. G. F. Meijs and E. Rizzardo, *J. Macromol. Sci., Macromol. Chem. Phys.*, C30(3&4), 305 (1990).
78. Y. Gnanou and P. Lutz, *Makromol. Chem.*, 190, 577 (1989).
79. F. R. Mayo and F. M. Lewis, *J. Am. Chem. Soc.*, 66, 1594 (1944).
80. M. Fineman and S. D. Ross, *J. Polym. Sci.*, 5, 332 (1981).
81. T. Kelen and F. Tudos, *J. Macromol. Sci. Chem.*, 9, 1 (1975).
82. I. Skeist, *J. Am. Chem. Soc.*, 68, 1781 (1946).
83. I. H. Spinner, B. C. Lu and W. F. Gaydon, *J. Am. Chem. Soc.*, 77, 2198 (1955).
84. V. E. Meyer and G. G. Lowry, *J. Polym. Sci.*, A3, 2843 (1965).
85. G. O. Schulz, R. J. Milkovich, *J. Polym. Sci., Polym. Chem. Ed.*, 22, 1633 (1984).
86. J. P. Kennedy and C. Y. Lo, *Polym. Bull.*, 8, 63 (1982).
87. J. P. Kennedy and C. Y. Lo, *ibid.*, 13, 343 (1985).
88. M. F. Farona and J. P. Kennedy, *Polym. Bull.*, 11, 359 (1984).

89. K. Ito, S. Yokoyama, F. Arakawa, Y. Yukawa, T. Iwashita and Y. Yamasaki, *Polym. Bull.*, 16, 337 (1986).
90. K. Ito, H. Tsuchida and T. Kitano, *Polym. Bull.*, 15, 425 (1986).
91. G. G. Cameron and M. S. Chisholm, *Polymer*, 26, 437 (1985).
92. G. G. Cameron and M. S. Chisholm, *Polymer*, 27, 1420 (1986).
93. Y. Tsukahara, M. Tanaka and Y. Yamashita, *Polym. J.*, 19, 1121 (1987).
94. Y. Tsukahara, H. Hayashi, X. L. Jiang and Y. Yamashita, *Polym. J.*, 21, 377 (1989).
95. F. Severini, M. Pegoraro and L. Saija, *Angew. Makromol. Chem.*, 133, 111 (1985).
96. K. Muhlbach and V. Percec, *J. Polym. Sci., Polym. Chem. Ed.*, 25, 2605 (1987).
97. T. Nishimura and T. Tsuruta, *Makromol. Chem., Rapid. Commun.*, 1, 573 (1980); *Makromol, Chem.*, 183, 29 (1982).
98. Y. Nabeshima and T. Tsuruta, *Polym. Prepr. Jpn.*, 36, 1496 (1987).
99. M. Arnold, W. Frank and G. Reinhold, *Polym. Bull.*, 24, 1 (1990); *Makromol. Chem.*, 192, 285 (1991).
100. V. Percec and J. H. Wang, *J. Polym. Sci., Polym. Chem., Ed.*, 28, 1059 (1990).
101. V. Percec, U. Epple, J. H. Wang and H. A. Schneider, *Polym. Bull.*, 23, 19 (1990).
102. J. P. Flory, "*Principles of Polymer Chemistry*", Cornell University Press, Ithaca, New York, 1953, Chap. 3.
103. S. Yamashita, K. Shibatani, K. Takakura and S. Imai, *Kobunsi Ronbunshu*, 39, 187 (1982).
104. K. Ito, Y. Masuda, T. Shintani, T. Kitano and Y. Yamashita, *Polym. J.*, 15, 433 (1983).
105. M. Niwa, M. Akahori and S. Nishizawa, *J. Macromol. Sci. Chem.*, A24, 1423 (1987).
106. S. Teramachi, A. Hasegawa, T. Matsumoto, Y. Tsukahara and Y. Yamashita, *Polym. Prepr. Japan*, 39, 4058 (1990).
107. Y. Tsukahara, Y. Okazawa, Y. Okamoto, S. Teramachi and Y. Yamashita, *Polym. Prepr. Japan*, 40, 182 (1991).
108. S. Teramachi, A. Hasegawa, T. Matsumoto, K. Kitahara, Y. Tsukahara, and Y. Yamashita, *Macromolecules*, 25, 4025 (1992).
109. J. Stejskal, D. Strakova, P. Kratochvil, S. D. Smith and J. E. McGrath, *ACS Polym. Prepr.*, 29, 116 (1988); *Macromolecules*, 22, 861 (1989).
110. R. Duran and P. Gramain, *Makromol. Chem.*, 188, 2001 (1987).
111. Y. Ohkatsu, M. Yokota and T. Kusano, *Makromol. Chem.*, 189, 755 (1988).
112. Y. Tsukahara, K. Tsutsumi, Y. Okamoto and S. Kojiya, *4th-International Polymer Conference*, Yokohama, p.195 (1992).
113. Y. Tsukahara, J. Inoue, S. Kojiya, K. Tsutsumi and Y. Okamoto, *Polym. Prepr. Japan*, 41, 4433 (1992).
114. Y. Tsukahara, K. Tsutsumi and Y. Okamoto, *Makromol. Chem., Rapid Commun.*, 13, 409 (1992).
115. H. Fujita, "*Polymer Solutions*", Elsevier, Amsterdam (1990).
116. G. C. Berry, *J. Polym. Sci.*, A2, 6, 1551 (1968).

117. I. Noda, T. Horikawa, T. Kato, T. Fujimoto and M. Nagasawa, *Macromolecules*, 3, 795 (1970).
118. G. C. Berry and T. A. Orofino, *J. Chem. Phys.*, 40, 1614 (1964).
119. H. Yamakawa, *"Modern Theory of Polymer Solutions"*, Harper and Row, New York, 1971.
120. I. Noda, T. Horikawa, T. Kato, T. Fujimoto and M. Nagasawa, *Macromolecules*, 3, 795 (1970).
121. P. Lang, W. Burchard, M. S. Wolfe, H. J. Spinelli and L. Page, *Macromolecules*, 24, 1306 (1991).
122. T. G. Fox and P. J. Flory, *J. Appl. Phys.*, 21, 581 (1950).
123. A. N. Gent and A. G. Thomas, *J. Polym. Sci., Polym. Phys. Ed.*, 10, 5201 (1972).
124. B. H. Bersted, *J. Appl. Polym. Sci.*, 24, 37 (1979).
125. D. T. Turner, *Polymer*, 23, 626 (1982).
126. W. W. Graessley, *Adv. Polym. Sci.*, 16, 1 (1974).
127. P. Prentice, *Polymer*, 24, 344 (1983).
128. S. Onogi, T. Masuda and K. Kitagawa, *Macromolecules*, 3, 107 (1970).
129. G. C. Berry and T. G. Fox, *Adv. Polym. Sci.*, 5, 261 (1967).
130. M. F. Refojo, *J. Appl. Polym. Sci.*, 9, 3163 (1965).
131. M. F. Refojo and H. Yasuda, *J. Appl. Polym. Sci.*, 9, 2425 (1965).
132. M. Macret and G. Hild, *Polymer*, 23, 81, 748 (1982).
133. J. D. Andrade, Ed., *"Hydrogelsfor Medical and Related Applications", Am. Chem. Soc., Symp. Ser.*, 31 (1976).
134. O. Kramer, Ed., *"Biologicaland Synthetic Polymer Networks"*, Elsevier Applied Science, London (1988).
135. H. Marushige, *"Taiden Boushizai"*, Saiwai Shobou, Tokyo (1968).
136. H. Morawetz, J. R. Cho and P. J. Gans, *Macromolecules*, 6, 624 (1973).
137. Y. Yamashita, Y. Tsukahara, in *"Modification of Polymers,"* C. Carraher and J. Moore, ed., Plenum Press, New York, (1983), p.131; Y. Tsukahara, K. Kohno, H. Inoue and Y. Yamashita, *Nippon Kagaku Kaishi*, (1985), 1070.
138. T. Kinoshita, A. Takizawa, Y. Tsujita, Y. Tsukahara and Y. Yamashita, *Kobunshi Ronbunshu*, 40, 493 (1983).
139. Y. Kawakami, T. Aoki and Y. Yamashita, *Polym. J.*, 18, 243 (1986).
140. T. Shiga, *Nippon Seirishi*, 44, 187 (1982).
141. T. Nakazaki, Y. Chujo, Y. Tsukahara, Y. Yamashita and T. Shiga, *Polym. Prep. Jpn.*, 33, 2103 (1984).
142. Y. Tsukahara, Y. Tsuruta, K. Kohno and Y. Yamashita, *Kobunshi Ronbunshu*, 47, 361 (1990).

# Macroinitiators Chapter 6

## MACROINITIATORS FOR CHAIN POLYMERIZATION

**Yusuf Yagci** and **Munmaya K. Mishra**

Department of Chemistry
Istanbul Technical University
Maslak, Istanbul - 80626
TURKEY

Polymers Research
Texaco Research Center
P.O.Box 509
Beacon, NY 12508, USA





Macromolecular Design: Concepts and Practice (Macromonomers, Macroinitiators, Macroiniferters, Macroinimers, Macroiniters, and Macroinifers) Pages 229 - 264

# 1. INTRODUCTION

In recent years macromolecular engineering has played a major role in designing tailor-made block and graft copolymers. The tailor-made polymers containing hetero-chains often exhibit unique properties which would not be otherwise shown by homopolymers. The precise control of various parameters with respect to structure, molecular weight, degree of polymerization, and uniformity have been achieved using various synthetic techniques. Some of the details are described in preceeding and following chapters.

Novel polymeric materials exhibiting specific types of performance are required for "high tech" applications, since available materials generally don't meet the requirements. Macromolecular targets reflecting improved properties can be achieved by several pathways: sequential addition of monomers to a living polymerization system; reaction between telechelic polymers; multi-mode polymerizations; and the use of macromonomers, macroiniferters, and macroinitiators, etc. Each of these methods has advantages and limitations. This chapter will describe some of the techniques of macroinitiation.

Let us first describe the term "macroinitiator." Macroinitiator is an abbreviation of macromolecular initiator and is based on the functional feature of the macromolecule. The functional feature of the macromolecule could be derived from deliberatly introduced groups that are capable of initiating chain polymerization or inherited during the synthesis of the original macromolecule. Selected functional groups of the macromolecule may be activated chemically, thermally, mechanochemically and by UV irradiation to yield active centers; i.e., radicals, cations or anions that initiate polymerization. The history of macroinitiators is about a century old. Polymeric compounds having an intramolecular peroxy group has been described in the literature as early as 1894 [1]. However, the concept of macroinitiation and the application of macroinitiators can be dated back to almost half a century. In 1954 Smets and Woodward [2,3] used the concept of macroinitiation and prepared AB type block copolymer using macroperoxy initiators. Polymer containing labile azo groups was described by Hill [4] as early as 1951. Similarly the concept of macroinitiators has been reported in living anionic [5,6] and cationic ring opening polymerizations [7-9] during the 50's and 60's respectively.

Macroinitiators are precursors for the synthesis of block and graft copolymers depending on the site of initiation, i.e., at the chain end or in the side chain, respectively. The general reaction steps are depicted in **Scheme 1**.

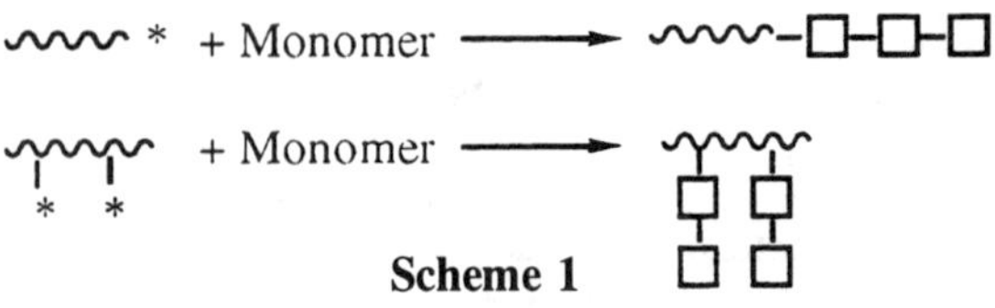


**Scheme 1**

The advantage of the macroinitiator technique is that the one component can be fully characterized prior to the preparation of block and graft copolymers. As expected, the final properties of block and graft copolymers generated by the use of macroinitiators depend on several factors: reactivity of functional groups to create initiating sites, position and number of functional groups, type of polymerization, termination mode of the polymerization, molecular weight of the macroinitiator, and the extent of molecular weight change during formation of the initiating sites.

Besides there use as precursors for block and graft copolymers, macroinitiators are often employed as conventional initiators. The potential advantages of macroinitiators over low molar mass initiators can be summarized as follows. First, low molar mass initiators suffer from high volatility. Secondly, initiation products which are not consumed during polymerization may cause some undesired properties in the final polymer such as undesired odor. However, the use of macroinitiators avoids these drawbacks. Macroinitiators will be discussed in the following sections according to the classical approach based on the type of active centers formed.

## 2. MACROINITIATORS FOR FREE RADICAL POLYMERIZATION

There have been various drawbacks which often retarded the practical application of macroinitiators for the synthesis of block and graft copolymers. These drawbacks essentially relate to homopolymerization, which generally proceeds in parallel with block and graft copolymerization, if these processess occur according to free radical mechanism. Therefore, the products should be purified by careful separation before characterization. On the other hand homopolymerization does not usually occur if macroinitiators function ionically. From the practical point of view, however, ionic macroinitiators are less attractive than radical macroinitiators, because radical polymerizations offer a wider choice of monomer combinations.

Block and graft copolymers resulting from free radical macroinitiators have an overall structure which depends on the mode of termination by either disproportionation and/or combination of the second monomer (**Scheme 2**).

### 2.1 Coordination-Type Macroinitiators

Polychelates which contain non-bonding atoms such as nitrogen and oxygen and complexing centers can act as macroinitiators for vinyl polymerization. Some natural and synthetic macromolecules or oligomers, such as starch, cellulose, silk, poly(vinyl alcohol), etc., initiate polymerization of vinyl monomers [10-16]. In these polymerizations, methyl methacrylate (MMA) was more reactive than any other vinyl monomer. Moreover, a slight amount of metallic ions, particularly cupric ions, was found [17] to enhance the reaction. For instance, polyamides such as nylon 3 or nylon 6 form bonds between the copper (II) ion, MMA monomer, water and non-bonding oxygen atom present in the polyamide (**Scheme 3**).

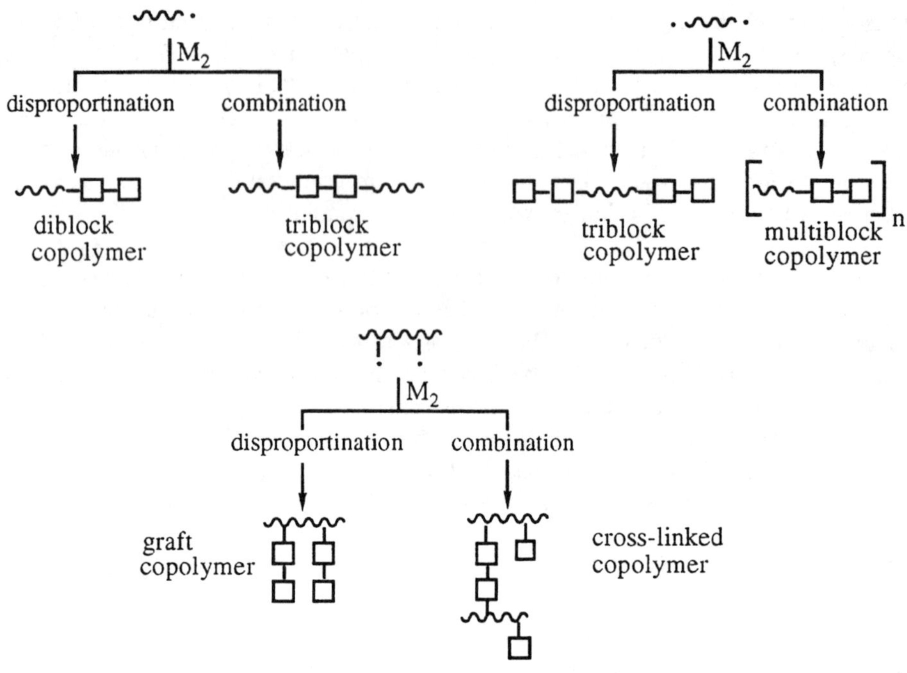


**Scheme 2**

**Scheme 3**

The rate of polymerization increased [18] with the molecular weight of the polyamide used in the system. This was attributed to the better complexing property of shorter chains than the longer ones. The former fills three or four coordination sites of copper (II) ion as shown in **Scheme 4**. It is, therefore, difficult to keep two vacant sites for MMA and water which is a crucial requirement for a successful polymerization.

When polyamides were used as coordination-type macroinitiators for the polymerization of MMA, the product contained up to about 50 % graft copolymer.

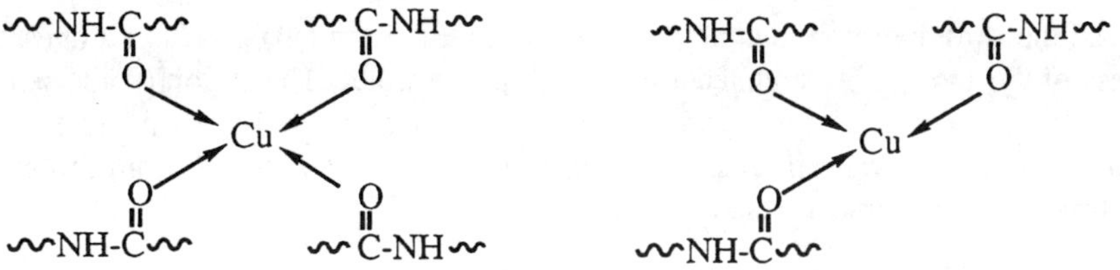

Scheme 4

Notably, the efficiency of grafting and the degree of MMA homopolymer did not depend on the chain length of the polyamide. Moreover, addition of urea increased [19,20] both the rate of polymerization and the efficiency of grafting, which was explained on the basis of conformational changes of the polyamide. Urea favours unfolding of the polyamide chain, hence grafting becomes more feasible.

Apart from the above polymers, anion-type macromolecular electrolytes, such as sodium salts of poly(alkene sulphonates) [21,22], poly(styrene sulphonate) [23], poly(ethylene phosphate) [24], poly(vinyl phenolate) [25], and copolymers of acrylates [26] as well as cation-type macromolecular electrolytes such as poly(vinyl benzyltrimethyl ammonium choloride) [27] initiate the polymerizations of vinyl monomers. In this case, the initiation occurs by hydrogen transfer from free monomer to the coordinated monomer. Typical mechanism for the poly(vinylbenzyl trimethyl ammonium choloride) / MMA system is depicted in **Scheme 5**.

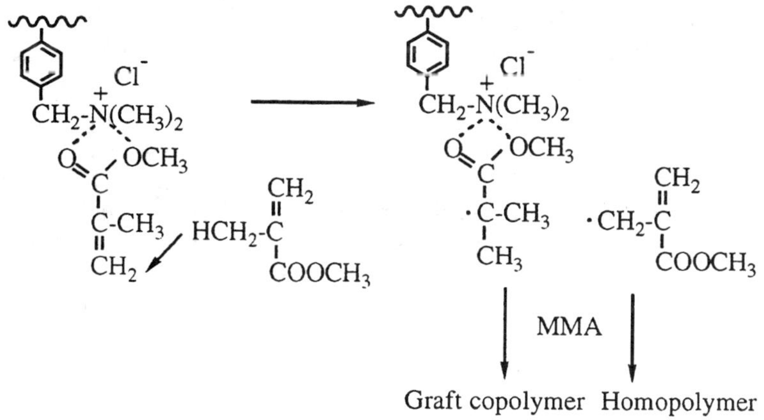

Scheme 5

In the case of styrene monomer, however, hydrogen abstraction occurs between $\pi$-complexed-monomer and the free monomer (**Scheme 6**).

Interestingly, styrene was found to be the most reactive monomer and acrylonitrile did not polymerize using this initiating system. The selectivity of vinyl monomers were explained [28] by the concept of hard and soft hyrdophobic areas and monomer. According to this concept the hard hydrophobic area incorporate the hard monomers, whereas the soft hydrophobic area prefers the soft monomers. The hardness and softness of the monomer is related to its solubility in water. The cationic-type polymer is considered to form hard hydrophobic areas since the polymer possesses 30% non-ammonium type repeating units. A good correlation between the conversions and solubility of the monomers was observed.

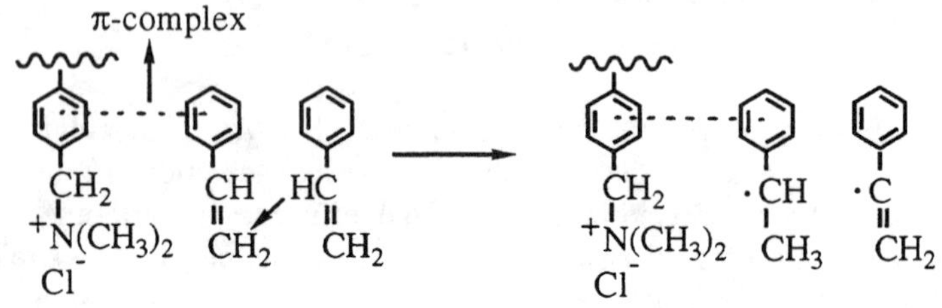

π-complex

$CH_2$ $^+N(CH_3)_2$ $Cl^-$ CH=$CH_2$ HC=$CH_2$ ⟶ $CH_2$ $^+N(CH_3)_2$ $Cl^-$ ·CH–$CH_3$ ·C=$CH_2$

Scheme 6

## 2.2 Mechano-Chemical Macroinitiators

Mechano-chemical macroinitiators are formed from the polymers which produce macroradicals when subjected to mechanical treatment. Unlike the other type of macroinitiators they function in their original structure and don't require incorporation of functional low molecular weight compounds or additives. In principle, most of the synthetic and natural polymers can act as initiators under mechano-chemical influence to yield block and graft copolymers. These processeses can be performed in standard equipment with low cost. Countering these advantages is the rather heterogeneous product formation from the chemical point of view. Mechano-chemical scission may occur at any position of the polymer. It is frequently difficult to produce pure block copolymers. Quite often graft copolymers are also produced as a result of formation of lateral macroradicals during secondary reactions. The homopolymerization also occurs via hydrogen transfer between macroradicals and monomer. The end products consist of a mixture of block, graft and homopolymers. Mechano-chemically induced block and graft copolymerization involves the following general steps (**Scheme 7**):

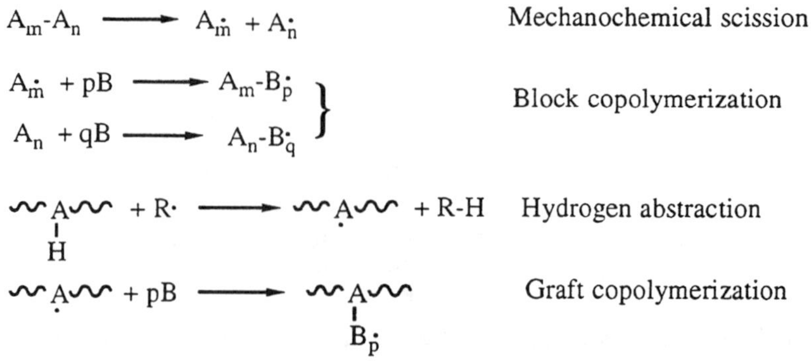

$A_m$-$A_n$ ⟶ $A_m^{\bullet}$ + $A_n^{\bullet}$ Mechanochemical scission

$A_m^{\bullet}$ + pB ⟶ $A_m$-$B_p^{\bullet}$
$A_n$ + qB ⟶ $A_n$-$B_q^{\bullet}$ Block copolymerization

~A(H)~ + R· ⟶ ~$\dot{A}$~ + R-H Hydrogen abstraction

~$\dot{A}$~ + pB ⟶ ~A($B_p^{\bullet}$)~ Graft copolymerization

Scheme 7

Many investigations on the synthesis of block and graft copolymers using natural polymers [29, 30], vinyl polymers [31-40] and heterochain polymers [41-50] as mechanochemical initiators have been performed. Although the mode of macroradical generation is the same in each case, the structure and properties of the copolymers obtained depends on the properties of the initial polymers and monomers, and the reaction conditions. By using various combinations and altering conditions, it is possible to regulate the properties and composition of the resulting products. For further information the readers may refer to the specialized books [51,52].

More recently, Oprea and Popa [53] reported the polymerization of methyl methacrylate initiated by mechano-chemically synthesized macroinitiators, i.e. acrylonitrile-based homopolymers and copolymers in water or dimethyl formamide (DMF). The polymers synthesized mechano-chemically exhibit a different structure than those obtained by classical methods. They bound metallic atoms either covalently or by coordination. Therefore, the initiation was presumed to occur via formation of a transition complex between metallic atom, trunk polymer, monomer and solvent (**Scheme 8**) in a similar manner to that described previously.

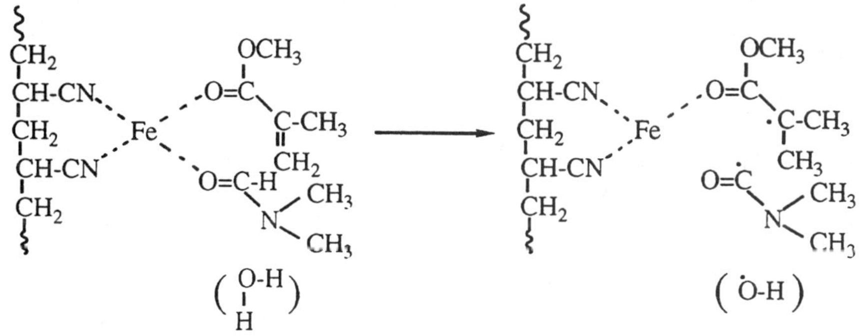

**Scheme 8**

The radicals formed from monomer are responsible for grafting whereas radicals formed from solvent (water or DMF) results in the formation of homopolymers. High-speed stirring or ultrasonic irradiation is another way to produce macroradicals for the synthesis of block and graft copolymers. Ultrasonic degradation of polyacrylamide, poly(vinylacetate), poly(methyl methactylate) and polystyrene solutions containing a second monomer has been demonstrated. It should be pointed out that "freeze-thaw",vapour-phase swelling"and "spark discharge"techniques, which also based on the macroradical generation, allow production of block copolymers.

## 2.3 Macrophotoinitiators

There has been much interest in macromolecules containing covalently bonded photoinitiating groups (macrophotoinitiators). Macrophotoinitiators may offer various advantages over the low molecular weight photoinitiators such as greater reactivity, low volatility and low migration. From the practical point of view, macrophotoinitiators are used as either photocrosslinking agents or precursors for block and graft copolymers. Macrophotoinitiators as curing agents have already been surveyed in two recent review articles [54,55] and are summarized in **Table 1**.

**Table 1**. Polymeric Photoinitiators (PPI) For UV Curing Applications

| PPI | Photosensitive Group | Type | References |
|---|---|---|---|
| Polybenzoin | Benzil | a, c | [56] |
| Poly(benzilketal) | Benzilketal | a, c | [57] |
| Poly(trichloro acetophenone) | Acetophenone derivative | a, b, c | [58] |
| Poly(benzoin acrylates) | Benzoin or benzoinmethylether | b, c | [59-61] |
| Poly(hydroxy alkylphenones) | Hydroxyalkyl phenone | b, c | [62-65] |
| Poly(vinyl or acrloyl benzophenone) | Benzophenone | c, d | [66-72] |
| Poly(acryloyl thioxanthone) | Thioxanthone | c, d | [73, 74] |
| Polysiloxanes | Benzoin ether, benzilketal acetophenone derivative | b, c | [75-77] |
| Polysilanes | Si-Si bond | a, c ,e | [78-81] |
| Polymeric iodonium salt | Iodonium salt | b, e | [82] |

a = Main chain photoinitiator
b = Side chain photoinitiator
c = Cleavage-type photoinitiator
d = H-Abstraction-type photoinitiator
e = Also cationic photoinitiator

The principal focus of this part will be to describe the synthesis of novel well defined block and graft copolymers by using macrophotoinitiators. Compared with most of the thermally induced macroinitiators, the radiation sensitive macroinitiators have the advantage of being applicable at low temperatures especially at room temperature. This is of prime importance with regard to the probability of occurence of side reactions such as homopolymerization which proceeds in parallel with block or graft copolymerization, if macroinitiators promoted thermally. Moreover, because of the selective absorptivity of certain groups (chromophores), it is possible to produce reactive sites at definite positions in the macromolecule [83].

Macrophotoinitiators possessing chromophoric groups either in the main chain or as pendant groups can be prepared in two ways:

- by synthesis and polymerization of monomers with photoactive groups
- by introduction of photoactive group into a polymer.

The latter case may be achieved in initiation or termination steps of a particular polymerization or reacting functional groups of a preformed polymer with functional groups of low moleculer weight compounds possessing also photoactive groups.

Macrophotoinitiators, analogous to the low molecular weight photoinitiators, are divided into two classes, according to their radical generation mechanism namely cleavage-type (Type I), **(Scheme 9)** and hydrogen abstraction-type (Type II), **(Scheme 10)** macrophotoinitiators.

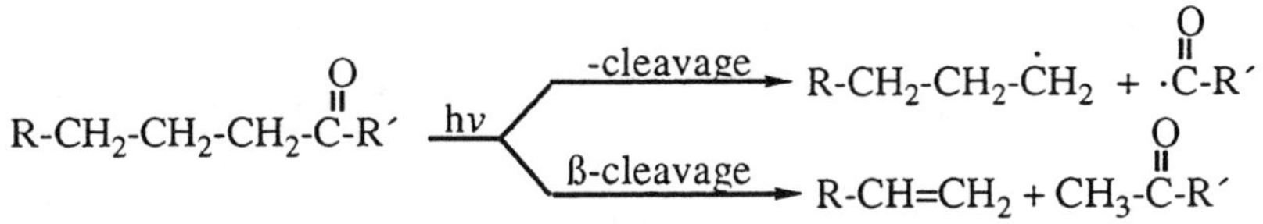

$$R\text{-}CH_2\text{-}CH_2\text{-}CH_2\text{-}\overset{O}{\overset{\|}{C}}\text{-}R' \xrightarrow{h\nu} \begin{cases} \xrightarrow{\text{-cleavage}} R\text{-}CH_2\text{-}CH_2\text{-}\dot{C}H_2 + \cdot\overset{O}{\overset{\|}{C}}\text{-}R' \\ \xrightarrow{\text{ß-cleavage}} R\text{-}CH{=}CH_2 + CH_3\text{-}\overset{O}{\overset{\|}{C}}\text{-}R' \end{cases}$$

**Scheme 9**

$$S \xrightarrow{h\nu} S^*$$

$$S^* + \underset{XH}{\sim\!\sim} \longrightarrow \dot{S}H + \underset{X\cdot}{\sim\!\sim}$$

$$\underset{X\cdot}{\sim\!\sim} + M \longrightarrow \underset{X\text{-}(M)_n\text{-}}{\sim\!\sim}$$

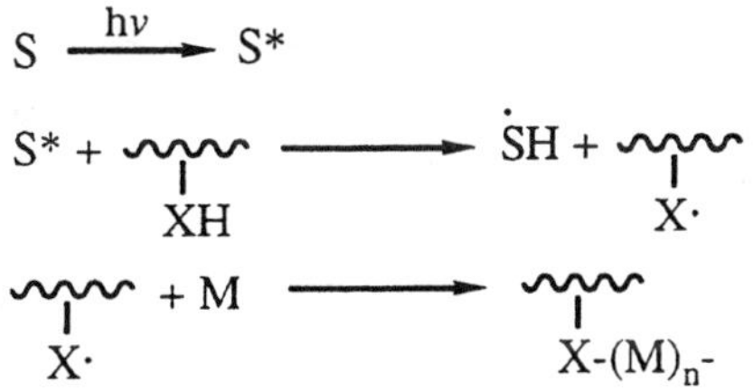

**Scheme 10**

$\alpha$-Cleavage is the most important reaction and leads to free radicals which are capable of initiating polymerization of appropriate monomers.

Type II macrophotoinitiators function in initiating process indirectly, as a result of hydrogen abstraction by electronically excited sensitizer molecules S* from lateral functional groups present in the macrophotoinitiator (**Scheme 10**). Obviously, when the hydrogen releasing groups are located at the chain ends, the resulting polymer will be a block copolymer.

### 2.3.1 Cleavage-type macrophotoinitiators

On the basis of $\alpha$-cleavage, Dhanras and Guillet [84] used poly(tetramethylene sebacate co-$\gamma$-ketopimalate) as a macrophotoinitiator according to the following reactions (**Scheme 11**). However, the resulting product was a mixture of unreacted polyester and block copolymer.

Kawai and coworkers [85-87] prepared similar carbonyl containing photoactive polystyrene and polyvinylchlorideby copolymerizing CO with corresponding monomers.

Irradiation of poly(styrene-co-carbonmonoxide) in the presence of methylmethacrylate resulted in a mixture containing up to 70 % by weight block copolymer according to fractionation, turbidity and elemental analysis (**Scheme 12**).

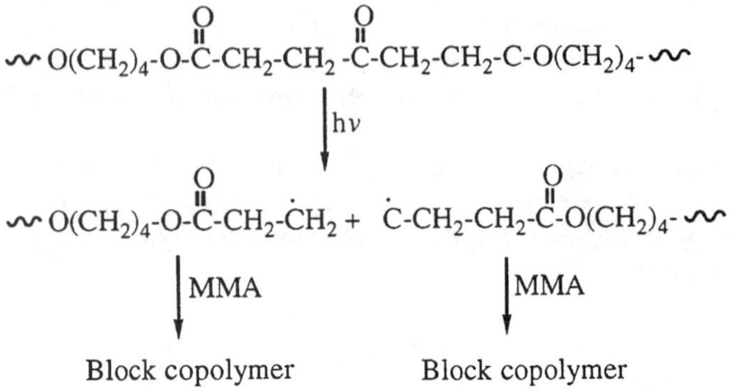

$$\sim O(CH_2)_4\text{-}O\text{-}\overset{O}{\overset{\|}{C}}\text{-}CH_2\text{-}CH_2\text{-}\overset{O}{\overset{\|}{C}}\text{-}CH_2\text{-}CH_2\text{-}C\text{-}O(CH_2)_4\text{-}\sim \xrightarrow{h\nu} \sim O(CH_2)_4\text{-}O\text{-}\overset{O}{\overset{\|}{C}}\text{-}CH_2\text{-}\dot{C}H_2 + \dot{C}\text{-}CH_2\text{-}CH_2\text{-}\overset{O}{\overset{\|}{C}}\text{-}O(CH_2)_4\text{-}\sim$$

MMA → Block copolymer; MMA → Block copolymer

**Scheme 11**

$$\sim CH_2\text{-}CH(C_6H_5)\text{-}\overset{O}{\overset{\|}{C}}\sim CH_2\text{-}CH(C_6H_5)\sim \xrightarrow{h\nu} \sim CH_2\text{-}\dot{C}H(C_6H_5) + \cdot\overset{O}{\overset{\|}{C}}\sim CH_2\text{-}CH(C_6H_5)\sim$$

MMA → St-MMA block copolymer

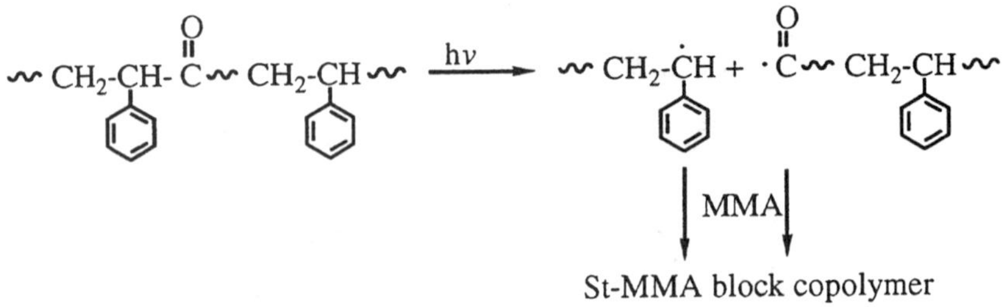

**Scheme 12**

When acrylic acid was used as the second monomer, the block copolymer yield was lower (50 % by weight). Carefully purified block copolymer after extraction with methanol was successfully tested as membranes in reverse osmosis experiments with aqueous NaCl solutions.

Polymers containing terminal photoactive benzoin ether groups have been synthesized by using azo-benzoin initiators by Onen and Yagci [88,89] (**Scheme 13**). Upon irradiation, benzoyl and alkoxy-benzyl radicals are produced, both capable of initiating polymerization to give mixtures of homopolymers and block copolymers [90] (**Scheme 14**).

A different sequence of the same procedure, i.e. photolysis first and thermolysis second, may also be used [91]. However, regarding formation of homopolymers the method in which thermolysis is employed first appears to be more suitable since benzoin groups are thermally stable and they remain unaffected for the second step. Moreover, azo-benzoin initiators allow [92] preparation of block copolymers with different chemical nature by using multi-mode polymerization approach (see Chapter 10).

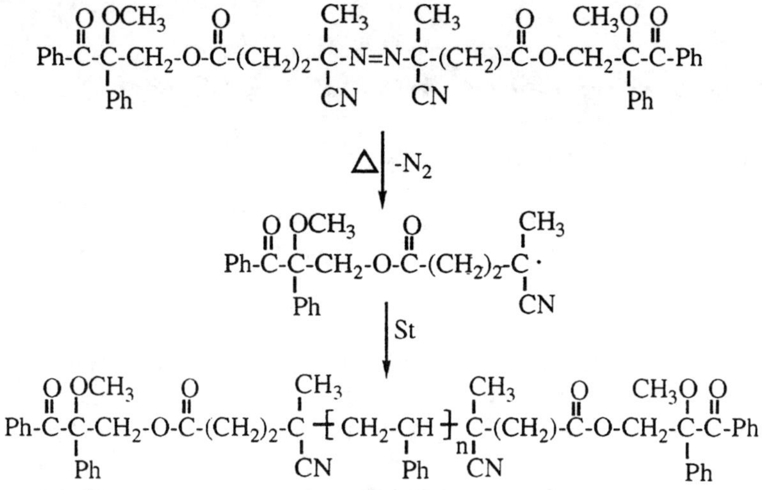

$$\text{Ph-}\overset{O}{\overset{\parallel}{C}}\text{-}\underset{Ph}{\overset{OCH_3}{C}}\text{-CH}_2\text{-O-}\overset{O}{\overset{\parallel}{C}}\text{-(CH}_2)_2\text{-}\underset{CN}{\overset{CH_3}{C}}\text{-N=N-}\underset{CN}{\overset{CH_3}{C}}\text{-(CH}_2)\text{-}\overset{O}{\overset{\parallel}{C}}\text{-O-CH}_2\text{-}\underset{Ph}{\overset{CH_3}{C}}\text{-}\overset{O}{\overset{\parallel}{C}}\text{-Ph}$$

$$\downarrow \Delta,\ -N_2$$

$$\text{Ph-}\overset{O}{\overset{\parallel}{C}}\text{-}\underset{Ph}{\overset{OCH_3}{C}}\text{-CH}_2\text{-O-}\overset{O}{\overset{\parallel}{C}}\text{-(CH}_2)_2\text{-}\underset{CN}{\overset{CH_3}{C}}\cdot$$

$$\downarrow St$$

$$\text{Ph-}\overset{O}{\overset{\parallel}{C}}\text{-}\underset{Ph}{\overset{OCH_3}{C}}\text{-CH}_2\text{-O-}\overset{O}{\overset{\parallel}{C}}\text{-(CH}_2)_2\text{-}\underset{CN}{\overset{CH_3}{C}}\text{-}\left[\text{CH}_2\text{-}\underset{Ph}{CH}\right]_n\text{-}\underset{CN}{\overset{CH_3}{C}}\text{-(CH}_2)\text{-}\overset{O}{\overset{\parallel}{C}}\text{-O-CH}_2\text{-}\underset{Ph}{\overset{CH_3}{C}}\text{-}\overset{O}{\overset{\parallel}{C}}\text{-Ph}$$

**Scheme 13**

$$\sim\text{PSt}\sim\text{CH}_2\text{-}\underset{Ph}{\overset{CH_3}{C}}\text{-}\overset{O}{\overset{\parallel}{C}}\text{-Ph} \xrightarrow{h\nu} \sim\text{PSt}\sim\text{CH}_2\text{-}\underset{Ph}{\overset{CH_3}{C}}\cdot + \cdot\overset{O}{\overset{\parallel}{C}}\text{-Ph}$$

$$\downarrow MMA \qquad\qquad \downarrow$$

PSt-b-PMMA HomoPMMA

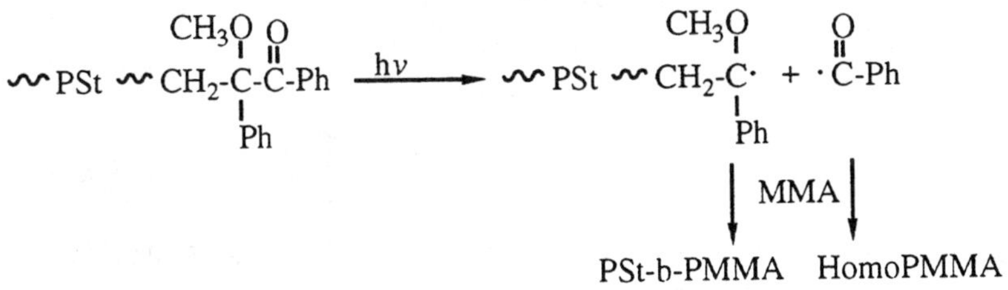

**Scheme 14**

Benzoin methyl ethers have been incorporated [93] also into a polycarbonate chain. The synthesis was achieved by performing the polycondensation of bisphenol A with phosgene in the presence of 4,4´-dihydroxybenzoin metylether as shown in **Scheme 15**.

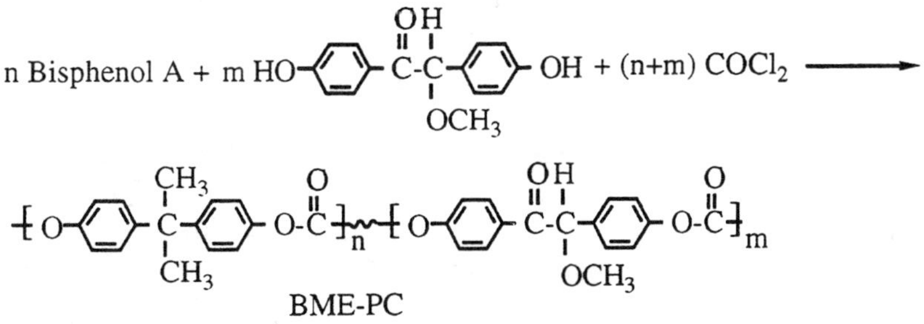

$$\text{n Bisphenol A} + \text{m HO-}C_6H_4\text{-}\overset{O}{\overset{\parallel}{C}}\text{-}\underset{OCH_3}{\overset{H}{C}}\text{-}C_6H_4\text{-OH} + (n+m)\ COCl_2 \longrightarrow$$

$$\left[\text{O-}C_6H_4\text{-}\underset{CH_3}{\overset{CH_3}{C}}\text{-}C_6H_4\text{-O-}\overset{O}{\overset{\parallel}{C}}\right]_n\sim\left[\text{O-}C_6H_4\text{-}\overset{O}{\overset{\parallel}{C}}\text{-}\underset{OCH_3}{\overset{H}{C}}\text{-}C_6H_4\text{-O-}\overset{O}{\overset{\parallel}{C}}\right]_m$$

BME-PC

**Scheme 15**

Irradiation of the resulting polycondensate (BME-PC) in the presence of methyl methacrylate resulted in block copolymer formation (**Scheme 16**).

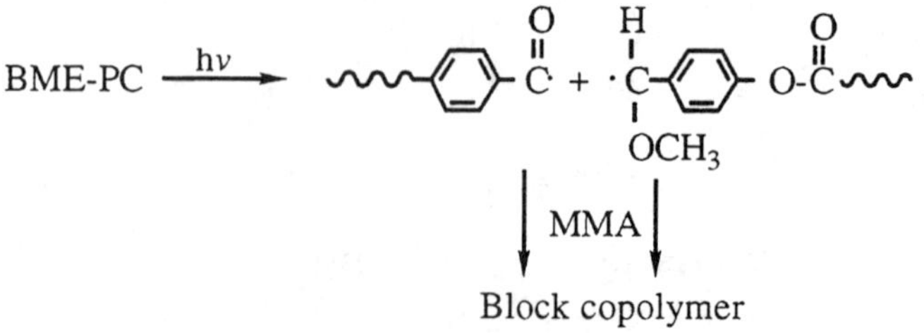

$$\text{BME-PC} \xrightarrow{h\nu} \sim\!\sim C_6H_4\text{-}\overset{O}{\overset{\|}{C}}\cdot \; + \; \cdot\underset{OCH_3}{\overset{H}{C}}\text{-}C_6H_4\text{-}O\text{-}\overset{O}{\overset{\|}{C}}\sim\!\sim$$

MMA

Block copolymer

**Scheme 16**

The applicability of this method was extended [94] to other vinyl monomers such as ethyl methacrylate and acrylonitrile. The obtained block copolymers exhibited intermediate properties with respect to the homopolymers.

Side chain macrophotoinitiators of benzoin type were synthesized by Kurusu et al [95].

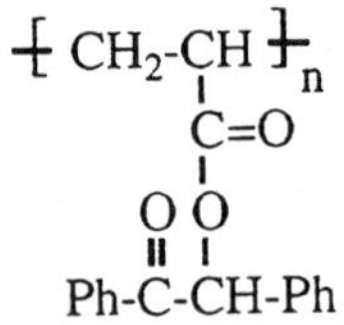

$$\left[ CH_2\text{-}CH \right]_n \quad CH\text{-}C(=O)\text{-}O\text{-}CH(Ph)\text{-}\overset{O}{\overset{\|}{C}}\text{-}Ph$$

Generally, acylation of the hydroxyl group of benzoin reduces the efficiency of $\alpha$-cleavage [96]. However, this is not the case when $CH_2$ spacer group is inserted between the benzoin moiety and the ester group.

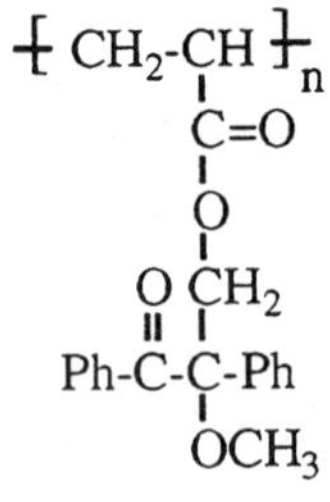

$$\left[ CH_2\text{-}CH \right]_n \quad CH\text{-}C(=O)\text{-}O\text{-}CH_2\text{-}C(Ph)(OCH_3)\text{-}\overset{O}{\overset{\|}{C}}\text{-}Ph$$

Indeed, poly($\alpha$-methylolbenzoin methylether acrylate) and related copolymers exhibited [97] a greater efficiency as side chain macrophotoinitiators for the polymerization of styrene and methyl methacrylate than poly(benzoin acrylate) and low-moleculer weight benzoin derivatives. In these studies photografting was accompanied with partial crosslinking.

Jackson and Sasse [98] reported grafting of triallylamine onto a copolymer of acrylic acid and allyl benzoinmethyl ether. The resulting graft copolymers were tested as thermally regenerable ion exchange resins. Notably little or no grafting resulted from irradiation of a copolymer of ABME and propyldiallylamine hydrochloride.

Lateral macroradicals can be formed if macrophotoinitiators containing carbonyl groups in side chains are irradiated (**Scheme 17**).

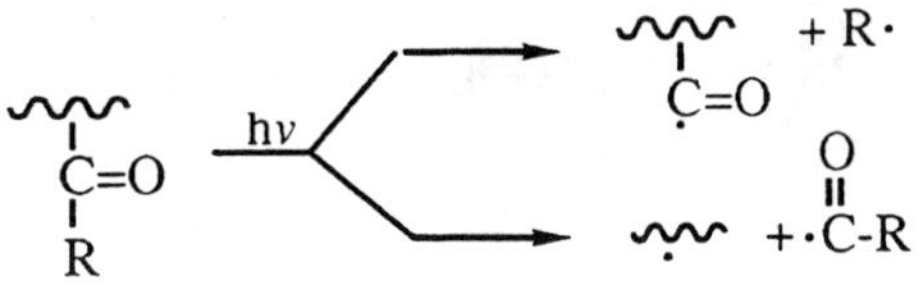

$$\sim\!\!\sim\!\!\underset{\underset{R}{|}}{C}\!=\!O \xrightarrow{h\nu} \begin{cases} \sim\!\!\sim\!\!\dot{C}\!=\!O + R\cdot \\ \sim\!\!\sim\!\!\cdot + \cdot\overset{O}{\overset{\|}{C}}\text{-}R \end{cases}$$

**Scheme 17**

These radicals are capable of initiating the polymerization of vinyl monomers. Homopolymers are also formed simultaneously with the graft copolymers, since the low molecular weight radicals are also reactive towards many monomers. The earliest example of this type of macrophotoinitiator was shown by Guillet and Norrish [99], who studied the photolysis of poly(methyl vinylketone) in dioxane solution containing acrylonitrile, vinyl acetate and methyl methacrylate (**Scheme 18**).

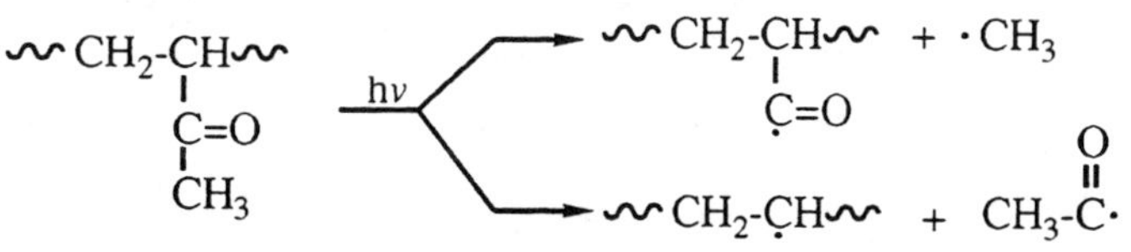

$$\sim CH_2\text{-}\underset{\underset{CH_3}{|}}{\underset{C=O}{|}}{CH}\sim \xrightarrow{h\nu} \begin{cases} \sim CH_2\text{-}\underset{\dot{C}=O}{\underset{|}{CH}}\sim + \cdot CH_3 \\ \sim CH_2\text{-}\dot{C}H\sim + CH_3\text{-}\overset{O}{\overset{\|}{C}}\cdot \end{cases}$$

**Scheme 18**

In addition to graft copolymer and homopolymer, acetaldehyde and methane were also formed indicating the readiness of low molecular weight radicals to participate in hydrogen abstraction reactions.

More recently, the graft copolymerization of styrene, methyl methacrylate, acrylonitrile and vinyl acetate onto poly(3-methyl-3-buten-2-one) (PMIK) and poly(4,4´-dimethyl-1-pentene-3-one) (PBVK), was reported [100].

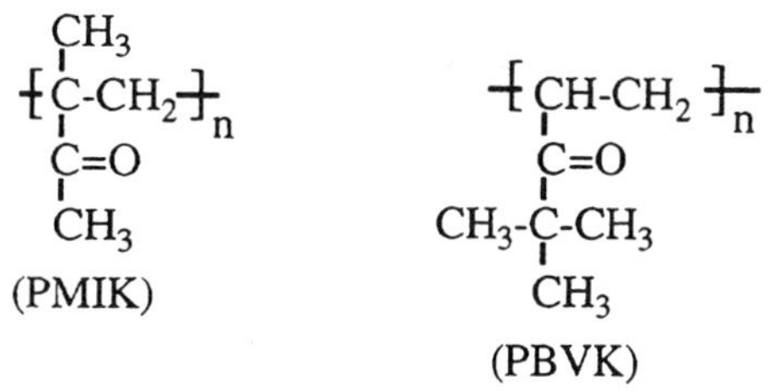

$$\left[\overset{CH_3}{\overset{|}{C}}\text{-}CH_2\right]_n \text{ with side group } C=O\text{-}CH_3 \quad \text{(PMIK)}$$

$$\left[CH\text{-}CH_2\right]_n \text{ with side group } C=O\text{-}C(CH_3)_3 \quad \text{(PBVK)}$$

(PMIK)

(PBVK)

Another photoinitiating groups, acyl oxime groups has been introduced [101] onto a polyester in the following manner (**Scheme 19**).

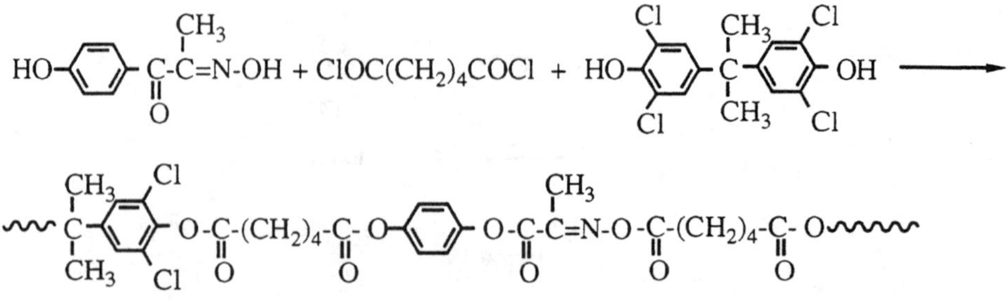

$$HO-C_6H_4-C(=O)-C(CH_3)=N-OH + ClOC(CH_2)_4COCl + HO-C_6H_2Cl_2-C(CH_3)_2-C_6H_2Cl_2-OH \longrightarrow$$

$$\sim\sim C(CH_3)_2-C_6H_2Cl_2-O-C(=O)-(CH_2)_4-C(=O)-O-C_6H_4-O-C(=O)-C(CH_3)=N-O-C(=O)-(CH_2)_4-C(=O)-O\sim\sim$$

Keto oxim ester polycondansate (KOS-PC)

**Scheme 19**

Irradiation of these polymers with 365 nm light in the presence of vinylmonomers such as styrene gave block copolymers (**Scheme 20**).

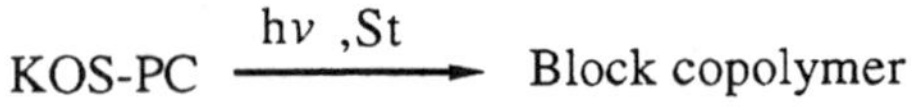

$$\text{KOS-PC} \xrightarrow{h\nu\text{ ,St}} \text{Block copolymer}$$

**Scheme 20**

Polymers containing N-nitroso groups are capable of acting as macrophotoinitiators [102] Treatment of various polyamides such as nylon 6,6,nylon 6,10 and polyurethanes with $N_2O_3$ affords N-nitroso polymers. Due to their property of undergoing a rearrangement on irradiation and subsequent expulsion of $N_2$, macroradicals are readily formed giving rise to the formation of block copolymers of polyamides with olefinic monomers. The proposed overall mechanism is presented in **Scheme 21**.

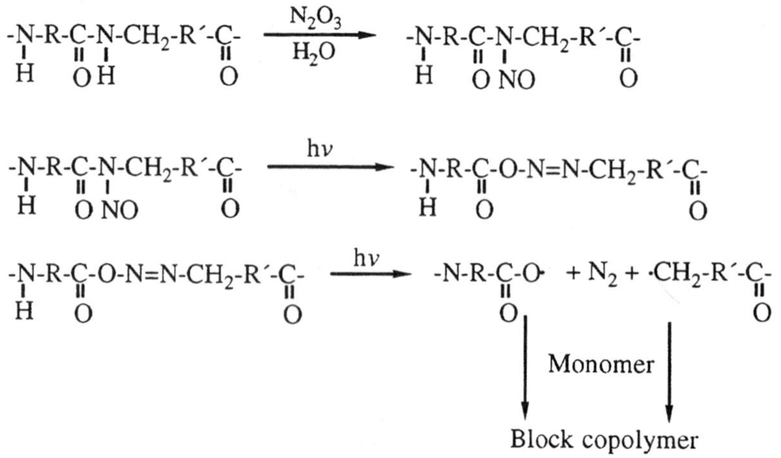

$$-N(H)-R-C(=O)-N(H)-CH_2-R'-C(=O)- \xrightarrow[H_2O]{N_2O_3} -N(H)-R-C(=O)-N(NO)-CH_2-R'-C(=O)-$$

$$-N(H)-R-C(=O)-N(NO)-CH_2-R'-C(=O)- \xrightarrow{h\nu} -N(H)-R-C(=O)-O-N=N-CH_2-R'-C(=O)-$$

$$-N(H)-R-C(=O)-O-N=N-CH_2-R'-C(=O)- \xrightarrow{h\nu} -N-R-C(=O)-O\cdot + N_2 + \cdot CH_2-R'-C(=O)-$$

$$\xrightarrow{\text{Monomer}} \text{Block copolymer}$$

**Scheme 21**

Notebly, macroradical generation involves two photon absorption. It was also pointed out [103] that care should be taken when poly(N-nitroso amides) are used as macroinitiators since they can *decompose violently under the influence of heat or light, especially at high contents of nitroso groups*.

Various organic disulfides including thiuram disulfides are also capable of acting as thermal and photochemical initiators for free radical polymerization. For example, when the polymerization of styrene was initiated by tetraethylthiuram disulfide, the polymer was found to contain terminal $Et_2N$-CS-S groups. These polymers are also photosensitive and could be photolysed in the presence of a reactive monomer to give block copolymers [104]. This route, i.e. using iniferter system, is treated in detail in the macroiniferter section (Chapters 11-13) by other authors and will not be discussed further here.

Peroxyesters containing triplet photosensitizer functionality are efficient initiators for the free radical polymerization of monomers not acting as triplet quenchers. Incorporated into polymer chains, it can serve as an initiator for photocrosslinking or photo grafting. For example, methylmethacrylate was photografted [105,106] onto polystyrene containing a low fraction of peroxybenzoate {p-(p-vinylbenzoyl) peroxybenzoate}, upon irradiation of the copolymer in methylmethacrylate at λ = 366 nm as illustrated below in **Scheme 22**. Homo-PMMA is also formed and the grafting efficiency increased with irradiation time and content of the peroxybenzoate monomer in the copolymer.

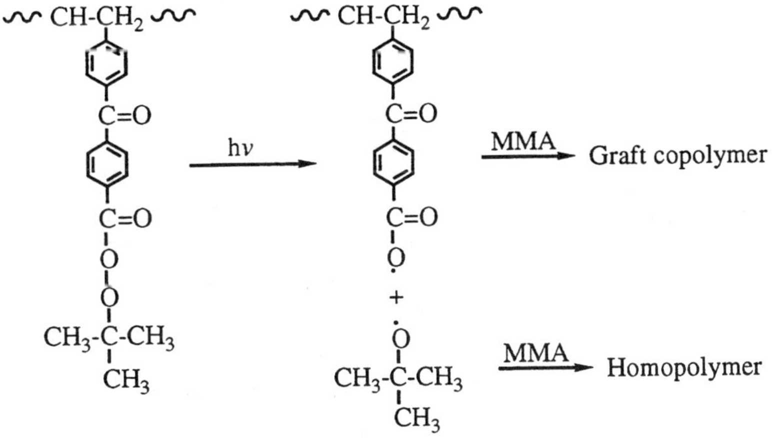

~ CH-$CH_2$ ~ (–C=O–phenylene–C(=O)–O–O–C($CH_3$)$_3$) —hν→ ~ CH-$CH_2$ ~ (–C=O–phenylene–C(=O)–O•) + •O–C($CH_3$)$_3$

—MMA→ Graft copolymer

—MMA→ Homopolymer

**Scheme 22**

Another type of macrophotoinitiator, polymers having halogen groups, also undergo cleavage. The earliest example of this type of initiator is brominated polystyrene reported by Jones et al. [107,108] about four decades ago. Brominated polystyrene was irradiated with a Mercury lamp at <310 nm in benzene solution containing MMA. The grafted material contained 1.2 to 5.9 branches per macromolecule, depending on the reaction conditions. The length of MMA branches was in the range of 1300-4500 repeating units. Similar method was also applied [109] to other brominated polymers. Brominated polyacrylamide and polyacrylonitrile served as the radical generating photoactive polymer to graft acrylamide and ethyleneglycol oligomers having methacrylate end groups, respectively. The latter product was aimed at the synthesis of antistatic materials.

### 2.3.2 H-Abstraction-type macrophotoinitiators

Block and graft copolymerization of vinyl monomers can be achieved by taking advantage of the fact that triplet states of aromatic ketones readily react with amines (**Scheme 23**).

$$>C=O \xrightarrow{h\nu} {}^{1}[>C=O]^{*} \longrightarrow {}^{3}[>C=O]^{*}$$

$${}^{3}[>C=O]^{*} + R^{1}\text{-}CH_2\text{-}NR^{2}_{2} \longrightarrow >\dot{C}\text{-}OH + R^{1}\text{-}\dot{C}H\text{-}NR^{2}_{2}$$

**Scheme 23**

Radicals formed by the reactions in **Scheme 23** can initiate the polymerization of vinyl compounds, and if the amine group is attached to a macromolecule, a block copolymer is formed (**Scheme 24**).

$$\sim\!\sim\dot{C}H\text{-}N(R_2)(R_2) + mB \longrightarrow \text{Block copolymer}$$

**Scheme 24**

Formation of homopolymer is expected to be negligible since the amine derived radicals are a better initiator than the semipinacol radical. The preparation of polystyrene with N,N´-dimethyl amino end groups from an azo-initiator is described by Yagci et al. [110].

$$(CH_3)_2N\text{-}CH_2\text{-}CH_2\text{-}O\text{-}\overset{O}{\overset{\|}{C}}\text{-}CH_2\text{-}CH_2\text{-}\underset{CN}{\overset{CH_3}{C}}\text{-}N{=}N\text{-}\underset{CN}{\overset{CH_3}{C}}\text{-}CH_2\text{-}CH_2\text{-}\overset{O}{\overset{\|}{C}}\text{-}O\text{-}CH_2\text{-}CH_2\text{-}N(CH_3)_2$$

Polymers functionalized by this method were irradiated at $\lambda = 350$ nm in the presence of 9-fluorenone and MMA to yield block copolymers (**Scheme 25**).

Kinstle and Watson [111] achieved the grafting of vinyl monomers using the same H-abstraction mechanism. They irradiated poly(styrene-co-p-N,N´-diethylamino styrene) in the presence of 2-ethylhexyl methacrylate (EHMA) and benzophenone at $\lambda = 350$ nm. In this case lateral macroradicals were formed by hydrogen abstraction from diethylamino groups by benzophenone triplets (**Scheme 26**).

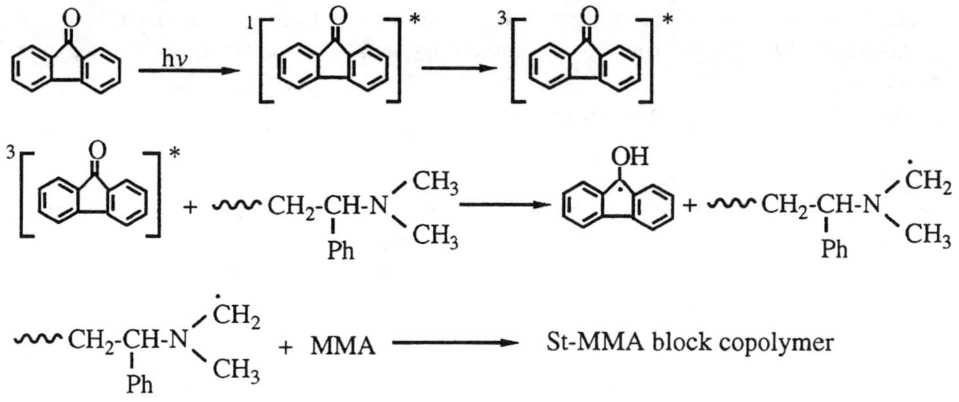

Scheme 25

Low monomer concentration and short irradiation time were chosen so that selectivity of reactions could be enhanced. Apart from graft copolymer, homo-PEHMA was also formed. However, the grafting efficiency was high: about 77 % of the converted monomer was grafted onto the backbone polymer. The rather low homopolymer yield was attributed again to the relatively inefficient initiating power of benzophenone ketyl radical.

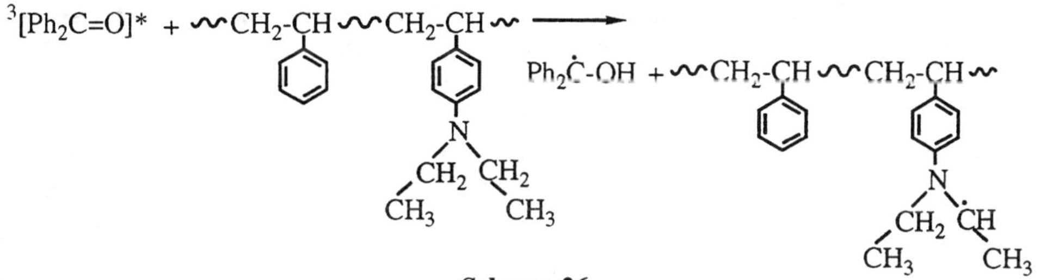

Scheme 26

### 2.3.3 Macrophotoinitiators with halogen-containing end groups

According to Bamford et al.[112], terminal carbon centered macroradicals are formed upon irradiation ($\lambda$ = ca.350-450 nm) of polymers having $-CX_3$, $-CHX_2$ or $-CH_2X$ end groups in the presence of dimanganese decacarbonyl $Mn_2(CO)_{10}$. Upon absorption of light, $Mn_2(CO)_{10}$ decomposes to $Mn(CO)_5$. The latter reacts with the terminal $CCl_3$ yielding macroradicals (**Scheme 27**).

$$Mn_2(CO)_{10} \xrightarrow{h\nu} 2Mn(CO)_5$$

$$-(-A-)_n-CCl_3 + Mn(CO)_5 \longrightarrow -(-A-)_n-\dot{C}Cl_2 + Mn(CO)_5Cl$$

Scheme 27

Radicals $\sim\sim\sim\sim An\sim\sim\sim\dot{C}Cl_2$ can initiate the polymerization of a second monomer B which is contained in the system. Notably, low molecular weight radicals are not formed and therefore homopolymerization cannot occur. Moreover, metal atoms don't become bound to the polymer in these processses.

Macrophotoinitiators with terminal halogen groups can be prepared in different ways. For example, a polymer with a terminal $CBr_3$ group is obtained if a vinyl monomer is polymerized in the presence of carbon tetrabromide as chain transfer agent. Using this polymer and the metal carbonyl compound, block copolymers are formed in the usual manner as illustrated below (**Scheme 28**).

$$-(\text{-A-})_n\text{-CBr}_3 + \text{Mn(CO)}_5 \longrightarrow -(\text{-A-})_n\text{-}\dot{\text{C}}\text{Br}_2 + \text{Mn(CO)}_5\text{Br}$$

$$-(\text{-A-})_n\text{-}\dot{\text{C}}\text{Br}_2 + m\text{B} \longrightarrow -(\text{-A-})_n\text{-CBr}_2\text{-(-B-)}_m^{\bullet}$$

$$-(\text{-A-})_n\text{-CBr}_2\text{-(-B-)}_m^{\bullet} \longrightarrow \begin{cases} -(\text{-A-})_n\text{-CBr}_2\text{-(-B-)}_{2m}\text{-(-A-)}_n\text{-} \\ 2\text{-(-A-)}_n\text{-CBr}_2\text{-(-B-)}_m \end{cases}$$

**Scheme 28**

The chemical nature of the block copolymer is determined again by the mode of termination. If the growing radicals terminates exclusivelyby combination, the product is a tri-block copolymer, whereas disproportionation leads to di-block structures. $-CBr_2-$ groups incorporated into the block copolymer normally do not initiate side reactions to an important extent, because the reactivity of halogen carrying group decreases in the order of $-CX_3 > -CX_2 > -CX$. An interesting application [113] of this procedure involves the synthesis of block copolymers in which each block was an alternating block copolymer:

poly-$(\text{-AB-})_n$-block-poly-$(\text{-CD-})_m$-block-poly-$(\text{-AB-})_n$-

The monomers used in this work were: styrene, butadiene, isoprene, methyl methacrylate and methacrylates.

The applicability of this method depends critically on the availability of chemical methods to prepare polymers with terminal halogen groups. Bamford and coworkers [114] showed that $CCl_3$ groups can be attached to polymers having terminal OH, $NH_2$ or COOH groups by reaction with tricholoroacetyl isocyanate (**Scheme 29**).

Isocyanation was successfully employed with dextran and poly(ethylene glycol). Recently an analogous procedure was used by Yagci et al. [115] to obtain a bifunctional initiator, 4,4-azobis(4-cyanopentane-trichloroacetylamide), ACPT (**Scheme 30**).

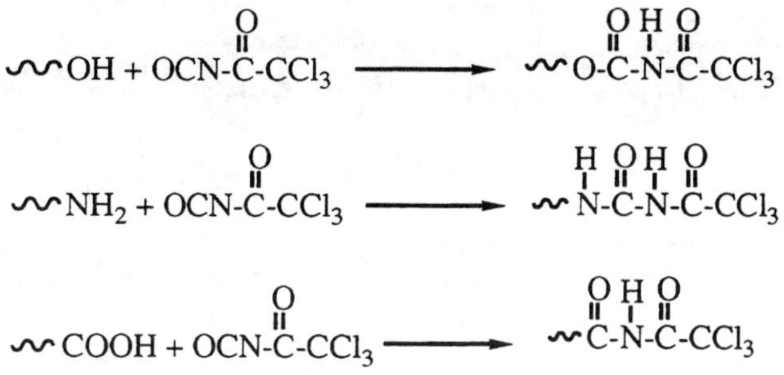

$$\sim\!\!\sim OH + OCN\text{-}C(=O)\text{-}CCl_3 \longrightarrow \sim\!\!\sim O\text{-}C(=O)\text{-}N(H)\text{-}C(=O)\text{-}CCl_3$$

$$\sim\!\!\sim NH_2 + OCN\text{-}C(=O)\text{-}CCl_3 \longrightarrow \sim\!\!\sim N(H)\text{-}C(=O)\text{-}N(H)\text{-}C(=O)\text{-}CCl_3$$

$$\sim\!\!\sim COOH + OCN\text{-}C(=O)\text{-}CCl_3 \longrightarrow \sim\!\!\sim C(=O)\text{-}N(H)\text{-}C(=O)\text{-}CCl_3$$

**Scheme 29**

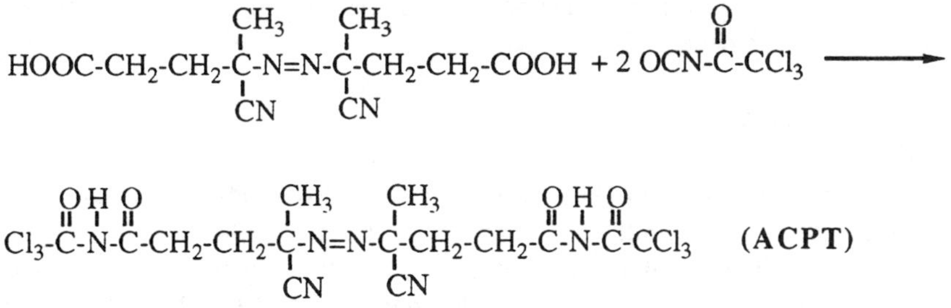

$$HOOC\text{-}CH_2\text{-}CH_2\text{-}C(CH_3)(CN)\text{-}N{=}N\text{-}C(CH_3)(CN)\text{-}CH_2\text{-}CH_2\text{-}COOH + 2\ OCN\text{-}C(=O)\text{-}CCl_3 \longrightarrow$$

$$Cl_3\text{-}C(=O)\text{-}N(H)\text{-}C(=O)\text{-}CH_2\text{-}CH_2\text{-}C(CH_3)(CN)\text{-}N{=}N\text{-}C(CH_3)(CN)\text{-}CH_2\text{-}CH_2\text{-}C(=O)\text{-}N(H)\text{-}C(=O)\text{-}CCl_3 \quad \textbf{(ACPT)}$$

**Scheme 30**

The initiator is at first thermally decomposed in the presence of styrene to obtain polystyrene containing terminal $CCl_3$ groups of the structure depicted below.

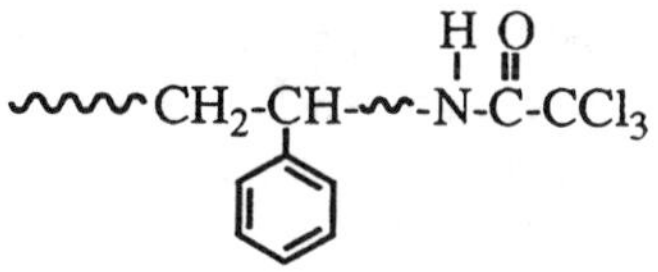

$$\sim\!\!\sim\!\!\sim CH_2\text{-}CH(C_6H_5)\text{-}\sim\!\!\sim\text{-}N(H)\text{-}C(=O)\text{-}CCl_3$$

This polymer was converted into a block copolymer by applying Bamford's method: the resulting polymer was mixed with methyl methacrylate and $Mn_2(CO)_{10}$ and block copolymer formation was induced by irradiating the system with UV light ( $\lambda = 436$ nm).

These results demonstrate that this method is a general technique for the preparation of block copolymers. Moreover, this method can be successfully employed in multi-mode polymerization as outlined in Chapter 10.

Bamford [116] also examined the utility of the method for graft copolymerization. In this case polymers possessing side chains with photoactive groups were employed (**Scheme 31**).

$$\underset{\substack{| \\ CBr_3}}{\sim\sim}\underset{\substack{| \\ CBr_3}}{\sim\sim} \xrightarrow[Mn_2(CO)_{10}]{h\nu} \underset{\substack{| \\ \cdot CBr_2}}{\sim\sim}\underset{\substack{| \\ CBr_3}}{\sim\sim} + Mn(CO)_5Br$$

**Scheme 31**

The grafting reactions leads to the synthesis of a network if combination of macroradicals is the predominant termination route. Network formation vs grafting of branches onto trunk polymers has been intensively studied using poly(vinyl trichloroacetate) as the trunk polymer. In addition, other trunk polymers were examined: polycarbonates and polystyrene with suitable functional groups, N-halogenated polyamides and polypeptides. Monomers used for these studies included styrene, methyl methacrylate and chloroprene [117-119].

The manganese carbonyl grafting method has also been applied to biopolymers [120] and polytetrafluoroethylene [121].

## 2.4 Redox Macroinitiators

During the last fifty years there had been a great deal of study on the use of redox systems; that is, systems containing both oxidizing and reducing agents, for the initiation of vinylpolymerization. Polymerizations initiated by the reaction between an oxidizing and a reducing agent are called redox polymerizations. These developments have been reported in many review articles by several authors [122-128].

Certain transition metals in their higher valence states alone or coupled with easily oxidizable organic substrates act as potential initiators for redox polymerization of vinyl monomers. The metal ions which have been used for initiating vinyl polymerization or graft copolymerization involving natural or synthetic polymers are chromium (VI), vanadium(V), cerium(IV), cobalt(III), ferric ion, permanganate, manganese(III), etc. [122-128]. When a polymeric substrate is used as a redox system in conjuction with metal ions for the polymerization of vinyl monomers to form graft coolymers, the radical generated on the polymer substrate acts as a macroinitiator. Due to the availability of several review articles in this area [122-128], this type of initiation will not be discussed in detail here. However, one example will be provided to introduce the concept in this chapter.

Of all the transitional metal ions, ceric salts have been used most extensively for oxidation studies of a multitude of organic as well as inorganic substrates. Many workers have reported the kinetics of vinyl polymerization of various monomers initiated by ceric ion redox systems. On account of its high grafting efficiency compared to other known redox systems, it has gained considerable importance in grafting vinyl monomers onto various natural and synthetic polymers. A large number of papers are available on graft copolymerization of vinylmonomers onto cotton and cellulose using Ce(IV). This technique is based on the fact that when cellulose is oxidized by ceric salts such as ceric ammonium nitrate or ceric ammonium sulfate, free radicals capable of initiating vinylpolymerization are formed on the cellulose according to the following reaction mechanism in **Scheme 32**. Although there are some doubts about the true mechanism of the radical formation, this mechanism leading to the formation of cellulose radicals has been published in the literature (refer to review articles [122-128] and references therein). Obviously, these cellulose radicals act as macroinitiators for the vinyl polymerization.

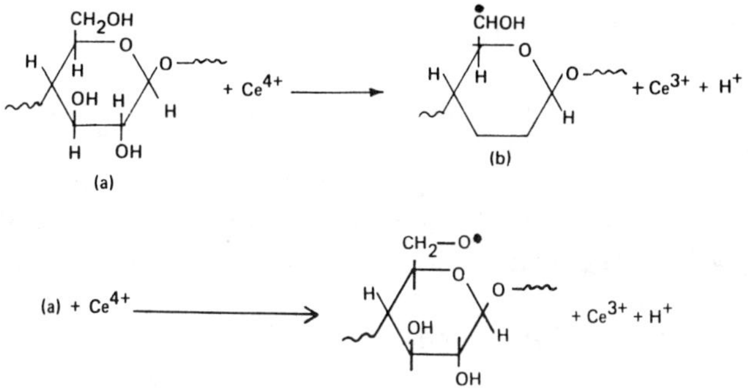

**Scheme 32**

## 2.5 Azo and Peroxy Macroinitiators

Azo and perxoy macroinitiators are described in detail by Ueda et al. and Nuyken et al. in Chapter 7 and 8 respectively in this book. Therefore they will not be included here.

# 3. MACROINITIATORS FOR CATIONIC POLYMERIZATION

Cationic polymerization is rather old [129] and after Friedel and Crafts initial work, cationic polymerization took a new dimension and some of the important publications [130-133], including those reporting on living cationic ring-opening of heterocyclic monomers [7-9], appeared during 1877-1982. The most recent development in cationic polymerization is the emergence of truly living polymerization of olefins leading to well-defined macromolecules. These milestones were reviewed elsewhere very recently [134].

It may be beneficial to define the "living polymers" and living polymerization where the concept of macroinitiation is evident. "Living polymers" have been defined [135] as systems in which macromolecules may spontaneously resume their growth whenever fresh monomer is supplied. The macromolecules act as initiators for further polymerization of fresh monomers. An operational definition of "livingpolymerization" was proposed by Szwarc [136] recently; e.g., a polymerization is living when the resulting polymer retains its integrity for a sufficiently long time to allow the operator to complete its task.

To demonstrate the existence of living carbocationic polymerizations, particularly those of olefins (isobutylene, styrene and its derivatives) of which the polymerization are often quite rapid, the incremental monomer addition (IMA) technique was developed [137,138]. The IMA technique has been described in detail elsewhere [134,139]. In the IMA technique, fresh monomers are added to the system at a predefined time interval and the living polymers act as macroinitiators for freshly added monomers in successive runs.

## 3.1 Synthesis of Block and Graft Copolymers

The understanding and controlling of elementary steps of carbocationic polymerization has helped to develop methods to design unique and useful block or graft copolymers using macroinitiator concept. The design of block/graft copolymers using the macroinitior concepts may be devided into two categories: 1) By controlled initiation and 2) By sequential living polymerization. These are described as follows.

### 3.1.1 Blocks by Controlled Initiation

The first block copolymer such as polystyrene-polyisobutylene (PSt-PIB) diblock was prepared by Peyrot et al. [140, 141] using a polyisobutylene macroinitiator. The synthesis involved three steps: (1) polymerization of isobutylene by the $C_6H_5CH_2Cl/Et_2AlCl$ initiating system, leading to the benzyl-headed polyisobutylene ($C_6H_5$-$CH_2$-PIB); (2) chloromethylation, yielding a PIB capped by a $ClCH_2C_6H_4CH_2$-group ($ClCH_2C_6H_4CH_2$-PIB); (3) polymerization of styrene by the PIB-p-$CH_2C_6H_4CH_2Cl/Et_2AlCl$ combination. However the products were not fully characterized.

Kennedy and Molby [142,143] prepared PSt-b-PIB using a polystyrene macroinitiator. Polystyrene containing a terminal tertiary bromine (PSt-Br) was prepared by the polymerization of styrene using 2-bromo-6-chloro-2,6-dimethyl heptane in conjuction with $Et_3Al$. Then the polymerization of IB by the PSt-Brt/$Et_2AlCl$ combination produced PSt-b-PIB. However, the polymerization of IB by PSt-Brt/$Et_2AlCl$ combination was not entirely free of chain transfer (some homo-PIB was also formed). The reaction steps may be represented as follows **(Scheme 33)**.

$$Cl\text{-}C(CH_3)_2\text{-}(CH_2)_3\text{-}C(CH_3)_2\text{-}Br \xrightarrow[Et_3Al]{Styrene} PSt\text{-}C(CH_3)_2\text{-}(CH_2)_3\text{-}C(CH_3)_2\text{-}Br$$

$$\xrightarrow[Et_2AlCl]{Isobutylene} PSt\text{-}C(CH_3)_2\text{-}(CH_2)_3\text{-}C(CH_3)_2\text{-}PIB$$

**Scheme 33**

Terminally chlorinated polystyrene and polyisobutylene were also used as macroinitiators for the polymerization of IB and St respectively. Kennedy et al. [144]

prepared terminally chlorinated PSt {sec-benzylic chlorine end group, PSt-$CH_2$-$CHCl(C_6H_5)$} or PIB (tert-chlorine-ended PIB) by initiating the polymerization of the respective monomers by a "$H_2O$"/$BCl_3$ combination. These terminally chlorinated polymers initiated the polymerization of IB or STY in the presence of $Et_2AlCl$ to produce PSt-b-PIB or PIB-b-PSt block copolymers. The block copolymers were also contaminated with homopolymers.

In another report Kennedy and Kurian [145] have reported the synthesis of poly(p-chlorostyrene)-block-polytetrahydrofuran block copolymer. The synthesis involved the living polymerization of p-chlorostyrene (p-ClSt) by the 1-chloro-2,4,4-trimethylpentane/$TiCl_4$ initiating system followed by the blocking of tetrahydrofuran (THF) from the sec-benzylic chlorine endgroups of poly(p-ClSt) in the presence of $AgPF_6$ in THF. Poly(p-ClSt) having a sec-benzylic chlorine end-group is used as the macroinitiator in conjunction with $AgPF_6$ for the THF polymerization to yield the corresponding block copolymer. The reaction may be represented by the following **Scheme 34**. The blocking efficiency was nearly 100%.

$$CH_3\text{-}C(CH_3)_2\text{-}CH_2\text{-}C(CH_3)_2\text{-}(CH_2\text{-}CH(C_6H_4Cl))_n\text{-}CH_2\text{-}CH(C_6H_4Cl)\text{-}Cl \xrightarrow[-AgCl]{THF,\ AgF_6}$$

$$CH_3\text{-}C(CH_3)_2\text{-}CH_2\text{-}C(CH_3)_2\text{-}(CH_2\text{-}CH(C_6H_4Cl))_{n+1}\text{-}(O\text{-}CH_2\text{-}CH_2\text{-}CH_2\ CH_2)_m\ OH$$

**Scheme 34**

Similarly Zsuga and Kennedy [146] have reported the synthesis of the block copolymers of THF and chlorostyrene {PTHF-Poly(p-ClSt)-PTHF}. In this case they have synthesized $\alpha,\omega$-dibenzylic chlorine-capped poly(chlorostyrene) by using 1,4-di(2-chloro-2-propyl) benzene and 1,3-di(2-chloro-2-propyl)-5-t- butylbenzene in conjunction with $BCl_3$. They have used $\alpha,\omega$-dibenzylic chlorine-capped poly(chlorostyrene) as the macroinitiator in the presence of $AgPF_6$ for the polymerization of THF for the synthesis of triblock copolymer.

Kennedy et al. [147, 148] had also prepared assymetric telechelic polyisobutylene carrying an olefinic head group and a tert.-chlorine end group by initiating the polymerization of IB with the $(CH_3)_2C{=}CHCH_2Cl/BCl_3$ combination. They used this polymer as a macroinitiator with $Et_2AlCl$ for the polymerization of styrene [149] and $\alpha$-methylstyrene [147, 148] to yield PIB-b-PSt or PIB-b-P$\alpha$MeSt. Although, chain transfer was evident during the polymerization of styrene, it was negligible during polymerization of $\alpha$-methylstyrene. In this manner they prepared clean PIB-b-P$\alpha$MeSt block copolymers containing $(CH_3)_2C=CH-CH_2$- headgroups.

In another approach Kennedy and Chen [150] have used an assymetric $\alpha,\omega$-dichloro telechelic (PIB); i.e, PIB capped by a neopentyl-chlorine headgroup and tert-chlorine tail group in conjunction with $Et_2AlCl$ to initiate the polymerization of $\alpha$-methylstyrene as represented in the following reaction, **Scheme 35**, to produce the PIB-b-P$\alpha$MeSt block copolymer.

Kennedy and Smith [149] had also demonstrated the synthesis of triblock copolymers using the concept of macroinitiators. In conjunction with $Et_2AlCl$, $\alpha,\omega$-di-tert.-chloro polyisobutylene ($Cl^t$-PIB-$Cl^t$) macroinitiator induced the polymerization of $\alpha,\omega$-methylstyrene at both ends of this telechelic prepolymer. Extraction of the crude product with a series of hydrocarbon solvents revealed a broad compositional distribution.

$$Cl\text{-}CH_2\text{-}C(CH_3)_2\sim\sim PIB\sim\sim CH_2\text{-}C(CH_3)_2\text{-}Cl \xrightarrow[Et_2AlCl]{\alpha\text{-methylstyrene}}$$

$$Cl\text{-}CH_2\text{-}C(CH_3)_2\sim\sim PIB\sim\sim CH_2\text{-}C(CH_3)_2\sim\sim poly(\alpha\text{-methylstyrene})$$

**Scheme 35**

The linear di-, tri- and three-arm star block copolymers of IB and $\alpha$-MeSt were synthesized in the presence of the proton trap 2,6-di-tert-butylpyridine(DTBP) [151]. The initiation of $\alpha$-methylstyrene by linear or three arm star PIB having one, two, or three tert-chlorine termine macroinitiator occured in conjunction with $SnCl_4$ coinitiator in presence of DTBP. The block copolymers were synthesized with 70% to > 95% efficiency.

Recently Liu et al. [152] have demonstrated the synthesis of poly(alkyl vinyl ether-2-ethyl-2-oxazoline) diblock copolymers. They have prepared monofunctional poly(alkyl vinyl ether) oligomers via a living cationic polymerization method using chloroethyl vinyl ether together with $HI/ZnI_2$ as the initiating system and lithium borohydride as the termination agent. Using the resultant chloroethyl ether functional oligomers in combination with sodium iodide as macroinitiators, 2-ethyl-2-oxazoline was polymerized to afford diblock copolymers. The reactions may be represented by the following **Scheme 36**.

No data on the blocking efficiency has been reported. However other results indicate that the blocking efficiency is very high. These polymers are being used to explore the molecular requirements for steric stabilization of alkoxide particles in polar solvents such as water and alcohol.

The different diblock and triblock copolymers described above are summarized in the following **Table 2**.

**TABLE 2** AB and ABA types blocks by controlled Initiation

| Block copolymer | Reference |
|---|---|
| Poly(isobutylene-*block*-styrene) | 140, 141, 144, 148 |
| Poly(styrene-*block*-isobutylene) | 142-144 |
| Poly(Isobutylene-*block*-α-methylstyrene) | 147, 148, 150, 151 |
| Poly(α-methylstyrene-*block*-isobutylene-*block*-α-methylstyrene) | 149, 151 |
| Radical[Poly(isobutylene-*block*-α-methylstyrene)]$_3$ | 151 |
| Poly(p-chlorostyrene-*block*-tetrahydrofuran) | 145 |
| Poly(THF-*b*-chlorostyrene-*b*-THF) | 146 |
| Poly(alkylvinylether-*block*-2-ethyl-2-oxazoline | 152 |

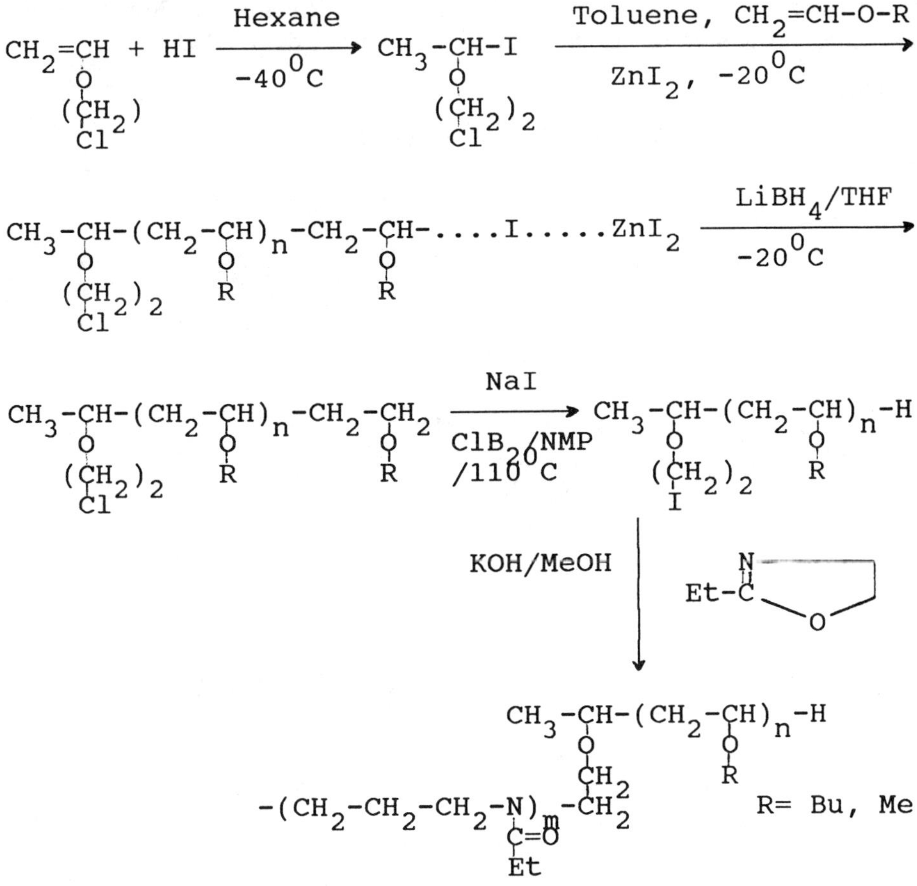

$$CH_2{=}CH(O(CH_2)_2Cl) + HI \xrightarrow[-40^0C]{Hexane} CH_3{-}CH(O(CH_2)_2Cl){-}I \xrightarrow[ZnI_2,\ -20^0C]{Toluene,\ CH_2=CH-O-R}$$

$$CH_3{-}CH(O(CH_2)_2Cl){-}(CH_2{-}CH(OR))_n{-}CH_2{-}CH(OR){-}....I.....ZnI_2 \xrightarrow[-20^0C]{LiBH_4/THF}$$

$$CH_3{-}CH(O(CH_2)_2Cl){-}(CH_2{-}CH(OR))_n{-}CH_2{-}CH_2(OR) \xrightarrow[ClB_2/NMP\ /110^0C]{NaI} CH_3{-}CH(O(CH_2)_2I){-}(CH_2{-}CH(OR))_n{-}H$$

$$\downarrow \text{KOH/MeOH}, \quad Et{-}C{=}N{-}CH_2CH_2{-}O \text{ (2-ethyl-2-oxazoline)}$$

$$CH_3{-}CH(O{-}CH_2{-}CH_2{-}(N(C{=}O\,Et){-}CH_2{-}CH_2{-}CH_2)_m{-}){-}(CH_2{-}CH(OR))_n{-}H$$

R= Bu, Me

**Scheme 36**

### 3.1.2 Blocks by Sequential Living Polymerization

Block copolymer synthesis by the sequential monomer addition "SMA"technique using livingpolymerization is one of the simplest and convenient methods. In this technique, block copolymers are prepared by first polymerizing monomer A to produce a living polymer (.......A)and then adding a second monomer B to the living polymer .......A*. The polymerization proceeds until monomer B is available. At this point the living polymer .....A......B*may be terminated to produce a diblock polymer or a third monomer C may be added to the living polymer sequence .........A......B*to produce $A_xB_yC_z$ triblocks, etc. If the first polymer segment is a living di- or triarm sequence then tri- or starblock copolymers, respectively will be produced. The general reaction pathways are presented in the following **Scheme 37**.

In the sequential monomer addition technique the living polymer sequence acts as the macroinitiator for the polymerization of the other monomers that has to be added in the sequential manner. The following **Tables 3 and 4** represent the different type of block copolymers prepared by the sequential monomer addition technique.

$$\ldots\ldots.A^* \xrightarrow{+B} \ldots.A\ldots..B^* \xrightarrow{+C} \ldots A\ldots B\ldots.C^*$$

quenching ↓

....A....B....C
(triblock copolymer)

$$^*A\ldots\ldots..A^* \xrightarrow{+B} {}^*B\ldots.A\text{--------}A\ldots\ldots B^*$$

quenching ↓

B-A-B
(triblock copolymer)

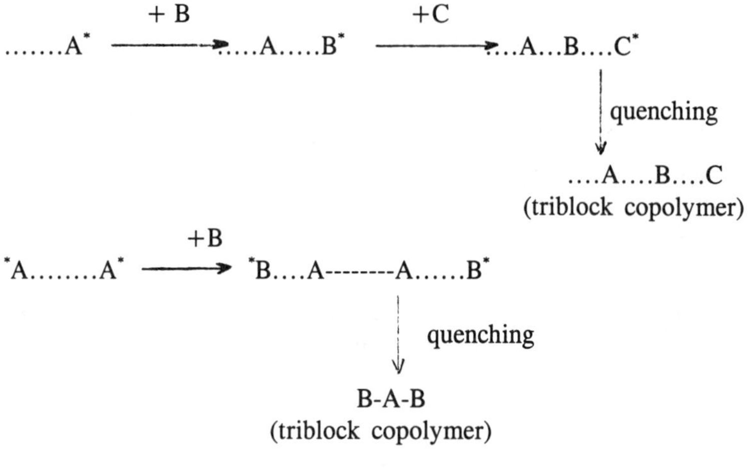

**Scheme 37**

## 4. MACROINITIATORS FOR ANIONIC POLYMERIZATION

Living polymerization is one of the best ways to prepare polymers with well-regulated chain lengths. A large variety of macromolecular structures; i.e., linear, cyclic, functional, block copolymers and graft copolymers are accessible via living polymerization techniques [170, 171]. Although several new types of polymerizations have been developed fulfilling partial or all the requirements of living conditions, anionic living polymerization is certainly the most established and well-studied method among the living polymerization systems.

**Table 3** AB Type Blocks by Sequential Monomer Addition.

| AB Blocks | References |
|---|---|
| Poly(styrene-*block*-isobutylene) | 153 |
| Poly(ethyl vinyl ether-*block*-N-vinyl-carbazole) | 154 |
| Poly(isobutyl vinyl ether-*block*-N-vinyl-carbazole) | 154 |
| Poly(p-methoxystyrene-*block*-isobutyl vinyl ether) | 155 |
| Poly(alkyl vinyl ether-*block*-methoxystyrene) (alkyl = methyl, ethyl, isobutyl) | 156 |
| Poly(cetyl vinyl ether-*block*-methyl vinyl ether) | 156 |
| Poly(cetyl vinyl ether-*block*-ethyl vinyl ether | 156 |
| Poly(methyl vinyl ether-*block*-cetyl vinyl ether) | 156, 157 |
| Poly(ethyl vinyl ether-*block*-cetyl vinyl ether) | 156 |
| Poly(isobutyl vinyl ether-*block*-cis(or trans)-ethyl propenyl ether) | 157 |
| Poly(cis(or trans)-ethyl propenyl ether-*block*-isobutyl vinyl ether | 157 |
| (CH2-CH)x (CH$_2$-CH)y with side groups OR and O–$CH_2$–$CH_2$–X; ($R = n\text{-}C_4H_9$, $i\text{-}C_8H_{17}$, $n\text{-}C_{16}H_{33}$; X = OH, $NH_2$, $NH_3^+Cl^-$ $CH_2COOH$, $CH_2COO^-Na^+$ ) | 158-160 |

Monomers bearing electron withdrawing substituents are best fitted for anionic living polymerization provided that they don't contain labile protons and highly nucleophilic functions which may react with propagating sites. However, a limited number of vinyl monomers are suitable for living anionic polymerization. The most suitable monomers include styrene and some derivatives, 1,3-butadiene, and isoprene. Vinylpyridines and alkyl methacrylates undergo anionic living polymerization under specific conditions [172-174]. Monomers possessing functional groups, which are reactive towards carbanionic species, can be protected [175,176]by using appropriate functions. After the monomer is polymerized anionically, deprotection is performed.

Obviously, SMA, in which monomers added in the order of increasing electroaffinity, is the most widely used method to prepare block copolymers via anionic living polymerization. This section aims by no means to review all published work and the reader's attention is directed to previous reviews [171, 177-180] devoted solely to anionic polymerization. Site transformation to anionic polymerization (multi-mode polymerization) is another versatile method to prepare well-defined block copolymers and is described in Chapter 10 of this book.

**Table 4** ABA type Blocks by Sequential Monomer Addition.

| ABA Blocks | References |
|---|---|
| Poly(styrene-*block*-isobutylene-*block*-styrene) | 161-164 |
| Poly(p-methylstyrene-*block*-isobutylene-*block*-p-methylstyrene) | 163 |
| Poly(t-butylstyrene-*block*-isobutylene-*block*-t-butylstyrene) | 163 |
| Poly(2,4,6trimethylstyrene-*block*-isobutylene-*block*-2,4,6 trimethylstyrene) | 164 |
| Poly(indene-*block*-isobutylene-*block*-indene) | 163 |
| Poly((t-butylstyrene-co-indene-*block*-isobutylene-*block*-t-butylstyrene-co-indene | 163 |
| Poly(p-chlorostyrene-*block*-isobutylene-*block*-p-chlorostyrene) | 165 |
| Poly(p-fluorostyrene-block-isobutylene-block-p-fluorostyrene) | 165 |
| Poly(trans-1,4-isoprene-b-isobutylene-b-trans-1,4-isoprene) | 166 |
| Poly(cy-isoprene-block-isobutylene-block-cy-isoprene)<br>cy = cyclized | 164, 166, 167 |
| Poly(α-methylstyrene-block-isobutyl vinyl ether-block- methylstyrene) | 168 |
| Poly(methyl vinyl ether-block-isobutyl vinyl ether-block-methyl vinyl ether) | 168 |
| Poly(methyl vinyl ether-block-cetyl vinyl ether-block-methyl vinyl ether) | 169 |
| Poly(cetyl vinyl ether-block-methyl vinyl ether-block-cetyl vinyl ether) | 169 |
| Poly(cetyl vinyl ether-block-ethyl vinyl ether-block-cetyl vinyl ether) | 169 |
| Radical [Poly(isobutylene-block-styrene)]$_3$ | 163, 165 |

# 5. MACROINITIATORS FOR GROUP TRANSFER POLYMERIZATION

As stated repeatedly, a major objective of synthetic polymer chemistry is the production of polymers with narrow molecular weight distribution and controlled chain length. About a decade ago, Webster and co-workers [181] reported a new living polymerization, named group transfer polymerization (GTP), for producing well-defined acrylic polymers by silyl keten acetal activation using nucleophilic [182] or electrophilic catalysts [183]. GTP involves generation of keten acetal at the chain end at each monomer unit addition (**Scheme 38**).

O, OMe + OSiMe$_3$, OMe ⟶ PMMA— OSiMe$_3$, OMe

**Scheme 38**

The major advantage of this polymerization over to the anionic living polymerization of acrylic esters is its applicability at room temperature or above. Generally, living anionic polymerization of methacrylates is conducted at low temperatures [184], e.g. at -60° C. At high temperatures, active centres react not only with monomer but also with polymer ester groups, which results in premature termination [185].

Preparation of block copolymers by GTP is restricted to methacrylate type monomers. Block copolymers can be prepared by utilizing monomers with appropriate reactivity ratios in a batch process or SMA technique. A wide range of block copolymers with AB or ABA type structures, using mono and difunctional initiators, respectively, were prepared and reviewed [186 - 189] elsewhere. Examples of block and graft copolymers of methacrylates with nonacrylic monomers using GTP in conjunction with other polymerization modes are dealt [190] in detail in Chapter 10.

In principle, any polymer, independent of polymerization mode, possessing terminal or side chain silylketen acetal groups would act as macroinitiators for GTP. Ester groups readily converted [191,192] to silylketen acetal macroinitiators by sequential treatment with lithium diisopropyl amide (LDA) and chlorotrimethylsilane (**Scheme 39**). Alternatively, termination of living GTP of methacrylate with suitable electrophilic reagents would yield polymers with a functionality for the second polymerization mode (**Scheme 40**).

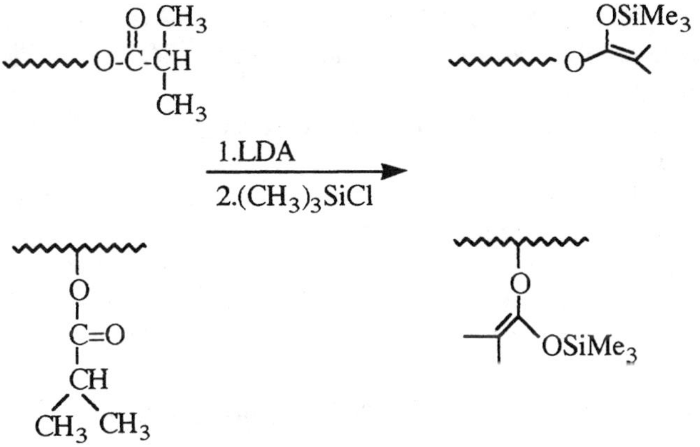

O CH$_3$
~~~O-C-CH
CH$_3$

OSiMe$_3$

1.LDA
2.$(CH_3)_3SiCl$

O
C=O
CH
CH$_3$ CH$_3$

O
OSiMe$_3$

**Scheme 39**

PMMA— OSiMe$_3$ OMe + E- F ⟶ PMMA — F

**Scheme 40**

Initiating functional groups can also be introduced in GTP by using a substituted initiator. For example, hydroxyl end groups are introduced [193, 194] into polymers by using initiators with the following structures and subsequently removing the trimethyl silyl group. Carboxylic, thiol and cyano groups can also be incorporated [193] by the same approach.

OSiMe$_3$
OSiMe$_3$

OSiMe$_3$
O OSiMe$_3$
~~~

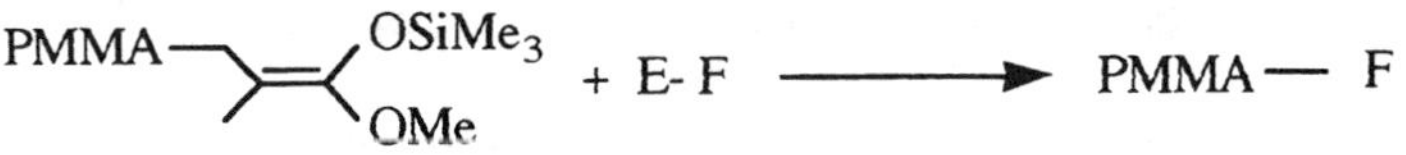

## REFERENCES

1. H. V. Pechmann and L. Vanino, *Ber.*, **27**, 1510 (1894)
2. G. Smets and A. E. Woodward, *J. Polym. Sci.*, **14**, 126 (1954)
3. A. E. Woodward and G. Smets, *J. Polym. Sci.*, **17**, 51 (1955)
4. J. W. Hill, U.S.Pat. 2,556,876,June 12, 1951
5. M. Szwarc, M. Levy and R. Milkovich, *J. Am. chem. Soc.*, **78**, 2656 (1956)
6. M. Szwarc, *Nature*, **178**, 1168 (1956)
7. D. Vofsi and A. V. Tobolsky, *J. Polym. Sci., Part A*, **3**, 3261 (1965)
8. M. Dreyfuss and P. Dreyfuss, *Polymer*, **6**, 93 (1965)
9. C. E. H. Bawn, R. M. Bell and A. Ledwith, *Polymer*, **6**, 95 (1965)
10. S.Kimura, T.Takitani and M.Imoto, *Bull.Chem.Soc.Jpn.*,**35**, 2012 (1962)
11. M.Imoto, K.Kondo and K.Takemoto, *Makromol.Chem.*, **89**, 165 (1965)
12. M.Imoto, K.Takemoto and K.Kondo, *Makromol.Chem.*,**98**, 74 (1966)
13. K.Takemoto, K.Kondo, T.Iwasaki and M.Imoto, *Makromol.Chem.*, **112,** 110 (1968)
14. M.Imoto, K.Takemoto, H.Azuma, N.Kita and M.Kondo, *Makromol. Chem.*, **107,** 188 (1967)
15. M.Imoto, K.Takemoto, A.Okura and M.Izubayashi, *Makromol. Chem.*, **113**, 111 (1968)
16. K.Takemoto and M.Imoto, *Cell.Chem.Techn.*, **4**, 347 (1969)
17. T.Ouchi, T.Nishimura and M.Imoto, *Kobunshi Ronbunshu.*, **32**, 196 (1975)
18. T.Ouchi, T.Nishimura and M.Imoto, *J.Polym.Sci.,Polym.Chem.Ed,* **14** 2695 (1976)
19. M.Iomoto, N.Sakade and T.Ouchi, *J.Polym.Sci.,Polym.Ed.,15*, 499 (1977)
20. T.Nishimura, T.Ouchi and M.Imoto, *Makromol.Chem.*, **177**, 1895 (1976)
21. M.Imoto, Y.Nakamura and T.Ouchi, *Bull.Chem.Soc.Jpn,*. **49**, 1342 (1976)
22. M.Imoto, T.Yamada, A.Tatsumi and T.Ouchi, *Nippon Kayaku Kaishi*, 1883 (1977)
23. M.Imoto, M.Sakae, H.Yamamoto and T.Ouchi, *Makromol.Chem,* **181,** 341 (1980)
24. T.Ouchi, H.Suzuki, T.Yamada and M.Imoto, *J.Makromol.Sci.,Chem,* **A12,** 1461 (1978)
25. T.Ouchi, M.Inamura, A.Kobayashi and M.Imoto, *J.Polym.Sci., Polym.Chem.Ed.*, **19,** 2561 (1981)
26. T.Ouchi, M.Inaba, K.Tadano and M.Imoto, *J.Polym.Sci.,Polym. Chem.Ed.*, **20**, 2089 (1982)
27. T.Ouchi, M.Inaba and M.Imoto, *Bull.Chem.Soc.Jpn.***55**,3229(1982)
28. T.Ouchi, T.Izumi, M.Inaba and M.Imoto, *J.Polym.Sci.,Polym.Chem. Ed.*, **21**, 2101 (1983)

29. W.Deters and H.DeHja-Chian, *Faserforsh.Textil.Tech.*,**14**, 58 (1963)
30. R.L.whister and J.L.Goatley, *J.Polym.Sci.*,**62**, 123 (1962)
31. H.Grohn and C.V.Oprea, *Rev.Roum.Chim.*, **11**, 1297 (1966)
32. H.Grohn and K.Bischof, *Plaste Kautsch,* **8**, 311 (1961)
33. K.Bischof and R.Korn, *Plaste Kautsh*, **10,** 28 (1963)
34. K.Bischof and R.Korn, *Plaste Kautsch,* **10**, 80 (1963)
35. A.M.Dubinskaya, P.Yu Butyagin and A.A.Berlin, *Dokl.Akad,Nauk SSSR* **159,** 595 (1964)
36. Yu.S.Simakov and N.K.Baramboim, *Nauchn.Tr.Mosk.Tekhnol.Inst. Legk.Prom.*, **36,** 59 (1969)
37. Yu.S.Simakov and N.K.baramboim, *Nauchn.Tr.Mosk.Tekhnol.Inst. Legk.Prom.*, **35**, 102 (1969)
38. V.G.Protasov and N.K.Baramboim, *Sov.Plast.*,**2**, 5 (1969)
39. H.Grohn, K.Bischof and H.Hensinger, *Plaste Kautsch*, **9,** 180 (1962)
40. A.Crusos, D.Feldman and C.Simionescu, *Rev.Roum.Chim.*,**13**, 1489 (1968)
41. N.K.Baramboim and Yu.S.Simakov, *Polym.Sci.USSR*, **8**, 255 (1966)
42. N.K.Baramboim and Yu.S.Simakov, *Nauchn.Tr.Mosk.Tekhn.Inst. Legk.Prom.*, **30**, 188 (1964)
43. H.Grohn and C.C.Oprea, *Rev.Roum.Chim.*, **9**, 757 (1964)
44. C.V.Oprea, C.Neguleanu and C.Simionescu, *Eur.Polym.J.*,**6**, 181 (1970)
45. C.V.Oprea, C.Neguleanu and C.Simionescu, *Plaste Kautsch.*, **17**, 9 (1970)
46. C.V.Oprea, C.Neguleanu and C.Simionescu, *Makromol.Chem.,* **126** 217 (1969)
47. C.Simionescu, C.Neguleanu and C.V.Oprea, *Makromol.Chem.*,**148**, 155 (1971)
48. C.Simionescu, C.Neguleanu and C.V.Oprea, *Bull.Inst.Politek Iasi*, **15**, 45 (1969)
49. C.Neguleanu, C.V.Oprea and C.Simionescu, *Makromol.Chem.*, **175,** 371 (1974)
50. C.Simionescu and C.V.Oprea, *Adv.Chem.Ser.*, **128,** 68 (1973)
51. N.K.Baramboim, "Mechanochemistry of polymers," (english translation by R.J.Moseley, Ed.W.F.Watson) Rubber and Plastic Res.Assoc. of Great Britan, Maclaren, 1964
52. A.Casale and R.S.Porter "Polymer Stress Reactions", Academy Press, New York, 1978
53. C.V.Oprea and M.Popa, *Polymer,* **26**, 452 (1985)
54. H.F.Gruber, *Prog.Polym.Sci.*,**17**, 953 (1992)
55. R.S.Davidson, *J.Photochem.Photobiol., Chem.*,**69**, 263 (1993)
56. M.A.Abd-Alla, *Makromol.Chem.*, **192**, 277 (1991)
57. R.Huesler, R.Kirchmayer, W.Rutsh, M.Rembold, Eur.Patent, EP. 0304 886 (1989)
58. R.H.Heiter, G.Rosen, US patent 3,992,363(1976)

59. H.J.Hageman, L.G.Jansen, *Macromol.Chem.*, **189**, 2781 (1988)
60. L.Angiolini, C.Carlini, *Chim.Ind.*, **72**, 124 (1990)
61. J.S.Shim, N.G.Park, U.Y.Kim, K.D.Ahn, *Polymer* (Korea), **8**, 34 (1984)
62. K.B.Hatton, E.Irvin, J.M.A.Walshie, A.Mollaband Eur.Pat.Appl. 302 831 (1987)
63. M.Kohler, J.Ohnemach, Polym, *Paint Colour J.*, **178**, 203 (1988)
64. R.Klos, H.Gruber and G.Greber *J.Macromol.Sci.Chem.*, **A28**, 925 (1991)
65. G.Li Bassi, L.Cadona, F.Broggi, Radcure Eur. 87 Conf.Proc., SME, Dearborn (1987)
66. C.Carlini, *Br.Polym.J.*, **18**, 236 (1986)
67. F.Ciardelli, G.Ruggierri, M.Aglietto, D.Angiolini, C.Carlini, G.Bianchi, G.Siccardi, G.Bigognou and L.Lioni, *J.Coat.Techn.*, **61**, 77 (1989)
68. C. Carlini and F.Gurzoni, *Polymer*, **24,** 599 (1983)
69. C. Carlini, L.Toniolo, P.A.Rolla, F.Barigeletti, P.Bortolus and L.Flamigni, *New Polym.Mater.*, **1**, 63 (1987)
70. G.Rehmer, A.Boettcher, G.Auchter, Eur.Pat.Appl. 377 199 (1989)
71. I.Lukac, C.H.Evans, J.C.Scaiano and P.Hrdlovic, *J.Polym.Sci. Polym.Chem.Ed.*, **28**, 595 (1990)
72. P.N.Green and W.A.Green, *Polym.Paint Colour J.*, **180,** 42 (1990)
73. F.Catalina, C.Peinado, E.L.Madruga, R.Sastre and J.L.Mateo, *J.Polym.Sci., Polym.Chem.Ed.*, **28**, 967 (1990)
74. N.S.Allen, F.Catalina, C.Peinado, R.Sastre, J.L.mateo and P.N.Green, *Eur.Polym.J.*, **23**, 985 (1987)
75. A.Shirata, US Pat. 4 391 963, 1982
76. L.F.Fabrizino, S.Q.S.Lin and A.F.Jakobine, US Pat. 4,536,265(1984)
77. S.Q.S.Lin and A.F.Jakobine, US Pat. 4,538,388(1984)
78. R.West, A.R.Wolf and D.J.Peterson, *J.Rad.Curing*, **13**, 35, (1986)
79. I.Kminek, Y.Yagci and W.Schnabel, *Polym.Bull.*, **29**, 277 (1992)
80. Y.Yagci, I.Kminek and W.Schnabel, *Eur.Polym.J.*, **28,** 387 (1992)
81. Y.Yagci, I.Kminek and W.Schnabel, *Polymer*, **34**, 426 (1993)
82. J.V.Crivello and J.L.Lee, US Pat. 4,780,511
83. Y.Yagci and W.Schnabel, *Prog.Polym.Sci.*, **15,** 551 (1990)
84. J.Dhanras and J.E.Guillet, *J.Polym.Sci. C*, **23**, 433 (1968)
85. W.Kawai, *J.Polym.Sci.Polym.Chem.Ed.*, **15**, 1479 (1974)
86. W.Kawai, *J.Macromol.Chem.*, **A11**, 1027 (1974)
87. W.Kawai and T.Ichiashi, *J.Polym.Sci.Polym.Chem.Ed.*, **12**, 1041 (1974)
88. A.Onen and Y.Yagci, *J.Macromol.Sci.Chem.*, **A27**, 743 (1990)
89. A.Onen and Y.Yagci, *Angew.Makromol.Chem.*, **181,** 191 (1990)
90. Y.Yagci, A.Onen, *J.Macromol.Sci.Chem.*, **A28**, 129 (1991)
91. Y.Hepuzer, M.Bektas, S.Denizligil, A.Onen and Y.Yagci *Macromol.Reports*, **A30**(suppl.1&2) 111, (1993)
92. Y.Yagci, A.Onen and W.Schnabel, *Macromolecules*, **24**, 4620 (1991)
93. G.Smets, *Polym.J.*, *17*, 153 (1985)
94. T.Doi and G.Smets, *Macromolecules*, **22,** 25 (1989)

95. Y.Kurusu, H.N.Shiyama and M.Okawara, *J.Chem.Soc.Jap.,Ind.Chem.Sec.*, **70**, 593 (1967)
96. H.Rudolph, H.J.Rosenkrannz and H.G.Heine, *Appl.Polym.Symp.*,**26**,157 1975)
97. K.D.Hahn, K.J.Ihn, I.C.Kwon,*J. Macromol. Sci. Chem.*, **A23**,355 (1986)
98. M.B.Jackson and W.H.F.Sasse,*J. Macromol. Sci. Chem.*, **A11**,1137 (1977)
99. J.E.Guillet and R.G.W.Norrish, *Proc.R.Soc.Lond.*, **A 233,** 172 (1956)
100. I.Naito, T.Ueki, S.Tabara, M.Tomiki and A.Kinoshita, *J.Polym.Sci., Polym.Chem.Ed.,* **24**, 875 (1986)
101. E.Lanza, H.Bergmans and G.Smets, *J.Polym.Sci.,Polym.Phys.Ed.*,**11**, 95 (1973)
102. H.Craubner, *J.Polym.Sci.,Polym.Chem.Ed.*,**20**, 1935 (1982)
103. H.Craubner, *J.Polym.Sci.,Polym.Chem.Ed.*,**18,** 2011 (1980)
104. M.Imoto, T.Otsu and J.Yonezawa., *Makromol.Chem.*, **36**, 93 (1960)
105. I.Gupta, S.N.Gupta and D.C.Neckers, *J.Polym.Sci.,Polym.Chem.Ed.*, **20**, 147 (1982)
106. D.C.Neckers, *J.Radiat.Curing,***10**, 19 (1983)
107. M.H.Jones, H.W.Melville, W.H.Robertson, *Nature,* **174**, 78 (1954)
108. M.H.Jones, *Can.J.Chem.*,**34**, 948 (1958)
109. M.L.Miller, *Can.J.Chem,* **36,** 309 (1958)
110. Y.Yagci, G.Hizal and U.Tunca, Polymer Commun., 31, 7 (1990)
111. J.F.Kinstle and S.L.Watson, J.Radiat.Curing, 2, 7 (1975)
112. For a review see: C.H.Bamford in New Trends in the Photochemistry of Polymers (N.S.Allen and J.F.Rabek, Eds.) p.129, Elsevier Applied Science Publishers, London (1985)
113. C.H.Bamford and X.Z.Han, *Polymer*, **22,** 1299 (1981)
114. C.H.Bamford, J.P.Middleton, K.D.Al-Lamee and J.Paprotny, *Br.Polym.J.*,**19**, 269 (1987)
115. Y.Yagci, M.Muller and W.Schnabel, *Macromol.Reports,* **A28** (suppl.1) (1991)
116. C.H.Bamford in Reactivity, Mechanism and Structure in Polymer Chemistry (A.D.Jenkins and A.Ledwith, Eds.) p.52 J.Wiley, New York (1974)
117. C.H.Bamford, G.C.Eastmond and D.White, *Polymer*, **10**, 771 (1969)
118. C.H.Bamford, G.C.Eastmond and D.White, *Polymer*, **12**, 241 (1971)
119. J.Ashworth, C.H.Bamford and E.G.Smith, *Pure Appl.Chem*, **30**, 25 (1972)
120. C.H.Bamford, I.P.Middleton and K.G.Al-Lamee, *Polymer*, **27**, 1981 (1986)
121. C.H.Bamford and S.U.Mullik,*Polymer*, **17**, 225 (1976)
122. P. L. Nayak and S. Lenka, *J. Macromol. Sci.-Rev. Macromol. Chem.*, **C19**(1), 83 (1980)
123. M. K. Mishra, *J. Macromol. Sci.-Rev. Macromol. Chem.*, **C19**(2), 193 (1980)
124. M. K. Mishra, *J. Macromol. Sci.-Rev. Macromol. Chem.*, **C22(3)**, 471 (1982-83)
125. S. Lenka, *J. Macromol. Sci.-Rev. Macromol. Chem.,* **C22**(2), 303 (1982-83)
126. A. K. Mukherjee and H. R. Goel, *J. Macromol. Sci.-Rev. Macromol. Chem.*, **C25**(1), 99 (1985)
127. R. K. Samal, P.K. Sahoo and H. S. Samantaray, *J. Macromol. Sci.-Rev. Macromol. Chem.*, **C26**(1), 81 (1986)
128. R. K. Samal and S. S. Giri, *J. Macromol. Sci.-Rev. Macromol. Chem.*, **C32**(1), 55 (1992)

129. C. Fridel and J. M. Crafts, *Compt. Rend,* **84,** 1392 (1877).
130. R. M. Thomas, I. E. Lightbown, W. J. Sparks, K. P. Frdick and E. V. Murphree, *Ind.Eng. Chem.*, **32**, 1283 (1940).
131. A. G. Evans, D. Holden, P. H. Plesch, M. Polanyi, H. A. Skinner and M. A. Weinberger, *Nature* (London), **157**, 102 (1946).
132. J. V. Crivello and J. H. W. Lam, *J. Polym. Sci., Polym. Symp.* **56,** 383 (1986).
133. J. P. Kennedy and E. Merechal, Carbocationic Polymerization, Wiley-Interscience, N.Y, 1982.
134. M. K. Mishra in "Frontiers in Macromolecular Science and Application", *Ind. J. of. Technol*, V **31** (4-6), 197 (1993).
135. M. Szwarc, *Adv. Polym. Sci.,* **49**, 7 (1993).
136. M. Szwarc, *Makromol. Chem., Rapid Communication.*, **13,** 141 (1992).
137. R. Faust and J. P. Kennedy, *J. Polym. Sci., Polym. Chem. Ed.*, **25,** 1847 (1987).
138. M. K. Mishra and J. P. Kennedy, *J. Macromol. Sci., Chem.*, **A24** (8), 933 (1987).
139. J. P. Kennedy and B. Ivan, Designed Polymers by Carbocationic Macromolecular Engineering, Hanser, Munich, 1991.
140. J. Peyrot, Ger. Pat, 2,161,859(1972)
141. Y. Joliveta J. Peyrot, Int. Symp. Cationic Polym., Ro, France, Abstract C18 (1973).
142. J. P. Kennedy and E. G. Melby, *Polym. Preprints,* **15,** 180 (1974)
143. J. P. Kennedy and E. G. Melby, *J. Polym. Sci., Polym. Chem. Ed.* **13**, 29 (1975).
144. J. P. Kennedy, S. C. Feinberg and S. Y. Huang, *J. Polym. Sci., Polym. Chem. Ed.*, **16**, 243 (1978).
145. J. P. Kennedy and J. Kurian, *Polym. Bull.*, **23**, 259 (1990).
146. M. Zsuga and J. P. Kennedy, *J. Polym. Sci., Polym. Chem.*, **29**(6), 875 (1991).
147. J. P. Kennedy, S. Y. Huang and R. A. Smith, *Polym. Bull.* **1**, 371 (1979).
148. J. P. Kennedy, S. Y. Huang and R. A. Smith, *J. Macromol. Sci., Chem.*, **A14**, 1085 (1980).
149. J. P. Kennedy and R. A. Smith, *J. Polym. Sci., Polym. Chem. Ed.*, **18**, 1539 (1980).
150. J. P. Kennedy and F. J. -Y. Chen, *Polym. Prep.*,**20**(2), 310 (1979).
151. J. P. Kennedy, S. C. Guhaniyogi and L. R. Ross, *J. Macromol. Sci., Chem.*, **A18,** 119 (1982).
152. Q. Liu, M. Konas, R. M. Davis and J. S. Riffle, *J. Polym. Sci., Polym. Chem. Ed.*, **31**, 1709 (1993).
153. M. Gyor, J. P. Kennedy, T. Kelen and F. Tudos, *J. Macromol. Sci. Chem.*, **A21**, 1311 (1984)
154. J. M. Rooney, D. R. Squire and V. W. Stannett, *J. Polym. Sci., Polym. Chem.*, **14**, 1877 (1976)
155. T. Higashimura, M. Mitsuhashi and M. Sawamoto, *Macromolecules*, **12**, 178 (1979)
156. M. Miyamoto, S. sawamoto and T. Higashimura, *Macromolecules*, **17**, 2228 (1984)

157. M. Sawamoto, K. Ebara, A. Tanizaki and T. Higashimura, *J. Polym. Sci., Polym. Chem.*, **24**, 2919 (1986)
158. S. Kanoka, M. Minoda, M. Sawamoto and T. Higashimura, *J. Polym. Sci., Polym. Chem.*, **28,** 1127 (1990)
159. S. Minoda, M. Sawamoto and T. Higashimura, *Macromolecules*, **20**, 2045 (1987)
160. M. Minoda, M. Sawamoto and T. Higashimura, *Macromolecules*, **23**, 1897 (1990)
161. Zs. Fodor, J. P. Kennedy, T. Kelen and F. Tudos, *J. Macromol. Sci. Chem.* A**24**, 735 (1987)
162. G. Kaszas, J. E. Puskas, C.C.Chen, J. P.Kennedy and W. Hager, *J. Polym. Sci., Polym. Chem.*, **29**, 427 (1991)
163. J. P. Kennedy, J. E. Puskas, G. Kaszas, W. G. Hager, US Pat., 4,946,899 (1990)
164. M. K. Mishra and A. Osman, US Pat., 4,943,616(1990)
165. J. P. Kennedy and J. Kurian, *J. Polym. Sci., Polym. Chem.*, **28**, 3725 (1990)
166. G. Kaszas, J. E. Puskas and J. P. Kennedy, *J. Appl. Polym. Sci.*, **39**, 119 (1990)
167. J. Puskas, G. Kaszas and J. P. Kennedy, *Org. Coat., Appl. Sci., Proc.*, **57,** 77 (1987)
168. M. Sawamoto and J. P. Kennedy, *J. Macromol. Sci.-Chem.*, **A18**, 1293 (1982)
169. M. Miyamoto, M. Sawamoto and T. Higashimura, *Macromolecules*, **18,** 123 (1985)
170. K. M. Martin, T. C. Ward, J. E. McGrath, Anionic Polymerization ACS Symp. Series 166 (1981).
171. P. Rempp, E. Franta, J. E. Herz, *Adv. Polym. Sci.*, **86**, 145 (1988).
172. M. Tadori and P. Sigwalt, *Eur. Polym. J.*, **9**, 1369 (1973).
173. A. H. Soum, C. F. Tien, T. E. Hogen-Esh, N.B.D. Accorso and M. Fontanille, *Makromol. Chem. Rapid. Commun.*, **4**, 243 (1983).
174. M. Van Beylen, S. Bywater, G. Smets, M. Szwarc and D. J. Worsfold, *Adv. Polym. Sci.*, **86**, 87 (1988).
175. A. Hirao and S. Nakahama, *Prog. Polym. Sci.*, **17**, 283 (1992).
176. S. Bywater, *Prog. Polym. Sci.*, **4**, 27 (1974).
177. P. Remp, Makromol. Chcm., *Macromol. Symp.*, **60,** 209 (1992).
178. S. Penczek, P. Kubisa and K. Matyjaszewski, *Adv. Polym. Sci.*, **37**, 1 (1980).
179. Y. Doi and T. Keii, *Adv. Polym. Sci.*, **73/74**, 201 (1986).
180. M. Morton, "Anionic Polymerization: Principles and Practice", Academic Press, New York, 1983.
181. O. W. Webster, W. R. Hertler, D. Y. Sogah, W. B. Farnham and T. V. Rajan Babu, *J. Am. Chem. Soc.*, **105**, 5706 (1983).
182. I. B. Dicker, G. M. Cohen, W. B. Farnham, W. R.Hertler, E. D. Laganis and D. Y. Sogah, *Macromolecules*, **23**, 4034 (1990).
183. W. R. Hertler, D. Y. Sogah, O. W. Webster and B. M. Trost, *Macromolecules*, **17**, 1415 (1984).
184. A. H. E. Müller in "Recent Advances in Anionic Polymerization", T. E. Hogen-Esch and J. Smid, Eds, Elsevier, New York, 1987, p.205.

185. K. Hatada, T. Kiyama, K. Fumikana, K. Onto and H. Yuki, *ACS Symp. Ser.*, **166**, 327 (1981).
186. O. W. Webster in "Encyclopedia of Polymer Science and Engineering" J.I. Krosehwitz, Ed., Wiley-Interscience, New York, 1987, Vol. 7, p. 580.
187. O. W. Webster, *Makromol. Chem. Macromol. Symp.*, **33**, 133 (1990).
188. G. C. Eastmond and O. W. Webster in "New Methods of Polymer Synthesis" J. R. Ebdon, Ed., Blacktie, Glaskow, U.K., 1991, p. 22.
189. W. J. Brittain, *J. Rubber Chem. Tech*, **64,** 580 (1992).
190. Y. Yagci and M. K. Mishra in "Macromolecular Design: Concept and Practice," M. K. Mishra, Ed., Polymer Frontiers International, Hopewell Jct., New York, Chapter 10 (1993), in press.
191. A. D. Jenkins, E. Tsartolia, D. R. M. Walton, J. H. Jenkins, P. Kratochyil and J. Stejskal, *Macromol. Chem.*, **191**, 2511 (1990).
192. W. G. Ruth, C. G. Moore, W. J. Brittain, J. Si and J. P. Kennedy, *Polymer Preprints*, **34**(1), 479 (1993).
193. D. Y. Sogah, W. R. Hertler, O. W. Webster and G. M. Cohen, *Macromolecules*, **20,** 1973 (1987).
194. J. M. DeSimone, A. M. Hellstern, E. J. Siochi, S. D. Smith, T. C. Ward, P. M. Gallagher, V. J. Krukonis and J. E. McGrath, *Makromol. Chem., Macromol. Symp.*, **32**, 21 (1990).

# Macroinitiators Chapter 7

## MACROINITIATORS INCLUDING SYNTHESES AND APPLICATIONS OF BLOCK COPOLYMERS DERIVED THEREFROM

**Akira Ueda** and **Susumu Nagai**

Department of Plastics
Osaka Municipal Technical
Research Institute
1-6-50, Morinomiya
Joto-ku, Osaka 536
JAPAN

Dept. of Chemical Sciences
and Technology
Kyushu University
6-10-1, Hakozaki
Higashi-ku, Fukuoka 812
JAPAN





Macromolecular Design: Concept and Practice (Macromonomers, Macroinitiators, Macroiniferters, Macroinimers, Macroiniters, and Macroinifers) Pages 265 - 312

# 1. INTRODUCTION

## 1. 1 Historical Survey

**a) MacroPeroxyInitiator:** History of macroinitiators started with the unexpected formation of a polymer peroxide. In 1894, Pechman and Vanino [1] reported that a compound having an intramolecular peroxy group was formed as they synthesized phthaloyl peroxide by reacting phthaloyl chloride with $Na_2O_2$. Shah et al. [2] in 1951, carried out the radical polymerization of styrene using the peroxide as an initiator and found that the average molecular weight of polymer increased with increase in conversion. This result suggested that the structure of the initiator was not an intramolecular peroxide, but a polymer peroxide as shown in Eq. 1.

$$n\ C_6H_4(COCl)_2 + n\ Na_2O_2 \longrightarrow \left[ -\overset{O}{\overset{\|}{C}}-C_6H_4-\overset{O}{\overset{\|}{C}}-O-O- \right]_n + 2n\ NaCl \qquad (1)$$

The first researchers who applied it for block copolymerization were Smets and Woodward [3,4]. They synthesized poly(styrene)-b-poly(methyl-methacrylate) (PS-b-PMMA) and poly(styrene)-b-poly(vinylacetate) (PS-b-PVAc) using the macroperoxyinitiator.

After a long interval, Tsuetkov et al. [5] had reported in 1969 the modification of aliphatic dicarboxylic acid. After one more decade, in 1978 a modification method of aliphatic dicarboxylic acid by diol to obtain a macroperoxy compound [6] was disclosed, which has been developed to "Modiper series," now commercialized by Nippon Oil & Fats Co. Ltd.

**b) MacroAzoInitiator:** Hill [7] disclosed a method of synthesizing a polymer having -N=N . Polyazine was prepared by polycondensation of hidrazine with aliphatic ketone followed by the addition of HCN and oxidation [7, 33].

$$CH_3-\overset{O}{\overset{\|}{C}}-(CH_2)_n-\overset{O}{\overset{\|}{C}}-CH_3 \xrightarrow{N_2H_2-H_2O} \xrightarrow{HCN} \xrightarrow{Oxidation} \left[ -\underset{CN}{\overset{CH_3}{C}}-(CH_2)_n-\underset{CN}{\overset{CH_3}{C}}-N=N- \right]_m \qquad (2)$$

In 1963, Smith [8] synthesized a series of high molecular weight azo compounds. However, his interest was mainly in the scission of the azo linkage of the compound and he only observed the decrease in viscosity of its solution under heating .

The earliest report on the synthesis of block copolymers via macroazoinitiator by Furukawa et al. [9] was published in 1967. They carried out polycondensation of 4,4'-azobis-4-cyanopentanol and polypropylene glycol,of which the latter was previously functionalized by toluene diisocyanate (3), (4).

$$HO\left(\underset{CH_3}{\overset{|}{C}}HCH_2O\right)_n H + 2\, OCN-\underset{CH_3}{C_6H_3}-NCO \longrightarrow$$

$$OCN-\underset{CH_3}{C_6H_3}-NHCOO\left(\underset{CH_3}{C}HCH_2O\right)_n CONH-\underset{CH_3}{C_6H_3}-NCO \tag{3}$$

$$HO(CH_2)_3\overset{CH_3}{\underset{CN}{C}}-N{=}N-\overset{CH_3}{\underset{CN}{C}}(CH_2)_3OH + OCN-\overset{CH_3}{C_6H_3}-NHCOO\left(\overset{CH_3}{C}HCH_2O\right)_n CONH-\overset{CH_3}{C_6H_3}-NCO$$

$$\longrightarrow \tag{4}$$

$$\left[O(CH_2)_3\overset{CH_3}{\underset{CN}{C}}-N{=}N-\overset{CH_3}{\underset{CN}{C}}(CH_2)_3OCONH-\overset{CH_3}{C_6H_3}\ NHCOO\left(\overset{CH_3}{C}HCH_2O\right)_n CONH-\overset{CH_3}{C_6H_3}\ NHCO\right]_m$$

The macroazoinitiator formed was then decomposed in the presence of vinyl monomers such as vinyl acetate, styrene, methyl methacrylate and vinyl chloride to obtain block copolymers. But, further development has not been made by them.

Nagai, Hidaka, and Ueda [10,11]have disclosed azo linked polyamide synthesized by interfacial polycondensation between 4,4'-azobis-4-cyanopentanoyl chloride (ACPC) and hexamethylenediamine (HMDA) as shown in Eq. 5.

$$n\ ClOCCH_2CH_2\overset{CH_3}{\underset{CN}{C}}-N{=}N-\overset{CH_3}{\underset{CN}{C}}CH_2CH_2COCl + n\ H_2N(CH_2)_6NH_2 \longrightarrow$$

$$\left[OCCH_2CH_2\overset{CH_3}{\underset{CN}{C}}-N{=}N-\overset{CH_3}{\underset{CN}{C}}CH_2CH_2CONH(CH_2)_6NH\right]_n + 2n\ HCl \tag{5}$$

Since then a series of studies has been continued by Ueda and Nagai et al. to develop various macroazoinitiators and block copolymers as exemplified by [12,29]. In another approach, Walz et al. in 1977 used 2,2'-azobisisobutyronitrile as a starting material for macroazoinitiators [13].

$$n\ NC-\underset{CH_3}{\overset{CH_3}{\underset{|}{\overset{|}{C}}}}-N{=}N-\underset{CH_3}{\overset{CH_3}{\underset{|}{\overset{|}{C}}}}-CN \quad + \quad n\ HO-R-OH$$

$$\xrightarrow{\text{Pinner synthesis}} \left( OC-\underset{CH_3}{\overset{CH_3}{\underset{|}{\overset{|}{C}}}}-N{=}N-\underset{CH_3}{\overset{CH_3}{\underset{|}{\overset{|}{C}}}}-CO-O-R-O \right)_n \tag{6}$$

Presently, the macroinitiators have become one of the unique key materials for producing various block copolymers via radical process, providing high-performance or high-functional materials with a variety of molecular-pair design never attained before.

## 1.2 Features of MacroInitiators

Features of macroinitiators will be described below in detail, but are shortly summarized here as follows:

1)A macroinitiator is characterisitic of having many or at least one scissile bond, such as -O O or N=N , in a macromolecule or an oligomer (Fig. 1).

2)It is stable at room temperature, but it is decomposable at elevated temperatures.

3)When it is decomposed by the scission of -O O or N=N , radical polymerization takes place in the presence of monomers.

4)If a second monomer is added to the above, the second monomer also polymerizes due to the presence of the residual scissile bonds, in the previously formed polymer chains, i.e.,yielding a block copolymer.

5)This kind of block copolymerization never takes place with ordinary initiators such as benzoyl peroxide or 2,2'-azobisisobutyronitrile.

Block copolymerization via living anionic polymerization is advantageous in comparison with radical process, since it gives block copolymers with well defined compositions and sharp molecular weight distributions. On the other hand, it has disadvantages of requiring extra purity of materials and solvents, extremely low temperature, besides the limited choice of monomers usable for the polymerization.

Conversely, block copolymerization via radical process is advantageous, in comparison with anionic process, since it is easy to handle and open to wider choice in coupling block chains, although the length of each chain segment and the total molecular weight of block copolymers are not strictly controllable.

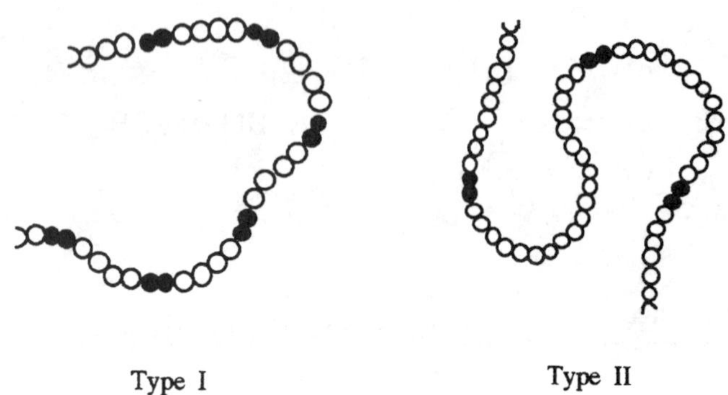

Type I Type II

Fig. 1 Simplified models of macroinitiator (●● scissile bond)

## 2. SYNTHESES OF MACROINITIATORS

### 2. 1 Materials

For materials of macroinitiator, it is essential to have a scissile bond capable of radical initiation, and to have bifunctionality at molecular ends for polycondensation or polyaddition. Materials mostly used are hydroperoxide [I] for macroperoxyinitiator, and 4,4'-azobis-4-cyanopentanoyl chloride (ACPC) [II] and 2,2'-azobisisobutyronitrile (AIBN) [III] for macroazoinitiator.

$$\text{H-O-O-H} \qquad [\text{I}]$$

$$\text{ClCOCH}_2\text{CH}_2\underset{\text{CN}}{\overset{\text{CH}_3}{\text{C}}}\text{-N=N-}\underset{\text{CN}}{\overset{\text{CH}_3}{\text{C}}}\text{CH}_2\text{CH}_2\text{COCl} \qquad [\text{II}]$$

$$\text{NC-}\underset{\text{CH}_3}{\overset{\text{CH}_3}{\text{C}}}\text{-N=N-}\underset{\text{CH}_3}{\overset{\text{CH}_3}{\text{C}}}\text{-CN} \qquad [\text{III}]$$

Other materials for macroperoxyinitiators are compounds having terminal peroxy-groups like [IV–VI] and peroxyisocyanate [VII].

Peroxy-compounds

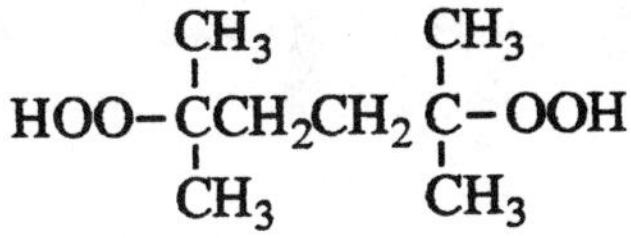

$$HOO-C(CH_3)_2CH_2CH_2C(CH_3)_2-OOH$$

[ IV ]

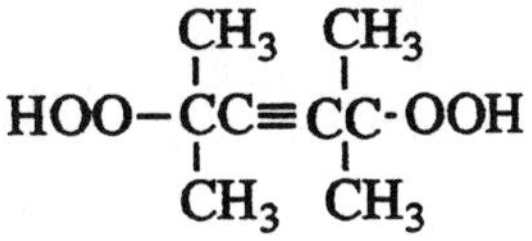

$$HOO-C(CH_3)_2C\equiv CC(CH_3)_2\cdot OOH$$

[ V ]

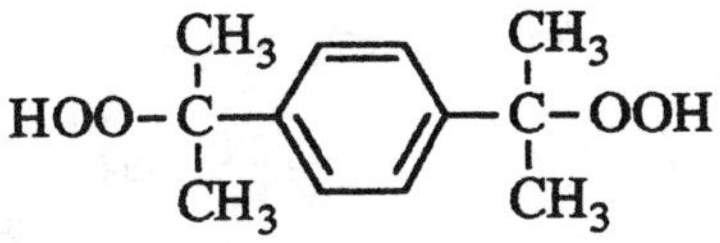

$$HOO-C(CH_3)_2-C_6H_4-C(CH_3)_2-OOH$$

[ VI ]

$$\text{OCN}(H_3C)(H_3C)_2\text{-cyclohexyl}-NHCOOOC(CH_3)_2CH_2CH_2C(CH_3)_2OOCONH-\text{cyclohexyl}(CH_3)_2(CH_3)\text{NCO}$$

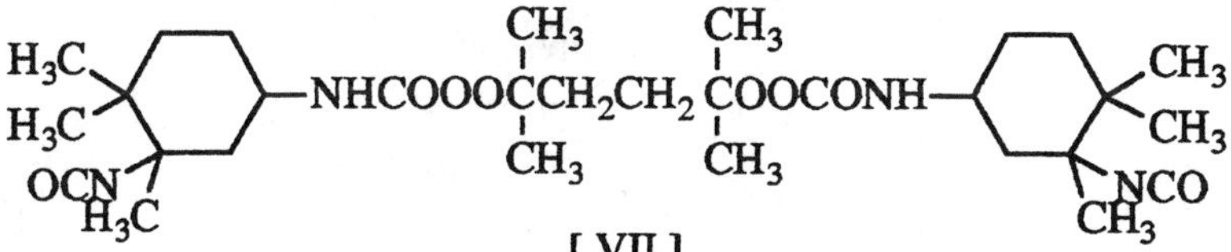

[ VII ]

For macroazoinitiators, azodiol compounds [VIII–XV] having terminal OH, azodiamine compounds [XVI–XXIV] having amino terminals, 4,4'-azobis-4-cyanopentanoic acid (ACPA) [XXV], which is the starting material of ACPC, and azodiisocyanate [XXVI] have been investigated.

Azo compounds :

Azo-diols

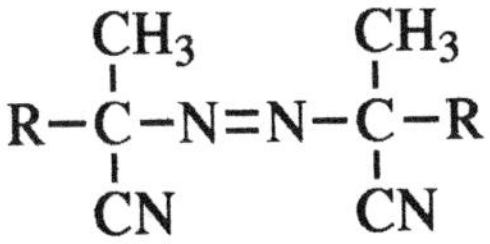

$$R-C(CH_3)(CN)-N=N-C(CH_3)(CN)-R$$

R: $-CH_2OH$ [VIII]

R: $-CH_2CH_2CH_2OH$ [IX]

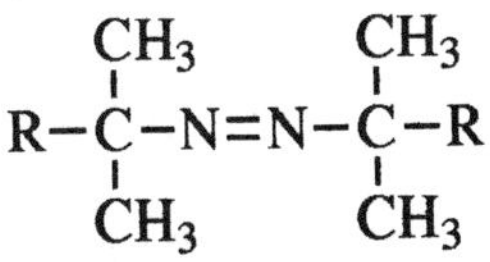

$$R-C(CH_3)_2-N=N-C(CH_3)_2-R$$

R: $-CONHCH_2CH_2OH$ [X]

R: $-CONHC(CH_3)(CH_2OH)_2$ [XI]

R: $-CONHC(CH_2OH)_3$ [XII]

R: $-C(=NH)NH-C_6H_4-OH$ [XIII]

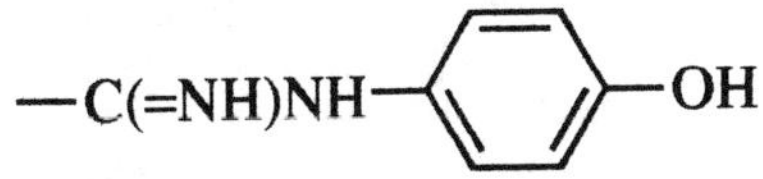

R : $-C(=NH)NH-C_6H_4-OH$ (3-hydroxyphenyl) [XIV]

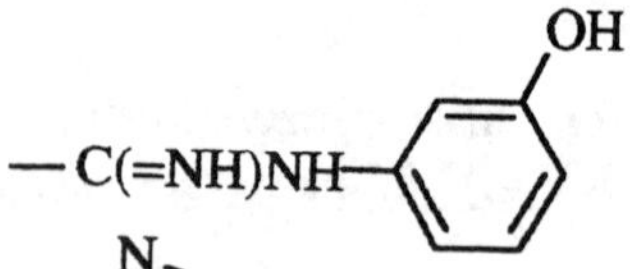

R : 2-imidazolinyl with $-CH_2CH_2OH$ at the 4-position ($-C(=N-CH_2-CH(CH_2CH_2OH)-NH)$) [XV]

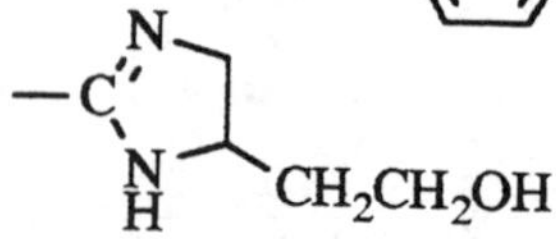

Azo-diamines

$$R-\underset{CH_3}{\overset{CH_3}{C}}-N{=}N-\underset{CH_3}{\overset{CH_3}{C}}-R$$

R : $-CONHNH_2$ [XVI]

R : $-CH_2CH_2CONHCH_2CH_2NH_2$ [XVII]

R : $-CONH-C_6H_4-NH_2$ (para) [XVIII]

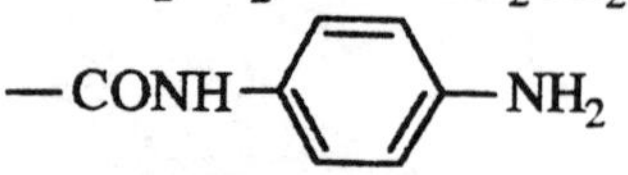

R : $-C(=NH)NH_2$ [XIX]

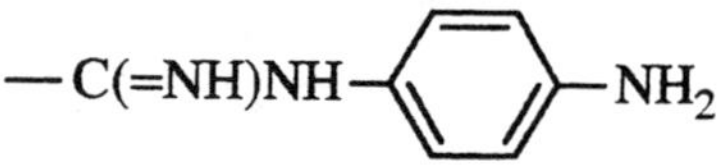

R : $-C(=NH)NH-C_6H_4-NH_2$ (para) [XX]

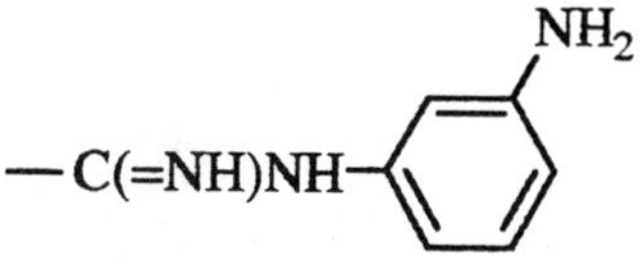

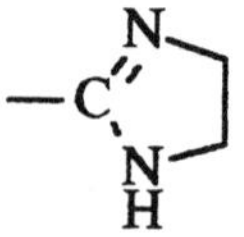

R : $-C(=NH)NH-C_6H_4-NH_2$ (3-aminophenyl) [XXI]

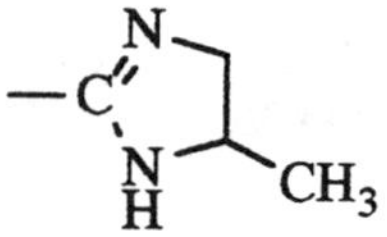

R : 2-imidazolinyl ($-C(=N-CH_2-CH_2-NH)$) [XXII]

R : 4-methyl-2-imidazolinyl ($-C(=N-CH_2-CH(CH_3)-NH)$) [XXIII]

R : 1,4,5,6-tetrahydropyrimidin-2-yl ($-C(=N-CH_2CH_2CH_2-NH)$) [XXIV]

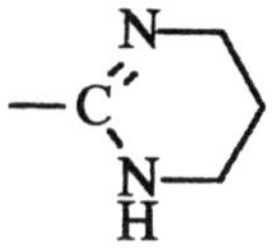

Azo-dicarboxylic acid

$$HOOCH_2CH_2-\underset{CN}{\overset{CH_3}{C}}-N{=}N-\underset{CN}{\overset{CH_3}{C}}-CH_2CH_2COOH \quad [XXV]$$

Azo-diisocyanate

$$OCNCH_2CH_2-\underset{CN}{\overset{CH_3}{C}}-N{=}N-\underset{CN}{\overset{CH_3}{C}}-CH_2CH_2NCO \quad [XXVI]$$

## 2. 2 Synthetic Methods

Macroinitiators are synthesized by using materials having both radical initiation ability and polycondensation or polyaddition functionality. Since radical initiators are sensitive and ready to decompose against heat and light, care should be taken to prevent the decomposition, and mild conditions must be kept throughout the processes in the preparation of macroinitiators.

Table 1 Synthetic Conditions of MacroPeroxyInitiators From Materials A and B

| Material A | Functional Group of A | Functional Group of B | Reaction Conditions | | | | Ref. |
|---|---|---|---|---|---|---|---|
| | | | Medium | Temp. °C | Catalyst | Time h | |
| HOOH [I] | –OOH | –COCl | Interface | 5 | NaOH | 0.7 | 14 |
| [IV] | –OOH | –COCl | Interface | 0–5 | KOH | 1 | 15 |
| [IV] | –OOH | –COCl | Benzene | 23–25 | Pyridine | 3 | 16 |
| [IV] | –OOH | –NCO | Toluene | r.t. | b) | 48 | 17 |
| [VII] | –NCO | –OH | DMF a) | 40 | b) | 80 | 18 |

a) Dimethylformamide b) Dibutyltin dilaurate

Optimum reaction conditions must be chosen, since reactivity between each pair of functional groups differs case by case as shown in Table 1 for macroperoxyinitiators, and in Table 2 for macroazoinitiators. Interfacial polycondensation, solution polycondensation or polyaddition are mostly taken place.

The reactions between -OH and COCl, and between COCl and -$NH_2$ to form ester and amide, respectively, proceed very quickly, while reactions between -OH and -NCO, CN and OH, $NH_2$ and OC O CO to form urethane, ester, imide, respectively, proceed very slowly. Any of these reactions are applicable to obtain macroinitiators preventing the decomposition of initiation sites of materials.

For example, an oligomer terminated with acid chlorides, obtained by reacting excess amount of adipoyl dichloride with triethyleneglycol, was further reacted with sodium peroxide and a macroperoxyinitiator (DP = 25) was prepared [27]. In another example, when ACPC was reacted with 1,6-hexanediol (HDO), a macroazoinitiator containing azo bond in every repeating unit (about 25) was obtained as shown in Eq. 7 and Fig. 2 [28].

Another type of macroazoinitiator has been developed by partly converting –NHCO to –N(NO) CO in polyamides and polyurethanes by $N_2O_3$, then converted to diazoesters (–N=N O CO ) by light or heat. These azo groups can be also used for radical initiation [30, 31].

Table 2 Synthetic Conditions of MacroAzoInitiators From Materials A and B

| Material A | Functional Group of A | Functional Group of B | Reaction Conditions | | | | Ref. |
|---|---|---|---|---|---|---|---|
| | | | Medium | Temp. °C | Catalyst | Time h | |
| ACPC [II] | –COCl | $-NH_2$ | Interface | r.t. | NaOH | 0 [e)] | 11 |
| ACPC | –COCl | $-NH_2$ | $CHCl_3$ | 30 | $Et_3N$ | 2 | 19 |
| Azodiamine | $-NH_2$ | –COCl | Interface | r.t. | NaOH | 1 | 1 |
| ACPC | –COCl | –OH | $CHCl_3$ | 30 | $Et_3N$ | 2 | 20 |
| AIBN [III] | –CN | –OH | $H_2SO_4$ | 0 | HCl | 20 | 21 |
| AIBN | –CN | –OH | $H_2SO_4$ | 0 | — | 1.5 | 22 |
| Azodiol | –OH | –COCl | $CHCl_3$ | 30 | $Et_3N$ | 2 | 23 |
| Azodiol | –OH | –NCO | THF [a)] | 25 | c) | 24 | 9 |
| Azodiamine | $-NH_2$ | -OC-O-CO- | MePydon [b)] | r.t. | — | 24 | 24 |
| ACPA [XXV] | –COOH | –OH | $CH_2Cl_2$ | r.t | $Et_3N$+ d) | 5–18 | 25, 26 |
| ACPA | –COOH | $-NH_2$ | $CH_2Cl_2$ | r.t. | $Et_3N$+ d) | 18 | 25, 26 |

a) Tetrahydrofuran
b) Methylpyrrolidone
c) Dibutyltin dilaurate
d) 1-Methyl-2-chloropyridinium iodide
e) Instantaneous

$$\text{n ClOCCH}_2\text{CH}_2\overset{\overset{\displaystyle CH_3}{|}}{\underset{\underset{\displaystyle CN}{|}}{C}}\text{-N=N-}\overset{\overset{\displaystyle CH_3}{|}}{\underset{\underset{\displaystyle CN}{|}}{C}}\text{CH}_2\text{CH}_2\text{COCl} + \text{n HO(CH}_2)_6\text{OH} \longrightarrow \quad (7)$$

$$\left[ \text{OCCH}_2\text{CH}_2\overset{\overset{\displaystyle CH_3}{|}}{\underset{\underset{\displaystyle CN}{|}}{C}}\text{-N=N-}\overset{\overset{\displaystyle CH_3}{|}}{\underset{\underset{\displaystyle CN}{|}}{C}}\text{CH}_2\text{CH}_2\text{COO(CH}_2)_6\text{O} \right]_n + 2n\,\text{HCl}$$

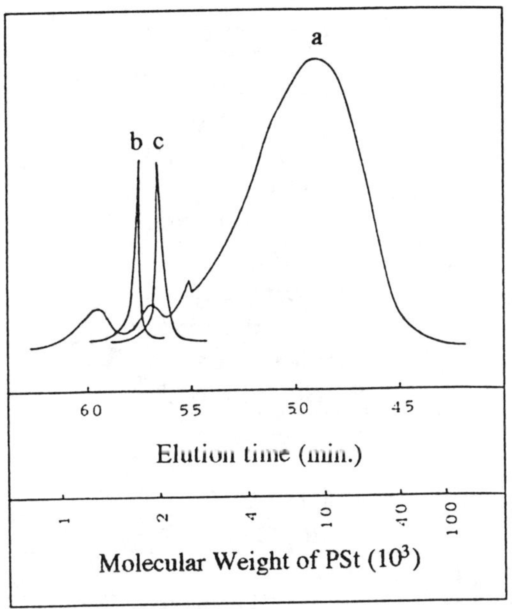


Fig. 2 GPC curve of a) Poly(hexamethylene-4,4'-azobis-4-cyanopentanoate) b) ACPA c) HDO

## 2. 3 Kinds of MacroInitiators

Macroinitiator for radical block copolymerization is classified to two types:

Type I: Macroinitiators consisted of repeated short units containing a scissile bond in every unit.

Type II: Macroinitiator consisted of repeated segments of a prepolymer linked with scissile bonds.

Macroperoxyinitiators hitherto prepared, so far as found in literatures, are summarized in Tables 3 and 4, and Macroazoinitiators are summarized in Tables 5 and 7.

Table 3 MacroInitiators (Type I ) Having Peroxy Groups Synthesized From Materials A, B and C

| Structure of Peroxy Group | Materials (A) | (B) | (C) | Ref. |
|---|---|---|---|---|
| —C(=O)—O—O—C(=O)— | [II] | phthaloyl dichloride | | 1 |
| | | adipoyl dichloride | | 32 |
| | | glutaroyl dichloride | | 5 |
| | | pimeloyl dichloride | | 5 |
| | | suberoyl dichloride | | 5 |
| | | azelaoyl dichloride | | 5 |
| | | sebacoyl dichloride | | 5 |
| | | undecanediacid dichloride | | 5 |
| | | decanediacid dichloride | | 5 |
| | | tridecanediacid dichloride | | 5 |
| —C(=O)—O—O— | [IV] | adipoyl dichloride | | 34 |
| | | tetrahydrophthaloyl dichloride | | 35 |
| | | 4-methyltetrahydro-phthaloyl dichloride | | 35 |
| | [V] | tetrahydrophthaloyl dichloride | | 35 |
| | | 4-methyltetrahydro-phthaloyl dichloride | | 35 |
| | [VI] | tetrahydrophthaloyl dichloride | | 35 |
| | | 4-methyltetrahydro-phthaloyl dichloride | | 35 |
| —C(=O)—O—O—C(=O)— | [I] | adipoyl dichloride | ethylenediol, propylenediol, 1,4-butanediol, diethyleneglycol, triethyleneglycol | 6, 14, 27, 36–38 |
| | | dodecanediacid dichloride | ethylenediol, diethyleneglycol | 6 |
| | | isophthaloyl dichloride | 2,2'-bis(4-hydroxy-phenyl)propane | 6, 38 |
| —O—O—C(=O)—NH— | [VII] | tolylenediisocyanate-butadiene-tolylendi-isocyanate | N, N'-bis(ß-hydroxyethyl)piperazine | 18 |

**a) MacroPeroxyInitiator:** Most of the macroperoxyinitiators belong to type I (Table 3) having active site in every short repeating unit. For the active site, peroxydiacyl group and peroxy ester group can be made by polycondensation. Some of the macroperoxyinitiators have peroxy diacyl group or urethane group modified by adding the third substance.

Only few examples of type II macroperoxyinitiator (Table 4), i.e. prepolymers linked by -O-O-, were reported using prepolymers such as poly(ethyleneglycol), poly(propyleneglycol) and poly(dimethylsiloxane).

Table 4 MacroInitiators (Type II) Composed of Prepolymer Linked with Peroxy Group

| Prepolymer | Peroxide | 3rd Component | Ref. |
|---|---|---|---|
| Polyethyleneglycol | [IV] | 4, 4'-methylene dicyclohexyl isocyanate<br>2, 2, 4-trimethylhexamethylene diiso-cyanate | 17, 42, 43 |
| Polypropyleneglycol | [IV] | 4, 4'-methylene dicyclohexyl isocyanate<br>2, 2, 4-trimethylhexamethylene diiso-cyanate | 42, 43 |
| Polydimethylsiloxane | [I] | phthaloyl dichloride<br>terephthaloyl dichloride | 44 |

**b) MacroAzoInitiator:** Both types I and II have been prepared. For type I having -N=N in every short repeating unit, various polyamides, polyesters, polyurethanes, and polyimides have been obtained (Table 5).

The type II macroazoinitiator containing -N=N which links prepolymers for final component of block copolymers is prepared by two routes:
One is *in situ* preparation of prepolymer. For example, copolycondensation of a mixture of ACPC and adipoyl chloride with hexamethylenediamine (HMDA) gives a copolyamide. If a very small amount of ACPC is used in feed, macroinitiator with long polyamide 6 6 segments linked by ACP unit is prepared (Table 6).

Another route is to use a ready-made oligomer having telechelic functionality for linking with scissile materials like ACPC. For oligomers, polyester, polyurethane, polybutadiene, polydimethylsiloxane, polycarbonate, polyethleneglycol, poly-propyleneglycol, polytetrahydrofuran, and polyolefin with telechelic functions have been available (Table 7).

Table 5 Macroinitiators (Type I) Having Azo Group in Every Short Repeating Unit Synthesized From Materials A and B

| (A) | (B) | Ref. |
|---|---|---|
| <Polyamide> | | |
| ACPC [II] | 1,2-ethylenediamine, 1,6-hexanediamine, 1,4-cyclohexanebismethylamine, 4,4'-diaminodibenzo-18-crown-6 | 8, 10, 11, 45–50 |
| Azodiamine [XVIII, XX, XXII] | sebacoyl dichloride, isophthaloyl dichloride | 51 |
| ACPA [XXV] | 1,6-hexanediamine, 4,4'-methylenedianiline | 25,26 |
| <Polyester> | | |
| ACPC | 1,6-hexanediol, 4,4'-isopropylidenediphenol, 2,2,3,3,4,4,5,5-octafluoro-1,6-hexanediol | 8, 10, 28, 52–55 |
| AIBN [III] | formaldehyde, 1,2-ethyleneglycol, 1,4-butanediol, 1,6-hexanediol, diethyleneglycol, triethyleneglycol, tetraethyleneglycol | 13, 21, 22, 56–61 |
| Azodiol [VIII–XIV] | sebacoyl dichloride, isophthaloyl dichloride | 23, 51, 62, 63 |
| ACPA [XXV] | 1,6-hexanediol, 4,4'-isopropylidenediphenol, 4,4'-sulfonyl diphenol, 3,3',5,5'-tetramethyl-4,4'-sulfonyl diphenol | 26, 64 |
| <Polyurethane> | | |
| Azodiol [VIII, X] | hexamethylenediisocyanate, tolylenediisocyanate | 23, 63, 65 |
| <Polyurea> | | |
| Azodiisocyanate [XXVI] | 1,2-ethylenediamine, 1,6-hexanediamine, diethylenetriamine | 50 |
| <Polyimide> | | |
| Azodiamine [XX] | 1,2,3,4-butanetetracarboxylic dianhydride | 24 |

Table 6 MacroInitiators (Type II) Having Block Components Synthesized by *in situ* Reaction of Azo Materials A with a Mixture of B and C

| (A) | (B) | (C) | Ref. |
|---|---|---|---|
| **<Polyamide>** | | | |
| ACPC [II] | 1,6-hexanediamine, 4,4'-diaminodibenzo-18-crown-6, 4,4'-isopropylidenediphenol | adipoyl dichloride, sebacoyl dichloride, succinyl dichloride, terephthaloyl dichloride | 10, 45, 46, 49, 66 |
| Azodiol [VIII, X, XIII] | isophthaloyl dichloride | 1,3-bis(4-aminophenoxy) benzene, 3,4'-diaminodiphenylether | 9, 62 |
| Azodiamine [XX, XXI, XXIII,XXIV] | isophthaloyl dichloride | 1,6-hexanediamine, isophoronediamine, 1,3-bis(4-aminophenoxy) benzene | 51 |
| **<Polyester>** | | | |
| Azodiol [VIII] | adipoyl dichloride | 3,3-dimethylolheptane | 62 |
| **<Polyarylate>** | | | |
| Azodiol [X, XIV] | isophthaloyl dichloride terephthaloyl chloride | 4,4'-isopropylidenediphenol | 51, 62 |
| Azodiamine [XX, XXII] | isophthaloyl dichloride terephthaloyl chloride | 4,4'-isopropylidenediphenol | 51 |
| **<Polyurethane>** | | | |
| Azodiol [VIII] | hexamethylenediisocyanate | neopentyl glycol, trimethylol propane | 65 |
| **<Polyimide>** | | | |
| Azodiamine[XXI] | 1,2,3,4-butanetetracarboxylic dianhydride, 3,3',4,4'-benzophenone tetracarboxylic dianhydride | 1,3-bis(4-aminophenoxy) benzene | 24 |

Table 7 Macroinitiators (Type II) Having Prepolymer Linked with Azo Group

| Prepolymer | End group | Azo compd. | The 3rd materials | Ref. |
|---|---|---|---|---|
| Polybutadiene | OH | ACPC [II] | | 20, 67 |
| Polyester | OH | ACPC | | 68–71 |
| F-polyester *1 | OH | ACPC | | 55 |
| Polyarylate | OH | ACPC | | 72 |
| Silicone | NH2 | ACPC | | 19, 73, 74, 88 |
| F-silicone *2 | NH2 | ACPC | | 75 |
| Silicone | OH | ACPC | | 76 |
| | OH | Azodiol [VIII] | hexamethylenediiso-cyanate | 77 |
| Polycarbonate | Clfmt *4 | Azodiol [VIII, X, XIII, XV] | sebacoyl chloride, isophthaloyl chloride, teraphthaloyl chloride | 63, 78 |
| | Clfmt *4 | Azodiamine [XIX, –XXIII] | | 78 |
| Polyurethane | OH | ACPC | | 68–71 |
| F-polyurethane *3 | OH | ACPC | | 55 |
| Polyethyleneglycol | OH | ACPC | | 53, 80–83 |
| | OH | AIBN | | 13, 84 |
| | OH | Azodiol [IX] | isophoronediisocyanate | 85 |
| | OH | ACPA [XXV] | | 25, 26 |
| | OH *5 | ACPC | | 87 |
| Polypropyleneglycol | OH | AIBN | | 13, 86 |
| | OH | Azodiol [VIII,IX] | hexamethylenediiso-cyanate,tolylenediiso-cyanate, xylenediiso-cyanate, isophoronedi-isocyanate, diphenyl-methylenediisocyanate | 9, 65, 85 |
| Polytetrahydrofuran | OH | AIBN | | 84 |
| | OH | Azodiol [IX] | diphenylmethylenedi-isocyanate | 9 |
| Polyolefin | OH | Azodiol [VIII–X] | tolylenediisocyanate, diphenylmethylenediiso-cyanate | 64, 65 |

*1–*3 Fluorinated Prepolymer *4 Chloroformate
*5 Polyethyleneglycol monomethylether

## 2. 4 Solubitity of MacroInitiators

One of the most important properties of a macroinitiator for block copolymerization is solubility in monomers. Homogeneous solution of monomers and macroinitiators is essential to proceed polymerization. In the case of macroperoxyinitiators, aliphatic peroxy compounds are more soluble in various monomers in comparison with aromatic ones (Table 8), and an important key point in successful development of "Modiper series", based on aliphatic macroperoxyinitiators, was their excellent solubility accompanied with controllability of peroxy content within its safe threshold [27].

Table 8 Solubility of MacroPeroxyInitiators (grams at 25 °C in 100 g of solvent)

| Materials / | Benzene | Toluene | Styrene | Vinyl acetate |
|---|---|---|---|---|
| A | 14.8 | 5.4 | 5.7 | 5.3 |
| B | 1.9 | 1.0 | 0.5 | 0.5 |

A: prepared from triethyleneglycol, adipoyldichloride, and $Na_2O_2$.
B: prepared from benzoyl chloride and $Na_2O_2$.

In the case of macroazoinitiators, macroazo amide compounds are soluble in few solvents such as DMF, *m*-Cresol. Further, monomers compatible to the solution are also limited. For example, poly(hexamethylene-4,4'-azobis-4-cyano-pentanamide) (cf. eq. 5) proved usable to polymerize styrene and methyl-methacrylate in DMF solution keeping homogeneous state, while other monomers caused precipitation of the macroinitiator and polymerization failed [10, 54, 88].

Macroazoinitiators of polyester type are advantageous in the view point of solubility and permit wide selection of solvents and monomers. Poly(hexamethylene-4,4'-azobis-4-cyanopentanoate) in benzene is kept homogeneous by addition of monomers such as styrene, methylmcthacrylatc, butyl acrylate and other kinds, and block copolymerization takes place smoothly [28].

For type II macroazoinitiators having long prepolymer structure linked with azo bonds, the solubility of the initiator depends on that of prepolymer since the latter dominates in the composition or volume fraction of the initiator.

# 3. DECOMPOSITION BEHAVIOR OF MACROINITIATORS

## 3. 1 Decomposition of MacroPeroxyInitiators

Decomposition behaviors of macroinitiators have been investigated in comparison with those of monomeric initiators. Decomposition of peroxy group in solution was measured by iodometry in terms of active oxygen content. Macroinitiator prepared by hydroperoxide [IV] and polyethyleneglycol (*MW* 4000), methylenedicyclohexylisocyanate proceeds its peroxy decomposition via 1st order reaction (Fig. 3, *Ea* = 123 kJ/mol) [42]. The rate constants of decomposition of both macro-initiators and monomeric ones are almost the same if the peroxy end structures are similar to one another as shown in Table 9. A macroinitiator obtained by reacting triethyleneglycol, adipoyl chloride and sodium peroxide behaves similarly.

Table 9 Decomposition Rate and Initiation Efficiency of MacroPeroxyInitiators in Comparison with Ordinary Initiators

| Materials | Solvent | Temp. °C | Decomp. Rate $10^{-5}$/sec. | Init. Eff. | Ref. |
|---|---|---|---|---|---|
| Polyethyleneglycol(MW4000)+ Methylenedicyclohexyliso-cyanate + peroxide [IV] | toluene | 80 | 2.0 | — | 42 |
| Bis(α)cumyl-N, N'-hexa-methylenediperoxycarbamate | toluene | 80 | 1.8 | — | 42 |
| Triethyleneglycol + adipoyldichloride + $Na_2O_2$ | benzene | 70 | 4.9 | 0.24 | 89 |
| Lauroyl peroxide | benzene | 70 | 4.4 | 0.52 | 89 |
| Benzoyl peroxide | — | 70 | — | 0.90 | 89 |

## 3. 2 Decomposition of MacroAzoInitiators

Azo decomposition in solid state has been measured by TG and DSC. A macroinitiator obtained by polycondensation of ACPC and hexamethylenediamine shows about the same decomposition temperature measured by TG as monomeric azo carboxylic acid [XXV] having identical azo group as shown in Fig. 4 [10, 11]. DSC provides more precise data of decomposition temperature.

Macroazoinitiators obtained from ACPC or azodiisocyanate [XXVI] by reaction with various diamines showed different activation energies from 100 to 139 kJ/mol depending on the morphology of polymers, although they have identical azo group [50].

In solution, decomposition begins at lower temperatures than in solid state. Fig. 5 shows the change in solution viscosity of poly (hexamethylene-4,4'-azobis-4-cyanopentanamide) at 50-70 oC due to decomposition [11].

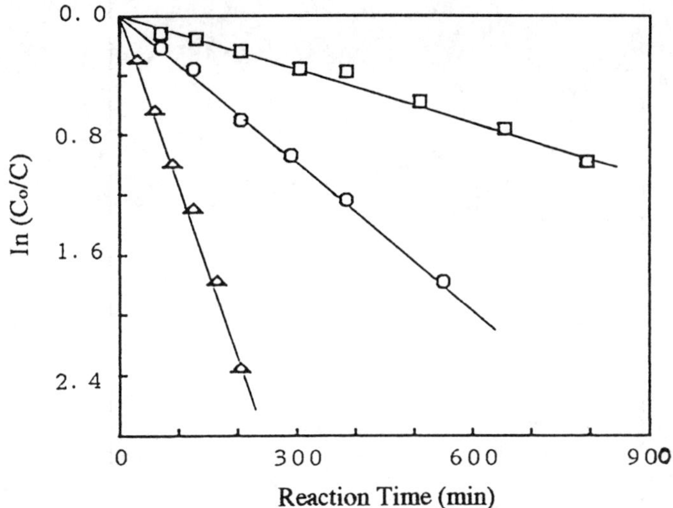


Fig. 3 Thermal decomposition rate indicating the first order reaction of a macroperoxy initiator prepared from hydroperoxide, polyethyleneglycol (*MW* 4000), and methylenedicyclohexylisocyanate [42] ( □ ; 80°C, ○ ; 90°C, △ ; 100°C )

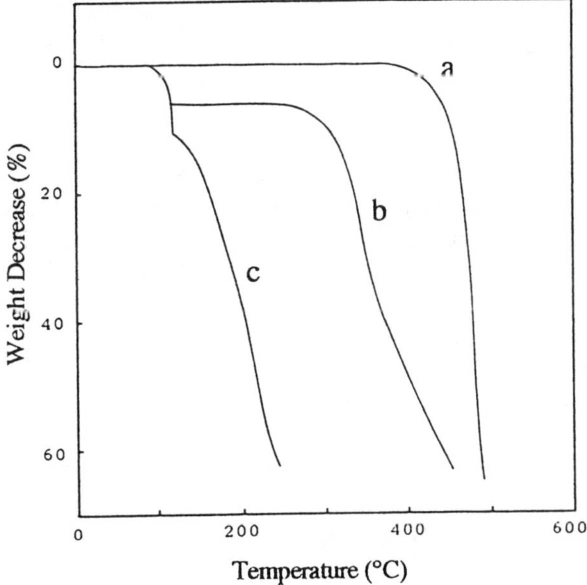


Fig. 4 Thermal decomposition behavior of a macroazoinitiator, observed by thermogravimetry in comparison with polyamide 66. a; PA 66, b; Poly(hexamethylene-4,4'-azobis-4-cyanopentanamide), c; ACPA

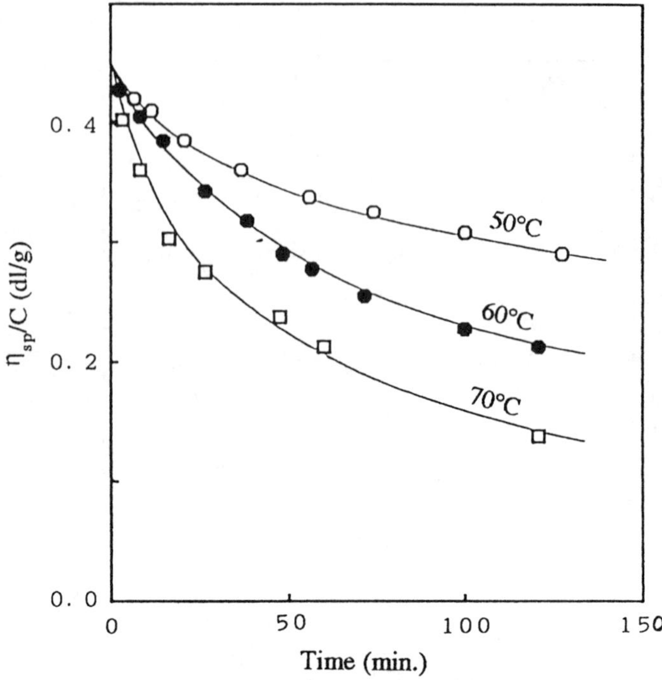


Fig. 5 Thermal degradation of a macroazoinitiator, poly(hexa methylene-4,4'-azobis-4-cyanopentanamide) in DMF solution, observed by viscometry at 50, 60 and 70°C

Decomposition temperature of azo group in solution has been measured by azotometry, UV absorption, or DSC analysis of the solution. The decomposition of azo group also proceeds via 1st order. Table 10 summarizes the data on the rate of decomposition, of macroinitiators and monomeric initiators. For ACPA analogues, macroinitiators show faster decomposition than monomeric ones, while AIBN analogues show reverse tendency. Length of prepolymers also affects on the decomposition, i.e., poly-ethyleneglycol with *MW* 7000 decomposes more slowly than that with *MW* 1000. Macroinitiators made from ACPC/terephthaloyl chloride with crown ether show almost the same activation energy of decomposition independent of poly-condensation composition i.e., 9097 kJ/mol, but a macroinitiator containing no terephthaloyl chloride unit shows a higher energy around 115 kJ (refer to 108 kJ for ACPA) [49]. Besides chemical structures, conformation or solid state might affect the decomposition behavior of macroinitiators.

# 4. ROUTES FOR SYNTHESIS OF BLOCK COPOLYMERS

## 4. 1 Characteristics of Polymerization Initiated with MacroInitiators

For radical polymerization initiated with an ordinary peroxy or azo compound at a given temperature, it is known that the overall rate of polymerization depends on the concentrations of monomer and initiator as exemplified in Figs. 6 and 7 [11]. The same dependency is held in the polymerization by a macroinitiator.

Table 10 Decomposition Rate and Initiation Efficiency of MacroAzoInitiators in Comparison with Ordinary Initiators

| Materials | Solvent | Temp. °C | Decomp. Rate $10^{-5}$/sec. | Init. Eff. | Ref. |
|---|---|---|---|---|---|
| ACPC + bisphenol-A | DMF | 60 | 2.1 | 0.28 | 52 |
| ACPA | DMF | 60 | 1.3 | 0.61 | 52 |
| Polyethyleneglycol (MW300) + AIBN | *o*-xylene | 85 | 27.5 | 0.3–0.4 | 13, 56 |
| AIBN | *o*-xylene | 85 | 21.1 | 0.57 | 13 |
| Polyethyleneglycol (MW1000) + ACPC | water | 70 | 3.8 | — | 79 |
| Polyethyleneglycol (MW7000) + ACPC | water | 70 | 3.2 | — | 79 |
| Crown ether + ACPC | DMSO | 70 | — | 0.13 –0.17 | 49 |
| Hexamethylenediamine + adipoyldichloride + ACPC | *m*-cresol | 60 | — | 0.15 | 45 |

However, a definite difference between the ordinary initiator system and macroinitiator system exists in molecular weight of polymer produced. While polymers obtained in the ordinary initiator system show almost the same average molecular weight independent of the conversion, polymers obtained by a macroinitiator show growth of average molecular weight with increase in conversion (Fig. 8) [11]. This occurs because every scissile bond in the macroinitiator molecule decomposes step by step, thus connecting multi-grown polymer chains. This multisteps initiation behavior of a macroinitiator can be utilized for block copolymerization.

## 4. 2 Routes for Block Formation

Macroinitiator is classified to two types as described in **2. 3** :
**Type I**: Macroinitiator consisted of repeated short units containing a scissile bond in the every unit.
**Type II**: Macroinitiator consisted of repeated segments of a prepolymer linked with scissile bonds.

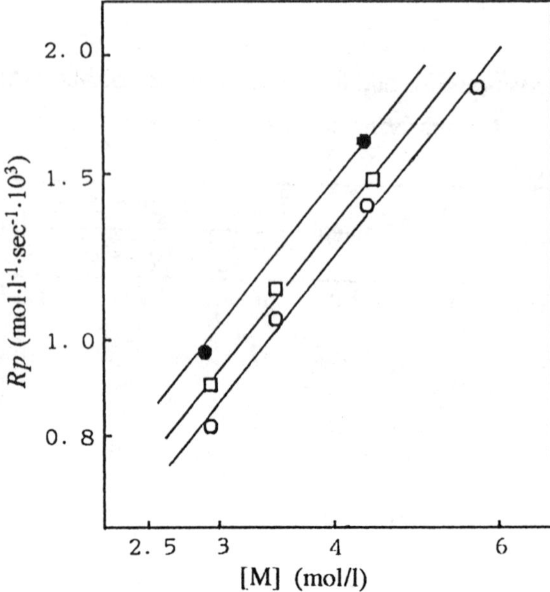


Fig. 6 Dependency of rate of polymerization of styrene on monomer concentrations: log $Rp$ vs. log [M] initiated by a constant concentration of poly(hexamethylene-4,4'-azobis-4-cyanopentanamide) (○), AIBN (□) and ACPA (●), respectively

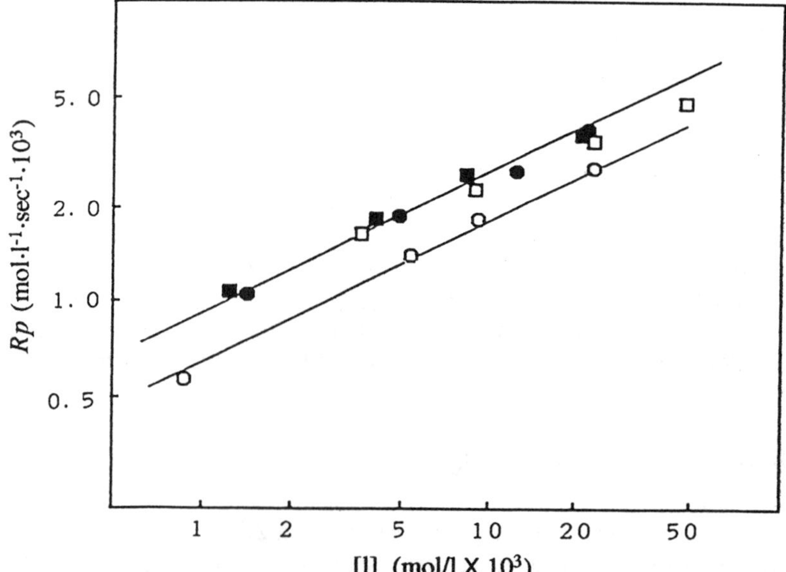


Fig. 7 Dependency of rate of polymerization of styrene on initiator concentrations: log $Rp$ vs. log [I] initiated by poly(hexamethylene-4,4'-azobis-4-cyanopentanamide) (○), ACPA•Me (■), AIBN (□) and ACPA (●)

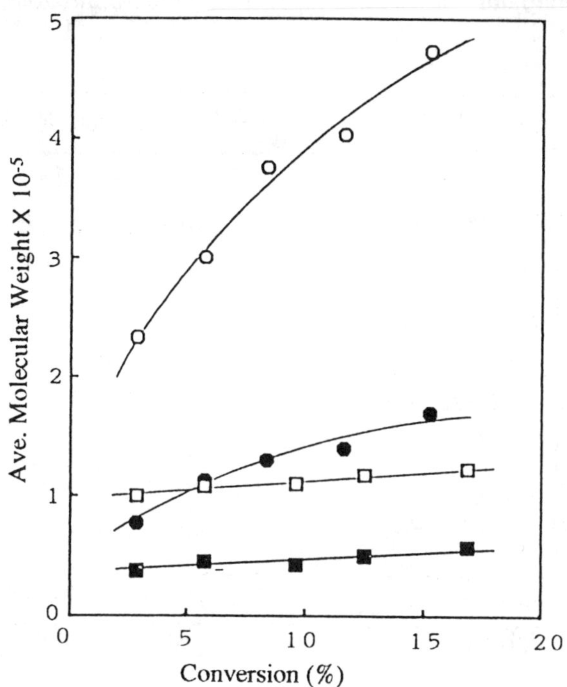


Fig. 8 Relationship between conversion and the average molecular weight of polystyrene

○$\overline{Mw}$ ●$\overline{Mn}$ : Macroazoinitiator system [M] = 4.34 mol/l, [I] =$5.21x10^{-5}$mol/l at $60^0$C in DMF

□$\overline{Mw}$ ■$\overline{Mn}$ : Macroazoinitiator system [M] = 4.41 mol/l, [I] =$4.41x10^{-5}$mol/l at $60^0$C in DMF

Fig. 9 shows a schematic explanation of the principle of synthesizing block copolymers with macroinitiators. In the case of type I macroinitiator, which has a scissile bond in every short repeating unit, polymer A segments are grown by the first step decomposition of the scissile bonds in the presence of monomer A. After some conversion, monomer A is purged and a new monomer B is charged. The polymer at this stage yet holds active scissile bonds on the chain, and considered a kind of macroinitiator, too. Then by the second step polymerization in the presence of monomer B initiated with the rest of scissile bonds, polymer B is grown connected with polymer A. This is a route for producing a block copolymer of A and B, via two step radical polymerization. Here, both A and B are addition polymers.

In the case of type II macroinitiator, the segment A (prepolymer) which had been already connected is considered a component of block copolymer to be synthesized.

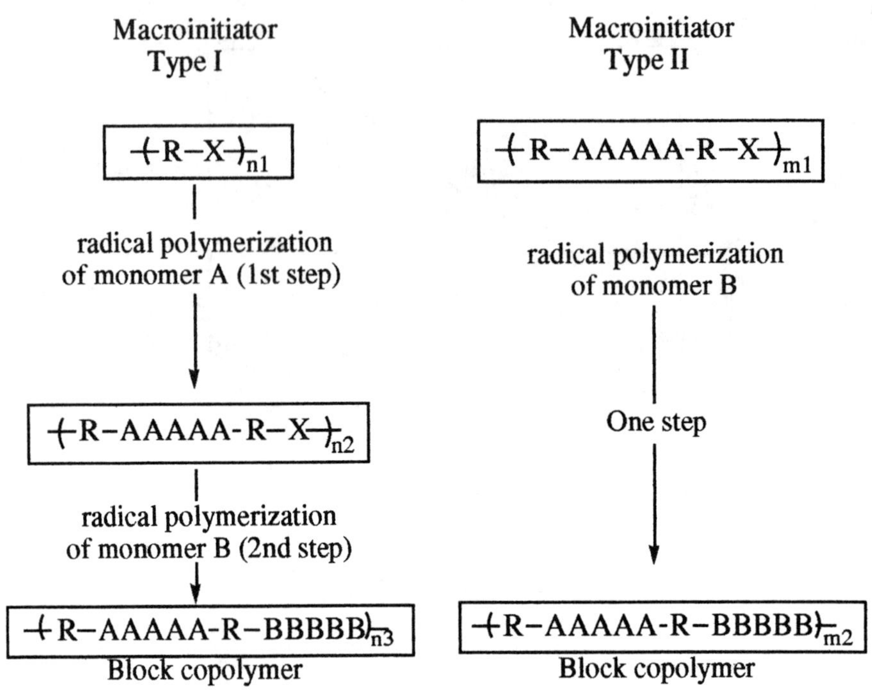


Fig. 9 The Principle of Block Copolymerization
X: active site for scission; R: aliphatic or aromatic hydrocarbons

By decomposing this initiator in the presence of a monomer B, new polymer of B is radically grown and connected with prepolymer A. This is cosidered to be a one step polymerization process for producing a block copolymers of A and B. This route opens the possibility of incorporating various kinds of prepolymer A, including those prepared by codensation polymerization, and ring opening polymerizations, etc.(refer Table 12) in the block copolymer with various kinds of addition polymers B.

If the polymerization step is multiplied, with feed of new monomer species, multi-component block copolymers can be derived. This kind of block copolymerization never occurs when ordinary benzoyl peroxide or 2,2'-azobisisobutyronitrile is used, since they have only one scissile bond and produce dead polymers after initiating radical polymerization of a monomer.

## 5. BLOCK COPOLYMERS PREPARED FROM MACROINITIATORS

Various block copolymers have been prepared by macroinitiators, and summarized in Tables 11 and 12.

By using macroperoxyinitiators, most block copolymers composed of a pair of addition polymers. In contrast, by using macroazoinitiators, abundant examples of block copolymers between addition polymers and condensation polymers have been reported.

Incorporation of condensation polymers as a member of block segment is one of the important characteristics of macroinitiator method of block copolymerization which is not attainable by ionic polymerization. Many engineering plastic materials, such as polyamides, polycarbonate, polybutyleneterephthalate, polyphenylene ether, polyphenylene sulfide, and liquid crystalline polymers, are made of condensation polymers, suggesting bright future of application of the macro-initiator method of block copolymerization involving them as a useful member.

Another characteristic of this macroinitiator method is to provide a variety of block copolymers composed of random copolymers used as a member. Many examples have been practiced in macroperoxy initiation systems applied for adhesives and coatings. Also, various monomers with functional groups such as sulfonic acid, quarternary ammonium salt, hydroxyl, carboxyl, acid amide, are available as a segment member. Further, a variety of multi-component block is also easy to derive by multi-step polymerization.

Table 11 Block Copolymers Prepared From MacroPeroxyInitiators

| Component 1 (Addition polymer) | Component 2 (Addition polymer) | Ref. |
|---|---|---|
| Polyacrylamide | Polyacrylate copolymer | 90, 93 |
| Polyacrylate copolymer | Polyacrylate copolymer | 90–95 |
| Poly(butylacrylate-co-acrylic acid) | Polyacrylate copolymer | 92 |
| Polyethylacrylate | Poly(hydroxyethylmethacrylate -co-ethylhexylmethacrylate) | 95 |
| Polyethylhexylacrylate | Polyacrylate copolymer | 95 |
| Polyethylhexylmethacrylate | Polyacrylate copolymer | 93, 95 |
| Poly(ethylhexylacrylate-co-hydroxyethylmethacrylate) | Polyacrylate copolymer | 95 |
| Poly(ethylhexylmethacrylate-co-butylmethacrylate) | Polyacrylate copolymer | 95 |
| Polyethylmethacrylate | Poly(hydroxyethylmethacrylate -co-butylacrylate) | 95 |
| Polydifluorochloroethylene | Polyethylmethacrylate | 60 |
| Polyperfluorooctylethylacrylate | Polybutylmethacrylate | 96 |

(continues)

| | | |
|---|---|---|
| | Poly(methylmethacrylate-co-butylacrylate) | 96 |
| | Poly(methylmethacrylate-co-hydroxyethylmethacrylate) | 96 |
| Poly(perfluorooctylethylacrylate -co-trifluoroethylmethacrylate) | Polyacrylate copolymer | 96 |
| Polyhydroxyethylacrylate | Polyacryl copolymer | 94, 97 |
| Polyhydroxyethylacrylate-co-N-methylol acrylamide) | Polyacrylate copolymer | 94 |
| Polyhydroxyethylacrylate-co-hydroxyethylmethyacylate) | Polyacrylate copolymer | 94 |
| Polyhydroxyethylacrylate-co-diethyleneglycol monomethacrylate) | Polyacrylate copolymer | 94, 97 |
| Polyhydroxyethylmethacrylate | Polyacryl copolymer | 94, 97, 98 |
| Polylaurylmethacrylate | Polyacrylate copolymer | 98 |
| Poly(laurylmethacrylate-co-acrylic acid) | Polyacrylate copolymer | 93, 95 |
| Polymethylmethacrylate | Polyvinylacetate | 37, 99 |
| | Polyhydroxyethylmethacrylate | 89 |
| | Polyperfluorooctylacrylate | 14 |
| | PFEA *1 | 60, 96 |
| Poly(methylmethacrylate-co-stearylmethacrylate) | Poly(stearylmethacrylate-co-EFS *2) | |
| Poly(methylmethacrylate-co-hydroxyethylmethacrylate) | Polyacrylic copolymers (contg. F) | 60 |
| Poly(methylmethacrylate-co-butylacrylate) | Poly(butylmethacrylate-co-PFOEA *3 ) | 96 |
| Poly(methylmethacrylate-co-ethylacrylate) | Poly(PFHEA *4 -co-PFOEA) | 60 |
| Poly(PFHVE) *5 | Poly(PEGMA *6 -co-propyl-methacrylate) | 60 |
| Polystearylmethacrylate | Polyacrylate copolymer | 90, 93, 95 |
| Poly(stearylmethacrylate-co-N-methylolacrylamide) | Polyacrylate copolymer (contg. F) | 60 |
| Polystyrene | Polyacrylonitrile | 100 |
| | Polybutylvinylether | 101 |
| | Poly(butyl acrylate-co-HE *7) | 102 |
| | Poly(butyl methacrylate-co-HE) | 102 |
| | Poly(2HPMQA) *8 | 89 |
| | Polyhydroxyethylmethacrylate | 89 |
| | Polymethylmethacrylate | 3, 4, 14, 99, 103–105 |
| | Poly(NaAMPS) *9 | 89 |
| | Polyvinylacetate | 4, 14, 37, 38, 94, 104, |

(continues)

| | | |
|---|---|---|
| | | 106 |
| | Poly(vinylacetate-co-AC *10) | 102 |
| | Poly(vinylacetate-co-methacrylic acid) | 36 |
| Poly(styrene-co-butylacrylate) | Polyacrylate copolymer | 92 |
| | Polyacrylate copolymer (contg. F) | 96 |
| Poly(styrene-co-methyl methacrylate) | Polyacrylate copolymer | 90 |
| Poly(styrene-PENS *11) | Poly(vinylacetate-co-butylacrylate) | 60 |
| Polyvinylacetate | Polyacrylate copolymer | 97 |
| | Polybutyl acrylate | 98, 107 |
| | Poly(butyl acrylate-co-ethylhexylacrylate) | 107 |
| | Polybutylmethacrylate | 90, 93 |
| | Poly(styrene-co-methacrylic acid) | 106 |
| | Poly(styrene-co-methylmethacrylate) | 36 |
| Polyvinylchloride | Poly(methylmethacrylate-co-MFS *12) | 96 |
| Poly(vinylchloride-co-vinylacetate) | Polyacrylate copolymer | 90, 93 |

| Component 1 (Addition polymer) | Component 2 (Condensation or ring opening polymer) | Ref. |
|---|---|---|
| Polymethylmethacrylate | Polydimethylsiloxane | 44 |
| Polystyrene | Polydimethylsiloxane | 44 |
| | Polyethyleneglycol | 42 |
| Polyvinylacetate | Polydimethylsiloxane | 44 |

*1 ß-Perfluorooctylethylacrylate
*2 N-ethyl-N-(ß-methacryloxyethyl)hexafluoroethylsulfonamide
*3 Perfluorodecylacrylate
*4 Perfluorooctylacrylate
*5 Perfluoroheptylvinylether
*6 Polyethyleneglycolmonomethylacrylate
*7 Methacryloyloxyethyl-*t*-butyl-peroxycarbonate
*8 2-Hydroxyl-3-methacryloxypropyltrimethylammonium
*9 2-Acrylamide-2-methylpropane sodium sulfonate
*10 Allyl-*t*-butylperoxycarbonate
*11 *p*-Perfluorononyl styrene
*12 N-methyl-N-(ß-methacryloxyethyl)-perfluorooctyl sulfonamide

Table 12 Block Copolymers Prepared by MacroAzoInitiators

| Component 1 (Addition polymer) | Component 2 (Addition polymer) | Ref. |
|---|---|---|
| Polyacrylonitrile | Polybutadiene | 20 |
| | Poly-1-vinyl-2-pyrrolidone | 20 |
| Polyacrylate copolymer | Polyacrylate copolymer | 53 |
| | Polyfluoroacrylate copolymer | 60, 86 |
| Polybutadiene | Polydimetylitaconate | 67 |
| Polymethylmethacrylate | Polybutadiene | 20 |
| | Polybutylacrylate | 28 |
| Polystyrene | Polyacrylonitrile | 11 |
| | Polybutadiene | 20 |
| | Polybutylacrylate | 51, 62 |
| | Polymethacrylic acid | 61 |
| | Polymethylacrylate | 57 |
| | Polymethylmethacrylate | 11, 21, 22, 88 |
| | Polymethylmethacrylate – Polybutylmethacrylate | 110 |
| | Poly-N-methyl-N-vinylacetate | 57 |
| | Polyolefin | 64 |
| | Poly(styrene-co-styrene Na sulfonate) | 109 |
| | Poly-1-vinyl-2-pyrrolidone | 88 |
| Poly(styrene-co-acrylonitrile) | Poly(styrene-co-butadiene) | 59 |
| Polyvinylacetate | Polyacrylamide | 56 |
| | Polybutadiene | 20 |

| Component 1 (Addition polymer) | Component 2 (Condensation or ring opening polymer) | Ref. |
|---|---|---|
| Polyacrylonitrile | Polyamide | 30 |
| Polyacrylate copolymer | Polyester | 68–71 |
| | Polyethyleneglycol | 115 |
| | Polyfluoroester | 55 |
| | Polysiloxane | 76, 77 |
| | Polyuretahene | 65, 68, 70, 71 |
| | Polyfluorourethane | 55 |
| Polymethylacrylate | Polyethyleneglycol | 82 |
| Polymethylmethacrylate | Polyamide | 11 |
| | Polyester | 66 |
| | Polyethyleneglycol | 82 |
| | Polyfluorosiloxane | 19, 75 |
| | Polypropyleneglycol | 9 |

(continues)

| | | |
|---|---|---|
| | Polysiloxane | 19, 73 |
| | Polyurethane | 30 |
| Poly(methylmethacrylate-co-vinylacetate) | Polyamide | 30 |
| Polystyrene | Polyamide | 11, 45, 49, 62 |
| | Polyarylate | 51, 72 |
| | Polycarbonate | 78, 111, 112 |
| | Polyester | 49 |
| | Polyethyleneglycol | 80, 82, 84, 85 |
| | Polyimide | 24 |
| | Polypropyleneglycol | 9, 85 |
| | Polysiloxane | 19 |
| | Polytetrahydrofuran | 9, 84 |
| Poly(styrene-co-acrylonitrile) | Polyamide | 30 |
| Poly(styrene-co-isobutyrene) | Polyamide | 30 |
| Polyvinylacetate | Polyethyleneglycol | 82 |
| | Polypropyleneglycol | 9 |
| | Polysiloxane | 19 |
| Poly(vinylacetate-co-acrylic acid) | Polyamide | 30 |
| Polyvinyl chloride | Polyethyleneglycol | 79, 81, 114 |
| | Polypropyleneglycol | 9 |
| | Polysiloxane | 113 |
| Poly(vinylisobutylether-co-vinylpropionate) | Polyamide | 30 |

## 6. INITIATION EFFICIENCY AND BLOCK EFFICIENCY

Initiation efficiency is a fraction of effective initiation of radical polymerization when scission of an initiator occurred. Block efficiency is a fraction of polymers connected as a component of block copolymers among all polymers formed.

Experimental data of initiation efficiency of macroinitiators are lower than those of monomeric initiators, as shown in Tables 9, 10 (**3. 1, 2**). It seems that radical pairs formed after the scission of a macroinitiator tend to recombine due to slow diffusion out of the cage. In the case of macroazoamide type initiators, many amide groups and solvent such as *m*-cresol also help lowering the initiation efficiency by large chain transfer reactivity.

Block efficiency, however, can be improved even if the initiation efficiency is low. Since a macroinitiator has many scission points to initiate polymerization, the low initiation efficiency of a scissile point can be covered, statistically, by effective block initiation of another scissile point in the same segment of the macroinitiator, resulting in improving practical block efficiency. At the same time, the most possible recombination of original segments after inefficient scission helps minimizing the fraction of single dead fragments of macroinitiators. Simply suppose that the initiation efficiency = 0.33 for a macroinitiator, then one should synthesize the initiator having at least 3 scissile points.

There is a factor which decreases the block efficiency. The larger the monomer transfer constant, the higher becomes the fraction of homopolymers produced independent of initiators. This is unavoidable in radical polymerization system, and situation differs, case by case, depending on the type of monomer, solvent, and temperature, etc.

In order to evaluate the block efficiency, one may carry out the decomposition of the macroinitiator in the presence of styrene. Polymerization of styrene takes place starting from the scission points of the initiator, and termination occurs by the recombination of growing radicals. After complete decomposition of all scissile points in the initiator, total length of the polymer, composed of linear combinations of ordinary polystyrene, is multiplied and proportional to the numbers of scissile points which involved in effective initiation (Fig. 10).

The plot of the molecular weight versus the monomer conversion provides information on the block efficiency of a macroinitiator (Fig. 11). Fig. 12 is another example, when 2nd step polymerization of styrene was initiated with polyvinylpyrrolidone containing scissile azo groups which had been prepared by polymerization of vinylpyrrolidone initiated with poly(hexamethylene-4,4'-azobis-4-cyanopentanoate) (Eq. 7), a typical macro-azoinitiator.

For effective block copolymerization, care should be taken to feed a considerable concentration of compatible monomers. Sometimes it needs a common solvent to keep a homogeneous solution. In the case of insufficient feed of monomers (Fig. 11), block efficiency became low due to insufficient initiation efficiency.

# 7. MOLECULAR DESIGN OF BLOCK COPOLYMERS

## 7.1 Types of Block Copolymers

According to the various types of combination of the block segments A and B, block copolymers are classified by three types: -(AB) n, ABA, and AB.

Types of block copolymers derived from macroinitiators are determined depending on the mechanism of termination of polymer radicals (Fig. 13). -(AB) n type block copolymers are formed, if polymer radicals terminate by recombination in both 1st and 2nd steps. If the polymer radicals terminate by disproportionation in both steps, AB type block copolymers will be formed. If the polymerization in the 1st step terminates by recombination, and in the 2nd step by disproportionation, or *vice versa*, ABA type block copolymers should be obtained, where, disproportional terminating polymer builds up segment A in both cases.

R N=N R N=N R N=N R + Styrene (S) + heat
*1 *2 *3
↓
R SSSSS • + • SSSSSSSS R N=N R N=N R + $N_2$
*2 *3
↓
R SSSSS - SSSSSSSS R N=N R N=N R
*2 *3
↓
R SSSSS - SSSSSSSS R SSSSSSS • + • SSSS R N=N R + $N_2$
*3
↓
R SSSSS - SSSSSSSS R SSSSSSS - SSSS R N=N R
*3
↓
R SSSSS - SSSSSSSS R SSSSSSS - SSSS R SSSSS • + • SSSS R + $N_2$
↓
R SSSSS - SSSSSSSS R SSSSSSS - SSSS R SSSSS - SSSS R

Fig. 10 Schematic expression of block homopolymerization initiated stepwise by macroazoinitiator having 3 scissile points

N=N : scissile point due to azo bond
SSS • : growing polystyrene radical
S–S : bond after recombination

In case of poly(methylmethacrylate)-b-poly(butylacrylate) block copolymer (PMMA-b-PBA), methylmethacrylate terminates mainly by disproportionation, while butylacrylate prefers to terminate by recombination. Fig. 14 shows experimental data on the change in the molecular weight of the block copolymers via two modes of polymerizations: (1) methylmethacrylate polymerized, followed by butylacrylate, (2) butylacrylate polymerized, followed by methylmethacrylate. In the course (1), drastic increase in molecular weight is observed during both 1st and 2nd step of the polymerization. However, in the course (2) the drastic increase in molecular weight by the n-butylacrylate polymerization in the 1st step was followed by a decrease by the polymerization of methylmethacrylate. The block copolymers obtained via both modes

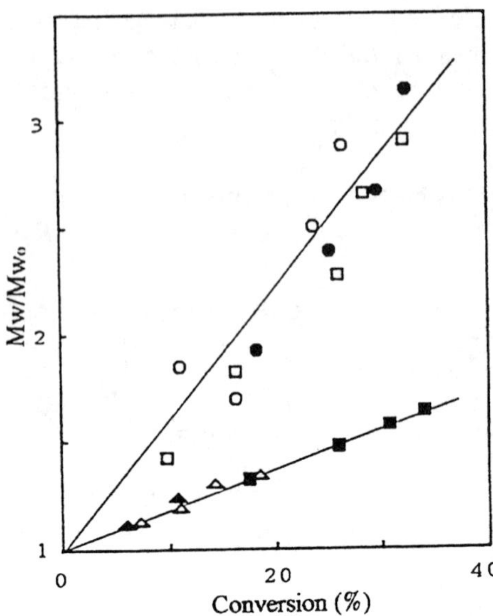


Fig. 11 Relationship between $Mw/Mw_o$ and conversion in the polymerization of styrene at 80°C in DMF initiated by macroazoinitiators prepared from ○ ; 2,2'-azobis(2-cyanopropanol) (ACPO) + sebacoyl chloride (SC), ● ; ACPC + HDO, □ ; 4,4'-azobis(4-cyanopentanol) (ACPEO) + SC, ■ ; AIBN, △ ; ACPO + TDI, ▲ ; NON

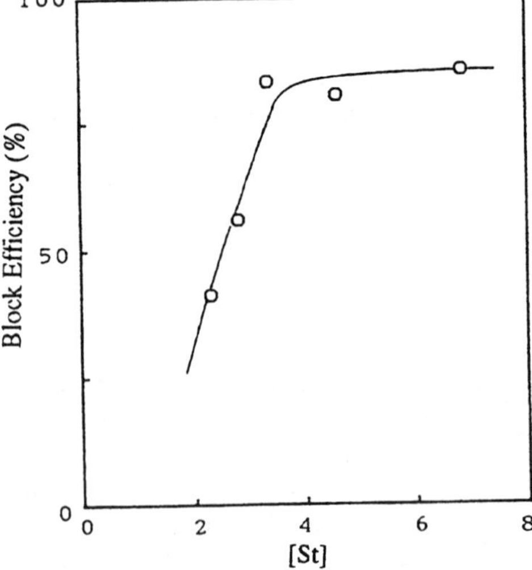


Fig. 12 Block efficiency plotted against monomer concentrations in the 2nd step polymerization of styrene connecting to polyvinylpyrrolidone (PVP). PVP was prepared by poly(hexamethylene-4,4'-azobis-4-cyanopentanoate) used as initiator.

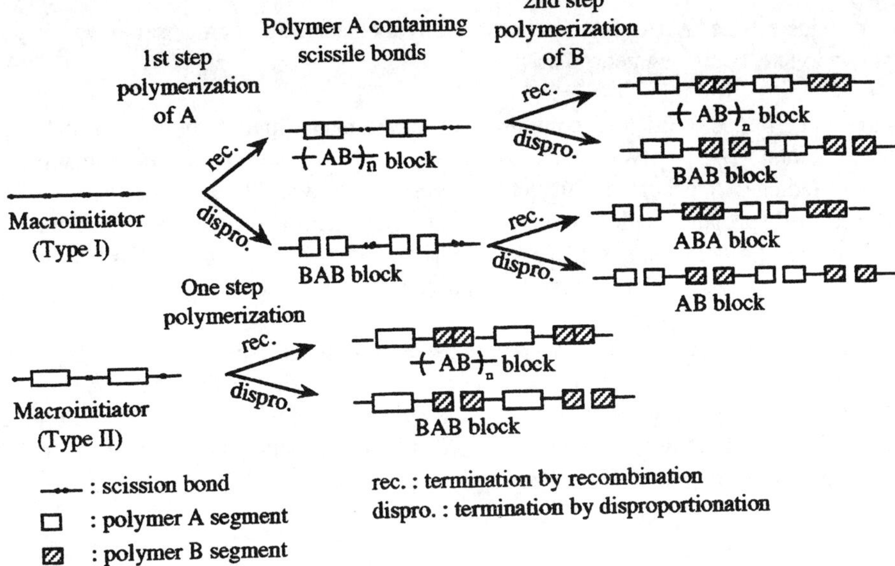


Fig. 13 Flow chart for producing different types of block copolymers depending on termination mechanisms

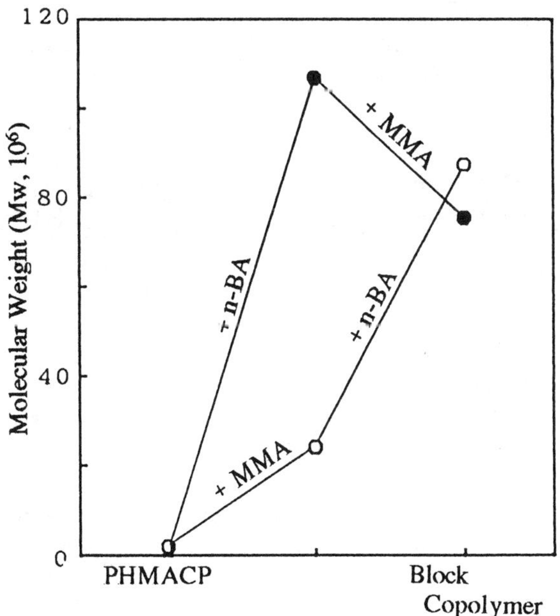


Fig. 14 Molecular weight change during 1st and 2nd steps of polymerization to prepare Poly(MMA-b-nBA)
○ 1st step : MMA 2nd step : n-BA,
● 1st step : n-BA 2nd step : MMA

are considered as ABA type, where methylmethacrylate being segment A and n-butylacrylate being segment B [54].

Using type II macroinitiator, composed of ready made prepolymers B linked with scissile points, $-(AB)_n$ type block copolymers can be obtained if recombination of growing radicals A occurs exclusively, whereas ABA type block copolymers can be obtained if disproportionation only occurs. In general, however, both recombination and disproportionation takes place at the same time, and ineffective initiation and chain transfer reaction also occur, thus the polymers obtained are of mixed types.

Some efforts have been made to prepare A–B type block copolymer. Starting from macroinitiator having several peroxy groups in a molecule, methylmethacrylate was firstly polymerized followed by styrene. The block copolymer was considered to be A–B type since its molecular weight accorded with the sum of molecular weight of each homopolymer (Table 13) [104].

Table 13 Molecular weights of extracted homopolymers and block copolymer

| Monomer ratio | $MW \times 10^{-3}$ a) | | |
|---|---|---|---|
| MMA/S (wt ratio) | PMMA | PS | Block copolymer |
| 3/7 | 387 | 379 | 660 |
| 5/5 | 291 | 267 | 565 |

a) By GPC calibrated with PSt.

A trial has been also made starting from type II macroazoinitiator having only one azo unit to successfully produce A–B type block copolymer (PMMA-b-PEG) [87].

## 7. 2 Control of Block Copolymer Compositions

Compositions of block copolymers can be controlled by the conditions of the second step polymerization. Polymerized amounts of monomers must be calculated by integrated equation of polymerization rate, but could be estimated simply by the rate equation for initial stage of polymerization.

In the case of preparing polyethyleneglycol-b-polystyrene (PEG-b-PS) initiated with poly(polyethyleneglycol-4,4'-azobis-4-cyanopentanoate) [80],the rate of polymerization of styrene (*Rp*) is expressed by Eq. 8,

$$\frac{100}{X} - 1 = \frac{k\,Rp}{[-N{=}N-]} \tag{8}$$

where x is the content of polyethyleneglycol in the block copolymer, and [-N=N-] is the concentration of azo group since $Rp = k\ [\text{-N=N-}]^{0.5}[M]^{1.2}$. Thus, Eq. 9 is derived as follows.

$$\frac{100}{X} - 1 = \frac{k\ [M]^{1.2}}{[\text{-N=N-}]^{0.5}} \tag{9}$$

This relation is shown in Fig. 15 and is utilized for designing the composition of block copolymers.

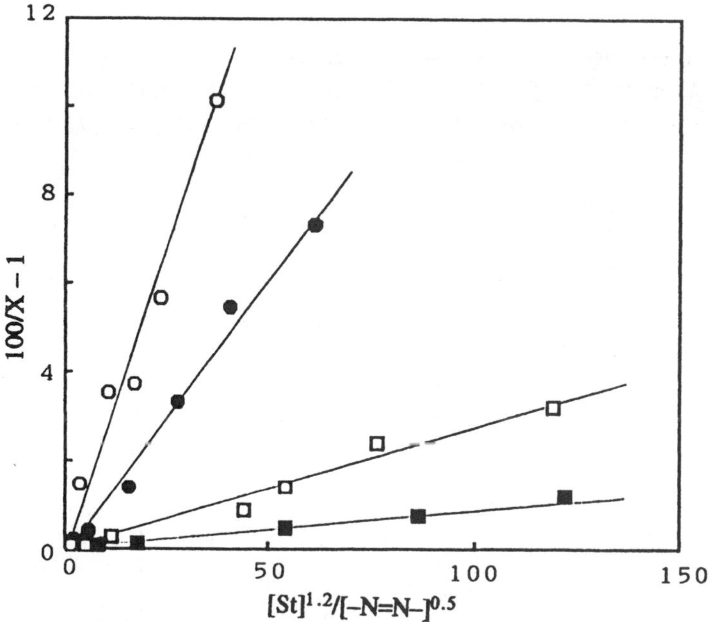


Fig. 15 Relationship between 100/X – 1 and $[M]^{1.2}/[\text{–N=N–}]^{0.5}$ in the preparation of PEG-b-PS

○ Macroazoinitiator (Type II) composed of PEG (*MW*: 600)
● Macroazoinitiator (Type II) composed of PEG (*MW*: 2000)
□ Macroazoinitiator (Type II) composed of PEG (*MW*: 6000)
■ Macroazoinitiator (Type II) composed of PEG (*MW*: 20000)

## 7. 3 Control of Segment Length

Block segment length is easily controllable in anionic livingpolymerization by changing monomer/initiator ratio. In radical polymerization, the degree of poly-merization is controlled by adding chain transfer agents. However, in the block copolymerization derived from macroinitiator, addition of chain transfer agents is not recommended, since they accelerate the increase of homopolymer fractions and decrease of block efficiency of the same time. Change of monomer/initiator ratio is rather desirable to control the block composition within a limited range.

Segment length of the macroinitiator type II, is expected to be unchanged after the incorporation in the final block copolymer. But, it often increases due to recombination just after the scission without effective initiation of polymerization. Segment length of PS block copolymer prepared by poly(ethyleneglycol-4,4'-azobis-4-cyanopentanoate) was evaluated, and Eq. 10 was derived.

$$\overline{MW} = k\ [\mathrm{St}]^2 / Rp \tag{10}$$

Here, rate of polymerization, Rp was proportional to $[-N=N\ ]^{0.5}[S]^{1.2}$. Therefore, [Picture] is expressed by $[S]^{0.8}/[-N=N-]^{0.5}$. Based on the equation, average molecular weight of polymer segment was controlled, ranging from 7,000to 220,000(Fig. 16) [80].

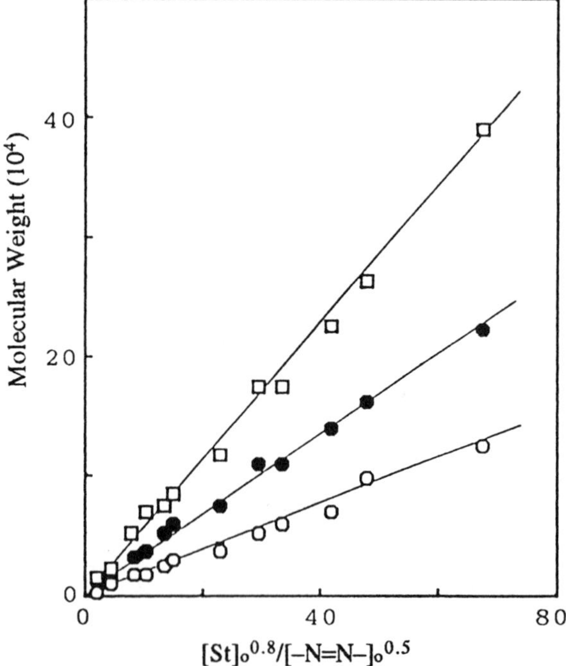


Fig. 16 Average molecular weight of PS segment in PEG-b-PS vs. $[S]^{0.8}/[-N=N-]^{0.5}$
○ $\overline{Mn}$, ● $\overline{Mw}$, □ $\overline{Mz}$

## 7. 4 Block Segments Composed of Random Copolymers

Preparation of copolymer-b-copolymer is considered a characteristics of radical process, providing modified properties by varying the copolymer compositions. Using a macroazoinitiator prepared from AIBN and hexanediol, a thermoplastic elastomer, P(St/Bd)-b-P(St/AN) has been synthesized. In P(St/Bd) matrix, P(St/AN) domains are dispersed, acting as physical crosslinkings. The thermal endurance and the modulus of elasticity are controlled [59] by changing the composition of P(St/AN) and P(St/Bd) respectively.

# 8. APPLICATIONS OF BLOCK COPOLYMERS

## 8. 1 Features of Block Copolymers

Block copolymers derived from radical polymerization via macroinitiators and anionic polymerization using the same monomer pairs, should basically have the same properties. But, there is an essential difference between them in the molecular weight distributions of block segments and of total copolymers. The radical process is inevitable to give narrow molecular weight distribution to block copolymers.

Block copolymers show both the properties identical to the component polymers. If the components are insoluble to each other, microphase separation occurs in the morphology of block copolymers due to agglomeration of each component. Electron microscopic observation provides that the microdomains (10–30 nm) of one component dispersed in matrix of another component. In measurement of dynamic viscoelastic properties, each component of block copolymers shows its *Tg*, elastic modulus change and peak of tan d in its respective region.

## 8. 2 Thermoplastic Elastomers

Application of block copolymers derived from macroinitiators for thermoplastic elastomers is a matter of great interest. Thermoplastic elastomers having PS as a hard segment and poly(oxytetramethylene)glycol as a soft segment [9], and S/AN copolymer as a hard segment and S/Bd copolymer as a soft segment [59], have been reported. In the latter, for improving the heat resistance and processability, *Tg* of the hard segment was controlled ranging from 90°C to 120°C by changing the composition of S/AN. For improving elasticity, the composition of S/Bd was changed.

Poly(dimethylitaconate)-b-poly(butadiene) is considered one of the block copolymers difficult to obtain by anionic process. It aims to improve the heat resistance of commercial SBS block copolymers by substituting for PS (*Tg* = 80°/110°C) segment by poly(dimethylitaconate) segment having *Tg* = 132–143°C [67].

The same block component combination with the lower content of the soft segment, serves as impact modifiers. Transparent high impact materials made of PS-b-PBd [61], and PMMA-b-PBA [108] were reported.

## 8. 3 Antishrinking Agents

Polystyrene-b-Polyvinylacetate (PS-b-PVAc) and PMMA-b-PVAc prepared by macroperoxyinitiators have been commercially available as antishrinking agents (low profile additives) for processing of FRP [36–38, 106]. In comparison with homopolymers of S, VAc, and MMA, the block copolymers show better compatibility as well as excellent effect of antishrinkage for unsaturated polyesters, resulting in high gloss products with smooth surfaces. PSt and PVAc segments were outside and inside of the micelles repectively when PS-b-PVAc was dispersed in a styrene solution. When this solution was mixed with unsaturated polyester, phase inversion took place, turning PVAc segments to outside. Again the phase inversion took place at the final stage of curing and PS-b-PVAc micelles with outside PSt segments separated as spheres. Voids formed in this process seemed to serve for prevention of shrinkage [106].

## 8. 4 Coatings

Block copolymers derived from macroinitiators are applicable to coatings. Various block copolymer systems dispersed in organic or aqueous media have been patented. They show excellent compatibility with media giving stable dispersion for coatings [53, 69, 95, 104]. The systems composed of condensation polymers such as polyurethane or polyester linked with acrylic copolymers are usable for coating materials which give balanced high performance of hardness, adhesiveness, dispersibility of pigments, staining resistance, weatherability, and low temperature resistance [55, 65, 68, 70]. Most of them are copolymerized with hydroxyl containing vinyl monomers, and added with melamine, polyisocyanate, amino resins, etc. for baking varnish. If they are further added with polydimethylsiloxane (PDMS), or fluorinated resins, improvement of performances are expected such as non-tackiness, water repellency, oil repellency, staining resistance, and weatherability [60, 76, 77, 86, 96].

## 8. 5 Film Modification

PMMA-b-PDMS film has been known to exhibit anisotropic surface property, i. e. different contact angles between one side and its backside when cast from chloroform solution on glass plate (Fig. 17). This phenomenon is considered to occur due to the difference in the amounts of plate-out of PDMS segments. The difference becomes significant by increasing PDMS segment length from 700, through 2100, to 4000. The kind of solvent for casting also affects on the diffusion behavior. Polar solvents such as methylethylketone and methylene-dichloride help PDMS segments to diffuse towards air side of cast film [74]. This effect is enhanced when fluorinated acrylic monomers and siloxanes having fluroalkyl side chains are used as components of block copolymers [75].

Diffusion of PDMS segments towards outside matrix gives drastic effects on adhesion property or peel strength of the matrix. Fig. 18 shows that the addition of only small amounts of the block copolymers into PMMA homopolymer are enough to cover the surface, giving excellent water repellency or non tackiness represented by drastic decrease in peel strength [39]. The film made of block copolymer of PDMS with polyvinylchloride (PDMS-b-PVC) shows excellent mold release property, weatherability and staining resistance [40].

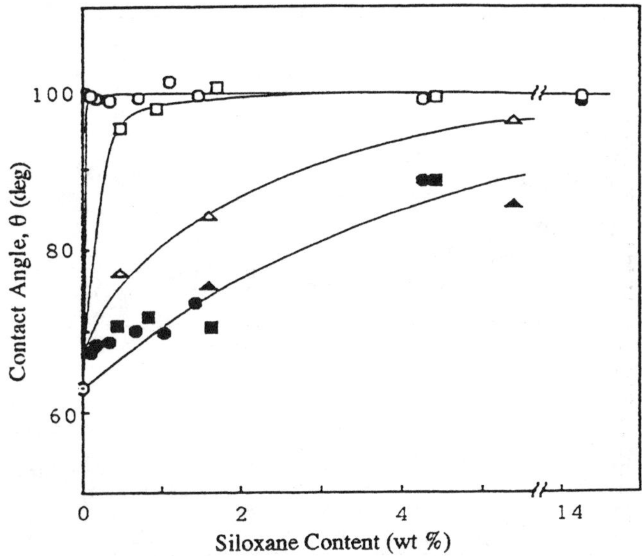


Fig. 17 Effect of blending PMMA-b-PDMS with various siloxane chain length (SCL) on the surface characteristics of PMMA cast film : water contact angle vs. siloxane content. air side surface : [○] SCL 4000 [□] SCL 2100 [Δ] SCL 770 glassplate side surface : [●] SCL 4000 [■] SCL 2100 [▲] SCL 770
casting solvent : chloroform

When block copolymer composed of PMMA-PDMS is blended to homoPMMA, the block copolymer tends to diffuse out the surface part of PMMA. The tendency is accelerated when PDMS segment is longer (Fig. 17). For giving oxygen permeability to PMMA, localization of PDMS segment on the surface of PMMA is not desirable. In contrast, block copolymer with shorter PDMS segment is more desirable so that it can be dispersed homogeneously inside the matrix (shorter PDMS segments prefer to disperse homogeneously inside the matrix). The oxygen permeability increased 2 to 3 times the original PMMA [41].

Surface of the film made of block copolymers of PS and hydrophilic polymers containing –OH, $SO_3Na$, $N^+(CH_3)_3$ shows water affinity. This characteristics become significant when the content of hydrophilic polymer segments is beyond 20%, where antifogging effect and antistatic effect on the film are expected. Block copolymers containing polyethyleneglycol (PEG) as a hydrophilic member have been also investigated. PVC-b-PEG dose not become cloudy when immersed in hot water [12]. It is of interest that PEG containing azo group is considered as a macroinitiating surfactant, usable in water dispersion system without emulsifiers, to obtain emulsifier-free coating materials resistant to water [115]. Polyacrylamide-b-PVAc is also cited to obtain emulsion dispersed not only in water but also in methanol [56, 58, 116].

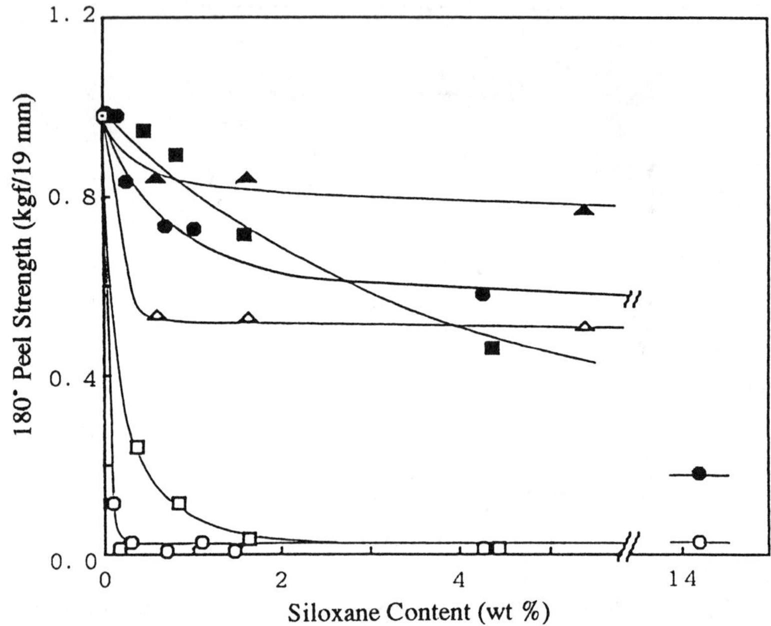


Fig. 18 Effect of blending PMMA-b-PDMS with various siloxane chain length (SCL) on the surface characteristics of PMMA cast film : 180° peel strength vs. siloxane content.
air side surface : ○ SCL 4000 □ SCL 2100 △ SCL 770
glassplate side surface : ● SCL 4000 ■ SCL 2100 ▲ SCL 770
casting solvent : chloroform

Shrinkable film made of PVC-b-PEG without plasticizer has been tested and extensibility due to flexible PEG segment was observed and antistatic property was also given (Fig. 19) [114].

## 8. 6 Compatibilizers

Compatibilizers for polymer alloys are considered as one of the most suitable applications for block copolymers. Polycarbonate-b-polystyrene (PC-b-PS) has been confirmed to be an effective compatibilizer for PC/PS blends. In comparison with a simple blend of each homopolymers, a modified blend added with the compatibilizer showed superior transparency and especially mechanical properties as shown in Table 14 [111].

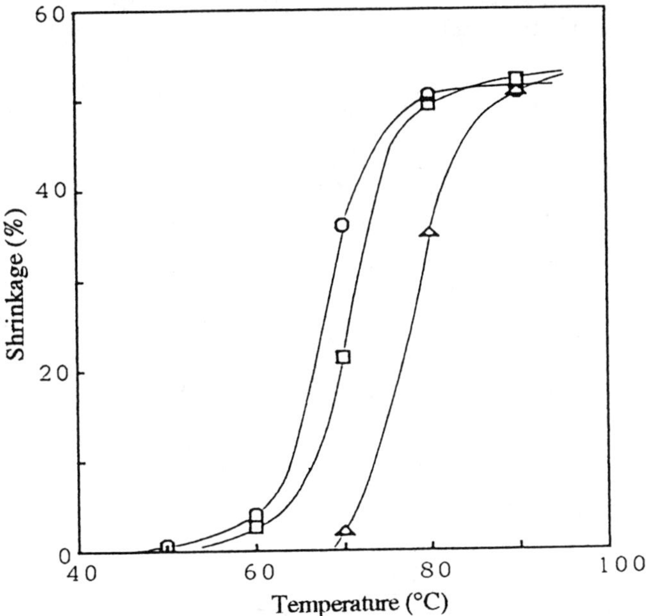


Fig. 19 Shrinkage behavior of stretched films : shrikage vs. temperature
○ PVC-b-PEG containing 6% PEG,
□ shrinkable film (plasticized PVC containing 7% DOP),
△ shrinkable film (non-plasticized PVC)

Polyarylate-b-Polystyrene (PAR-b-PS) has been prepared using PAR derived from condensation of bisphenol A with a mixture of terephthalic acid/isophthalic acid (1/1). When it is added from 1 to 30% into a blend of PAR/PS (7/3), dispersed domain size becomes small by increasing the added amount, and tensile modulus of the blend was proportionally improved at the same time (Fig. 20) [72].

## 8. 7 Graft Copolymers

Effective graft copolymerization of polyolefins via a block copolymer having peroxy side chains has also been reported, where monomers containing peroxy group was copolymerized most effectively with soft vinyl acetate [102].

A grafting reaction during injection molding has been tried, where molding compound is a blend of ethylene-propylene copolymers (10 parts) and poly-hydroxyethyl methacrylate (1 part) polymerized with a mixture of benzoyl peroxide and macroperoxy initiator. If a suitable amount of macroperoxy group is left active, effective grafting occurs, and the moldings show excellent adhesive property when coated with polyurethane indicating peel strength of 450 g/cm in contrast with 10 g/cm for ethylene-propylene copolymer itself.

Table 14 Properties of PC–b–PS

| Materials | Molding temp. °C | Tensile strength $kg/cm^2$ | Tensile modulus $kg/cm^2$ | Flexural strength $kg/cm^2$ |
|---|---|---|---|---|
| PC–b–PS * | 230 | 680 | 24,200 | 900 |
| PC | 320 | 630 | 18,000 | 970 |
| PS | 210 | 530 | 26,200 | 880 |
| PC/PS blend * | 230 | 410 | 23,500 | 660 |

* PC/PS = 15 w/75 w

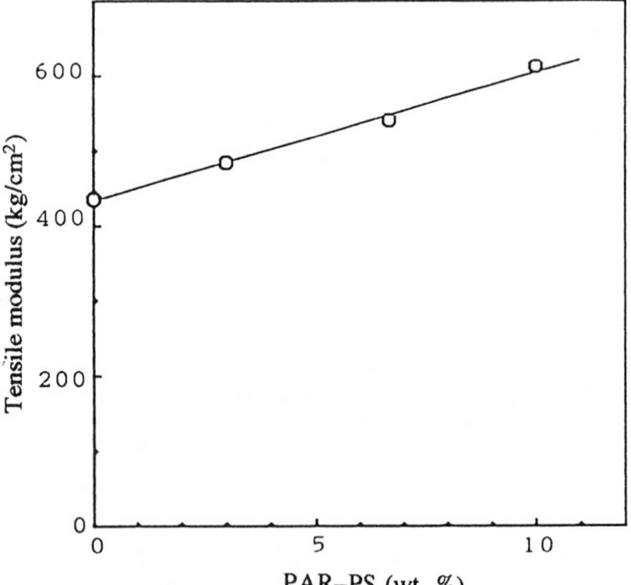


Fig. 20 Improvemant of the tensile modulus of PAR/PS (7/3) blend by the addition of PAR-b-PS

## 8. 8 Other Applications

When a macroperoxyinitiator having long side alkyl chains ($C_{20}$) was used for polymerization of MMA or S, the polymers showed high impact strength [117, 118]. Also, the same macroperoxyinitiator was utilized for polymerization of S or S/AN mixture in the presence of PBd to give high impact plastic materials with high gloss [119, 120].

Polymers with a wide range of molecular weight distribution are obtained using a mixture of macroinitiator and coventional initiator, since high molecular weight polymers are produced via initiation through the macroinitiator. Improvement of mechanical strength of moldings, rapid molding cycle, improved foam-moldability was achieved [121, 122].

Development of heat sensitive micells including ink was reported. The wall of micells is composed of azo containing crosslinked polyamide prepared by interfacial polycondensation under vigorous stirring. Decomposition of azo group by heat induces the evolution of ink. Heat sensitive recording paper with excellent resolution was made [50].

# 9. CONCLUSION

Macroinitiators for radical block copolymerization have scissile or labile bonds in their main chains, and are capable to link polymers linearly. In this chapter, their features, classfication, synthesis and decomposition behaviors have been reviewed, and then preparations, molecular design and applications of block copolymers derived from the macroinitiators have been described.

It is an advantage of macroinitiators to be able to choose various components for preparing block copolymers among both addition polymers and condensation polymers. This method seems rather easy to carry out than anionic process, although block copolymers have wide distribution in composition. The review by Nuyken et al. [33] and orther chapters of this book provide information of macroinitiators having scissile bonds in their side chains, which are expected to derive graft polymers.

It is our hope that this article will further stimulate interest in this field of polymer science and accelerate the development of new and high functional materials as well as progress of polymer technology.

## Acknowledgment

Finally we wish to express our thanks to Miss Michie Yoshida (Kyushu Univ.) for her sincere cooperation in completion of this article.

## REFERENCES


1. H. V. Pechmann and L. Vanino, *Ber.*, **27**, 1510 (1894).
2. H. A. Shah, F. Leonard and A. V. Tobolsky, *J. Polym. Sci.*, **7**, 537 (1951).
3. G. Smets and A. E. Woodward, *J. Polym. Sci.*, **14**, 126 (1954).
4. A. E. Woodward and G. Smets, *J. Polym. Sci.*, **17**, 51 (1955).
5. M. S. Tsvetkov, R. F. Markovskaya and A. A. Sorokin, *Vysokomol. Soedin. Ser. B*, **11**, 519 (1969), (*Chem. Abst.*, **71**, 102412u (1969)).
6. T. Komai and M. Matsushima, *Japan Kokai* S53-149918 (1978).
7. J. W. Hill, *U. S. Pat.*, 2556876 (1951); *Chem. Abst.*, **45**, 9915f (1951).
8. D. A. Smith, *Makromol. Chem.*, **103**, 301 (1967).
9. J. Furukawa, S. Takamori and S. Yamashita, *Angew. Makromol. Chem.*, **1**, 92 (1967).
10. S. Nagai, Y. Hidaka and A. Ueda, *Japan Kokai*, S49-17895 (1974); *Chem. Abst.*, **81**, 92227c (1974).
11. A. Ueda, Y. Shiozu, Y. Hidaka and S. Nagai, *Kobunshi Ronbunshu*, **33**, 131 (1976).
12. S. Nagai, A. Ueda, H. Inoue, Y. Kita, K. Matsukawa, A. Matsumoto, M. Tanaka and T. Harada, *SPE Antec 88 Technical Papers*, 1849 (1988).
13. R. Walz, B. Bömer and W. Heitz, *Makromol. Chem.*, **178**, 2527 (1977).
14. T. Yamamoto, K. Aoshima, H. Ohmura, Y. Moriya, N. Suzuki and Y. Oshibe, *Polymer*, **32**, 19 (1991).
15. B. Hazer, *J. Polym. Sci.*, *Polym. Chem.*, **25**, 3349 (1987).
16. K. Masa, *Japan Kokai*, H3-62813 (1991).
17. B. M. Baysal, W. T. Short and A. V. Tobolsky, *J. Polym. Sci.*, *Polym. Chem.*, **10**, 909 (1972).
18. Y. Yagci, H. Yildirim, E. Altun and B. M. Baysal, *J. Appl. Polym. Sci.*, **44**, 367 (1992).
19. H. Inoue, A. Ueda and S. Nagai, *J. Polym. Sci.*, *Polym. Chem.*, **26**, 1077 (1988).
20. A. Ueda and S. Nagai, *Kobunshi Ronbunshu*, **43**, 97 (1986).
21. C. O. Oppenheimer and W. Heiz, *Angew. Makromol. Chem.*, **98**, 167 (1981).
22. Y. Yagci and S. Denizligil, *Eur. Polym. J.*, **27**, 1401 (1991).
23. A. Ueda, Y. Agari, S. Nagai, N. Minamii and T. Miyagawa, *Kobunshi Ronbunshu*, **47**, 321 (1990).

24. T. Ohta and S. Sano, *Japan Kokai*, H1-223127 (1989).
25. K. Matsukawa, A. Ueda, H. Inoue and S. Nagai, *Chem. Express*, **2**, 33 (1987).
26. K. Matsukawa, A. Ueda, H. Inoue and S. Nagai, *J. Polym. Sci., Polym. Chem.*, **28**, 2107 (1990).
27. Y. Oshibe and T. Yamamoto, *Kobunshi Ronbunshu*, **44**, 73 (1987).
28. A. Ueda and S. Nagai, *Kobunshi Ronbunshu*, **44**, 469 (1987).
29. A. Ueda, *Bulletin of Osaka Municipal Technical Research Institute*, **84**, 1–139 (1989).
30. H. Craubner, *J. Polym. Sci., Polym. Chem.*, **18**, 2011 (1980).
31. H. Craubner, *J. Polym. Sci., Polym. Chem.*, **20**, 1935 (1982).
32. N. S. Tsvetkov and M. B. Fagarash, *Vyskomol. Soedin*, **5**, 1002 (1963); *Chem. Abst.*, **59**, 7651a (1963).
33. O. Nuyken and R. Weidner, *Adv. Polym. Sci.*, **73–74**, 145 (1986).
34. B. Hazer, *J. Polym. Sci., Polym. Chem.*, **25**, 3349 (1987).
35. O. Suyama, K. Taura and M. Kato, *Japan Kokai*, H3-252413 (1991).
36. K. Fukushi and A. Hatachi, *Japan Kokai*, S58-189214 (1983).
37. M. Nakayama, M. Matsushima, S. Banno and N. Kanazawa, *Japan Kokai*, S56-79113 (1981).
38. K. Fukushi, *Japan Kokai*, S60-217222 (1985).
39. H. Inoue, A. Matsumoto, K. Matsukawa, A. Ueda and S. Nagai, *J. Appl. Polym. Sci.*, **41**, 1815 (1990).
40. H. Murakami, S. Hirai, Y. Shikoku, A. Ueda, H. Inoue and S. Nagai, *Japan Kokai*, H4-39315 (1992).
41. H. Inoue, K. Matsukawa, A. Ueda and S. Nagai, *Kagaku to Kogyo (Osaka)*, **64**, 106 (1990).
42. I. Yilgör and B. M. Baysal, *Makromol. Chem.*, **186**, 463 (1985).
43. H. Yildirim and B. M. Baysal, *Angew. Makromol. Chem.*, **181**, 103 (1990).
44. T. Saigou and M. Nakayama, *Japan Kokai*, S63-57644 (1988).
45. A. Ueda and S. Nagai, *J. Polym. Sci., Polym. Chem.*, **22**, 1611 (1984).
46. A. Ueda and S. Nagai, *J. Polym. Sci., Polym. Chem.*, **22**, 1783 (1984).
47. Y. Yagci, U. Tünca and N. Bicak, *J. Polym. Sci., Polym. Lett.*, **24**, 491 (1986).
48. Y. Yagci, U. Tünca and N. Bicak, *J. Polym. Sci., Polym. Lett.*, **24**, 49 (1986).

49. U. Tünca and Y. Yagci, *J. Polym. Sci., Polym. Chem.*, **28**, 1721 (1990).
50. O. Nuyken, J. Dauth and W. Pekruhn, *Angew. Makromol. Chem.*, **187**, 207 (1991).
51. T. Ohta and S. Sano, *Japan Kokai*, H1-201331 (1989).
52. M. H. George and J. R. Ward, *J. Polym. Sci., Polym. Chem.*, **11**, 2909 (1973).
53. H. Kinoshita, G. Iwamura, A. Shozi, S. Nagai and A. Ueda, *Japan Kokai*, H1-254778 (1989).
54. A. Ueda and S. Nagai, *Kagaku to Kogyo (Osaka)*, **64**, 446 (1990).
55. H. Kinoshita, H. Kumada, G. Iwamura, A. Shozi, S. Nagai and A. Ueda, *Japan Kokai*, H2-167375 (1990).
56. H. Dicke and W. Heitz, *Makromol. Chem. Rappid Commun.*, **2**, 83 (1981).
57. P. S. Anand, H. G. Stahl and W. Heitz, *Makromol. Chem.*, **183**, 1685 (1982).
58. H, Dicke and W. Heitz, *Coll. Polym. Sci.*, **260**, 3 (1982).
59. X. Y. Qiu, W. Ruland and W. Heitz, *Angew. Makromol. Chem.*, **125**, 69 (1984).
60. Y. Oshibe, T. Yamamoto, M. Nakayama and H. Ohmura, *Japan Kokai*, S60-36517 (1985).
61. F. Tianru, X. Suxian and L. Yan, *Polym. Bull.*, **22**, 317 (1989).
62. T. Ohta and S. Sano, *Japan Kokai*, H1-138224 (1989).
63. A. Ueda, Y. Agari, S. Nagai, N. Minamii and T. Miyagawa, *Chem. Express*, **4**, 193 (1989).
64. T. Ohta, *Japan Kokai*, H1-138207 (1989).
65. H. Kinoshita, *Japan Kokai*, H1-245066 (1989).
66. C. I. Simionescu, *Eur. Polym. J.*, **23**, 921 (1987).
67. J. M. G. Cowie and M. Y. Pedram, *Briti. Polym. J.*, **16**, 127 (1984).
68. G. Iwamura, H. Kinoshita, A. Shozi, S. Nagai and A. Ueda, *Japan Kokai*, H1-252611 (1989).
69. G. Iwamura, H. Kinoshita, A. Shozi, S. Nagai and A. Ueda, *Japan Kokai*, H1-254703 (1989).
70. G. Iwamura, H. Kinoshita, A. Shozi, S. Nagai and A. Ueda, *Japan Kokai*, H1-254774 (1989).
71. H. Kumada, N. Harui, G. Iwamura, A. Shozi, S. Nagai and A. Ueda, *Japan Kokai*, H2-229873 (1990).
72. T. O. Ahn and H. M. Jeong, *Polym. Preprints, Japan*, **39**, 25 (1990).

73. H. Inoue, S. Nagai and A. Ueda, *Japan Kokai*, S61-252230 (1986).
74. H. Inoue, A. Ueda and S. Nagai, *J. Appl. Polym. Sci.*, **35**, 2039 (1988).
75. H. Inoue, A. Matsumoto, K. Matsukawa, A. Ueda and S. Nagai, *J. Appl. Polym. Sci.*, **40**, 1917 (1990).
76. H. Kumada, N. Harui, G. Iwamura, A. Shozi, S. Nagai and A. Ueda, *Japan Kokai*, H1-254775 (1989).
77. G. Iwamura, H. Kumada, H. Kinoshita and A. Shozi, *Japan Kokai*, H2-67381 (1990).
78. T. Ohta, S. Sano and R. Kozika, *Japan Kokai*, H1-236233 (1989).
79. J. J. Laverty and Z. G. Gardlund, *J. Polym. Sci., Polym.Chem.*, **15**, 2001 (1977).
80. A. Ueda and S. Nagai, *J. Polym. Sci., Polym. Chem.*, **24**, 405 (1986).
81. Y. Kita, A. Ueda, T. Harada, M. Tanaka and S. Nagai, *Chem. Express*, **1**, 543 (1986).
82. A. Ueda and S. Nagai, *J. Polym. Sci., Polym. Chem.*, **25**, 3495 (1987).
83. J. C. Saramone, W. P. Liao, A. C. Watterson and C. C. Olson, *Polym. Prepr.*, **29** (**1**), 275 (1988).
84. R. Walz and W. Heitz, *J. Polym. Sci., Polym. Chem.*, **16**, 1807 (1978).
85. H. Kürük, A. B. Özdemir and B. M. Baysal, *J. Appl. Polym. Sci.*, **31**, 2171 (1986).
86. T. Fuyuki, A. Murakami, K. Shinkai, M. Ishimoto, T. Yamamoto, *Japan Kokai*, S61-95081 (1986).
87. S. Nagai, H. Terada and A. Ueda, *Chem. Express*, **8**, 157 (1993).
88. A. Ueda, Y. Agari, K. Kosai, N. Nishioka and S. Nagai, *Chem. Express*, **7**, 725 (1992).
89. H. Ohmura, H. Dohya, Y. Oshibe and T. Yamamoto, *Kobunshi Ronbunshu*, **45**, 857 (1988).
90. H. Ohmura and M. Nakayama, *Japan Kokai*, S56-93722 (1981).
91. H. Ohmura and M. Nakayama, *Japan Kokai*, S56-76423 (1981).
92. H. Ohmura and M. Nakayama, *Japan Kokai*, S56-133312 (1981).
93. H. Ohmura and M. Nakayama, *Japan Kokai*, S56-93723 (1981).
94. H. Ohmura and M. Nakayama, *Japan Kokai*, S56-55415 (1981).
95. H. Ohmura and M. Nakayama, *Japan Kokai*, S55-142016 (1980).
96. Y. Hiramatsu, M. Tokieda, A. Kato, T. Yamamoto, *Japan Kokai*, S60-228575 (1985).
97. H. Ohmura and M. Nakayama, *Japan Kokai*, S56-50908 (1981).
98. T. Sanui and M. Matsushima, *Japan Kokai*, S55-7855 (1980).

99. M. Matsushima and M. Nakayama, *Japan Kokai*, S55-75414 (1980).
100. M. S. Tsvetkov and R. F. Markovskaya, *Izv. Vyssh. Ucheb. Zaved., Khim. Khim. Tekhnol.*, **11**, 936 (1968); *Chem. Abst.*, **70**, 29399t (1969).
101. Y. Yagci, M. Acar, G. Hizal, H. Yildirim and B. Baysal, *Angew. Makromol. Chem.*, **154**, 169 (1987).
102. N. Suzuki, Y. Moriya and T. Yamamoto, *Kobunshi Ronbunshu*, **44**, 89 (1987).
103. T. Sugimura, I. Sudo and Y. Minoura, *Kogyo Kagaku Zasshi*, **69**, 718 (1966).
104. N. Suzuki, Y. Moriya and T. Yamamoto, *Kobunshi Ronbunshu*, **44**, 81 (1987).
105. W. Kawai, *J. Polym. Sci., Polym. Chem.*, **15**, 1479 (1977).
106. K. Fukushi, Y. Moriya and T. Yamamoto, *Kobunshi Ronbunshu*, **44**, 97 (1987).
107. H. Ohmura, *Japan Kokai*, S58-103515 (1983)
108. A. Ueda and S. Nagai, *Kagaku to Kogyo (Osaka)*, **66**, 385 (1992).
109. A. Ueda, Y. Agari and S. Nagai, *Polym. Preprints, Japan*, **41**, 390 (1992).
110. B. Hazer, *Angew. Makromol. Chem.*, **129**, 31 (1985).
111. K. Takahashi and K. Kawasaki, *Japan Kokai*, S59-27908 (1984).
112. T. Ohta, S. Sano and R. Kozika, *Japan Kokai*, S64-236234 (1989).
113. H. Murakami, S. Hirai, Y. Shikoku, A. Ueda, H. Inoue and S. Nagai, *Japan Kokai*, H2-39315 (1990).
114. H. Murakami, S. Hirai, Y. Shikoku, A. Ueda, H. Inoue and S. Nagai, *Japan Kokai*, H2-269744 (1990).
115. Y. Hasegawa, H. Kinoshita, G. Iwamura, S. Nagai and A. Ueda, *Japan Kokai*, H1-254780 (1989).
116. W. Heitz, C. Oppenheimer, P. S. Anand and X. Y. Qiu, *Makromol. Chem. Suppl.*, **6**, 47 (1984).
117. H. Suezawa and Y. Masuda, *Japan Kokai*, H3-84008 (1991).
118. A. Tohara and K. Masa, *Japan Kokai*, H3-111434 (1991).
119. K. Masa, H. Kasahara and Y. Ohtani, *Japan Kokai*, H3-63813 (1991).
120. H. Ishikawa, H. Kasahara and Y. Ohtani, *Japan Kokai*, H3-70717 (1991).
121. K. Masa, *Japan Kokai*, H4-55441 (1992).
122. K. Masa, *Japan Kokai*, H4-55409 (1992).

# Macroinitiators — Chapter 8

## POLYMERIC AZO INITIATORS

**Oskar Nuyken and Brigitte Voit**

Technische Universität München
Institut für Technische Chemie
Lehrstuhl für Makromolekulare Stoffe
Lichtenbergstr. 4
D-85747 Garching, GERMANY





Macromolecular Design: Concept and Practice (Macromonomers, Macroinitiators, Macroiniferters, Macroinimers, Macroiniters, and Macroinifers) Pages 313 - 358

# 1. INTRODUCTION

In the field of polymeric initiators the use of the azo function is probably the best known and the oldest method. This is due to the long tradition of azo compounds as initiators in free radical polymerization. Studies on thermal decomposition of azobisisobutyronitrile (AIBN) were reported in 1949 [1] and as early as 1951, polymers containing labile azo groups can be found in the patent literature [2]. Up to now a broad variety of azo containing polymers were synthesized and several reviews have been published on this subject [3,4,5].

The "azo polymers" can be formally classified according to their properties and the location of the azo function. First we have to distinguish labile and inert azo groups. Of course this is a question of an exact definition. Under certain conditions every azo function will be destroyed. In principle, it is possible to cleave the azo group by heat [3,6], by irradiation with UV-light [6] and even by bacterial reduction [7]. In this article we will talk only about polymeric azo initiators, indicating polymers with a thermally labile azo functionality which decomposes under moderate temperature conditions according to the scheme below. This restriction excludes polymers with azo benzene units which have not been used as polymeric initiators but as NLO, liquid crystalline and biodegradable polymers [3,4].

$$R_1{-}N{=}N{-}R_2 \xrightarrow{\Delta T} R_1^{\bullet} + R_2^{\bullet} + N_2$$

$$(R_1 = R_2 \neq Ph)$$

The cleavage of the azo functionality by UV-light at low or moderate temperatures also provides radicals that have initiating ability. However, this reaction has found only little interest for polymeric initiators due to its low quantum yield [6] and possible side reactions.

The thermally labile azo polymers can be further classified into polymers with the azo function in the main chain or in the side chain. We will first give an overview over the synthetic routes to main chain macroazoinitiators and then discuss the side chain azo polymers in detail. In the last part, possible applications of these polymers will be reviewed.

# 2. MAIN CHAIN POLYMERIC AZO INITIATORS

Main chain macroinitiators are characterized by one or more azo functionalities which are part of the main chain. They can be distinguished as follows:

The polymer chains with exactly one azo group in the middle (type A). Type B1 and type B2 contain several azo units, where B1 is the homopolymer with an azo repeating unit and B2 represents the copolymers with a more or less controlled amount of azo functions statistically distributed along the chain. It is further possible to introduce exactly one (type C1) or two (type C2) azo end groups into the polymer.

~~~~N=N~~~~ type A

~~~N=N~~~N=N~~~N=N~~~ type B1 and B2

R—N=N~~~~ type C1

R—N=N~~~~N=N—R type C2

Since other authors in this book have reviewed macroinitiators in general and blockcopolymers in detail which were made often from main chain polymeric azo initiators, in this chapter we will focus on the different synthetic methods used for the preparation of various kinds of macroazoinitiator structures.

The largest groups of main chain azo polymers are those of type B1 and B2 with several azo functions. As early as 1951 [2] these polymers were synthesized by the reaction of aliphatic diketones **1** with hydrazine to polyazines **2**. In a second step, which leads us to a polymer modification reaction, HCN was added to the azine and by oxidation with 1,3-bis(chloromethyl)hydantoin a polymer **4** with an azo functionality in each repeating unit was obtained.

$$CH_3-\overset{O}{\overset{\|}{C}}-(CH_2)_{n>3}-\overset{O}{\overset{\|}{C}}-CH_3 \xrightarrow{N_2H_4 \times H_2O} \left[=\overset{CH_3}{\overset{|}{C}}-(CH_2)_{n>3}-\overset{CH_3}{\overset{|}{C}}=N-N=\right]_n$$

**1** **2**

$$\xrightarrow{HCN} \left[-\underset{CN}{\overset{CH_3}{C}}-(CH_2)_{n>3}-\underset{CN}{\overset{CH_3}{C}}-\underset{H}{N}-\underset{H}{N}-\right]_n \quad \mathbf{3}$$

$$\xrightarrow{Ox.} \left[-\underset{CN}{\overset{CH_3}{C}}-(CH_2)_{n>3}-\underset{CN}{\overset{CH_3}{C}}-N=N-\right]_n \quad \mathbf{4}$$

In this early patent the labile azo functionality was already used as radical source for block copolymerization - the main application of all main chain polymeric azo initiators.

The polymer modification reactions, especially those involving several reaction steps, are usually not the most efficient ways to obtain clean products. Therefore only one more example should be mentioned here for the synthesis of azo polymers. Craubner et al. [8] have described the synthesis of poly(N-nitroso-acylamines) **6** from polyamide **5**.

On irradiation, the poly(N-nitroso-acylamine) rearranges to form azo functions **7** which were photolytically further decomposed (without isolation) to polymeric radicals **8** - the initiators for blockcopolymerization. This is also one of the rare examples where the macroazoinitiators were photolytically and not thermally activated.

$$\left[ R-\overset{O}{\overset{\|}{C}}-\underset{H}{\underset{|}{N}} \right]_n \xrightarrow{N_2O_3} \left[ R-\overset{O}{\overset{\|}{C}}-\overset{NO}{\overset{|}{N}} \right]_n \xrightarrow{h\nu} \left[ R-\overset{O}{\overset{\|}{C}}-O-N=N \right]_n$$

5 → 6 → 7 →(hν) ∿∿∿R· + $N_2$ + $CO_2$ (8)

A better alternative to the polymer modification and a very often used pathway to introduce the azo function into the polymer, is the polycondensation of AIBN derivatives containing acid, acid chloride or alcohol groups with diamines, dialcohols, diacids or bisacid chlorides, respectively. A variety of bifunctional azo compounds which are suitable for polycondensation can be found in several patents [9, 10].

Polyamides and polyesters **10** are the result from the direct reaction of 4,4-azobis-(4-cyanopentanoic acid) (ACPA) **9** with e.g. poly(ethylene glycols) or 1,6-hexamethylene diamine using mineral acids [10] or more recently [11] 1-methyl-2-chloropyridinium iodide as promotor. Matsukawa et al. [11] studied this reaction with several diols and diamines. It was possible to carry out the reaction at ambient temperature to avoid the cleavage of the azo function. But the degree of polymerization in this reaction could not be raised above 10.

$$\left[ HO-\overset{O}{\overset{\|}{C}}-CH_2-CH_2-\underset{CN}{\underset{|}{\overset{CH_3}{\overset{|}{C}}}}-N= \right]_2 + HO-PEG-OH \xrightarrow{catalyst}$$

9

$$\left[ \overset{O}{\overset{\|}{C}}-CH_2-CH_2-\underset{CN}{\underset{|}{\overset{CH_3}{\overset{|}{C}}}}-N=N-\underset{CN}{\underset{|}{\overset{CH_3}{\overset{|}{C}}}}-CH_2\cdot CH_2-\overset{O}{\overset{\|}{C}}-O-PEG-O \right]_n$$

10

A very versatile and successful method is to convert ACPA **9** into the more reactive acid chloride **11** (ACPC) which can be condensed with all types of diols [12,13] or diamines [14] to polyesters and polyamides **12** that was first described by Smith [15].

$$\left[ Cl-\overset{O}{\overset{\|}{C}}-CH_2-CH_2-\underset{CN}{\underset{|}{\overset{CH_3}{\overset{|}{C}}}}-N= \right]_2 + NH_2-R-NH_2 \longrightarrow$$

11

$$\left[ \overset{O}{\overset{\|}{C}}-CH_2-CH_2-\underset{CN}{\underset{|}{\overset{CH_3}{\overset{|}{C}}}}-N=N-\underset{CN}{\underset{|}{\overset{CH_3}{\overset{|}{C}}}}-CH_2\cdot CH_2-\overset{O}{\overset{\|}{C}}-\underset{H}{\underset{|}{N}}-R-\underset{H}{\underset{|}{N}} \right]_n$$

12

Such polymer formation can also be achieved by interfacial polycondensation under mild conditions and at low temperatures. An interesting application of the interfacial polycondensation is the synthesis of microcapsules. Several crosslinked thermo- or photolabile capsule shells have been synthesized using e.g. **11, 11a, 11b**, or **11c** in combination with trifunctional amines [15a]. Compared to the AIBN-derivatives **11, 11a,** and **11b** the aryl-alkyl azo compound **11c** is thermally quite stable but can be readily cleaved by irradiation with UV-light. Even though these polymers belong to the class of main chain azo polymers, they were not applied as polymeric initiatitors but in thermo printing techniques. Encapsulated printing ink can be released through thermal treatment of the azo group containing microcapsules [15a].

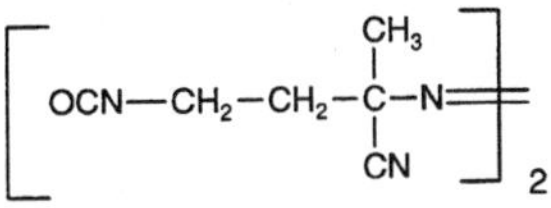

$$\left[ OCN-CH_2-CH_2-\underset{CN}{\overset{CH_3}{C}}-N= \right]_2$$

**11a**

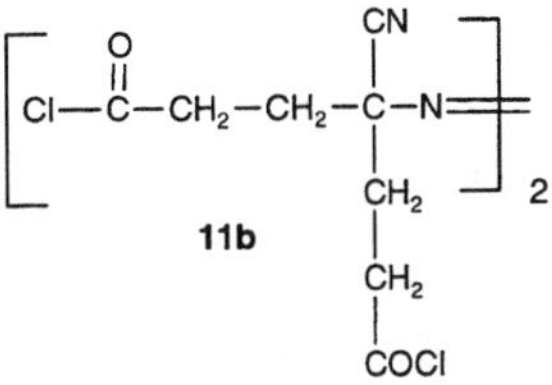

$$\left[ Cl-\overset{O}{\overset{\|}{C}}-CH_2-CH_2-\underset{CH_2-CH_2-COCl}{\overset{CN}{C}}-N= \right]_2$$

**11b**

$$Cl-\overset{O}{\overset{\|}{C}}-C_6H_4-N=N-\underset{CN}{\overset{CH_3}{C}}-(CH_2)_2-\overset{O}{\overset{\|}{C}}-Cl$$

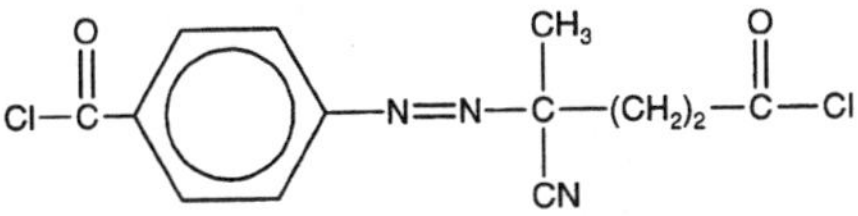

**11c**

Several water soluble azo polymers containing poly(ethylene glycol) units of different length **13** with DP up to 21 and relatively broad molecular weight distributions, prepared by the Smith method [15] are reported [16-18].

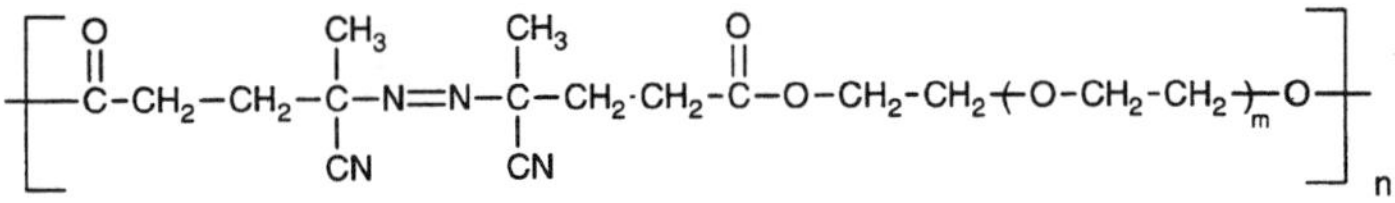

$$\left[ \overset{O}{\overset{\|}{C}}-CH_2-CH_2-\underset{CN}{\overset{CH_3}{C}}-N=N-\underset{CN}{\overset{CH_3}{C}}-CH_2\cdot CH_2-\overset{O}{\overset{\|}{C}}-O-CH_2-CH_2\left(O-CH_2-CH_2\right)_m O \right]_n$$

**13**

Yagci et al. [19] condensed 4,4 azobis(4-cyanopentanoylchloride) (ACPC) **11** with cis or trans-4,4'-diaminodibenzo-18-crown-6 (DADC) **14** or its KI-complex by interfacial [19] and low temperature solution polycondensation [20].

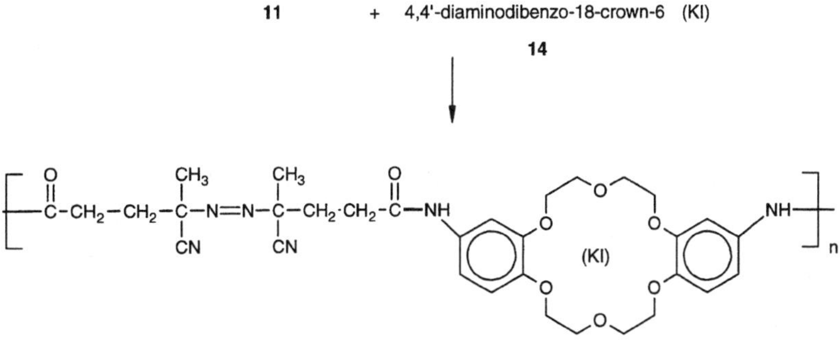

**11** + 4,4'-diaminodibenzo-18-crown-6 (KI)

**14**

↓

$$\left[ \overset{O}{\overset{\|}{C}}-CH_2-CH_2-\underset{CN}{\overset{CH_3}{C}}-N=N-\underset{CN}{\overset{CH_3}{C}}-CH_2\cdot CH_2-\overset{O}{\overset{\|}{C}}-NH-\text{dibenzo-18-crown-6 (KI)}-NH \right]_n$$

**15**

In some of Yagci's examples [19,20],**11** was substituted partly by an azo free bisacid chloride to obtain copolymers with lower azo content but a high amount of crown ether units which could be used for the binding of ions. Copolycondensation with azo free units was used in many of the here mentioned synthetic methods for adjusting the azo units per polymer chain to the latter application.

The reaction of amino [21] or hydroxy [21a] functionalized oligosiloxanes with **11** in $CHCl_3$ solution yields poly(siloxanes) **16** having thermally labile azo groups. Degrees of polymerization up to 32 are reported [21].

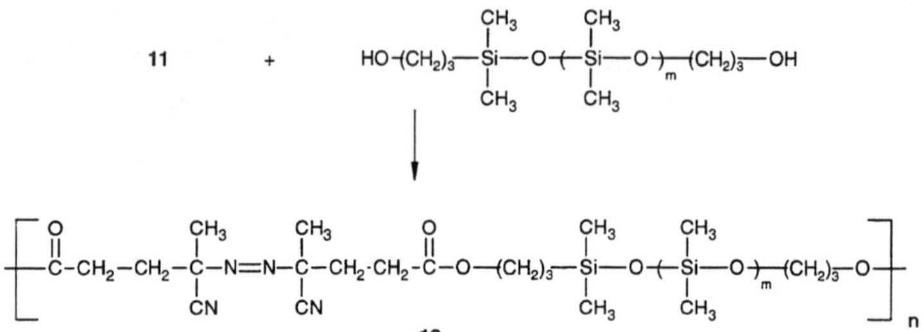

Akar et al. [22] introduced the azo groups in a single polymer modification step. A cyclohexane-formaldehyde resin (CF) **17**, containing free OH groups, was reacted with **11**. The bifunctional azo compounds were bound by ester linkage to the hydroxy groups of the resin (structure **18**). The ratio ACPC/OH was kept low, the resulting polymer resin was therefore still soluble and used as polymeric initiator for styrene polymerization.

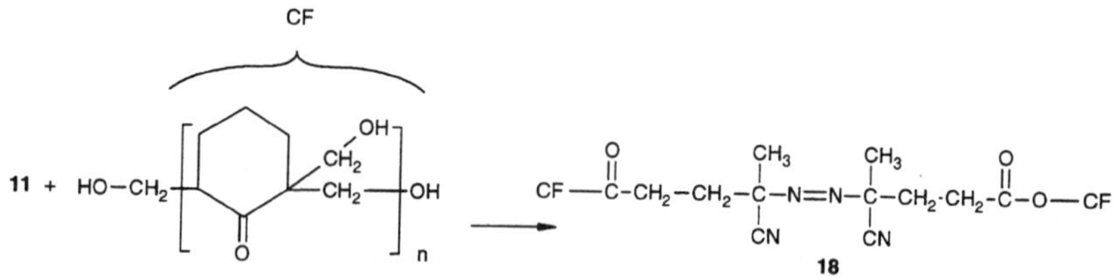

Beside the acid derivative of AIBN also diol derivatives, e.g 4,4'-azobis(4-cyanopentanol) **20**, have been used in the preparation of macroinitiators. The azo diols were added e.g. to hexamethylene diisocyanate [23] or to poly(oxypropylene glycol) (HO-PPG-OH) which was functionalized with diisocyanate **19** to yield "macroazonitrile" **21** [24] with high intrinsic viscosities ($[\eta] > 0.75$) and up to 20 azo units per polymer chain.

$$[HO-(CH_2)_3-C(CH_3)(CN)-N=]_2 \; (\textbf{20}) + OCN-Ar-NH-C(=O)-O-PPG-O-C(=O)-NH-Ar-NCO \; (\textbf{19}) \longrightarrow$$

$$[-O-(CH_2)_3-C(CH_3)(CN)-N=N-C(CH_3)(CN)-(CH_2)_3-O-C(=O)-NH-Ar-NH-C(=O)-O-PPG-O-C(=O)-NH-Ar-NH-C(=O)-]_n \; (\textbf{21})$$

Macroinitiators can also be synthesized by direct reaction of AIBN **22** with diols in an acid medium at low temperatures (Pinner reaction) [25]. Heitz et al. [26,27] studied this reaction in detail with different diols and poly(ethylene glycols) yielding polymers with DP = 5-10. Only after the partial decomposition of these oligomers **23** in the presence of a monomer like styrene (combination termination preferred) a new type of polymeric azo initiators - now based on polystyrene - was achieved [28 - 30].

$$CH_3-C(CH_3)(CN)-N=N-C(CH_3)(CN)-CH_3 \; (\textbf{22}) + HO-R-OH \xrightarrow[2.\,+\,H_2O]{1.\,+\,HCl/benzene} [-C(=O)-C(CH_3)_2-N=N-C(CH_3)_2-C(=O)-O-R-O-]_n \; (\textbf{23})$$

The preparation of poly(azoamides) **24** from AIBN **22** and formaldehyde in the presence of sulfuric acid is described by Yagci et al. [31,32]. Soluble polyamides with $\eta_{sp/c} = 2.3$ dl/g could be obtained.

$$CH_3-C(CH_3)(CN)-N=N-C(CH_3)(CN)-CH_3 \; (\textbf{22}) + CH_2O \xrightarrow{H_2SO_4} [-N(H)-C(=O)-C(CH_3)_2-N=N-C(CH_3)_2-C(=O)-N(H)-CH_2-]_n \; (\textbf{24})$$

There are, obviously, many possibilities to synthesize macroinitiators with several azo functions in the main chain. Similarly there are numerous methods available to introduce exactly one central azo group (type A).

The above mentioned Pinner reaction [33] and the condensation reaction of ACPC **11** (34) yield in polymers with one azo group in the middle and hydroxyl end groups by keeping the ratio azo compound/poly(ethylene glycol) at 1:2. The molecular weight of these polymers **25** is fully determined by the molecular weight of the reacted PEG or any other hydroxy functionalized telechelics which represent the diol component. Hazer [34] further functionalized these macroinitiators with methacryloyl chloride to obtain "macroinimers" **26**, a polymeric initiators with monomer functionalities as end groups.

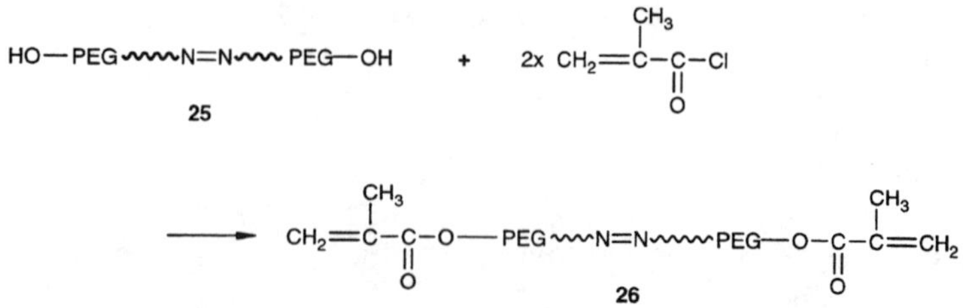

$$HO-PEG\sim\sim N{=}N\sim\sim PEG-OH \quad + \quad 2x\; CH_2{=}C(CH_3)-C(=O)-Cl$$

25

$$\longrightarrow CH_2{=}C(CH_3)-C(=O)-O-PEG\sim\sim N{=}N\sim\sim PEG-O-C(=O)-C(CH_3){=}CH_2$$

26

Bifunctional initiator systems which are not based on radical decomposition but include a functional azo compound are an excellent starting material to introduce the azo group in the middle of the polymer chain. It is known, for example, that certain ceric salts form effective redox initiator systems in the presence of organic reducing agents such as alcohols [35]. Yagci et al. [36] therefore used ceric ammonium nitrate in combination with 4,4'-azobis(4-cyanopentanol) **20** as redox initiator for the polymerization of acrylamide. The termination mechanism, which was exclusively linear, offered one azo group per chain (polymer **27**) [36 - 38].

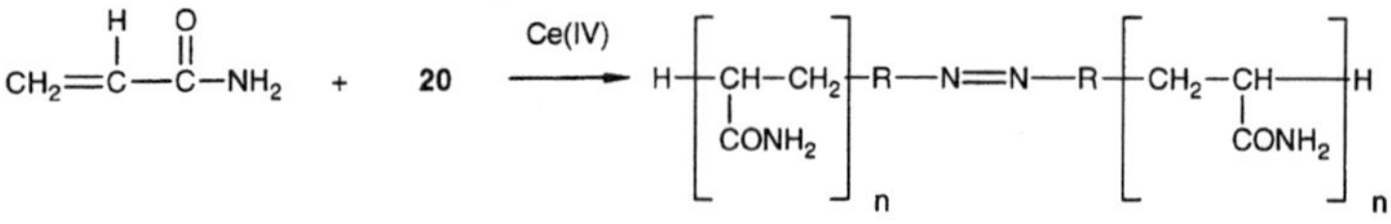

$$CH_2{=}CH-C(=O)-NH_2 \quad + \quad 20 \xrightarrow{Ce(IV)} H\!\left[CH(CONH_2)-CH_2\right]_n\!-R-N{=}N-R-\!\left[CH_2-CH(CONH_2)\right]_n\!H$$

27

A different redox initiator was reported by Araki et al. [39]. In the presence of $CuCl_2$ a bifunctional azo compound with imino groups **28** initiates the polymerization of a vinyl monomer $M_1$ (polymer **29**). Chain termination by radical combination leads to several azo groups in the polymer. A linear termination mechanism guarantees again, one central azo function.

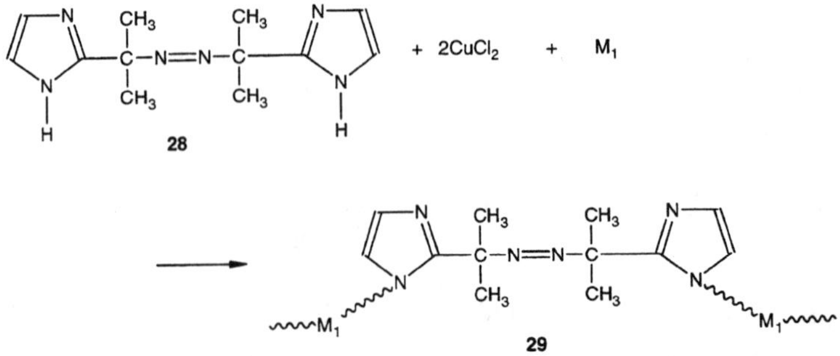

$$\text{(imidazolinyl)}-C(CH_3)_2-N{=}N-C(CH_3)_2-\text{(imidazolinyl)} \quad + \quad 2CuCl_2 \quad + \quad M_1$$

28

$$\longrightarrow \sim\sim M_1\sim\sim N\text{(imidazolinyl)}-C(CH_3)_2-N{=}N-C(CH_3)_2-\text{(imidazolinyl)}N\sim\sim M_1\sim\sim$$

29

Acrylonitrile was polymerized by an anionic insertion process by adding to a solution containing the adduct of the reaction of tetrakis(dimethylamino)-titanium and AIBN **30** [40]. An azolinked polyacrylonitrile **31** was obtained in this fast polymerization process.

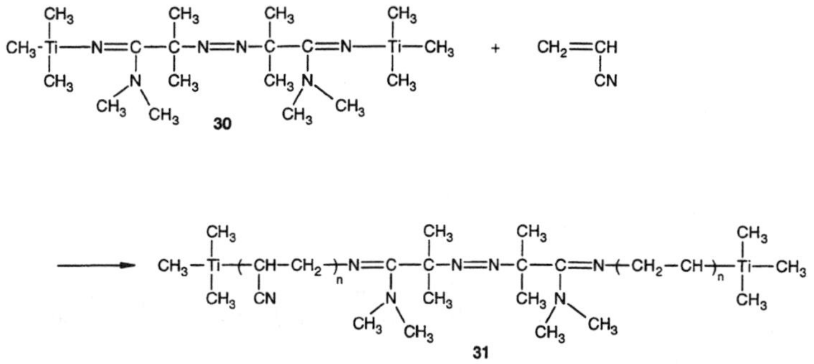

A cationic initiation process was used to polymerize THF with a central azo linkage [41,42]. The azo bisacid chloride **11** in the presence of silver salts ($AgSbF_6$ or $AgBF_4$) forms a difunctional azo-oxocarbenium initiator which can start the cationic polymerization of heterocycles like THF:

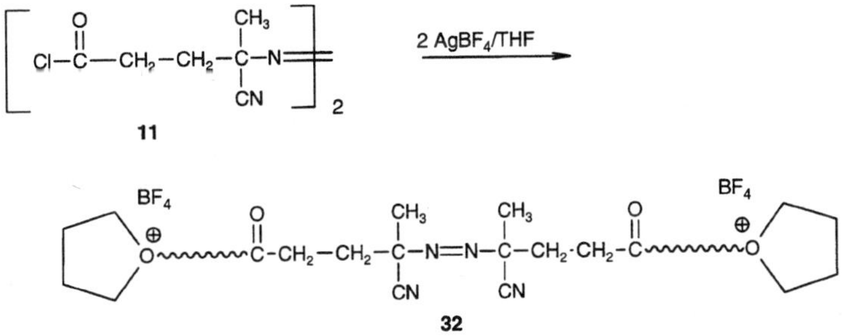

The reported kinetic results [41] indicate the absence of termination reactions as it was found earlier for other oxocarbenium salt initiating systems [43]. The best results were obtained with $AgSbF_6$. The molecular weights of the azo containing poly(tetrahydrofuran)s **32** were in the range of 3500 - 34000 g/mole with narrow molecular weight distributions. Macroinitiators via living cationic polymerization were also synthesized using the vinylether/HI/coinitiator system [44]. An azo compound functionalized with two vinylether groups **33** forms the initiator adduct with HI/coinitiator (tetrabutylammonium perchlorate, TBAP) and starts the polymerization. The resulting azolinked poly(vinylether)s **34** have molecular weights in the range of 1300-16600 g/mole and narrow molecular weight distributions (Mw/Mn = 1.11-1.38).

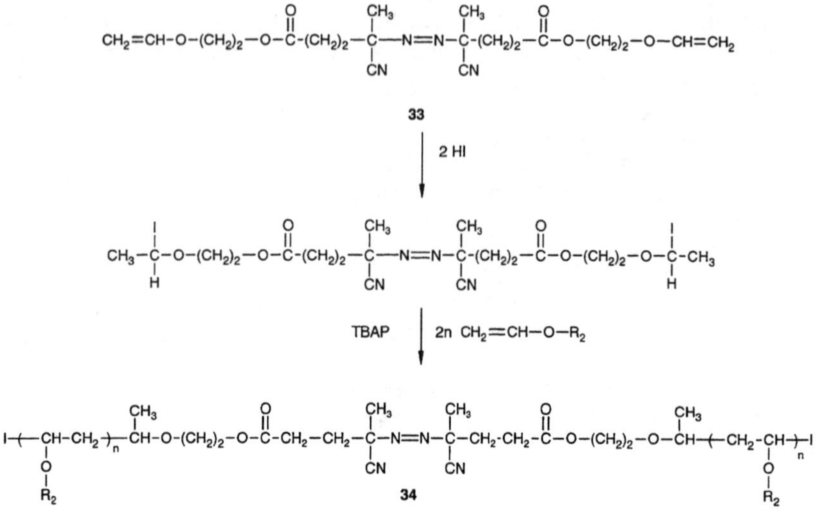

An interesting variation of the anionic polymerization technique was applied for the polymeric initiator **36**. An N-carboxy anhydride is polymerized in the presence of 2,2 -azodimethyl propiohydrazide **35** to yield polypeptide containing a central azo group [45]:

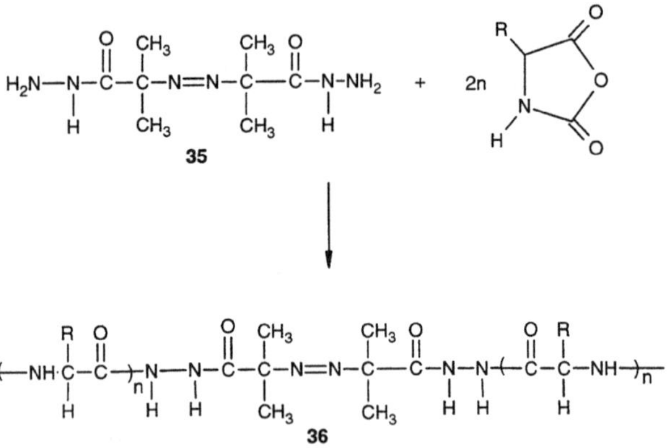

The reaction of two living polystyrene chains made by anionic polymerization with the azo bisacid chloride **11** leads to an azolinked polystyrene **37**. This dimerization of PSLi was more successful when PSLi was reacted with 1,1-diphenyl ethylene prior to the coupling reaction with the azo compound [46].

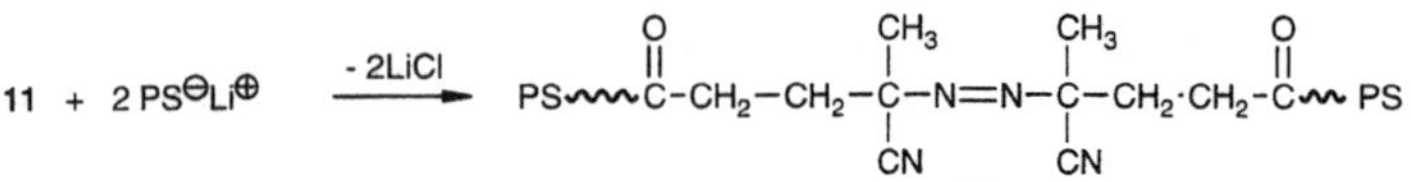

A similar coupling product was obtained by the addition of living polystyrene to the nitrile group of AIBN [47]. If $PS^{\ominus}$ was first derivatized with α-methyl styrene or 1,1-diphenyl ethylene, the dimer with a central azo group was isolated. However, the underivatized terminal styryl anion can form a pseudo cycle with the azo compound and the addition of $PS^{\ominus}$ to the second nitrile group is no longer possible. One obtains PS with one azo end group **38**. In some later work [48] it was shown that this addition process finally leads to the elimination of the nitrile group, especially in the case of high coupling efficiency with the derivatized polystyrene.

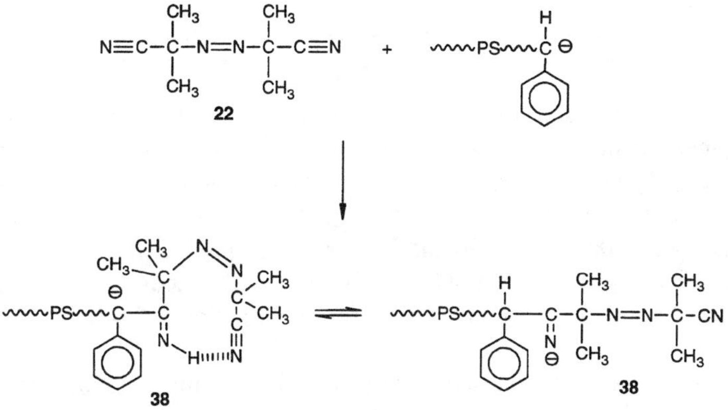

$$N{\equiv}C{-}C(CH_3)_2{-}N{=}N{-}C(CH_3)_2{-}C{\equiv}N \quad (22) \quad + \quad \sim\!PS\!\sim C^{\ominus}H(C_6H_5)$$

**22**

**38** ⇌ **38**

Termination of a living anionic polymerization with a monofunctional azo compound should always lead to polymeric azoinitiators with terminal azo functions (type C1 or type C2, depending on the initiation of the living polymerization). A similar product (type C2) **40** is achieved by the reaction of a OH difunctional polysiloxane or poly(ethylene glycol) with a monofunctional azo acid chloride **39** [49].

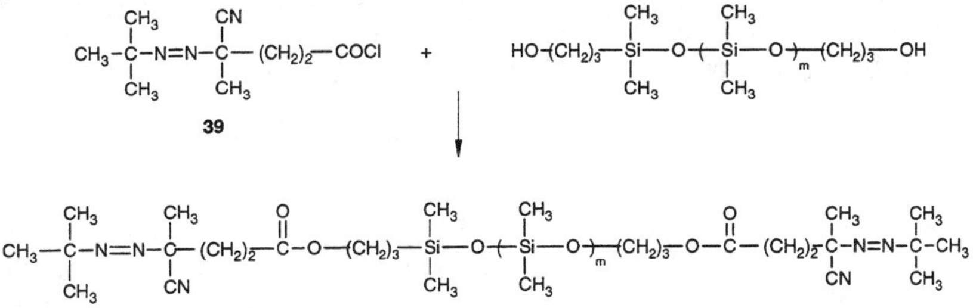

$$CH_3{-}C(CH_3)_2{-}N{=}N{-}C(CN)(CH_3){-}(CH_2)_2{-}COCl \quad + \quad HO{-}(CH_2)_3{-}Si(CH_3)_2{-}O{-}[Si(CH_3)_2{-}O]_m{-}(CH_2)_3{-}OH$$

**39**

**40**

Azo compounds with two azo functions of different stability were used to introduce azo end groups into a polymer [50,50a]. The more labile function starts the free radical polymerization of a vinyl monomer while leaving the remaining more stable azo function intact and connected to the polymer. Termination of the polymerization by recombination leads such to macroinitiators with two terminal initiator functions, termination by transfer results in a monofunctional macroinitiator. Difunctional polystyrene and polyisoprene macroinitiators could be synthesized by this method [50a].

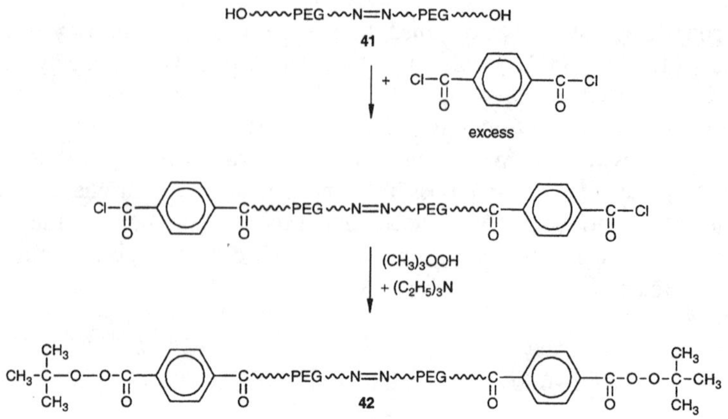

HO〰PEG〰N=N〰PEG〰OH
**41**

+ Cl–C(=O)–C$_6$H$_4$–C(=O)–Cl
excess

Cl–C(=O)–C$_6$H$_4$–C(=O)〰PEG〰N=N〰PEG〰C(=O)–C$_6$H$_4$–C(=O)–Cl

$(CH_3)_3OOH$
+ $(C_2H_5)_3N$

$CH_3$–C($CH_3$)$_2$–O–O–C(=O)–C$_6$H$_4$–C(=O)〰PEG〰N=N〰PEG〰C(=O)–C$_6$H$_4$–C(=O)–O–O–C($CH_3$)$_2$–$CH_3$
**42**

In some cases it is useful to have initiator functions of different stability in one single polymer chain. This allows a stepwise formation of multiblockcopolymers. Hazer et al. [51] have synthesized macroinitiators with peroxy and azo groups. Diisocyanate terminated poly(peroxycarbamate) (Mn ~ 2300) was reacted with a polyazoester (Mn ~ 800) made by the Pinner reaction of AIBN with ethylene glycol. Furthermore, an azolinked PEG **41**, synthesized according to Pinner, was functionalized with two peroxy end groups **42** [52] as shown above.

We hope that we have given a satisfactory overview on the possible synthetic routes to main chain polymeric azo initiators. For characterization purpose all commonly known methods have been used: NMR, IR, UV and DSC (differential scanning calorimetry, determination of azo content and the thermal decomposition), GPC (gel permeation chromatography, molecular weight determination), viscosity measurements and so on. Even though there is a great interest in these polymers at the university and industrial research organisations and the syntheses are very versatile, no commercially valuable products seem to be the results of these efforts up to now. The main application, but still on laboratory scale, is the use as starting material for blockcopolymerization. More details on this will be given in section 4.1.

## 3. SIDE CHAIN POLYMERIC AZO INITIATORS

Beside the important group of the **main chain** polymeric azo initiators, polymers with azo functions in the **side chain** found early interest (1956, Hahn and Fischer [53]).The main intention for the synthesis of these polymeric initiators was, analog to the potential radical blockcopolymerization using the main chain initiators, the possibility of radical grafting from the polymer backbone.

Side chain polymeric azo initiators (SCPAI) of the general structure A (again the restriction to labile azo functions applies) can be synthesized mainly via three methods: polymer analogous reaction, copolymerization and copolycondensation. Ionic polymerization techniques play a minor role up to now and only one example will be mentioned under the copolymerization techniques.

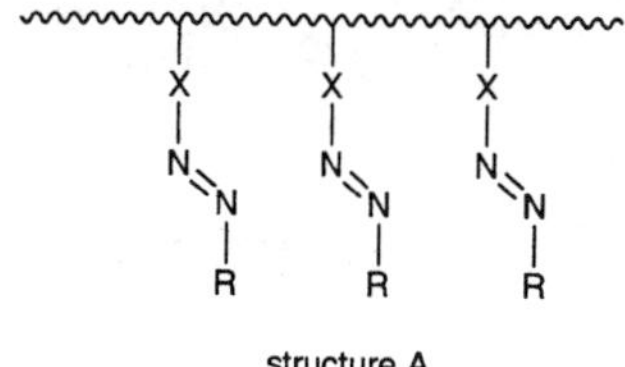


structure A

We will see that the azo function in structure A can be directly bound onto a monomer unit (X = bond) or will be introduced over a spacer linkage (X = aromatic, aliphatic, heteroatom). The nature of the polymer backbone can vary widely and ter- or even tetrapolymerization (polymer backbone composed of four different structural units) was applied.

Restrictions are given through the thermal lability of the azo functions. Reaction conditions had to be chosen carefully to avoid undesired cleavage of the azo group. However, variation of R in structure A opens the possibility to tailor-made the thermal stability and the initiator reactivity of the formed polymer radical, similar to AIBN or almost as stable and unreactive as azo benzene. Therefore the limitation to AIBN-derivatives as used mainly in the main chain azo polymers no longer applies. This requires in many cases an intensive study on the thermal behavior and the initiator efficiency of the used azo structures. The synthetic approaches, results, and thermal behavior of the SCPAI's will be discussed in detail in this chapter. Interesting examples of potential applications, not only grafting, will be given in section 4.

## 3.1 Synthesis via Polymer Analogous Reaction

Even though many important technical products are the result of successful polymer modification via polymer analogous reaction, this technique still battles with its bad reputation. Slow reaction, low conversion, the difficulties to remove side products, and often heterogeneous products repel many synthetic polymer chemists to use this pathway to new products. However, for the introduction of azo functions as side groups in a polymeric chain, this was the method used in one of the first reported attempts. Hahn and Fischer [53] converted pendant acetylamino groups on a polymer chain into azo groups (structure **43**) via nitrosation and rearrangement. First, starting polymer was poly-p-aminostyrene and its copolymers with methyl-methacrylate (MMA). Later 4-acetylaminostyrene was copolymerized directly with MMA to reduce the number of polymer analogous reaction steps.

The azo polymer **43** is thermally quite instable and therefore could not be isolated and purified. In situ grafting with MMA or acrylonitrile (AN), resulted crosslinked products [53] in many cases. This proves that several reactions steps carried out at a polymer backbone can lead to undefined products and side reactions.

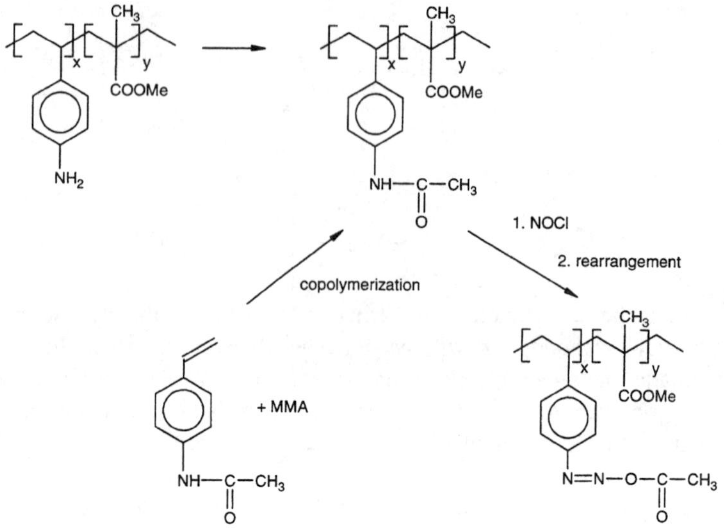

43

Vollmert and Bolte [54] functionalized OH-groups of poly(butylacrylate-co-butanediolmonoacrylate) with 4,4'-azobis(4-cyanopentanoic acid) **9** to form polymer **44**.

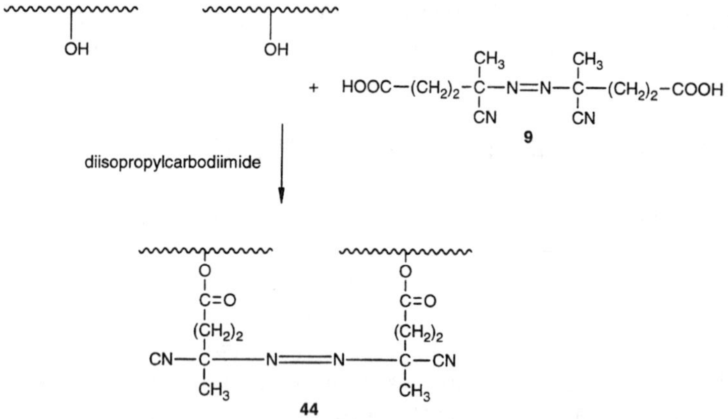

44

Due to the bifunctionality of the azo compound **9**, mostly inter- and not intramolecular reaction took place, and therefore crosslinking was observed. Nevertheless, grafting onto the in monomer 2 swollen gel was possible. However, from the technical standpoint, grafting from insoluble polymers is impractical.

Also, a bifunctional azo initiator of the structure **45** was used by Gupta et al.[55] in a reaction with a polymeric acid chloride yielding insoluble products after grafting. Dekking [56], however, described the successful grafting with the polymeric initiator **46** which contained **45** complexed to pendant acid moieties.

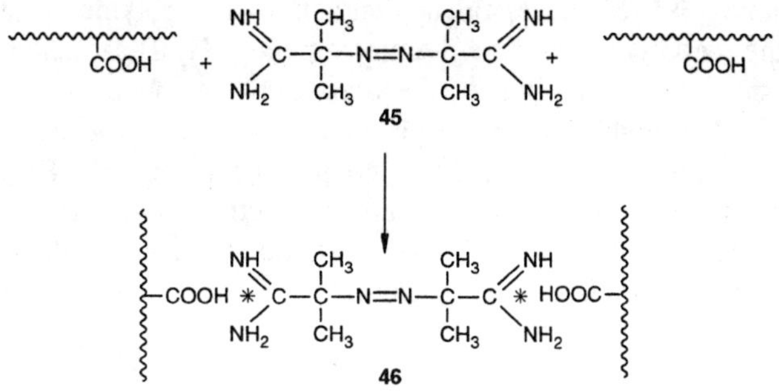

45

46

A similar complex formation was observed for the free acid groups of γ-poly(glutamic acid) with the difunctional compound 2,2'-azobis(2-(hydroxymethyl)propionitrile) (analogous to structure **20**) and soluble products were reported [56a] when less than 0.25mmole azo functions per g polymer were incorporated. Soluble products can be obtained more reliable by the single step polymer analogous reaction of a monofunctional azo compound like **39** with hydroxy group containing backbone or OH-terminated butadiene rubber [57].

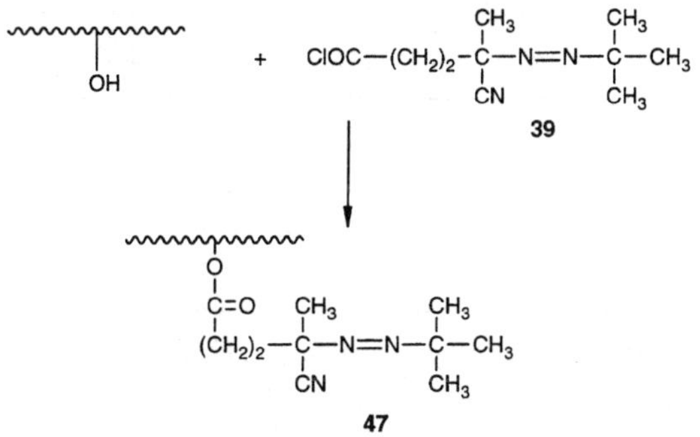

39

47

The reaction of a polymeric acid chloride e.g. poly-(methylmethacrylate-co-methacryloyl chloride) **48**, with an amino or hydroxy monofunctional azo compound **49** to structure **50** is similarly straight forward [58]. The composition of the prepolymer **48** was varied from 10-50 mol% acid chloride groups. In all cases soluble products could be isolated after modification. However, the azo content in the product was generally lower than expected.

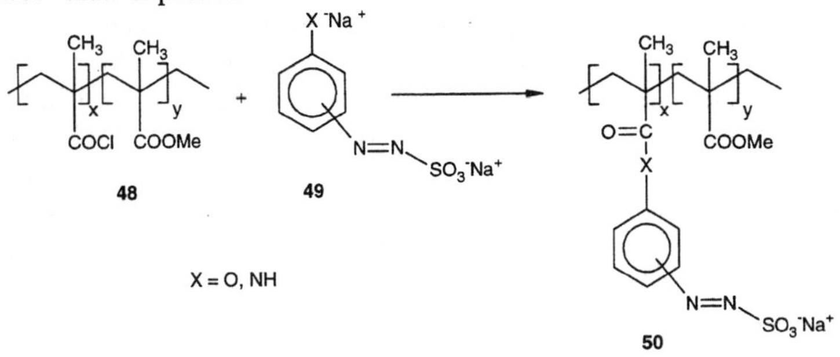

48 49

X = O, NH

50

Another example [59] for successful application of the polymer analogous reaction starts from the copolymer styrene/3-aminostyrene **51**. Diazotation of the amino function and subsequent coupling of the diazonium salt with methylmalonodinitrile could be achieved in high yields (Tab.1). By this method up to 35 weight % of azo containing units could be introduced in the polymer chain **52**. The polymers were soluble and the number of unconverted amino functions was very low. However, the prepolymer having higher than 50 mol% amino content resulted in crosslinked products after modification due to some intermolecular coupling reactions of the intermediate diazonium salts.

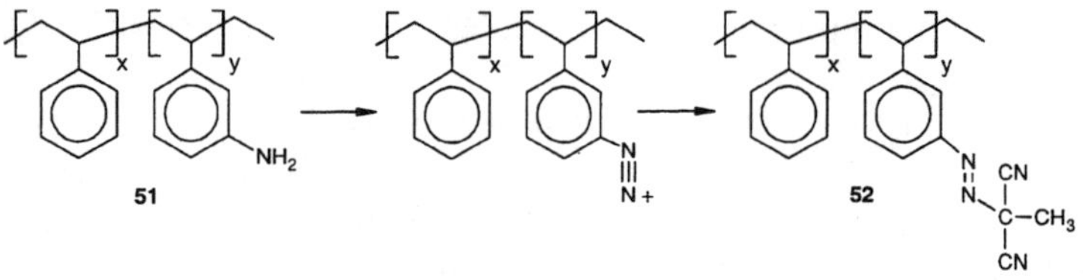

**Table 1** Results for the modification of polymer **51**

| amino content of **51** in wt.% | azo content in **52** in wt.%[a)] | solubility |
|---|---|---|
| 5.7 | 4.2 | soluble |
| 11.3 | 10.6 | soluble |
| 32.9 | 35.0 | soluble |
| 53.4 | 47.6 | insoluble |

*a) determined by differential scanning calorimetry (DSC)*

Furthermore the GPC-traces of the soluble products showed evidence of a very small amount of polymer chains with double the molecular weight which is again indication for intermolecular coupling reaction. However, at this low extent, the intermolecular reaction did not lead to insolubility.

These last two examples reveal quite well the problems of polymer modification reactions: side reactions, low conversion, crosslinking. Even single step reactions of functional groups with high reactivity can prove problematic. Nevertheless, polymer modification gives access to a variety of polymer structures with azo side groups which can not be synthesized by other polymerization techniques. And, as we will see later, polymers with azo contents of 10mole% or higher are very difficult to achieve by radical techniques due to the retarding influence of the azo function itself. Via polymer modification the control of the molecular weight is given through the prepolymer and therefore independent of the azo content.

## 3.2 Synthesis via Copolymerization

The polymer modification reaction, though at least partly successful and in many cases unavoidable, is dominated by copolymerization of azo group containing monomers in the synthesis of SCPAI's. The copolymerization leads to well-defined linear products which can be easily purified. The main challenge, however, is the synthesis of appropriate monomers.

### 3.2.1 Azo Monomers

Early on, difunctional symmetrical azo monomers derived from AIBN were successfully applied in radical co- and ter-polymerization yielding mainly crosslinked products, for example **54**, onto which MMA was grafted radically [60].

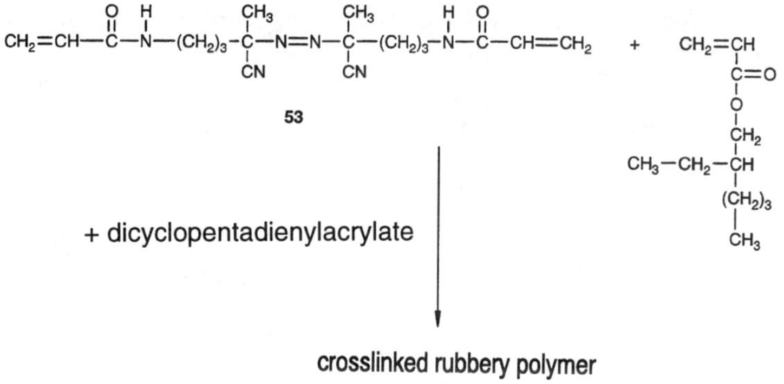

$CH_2{=}CH{-}C(O){-}NH{-}(CH_2)_3{-}C(CH_3)(CN){-}N{=}N{-}C(CH_3)(CN){-}(CH_2)_3{-}NH{-}C(O){-}CH{=}CH_2$ + $CH_2{=}CH{-}C(=O){-}O{-}CH_2{-}CH(CH_2CH_3){-}(CH_2)_3{-}CH_3$

53

+ dicyclopentadienylacrylate

↓

crosslinked rubbery polymer

54

Compound **55**, on the contrary, was copolymerized with isobutylvinylether cationically with the initiator $BF_3$ x $Et_2O$ [54]. This is one of the very rare examples where ionic initiation played a role for the synthesis of SCPAI's. Again, no soluble products could be obtained due to the difunctionality of the azo monomer. Monomers of this type are available via the synthetic pathway generally leading to AIBN derivatives.

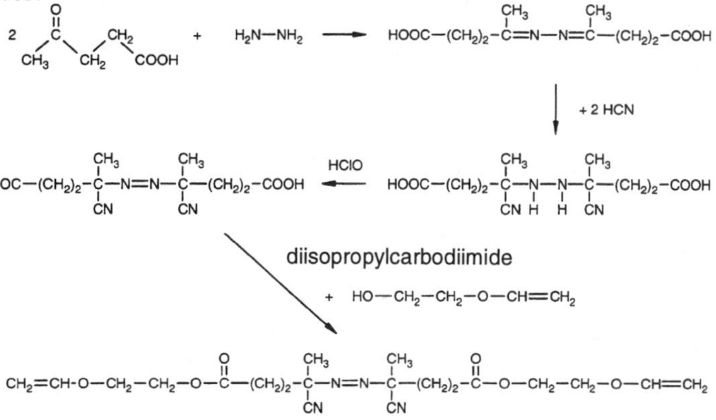

2 $CH_3{-}C(O){-}CH_2{-}CH_2{-}COOH$ + $H_2N{-}NH_2$ ⟶ $HOOC{-}(CH_2)_2{-}C(CH_3){=}N{-}N{=}C(CH_3){-}(CH_2)_2{-}COOH$

↓ + 2 HCN

$HOOC{-}(CH_2)_2{-}C(CH_3)(CN){-}NH{-}NH{-}C(CH_3)(CN){-}(CH_2)_2{-}COOH$

⟵ HClO

$HOOC{-}(CH_2)_2{-}C(CH_3)(CN){-}N{=}N{-}C(CH_3)(CN){-}(CH_2)_2{-}COOH$

diisopropylcarbodiimide + $HO{-}CH_2{-}CH_2{-}O{-}CH{=}CH_2$

↓

$CH_2{=}CH{-}O{-}CH_2{-}CH_2{-}O{-}C(O){-}(CH_2)_2{-}C(CH_3)(CN){-}N{=}N{-}C(CH_3)(CN){-}(CH_2)_2{-}C(O){-}O{-}CH_2{-}CH_2{-}O{-}CH{=}CH_2$

55

The unsymmetrical difunctional azo compound **56** was used in radical copolymerization with styrene [61]. Due to strong differences in reactivity of the two unsaturated groups in **56**, mainly the vinyl group was incorporated in the polymer backbone and the allyl function remained unreacted. Therefore, soluble products have been obtained as long as the azo monomer concentration was kept below 2mole%.

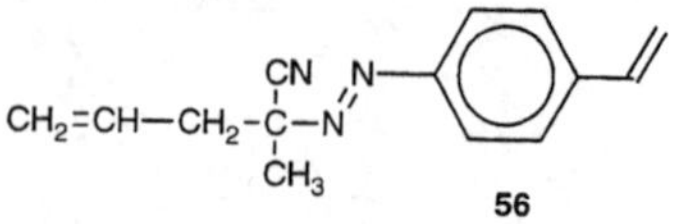

**56**

While difunctional azo monomers like **53** or **55** easily result in crosslinked polymers, a variety of monofunctional azo compounds, e.g. **57** and **58**, described in the patent literature [62], lead to linear, soluble products in the copolymerization with styrene.

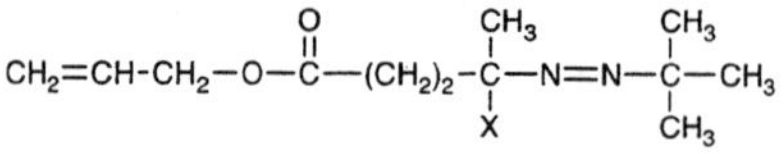

$CH_2{=}CH{-}CH_2{-}O{-}C(O){-}(CH_2)_2{-}C(CH_3)(X){-}N{=}N{-}C(CH_3)_2{-}CH_3$

**57**

X = CN, $OCH_3$, O-Ph

$CH_2{=}C(CH_3){-}C(O){-}O{-}CH_2{-}CH_2{-}O{-}C(O){-}(CH_2)_2{-}C(CH_3)(X){-}N{=}N{-}C(CH_3)_2{-}CH_3$

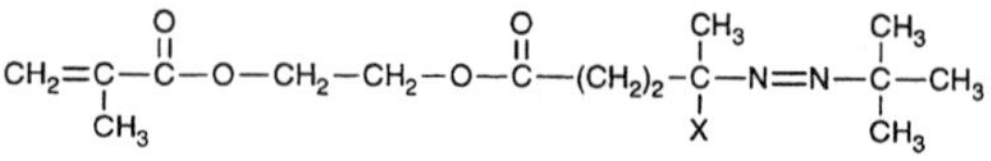

**58**

Since these initial reports on radical copolymerization of azo monomers, a broad range of suitable monomers have been described [3 - 6], some shown below.

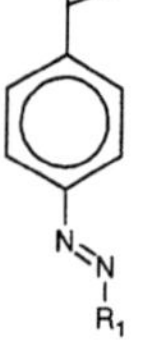

**59** 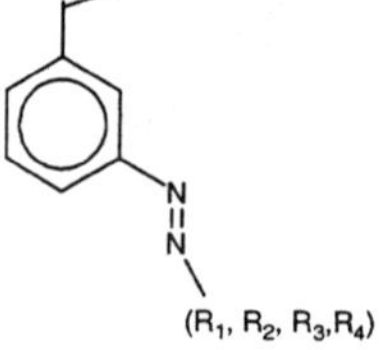 **60** 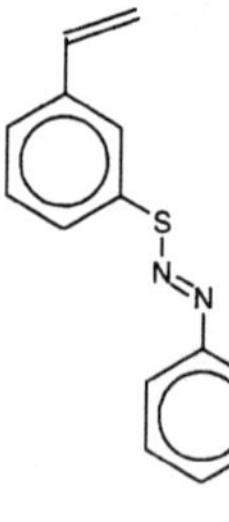 **61** (lit.64) **62** (lit.65)

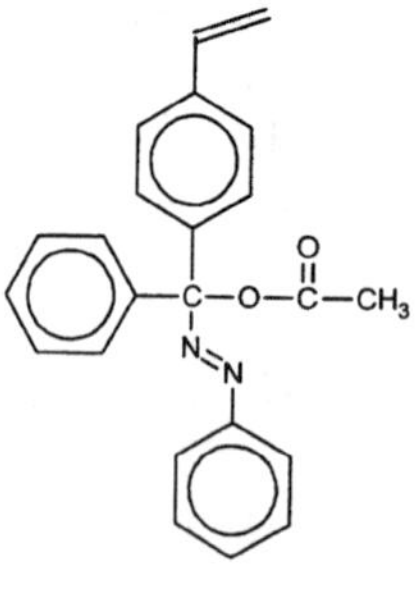

($R_1$, $R_2$, $R_3$, $R_4$)

$R_1$ = $C(CN)_2(CH_3)$ (lit.63)

$R_2$ = S-Ph (lit.64b)

$R_3$ = $-C(R_y)(R_x){-}O{-}C(O){-}CH_3$ (lit.65)

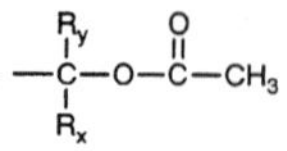

$R_x$, $R_y$ = $CH_3$ or $R_x$ = Ph, $R_y$ = $CH_3$ or $R_x$, $R_y$ = Ph

$R_4$ = $SO_3^-Na^+$ (lit.66)

Starting material for the azo monomers **56**, **59** to **62**, are styrene derivatives: para- or meta-aminostyrene, 3-mercaptostyrene, 4-(phenylketone)-styrene, and 3-hydrazostyrene. Hydrazin derivatives proved useful as starting material for the synthesis of many azo compounds, including **57**, **58**, **60($R_3$)**, and **62**, following the general scheme below:

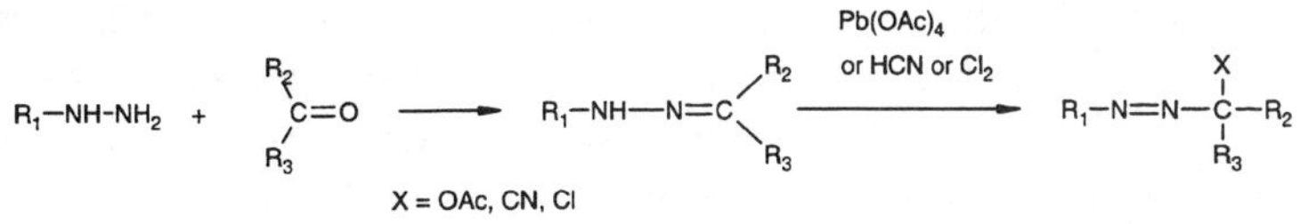

$$R_1{-}NH{-}NH_2 + \overset{R_2}{\underset{R_3}{C}}{=}O \longrightarrow R_1{-}NH{-}N{=}C\begin{smallmatrix}R_2\\R_3\end{smallmatrix} \xrightarrow[\text{}]{Pb(OAc)_4 \text{ or } HCN \text{ or } Cl_2} R_1{-}N{=}N{-}\overset{X}{\underset{R_3}{C}}{-}R_2$$

X = OAc, CN, Cl

On the other hand, the coupling reaction of an aromatic diazonium salt with a compound containing an acidic hydrogen was used for the synthesis of **56**, **60($R_1$,$R_2$,$R_4$)**, and **61**. The relatively simple one-step diazonium salt coupling reaction allows isolation of azo monomers in high yields.

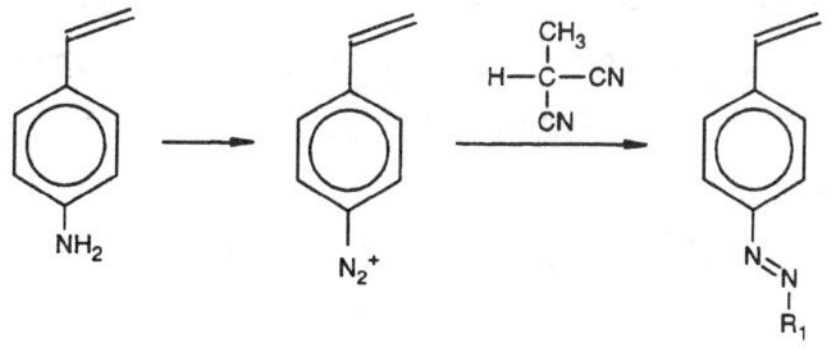

NH$_2$ → N$_2^+$ → (H–C(CH$_3$)(CN)–CN) → N=N–R$_1$

59

N$_2^+$ + SH → S–N=N

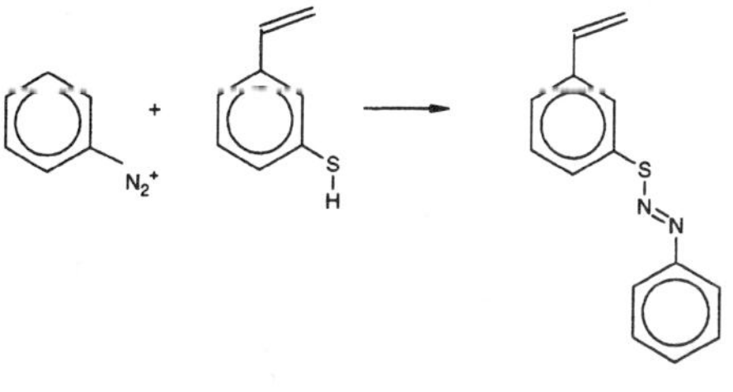

61

The styrene derivatives **56**, **59-62** represent only one class of radically polymerizable azo monomers. Other examples with methylmethacrylate or allyl groups are shown in structure **63** [67] and **64** [61].

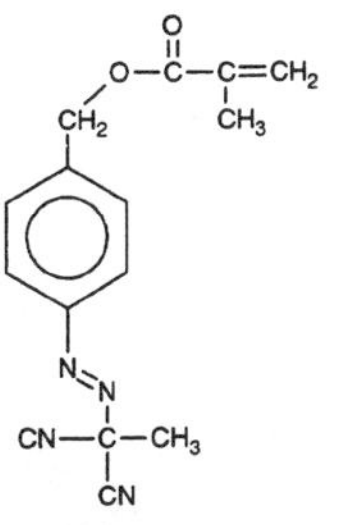

$CH_2{-}O{-}C(=O){-}C(CH_3){=}CH_2$; N=N; CN–C–CH$_3$; CN

63

Br; N=N; CN–C–CH$_3$; CH$_2$; CH; CH$_2$

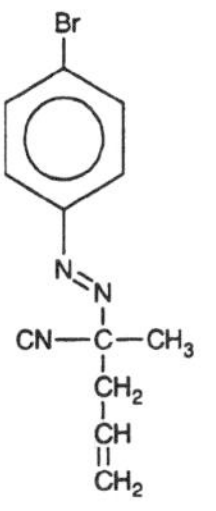

64

Monomer **64** could be obtained directly by the diazonium salt coupling of 4-bromo-aniline with 4,4-dicyano-1-butene; for **63**, the azobenzylalcohol, synthesized analogous as shown for **59**, was esterified with methacryloyl chloride.

### 3.2.2 Thermal Characteristics of Azo Monomers

One of the criteria for the choice of an azo monomer is the thermal stability or lability of the azo function. In the case of main chain polymeric azo initiators (MCPAI's), AIBN like structures predominate with therefore well known thermal characteristics. However, their initator efficiency is quite often much lower than for their low molecular weight models, but still high enough to allow the synthesis of blockcopolymers.

As seen in Section 3.2.1, the structures for azo monomers vary widely. First, there are again symmetrical AIBN derivatives. The thermolysis of **55** was studied by Nuyken et al. [44]: a rate constant of $k = 2.38 \times 10^{-4} s^{-1}$ at 70°C with an activation energy of 147 KJ/mole was found, a behavior very similar to AIBN ( $k = 1.25 \times 10^{-4} s^{-1}$, $E_a = 145$ KJ/mole). However, thermolysis of unsymmetrical azo compounds had to be studied in detail [61,64,64a,68,70,73]. In [5] the dependence of the thermal stability of the azo monomers on the substituents on the azo function is summarized and the results will be shown here in Table 2.

**Tab.2** Effects of varying the substituents on the thermolysis of azo compounds of the type $R_2$-$C_6H_4$-N=N-$R_1$ [5]

| $R_1$ | $R_2$ | T/°C ($t_{1/2}$ = 1h) | solvent |
|---|---|---|---|
| $C(CH_3)(CN)_2$ | 4-$OCH_3$ | 103 | 1,2-dichlorobenzene |
| | 4-$CH_3$ | 93 | 1,2-dichlorobenzene |
| | H | 93 | 1,2-dichlorobenzene |
| | 4-$NO_2$ | 95 | 1,2-dichlorobenzene |
| | 3-CH=$CH_2$ | 89 | toluene |
| | 2,6-$CH_3$ | 67 | 1,2-dichlorobenzene |
| S-$C_6H_5$ | 3-CH=$CH_2$ | 76 | toluene |
| $C(OCOCH_3)(Ph)_2$ | H | 126 | phthalic acid dibutyl ester |
| $C(OCOCH_3)(CH_3)(Ph)$ | H | 176 | phthalic acid dibutyl ester |
| $C(OCOCH_3)(CH_3)_2$ | H | 235 | phthalic acid dibutyl ester |
| AIBN | | 73 | toluene |

Instead of the azo monomers often model compounds without a polymerizable group were studied to exclude side reactions during kinetic measurements of the thermal decomposition. Thermal stability was measured in solution followed by

UV-spectroscopy and via DSC (differential scanning calorimetry) measurements. The latter allows the analysis of the solid compound without the effect of solvent and needs only a very small amount of substance for a complete kinetic analysis. The maximum of the exothermic peak represents the temperature of maximum decomposition and kinetic analysis of the area under the DSC-trace provides information on rate constants and activation energy (details in ref. 5 and 65). DSC proved very helpful in the analysis of azo group containing crosslinked products to determine the azo content and the thermal decomposition behavior.

The measurements in solution showed the predicted dependance of the thermolysis on the solvent. The decomposition is faster in polar solvents, but the effect is very small due to the radical character of the decomposition mechanism [5, 69]. The thermal decomposition can also be followed by other methods, e.g. $N_2$-volumetry, colorimetry or radical scavenging, however, UV-spectroscopy and DSC are in most cases very efficient and reliable.

In Tab.2, one can see the broad temperature range of the thermal decomposition of azo model compounds and monomers. Some are as easily cleaved as AIBN, others are still quite stable above 200°C. An other example for a very stable azo compound is structure **60($R_4$)**, the azo sulfonate. No rate constants could be determined for this compound due to very sharp DSC-traces, but the decomposition maximum lies at about 240°C [70]. This broad range of thermal stability allows the free choice of the polymerization and grafting temperature. Furthermore other applications of the resulting SCPAI s for the synthesis of core-shell polymers and even for reactive blending and selfcrosslinking networks, where higher temperatures will be applied, are made possible.

It should not be ignored that these azo compounds are also photochemically labile. However, the photochemical decomposition reaction, especially in solution, is no longer first order and side reactions beside the homogeneous breaking of the N,N-double bond can occur. Nevertheless, an application for selfcrosslinking, for example, is possible. Some studies showed that thermolysis in combination with UV-irradiation leads to a decomposition approximately 100°C below the original thermolysis temperature [71]. This decrease in thermal stability is due to the photochemical formation of the labile cis-isomer from the rather stable trans-isomer. By ESR-studies it was proven that radicals are formed in the photolysis of azo compounds [72].

### 3.2.3 Copolymerization

One advantage of the polymer modification reaction as synthetic route to SCPAI's is the fact that polymer backbone and structure of the azo compound can be designed independently. Normally, molecular weight and structure of the backbone is given by the prepolymer and does not change during modification.

In the free radical polymerization technique the azo monomer can have a strong effect on the copolymerization rate and the molecular weight obtained. Furthermore, the incorporation of the azo compound is dependent on the copolymerization parameters and therefore strongly on the comonomer. Moreover, azo monomers can act as initiators in copolymerization at temperatures close to their decomposition temperatures.

This last aspect requires polymerization techniques where the azo function of the monomer will not decompose. Surprisingly, this was neglected in the first copolymerization attempts of AIBN derivatives. The copolymerization was carried out at 60 or 70°C for several hours [60,62]. Cleavage of part of the azo functions can lead to crosslinking and branching. Since the products of Naarmann et al.[60] and MacLeay et al. [62] were insoluble due to the difunctionality of the azo monomer or due to added crosslinking agents, the negative effect of the partly decomposed azo monomer was not recognized.

Thermal decomposition of the azo monomer can be avoided when redox initiation in emulsion polymerization at room temperature or below is applied. A suitable system consists of water as dispersing medium, monomers (about 30wt%), Rongalit®, $FeSO_4x7H_2O$, ethylenediamine tetraacetate, sodiumhydrogenphosphate puffer and methanehydroperoxide as redox initiator at 5°C [63,64a,73]. However, polymerization in solution or in bulk is possible, too [64-66].

Copolymerization behavior or copolymerization parameter were studied in detail for several of the azo monomers described in section 3.2.1. For the styrene derivatives **59** and **60(R$_1$)**, it was found that the position of the azo function *para* to the vinyl group causes rather non ideal incorporation of the azo monomer and a strong decrease of the polymerization rate [61]. Azo function in *meta* position still slows the polymerization reaction down to a certain degree, but the incorporation of the azo monomers is ideally statistically [63]. This is due to the almost negligible effect of the azo group in meta position on the electronegativity of the vinyl group and therefore on its reactivity. Tab.3 shows some results of the copolymerization of **60(R$_1$)** with styrene.

The retarding effect of the azo unit on the radical polymerization rate is common for almost all of the described azo monomers and can be explained by the formation of hydrazyl radicals of structure **65**, leading to transfer and finally even to termination reactions. Also, this explains the decrease of molecular weight with increasing azo monomer concentration in the reaction mixture. Evidence for the hydrazyl radicals was found in ESR measurements [74]. Therefore using the styrene derivatives as monomers, the azo content in the monomer composition had to be kept rather low (below 5mole%) to achieve reasonable products.

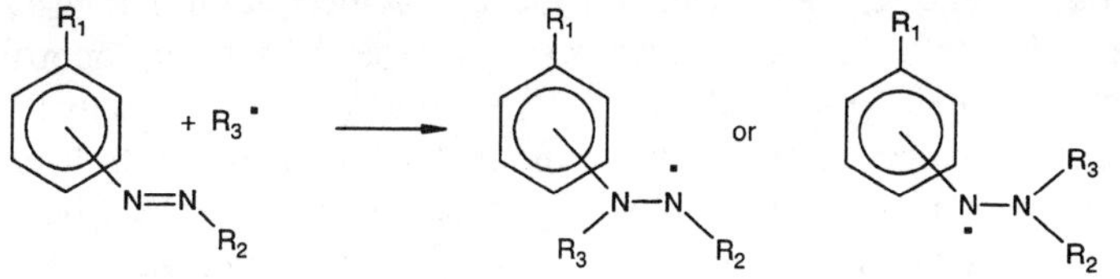

**Tab.3** Results of the emulsion copolymerization of $60(R_1)$ and styrene at 5°C [63]

| azo content[a)] [mole%] monomer mixture | copolymer | $R_p$ [b)] [%/h] | $M_n \times 10^{-3}$ [g/mole][c)] |
|---|---|---|---|
| 0 | 0 | 68.00 | 500 |
| 0.47 | 0.61 | 2.60 | 74 |
| 0.94 | 1.14 | 2.30 | 69 |
| 1.88 | 2.30 | 1.06 | 54 |
| 4.10 | 4.60 | 0.65 | 33 |

*a) by UV-analysis*
*b) conversion to polymer per hour in %*
*c) by membrane osmometry*

The above considerations for the copolymerization behavior are valid only for the copolymerization of the styrene derivatives **59-62** with styrene. Similar ideal copolymerization behavior was found for the copolymerization of the MMA-derivative **63** with MMA, with a retarding effect much less pronounced in respect to polymerization rate and molar masses than found for the styrene derivatives [67].

But in many cases, the copolymerization of chemically different monomers is very interesting. Therefore, attempts have been made to find an universal way for selecting monomers which copolymerize "ideally". It is known that the classical Mayo-Lewis equation turns into $[M_1] : [M_2] = [m_1] : [m_2]$ for $r_1 = r_2 = 1$. However, systems with $r_1 = r_2 = 1$ are rare. Under conditions applied for the design of polymeric azo initiators ( $[M_1] : [M_2] < 0.01$, $M_1$ = azo monomer) not both parameters but only one must be equal to one ($r_2 = 1$). For a given $M_1$, it is now possible to find monomers $M_2$ which confirm "ideal" copolymerization. $M_2$ is then selected on the basis of $Q_2$ and $e_2$ values calculated from $r_2 = 1 = f(Q_2, e_2, e_1, Q_1)$.

The potential of this concept is proven for the copolymerization of **63** with MMA, MAN, and styrene. Ideal copolymerization was observed for **63**/MMA and **63**/MAN but not for **63**/styrene, which was in good agreement with the experimental and calculated values (for more details compare [5] and [65]).

The allyl azo monomer **64** was copolymerized only with styrene [61]. A pronounced retarding effect of the azo monomer on the copolymerization rate was found and the incorporation of the azo monomer was, as expected, underproportional, since the conjugated vinyl group in styrene is much more reactive in radical polymerization compared to the allyl group of the azo compound. In contrast, the azo monomers **60($R_3$)** and **62** were overproportionally incorporated, but no negative effect on polymerization rate and molecular weight was observed. From the patent literature only some data could be obtained for the copolymerization of **57** and **58**. Both were copolymerized with styrene and **58** furthermore with MMA and ethylacrylate. Ideal behavior was not observed in any case.

All the cited polymers were usually characterized by common technique such as NMR, IR and elemental analysis to prove the structure, UV-spectroscopy and DSC to determine the azo content, and GPC (gel permeation chromatography), viscosimetry and membrane osmometry for the molecular weight determination. For the design of polymeric azo initiators via copolymerization several aspects have to be considered:

1. "Ideal" copolymerization, which allows complete conversion of the relative valuable azo monomer, is favorable in order to avoid additional inhomogenity as a function of conversion.
2. Azo compounds tend to show retarding effect to polymerization rate and molar masses. Therefore they should be applied in low concentrations or they should contain bulky substituents.
3. Due to thermo- and photosensitivity of azo monomers one is not completely free in choice of reaction conditions. Redox polymerization in emulsion at low temperatures is the method of choice in many cases.

Apart from these restrictions, copolymerization seems to us still one of the best methods for the synthesis of polymeric azo initiators.

### 3.2.4 Terpolymerization

The emulsion polymerization technique described above was also used for the synthesis of ter- and tetrapolymers (4 different monomers used in radical copolymerization). Main driving force for these activities was the synthesis of macroinitiators with a certain amount of unsaturated bonds and the formation of azo containing networks. Monomers **60** and **63** were mainly used for that purpose.

The styrene derivatives **60** [64a,73,75,76] were copolymerized with styrene and butadiene, and in some cases divinyl benzene was added as crosslinking agent. The ratio styrene/ butadiene was kept close to the "cold rubber" composition (1:3 by weight) [73]. Styrene was partly replaced by the azo monomers (0.5 - 5mole%) [64a,73]. The copolymerization of styrene and butadiene is known to be almost ideal. Since the azo monomer **60** behaved like styrene, it did not disturb this systeme and was incorporated ideally even at low conversion (Tab.4).

soluble terpolymer 66

azo group containing network 67

The incorporation of divinyl benzene leads to crosslinked products **67** [75]. Since the polymerization was carried out in emulsion, well-defined azo group containing beads were obtained which can be used as functionalized cores in core-shell synthesis or for reactive blending (section 4). The divinyl benzene content was varied from 0.5 - 2mole% [75].

**Tab.4** Results of the terpolymerization styrene/butadiene/**60($R_3$)** in emulsion ($R_x$ and $R_y$ as shown in structure **60**, section 3.2.1)

| azo monomer **60($R_3$)**, $R_x$, $R_y$ | azo content[a)] [mmole/100g] monomer mix. | copolymer | conversion [ %] | $R_p$[b)] [ %/h] | $M_{GPC}$ x $10^{-3}$ [g/mole][c)] |
|---|---|---|---|---|---|
| $R_x$, $R_y$ = $CH_3$ | 10.0 | 10.7 | 24.8 | 5.0 | 81 |
| $R_x$=$CH_3$,$R_y$=Ph | 10.0 | 10.7 | 36.5 | 4.1 | 54 |
| $R_x$,$R_y$ = Ph | 10.0 | 9.4 | 54.7 | 10.4 | 81 |

*a) by UV-analysis*
*b) conversion to polymer per hour in %*
*c) by GPC in THF, calibrated with polystyrene standards*

Similar to the procedure described above for butadiene/styrene/azo, a butadiene/acrylonitrile (2:1) rubber **68** with azo side groups was synthesized [75]. Again, addition of small amounts of divinyl benzene led to well-defined precrosslinked beads (structure **69**). Although the terpolymerization with butadiene, acrylonitrile and **60** was not ideal, the azo monomer was fully incorporated at high conversion, as expected. Since the azo content in the monomer mixture was usually very low, no danger of block formation in the polymer chains was given.

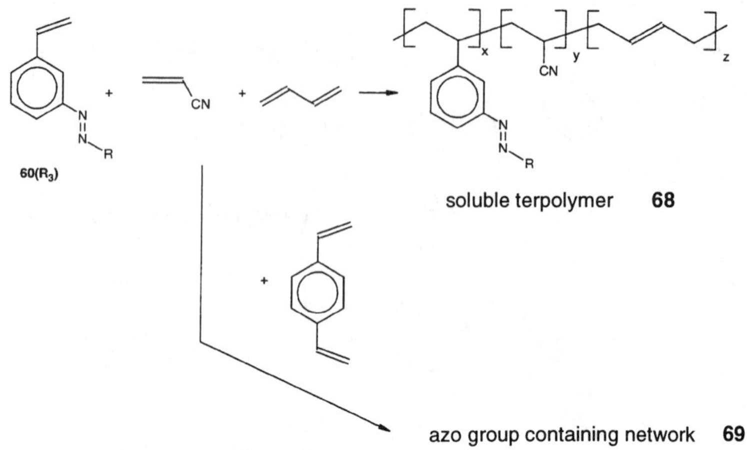

60(R$_3$)

soluble terpolymer **68**

azo group containing network **69**

As third example the terpolymerization of 3-vinylphenyl azo monomers **60(R$_1$)**, **60(R$_2$)** or **60(R$_3$)** (0.3-1.25mole%) with MMA and butanediol-1,4-dimethacrylate (BDDM) as crosslinking agent (0.5mole%) should be mentioned. Crosslinked beads were obtained through emulsion polymerization at 45°C. Part of the azo monomer **60(R$_1$)** decomposed under these conditions and the azo monomer incorporation was overproportional [77]. Results closer to ideal terpolymerisation could be observed in the system MMA, BDDM and the methacrylate functionalized monomer **63** (structure **70**), since the reactivities of the polymerizable groups were more or less equal and the polymerization temperature was kept below 25°C [78,79].

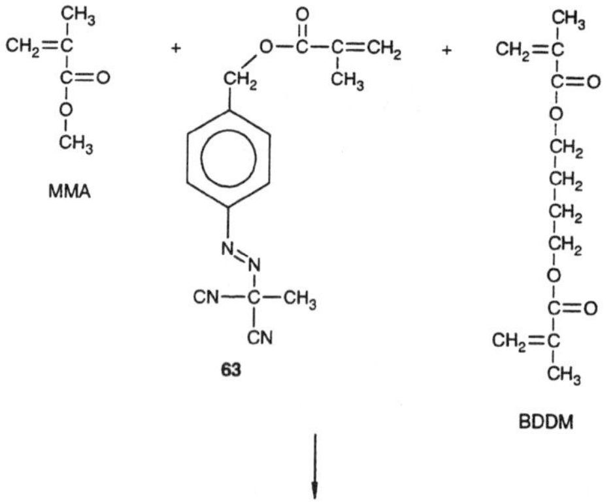

MMA + **63** + BDDM

azo group containing network **70**

Interesting products were obtained after the incorporation of two azo monomers of different thermostability into a styrene/butadiene copolymer [76]. The copolymerized azo monomers **60(R$_3$)** b ($R_x = CH_3, R_y = Ph$, $T_{dec.} = 168$°C) and **60(R$_3$)** a ($R_x = R_y = HCH_3$, $T_{dec.} = 225$°C) could be decomposed stepwise, separatly at different temperatures, which allows, for example, a grafting procedure initiated by the more labile azo function and subsequent network formation using the remaining more stable azo groups.

Characterization of the soluble ter- and higher copolymers was identical to the copolymers described before. However, the azo group containing networks were characterized mainly by DSC measurements and nitrogen analysis for determination of the azo content [75,76] and by swelling experiments. The latter gives the main network characteristics concerning crosslinking degree and the behavior of the network in different solvents [77].

## 3.3 Synthesis via Polycondensation

The radical copolymerization proved very versatile in the synthesis of side chain polymeric azo initiators. However, there are still some limitations to the azo content and to the structure of the polymer backbone. High molecular weight and high azo content can not be obtained at the same time. The restriction of molecular weight can be overcome by polymer modification reactions as shown earlier. Interesting backbone structures such as polyesters, polyethers, polycarbonates, polyamides, and so on, are possible to synthesize by polycondensation.

Polycondensation also allows and even requires higher azo content in the polymer chain. Since the molecular weights obtained in condensation are usually below 20,000, an higher azo content in the monomer composition is necessary to achieve at least one azo function per chain. Incorporation of high amounts of azo functions should be no problem in condensation reaction, as many synthetic approaches to main chain azo macroinitiators are based on it.

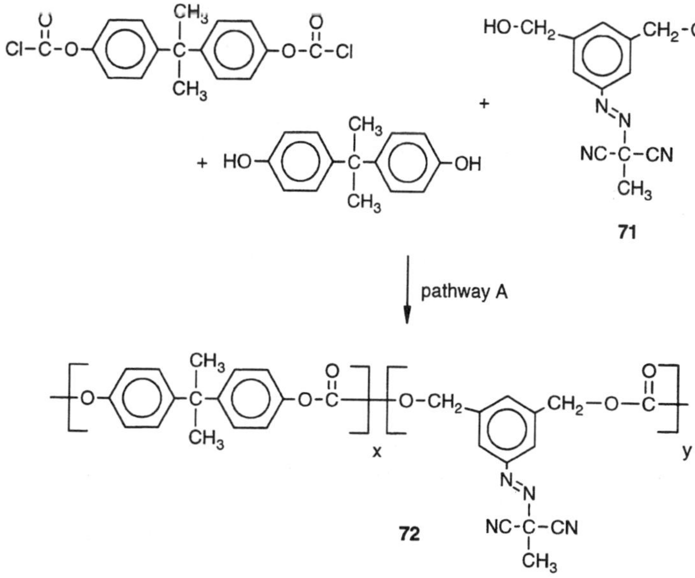

Cl-C(=O)-O-⟨C₆H₄⟩-C(CH₃)₂-⟨C₆H₄⟩-O-C(=O)-Cl + HO-⟨C₆H₄⟩-C(CH₃)₂-⟨C₆H₄⟩-OH + HO-CH₂-⟨C₆H₃(N=N-C(CH₃)(CN)-CN)⟩-CH₂-OH (**71**)

↓ pathway A

$[$-O-⟨C₆H₄⟩-C(CH₃)₂-⟨C₆H₄⟩-O-C(=O)-$]_x$ $[$-O-CH₂-⟨C₆H₃(N=N-C(CH₃)(CN)-CN)⟩-CH₂-O-C(=O)-$]_y$

**72**

Not many of the above listed structures, possible through polycondensation, are actually realized for SCPAI's. Some of the important techniques for condensation, for example reaction in bulk at high temperatures or polycondensation under very basic conditions, can not be applied in the presence of the labile azo compounds. Detailed investigations have shown that the solution polycondensation of bisphenol A, phosgene, and the difunctional hydroxy compound (3,5-dihydroxymethylphenyl)azo-methylmalonodinitrile **71** is suitable for the synthesis of polymeric azo initiators [80,81].

The azo compound **71** shows a thermal behavior similar to **60($R_1$)** (Table 2, section 3.2.2) with rate constant $k = 1.3 \times 10^{-5}\ s^{-1}$ at 70°C in 1,4-dioxane [80]. The incorporation in the polycondensate **72** was complete, since the conversion was driven very close to 100%. The azo content and the composition of the copolycondensate [80,81] was analyzed by UV-derivative-spectroscopy [82], DSC measurements, $N_2$-analysis, and by proton NMR.

**Tab.5** Results of the polycondensation of **71** with bisphenol A and phosgene [81-83] (pathway B)

| polymer | $M_n$[g/mole][a)] | azo content [mole%][b)] | functionality[c)] |
|---|---|---|---|
| homo-polycarbonate | 70,000 | - | - |
| **72a** | 5,100 | 8.9 | 1.8 |
| **72b** | 5,400 | 4.3 | 0.9 |
| **72c** | 7,800 | 14.4 | 4.5 |
| **72d** | 8,000 | 19.3 | 6.0 |
| **72e** | 20,600[d)] | 5.7[e)] | 3.2 |

*a) by vapor pressure osmometry*
*b) by UV-derivative-spectroscopy*
*c) average number of azo groups per polymer chain calculated from $M_n$ and the azo content*
*d) by viscosimetry*
*e) by $N_2$-analysis*

The polycondensation could be carried out at room temperature either by condensation of the bischloroformate of bisphenol A and **71** (pathway A) or by direct condensation of bisphenol A, **71**, and phosgene in pyridine (pathway B). Molecular weights obtained varied from 2,500 to 3,900 g/mole in pathway A (analyzed by vapor pressure osmometry) and from 5,100 to 20,600 in pathway B (analyzed by viscosimetry). The azo content was kept between 5 - 22 mole% to achieve an average of 1-6 azo groups per polymer chain.

The reducing effect of **71** on the molecular weight is remarkable (Tab. 5) and can not be explained easily. Molecular weight of 70,000 could be obtained for the azo free polycarbonate under the same conditions. One possible explanation is the lower reactivity of the azo compound which can lead to longer reaction time. But even at the lowest azo content **(72b)**, the molecular weight droped drastically and it could be shown that in the presence of the unfunctionalized model azo compound to **71**, (3,5-dimethylphenyl)-azo-methylmalonodinitrile, the molecular weight of the homo-poly(bisphenol-A-carbonate) decreases as well [83]. This effect is not yet well understood. Polycarbonate **72** was used very successfully in many grafting experiments which will be described in section 4.

# 4. APPLICATION OF POLYMERIC AZO INITIATORS

## 4.1 Block Copolymers

Undoubtly, the most important application of **main chain** polymeric initiators is the synthesis of blockcopolymers. Usually control of the product, meaning segment length and architecture, is an absolute necessity for the use of the full potential of blockcopolymers. Anionic polymerization technique proved to be ideal for controlled block synthesis. However, this technique is limited to special monomers and the high sensitivity toward impurities restricts its application. The alternative, blockcopolymerization by radical techniques, shows advantages in regard to a wide choice of monomer combination and low sensitivity toward impurities. However, the initiation of radical polymerization and various transfer as well as termination reactions all lead to a variety of products and, furthermore, inevitably to more or less homopolymer which has to be separated.

Despite these problems the advantages predominate and the efforts to synthesize blockcopolymers via radical techniques continue. One effective method is to introduce one or more initiator functionionalities (e.g. azo groups) into or at the end of the main chain. An overview on the synthetic approaches to main chain polymeric azo initiators is given in Section 2.

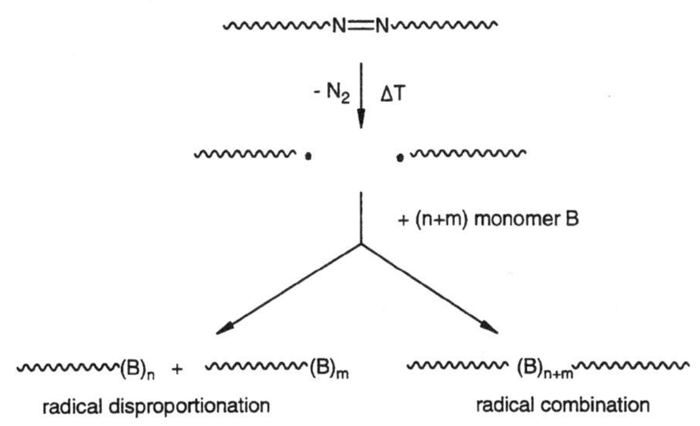


**Scheme 4.1**

Using polymeric initiators, the synthesis of block copoymers is straight forward. A thermal labile azo group along the polymer backbone is cleaved thermally or photochemically to yield a polymer radical which, in the presence of another monomer, initiates block copolymerization. According to the above scheme AB or ABA blockcopolymers are obtained, depending on the prefered type of termination reaction (disproportionation or recombination) .

In the case of several azo groups distributed along the prepolymer chain (type B), cleavage, polymerization and subsequent combination termination results mainly in $(AB)_n$-type stuctures. However, disproportionation leads to BAB- or AB-structures. Azo groups located at the chain end of the prepolymers (type C) can result in radicals not bound to a polymer chain. These radicals can start homopolymerization of the second monomer which should be avoided. Obviously, increasing the number of azo groups per chain reduces the relative effect of azo end groups. Also, it is possible to introduce an azo compound at the chain end with the radical R·, not bound to the chain after thermolysis, not effective as radical initiator. However, more reliable is the introduction of exactly one azo group in the middle of the prepolymer (type A). Its cleavage always results in two polymer radicals as idealized in **Scheme 4.1.**

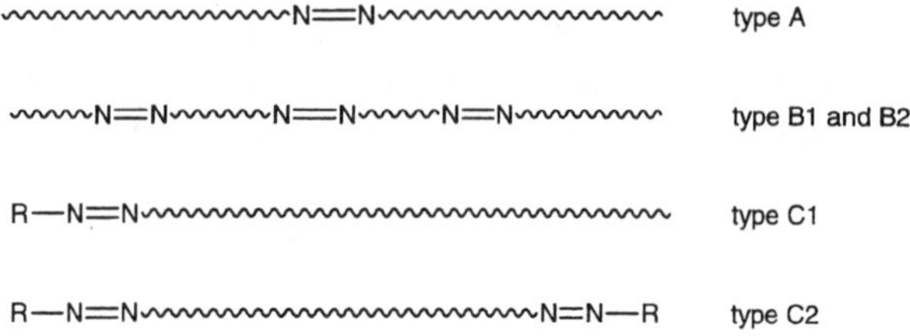


The decomposition kinetics is an important characteristic of the polymeric azo compounds for their subsequent use as initiators of free radical polymerization. Most of the prepolymers discussed in chapter 2 contain azo units derived from AIBN. The decomposition therefore follows first order kinetics and the rate constants are generally only slightly smaller than for AIBN [33]. However, the efficiency of the polymeric initiators seems to be considerably lower than for the low molecular weight azo initiators.

Furukawa et al. [24] studied the blockcopolymerization of styrene in the presence of an azo group containing poly(ethylene glycol). A decrease of initiator efficiency from 1 to 1/3 was noticed. Reasons for the low efficiency may be recombination or deactivation of the polymer radicals due to cage effect and chain transfer reactions. Therefore, a number of at least **6 to 20 azo units** per prepolymer chain was considered enough to produce mostly the biradicals desired for block copolymerization [24].

The examples of blockcopolymerizations actually carried out with polymeric azo initiators are numerous and cover almost all possible combinations of monomers.

Most common is the combination of a polyester or poly(ethylene glycol) prepolymers with styrene [17, 18, 24, 26, 27, 34], MMA [11, 17, 25], vinylchloride [16] or vinylpyrrolidone [25]. Using partial decompositon of the prepolymer in the presence of monomer 1 and subsequent decomposition of the remaining azo functions in the presence of monomer 2 resulted in styrene/MMA, styrene/ methylacrylate, methylacrylate/vinylpyrrolidone, or acrylamide/vinylacetate copolymers [28-30]. **Scheme 4.2** shows an example for the stepwise block copolymerization of styrene with MMA. Some of the above mentioned examples show hydrophilic/hydrophobic blockcharacter and are therefore of interest as emulsifier [25].

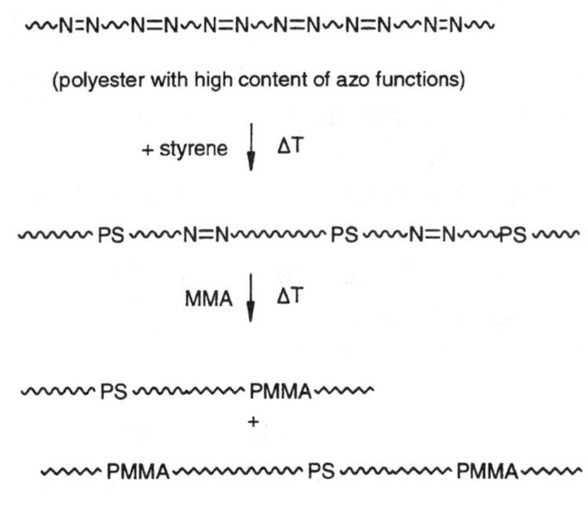


**Scheme 4.2**

Craubner [8] combined polyamide with acrylic acid, acrylonitrile, MMA and vinylacetate via photochemical rearrangement reaction and azo functions formed as intermediates. Polysiloxanes could be blockcopolymerized with a variety of monomers such as MMA, styrene, acrylonitrile [21,21a], via redox initiation mechanism acrylonitrile was combined with styrene [36] and acrylamide with acrylonitrile [38] as well as styrene with MMA and *vice versa* [37, 39], and anionic insertion reactions led to acrylonitrile/MMA block copolymers [40].

The list of examples is by far not complete and many details can be given on the copolymerization conditions and on the properties of the resulting blockcopolymers. However, it was our purpose only to demonstrate the potential of main chain polymeric azo compounds examplified by selected reactions.

As shown, MCPAI's are quite suitable for the synthesis of blockcopolymers via radical initiation. This opens the possibility to combine the properties of a variety of immiscible polymers without losing their original nature. In contrast to two-phase physical blends, covalent bonds between the phases improve mechanical strength and the formation of very small phases allows transparency of the products.

Blockcopolymers can also act as emulsifiers for mixtures of the two polymers from which the blockcopolymer was formed. Although blockcopolymers made via radical techniques are less well defined than those from anionic polymerization, the field of application is broad and the synthesis affordable and less sensitive to impurities.

There are other possible applications for labile **main chain** azo polymers. Labile implies that the azo function can easily be cleaved by UV-light or thermal treatment under degradation of the polymer backbone. Advantage was already taken of this property in the synthesis of labile microcapsule shell polymers [15a] and it might open the pathway to the group of degradable polymers.

## 4.2 Graft Copolymers

The most important application of the **side chain** polymeric azo initiators (SCPAI), analogous to the **main chain** azo polymers, is their use for radical grafting from a polymer backbone. Different synthetic approaches can lead to graftcopolymers [84]: copolymerization of a macromonomer with common monomers (PW1), polymer analogous reaction of a functional prepolymer with functional groups at a polymer backbone (grafting onto, PW2), and the grafting from, meaning the initiation of the polymerization of a second monomer directly from the backbone (PW3). These methods are summarized in **Scheme 4.3.**

The three pathways have advantages and disadvantages which have to be considered before one of these is chosen. In the grafting of functionalized polymers onto an existing backbone (PW2) the immiscibility of the two polymer chains, $P_1$ and $P_2$, may cause serious problems. The resulting phase separation leads to very low conversion of the coupling reaction. Even in homogeneous reaction medium it can occur that the reactivity and the accessibility of the functional groups connected to a polymer chain are quite reduced compared to low molecular weight models.

Differences in reactivity of the polymerizable groups in pathway 1 often result in slow and incomplete incorporation of the macromonomer during copolymerization. Furthermore, molar masses are quite often below the values observed for common copolymers.

In many cases the third method, "grafting from", is the most promising and the most generally applicable for a variety of monomers. Cationic polymerization can be initiated, for example, from poly(vinyl chloride) with $AlR_2Cl$ (carbocation formation) [85] or from pendant epoxy groups through ring opening polymerization. Pendant e.g. peroxy [86] or hydroxy groups allow radical graft copolymerization, initiated by heat or by $Ce^{4+}$ addition [87]. In the radical "grafting from", the role of polymeric azo initiators is rather important, at least for research purpose. The pendant azo function can be more or less easily cleaved, depending on the structure of the azo compound (compare 3.2.1), by heat or UV-irradiation to form radicals located at the polymer backbone and free radicals $R^{\bullet}$.

copolymerization of macromonomers (PW1):

initiator

$+ M_1$

$P_1$

$P_2$

grafting onto (PW2):

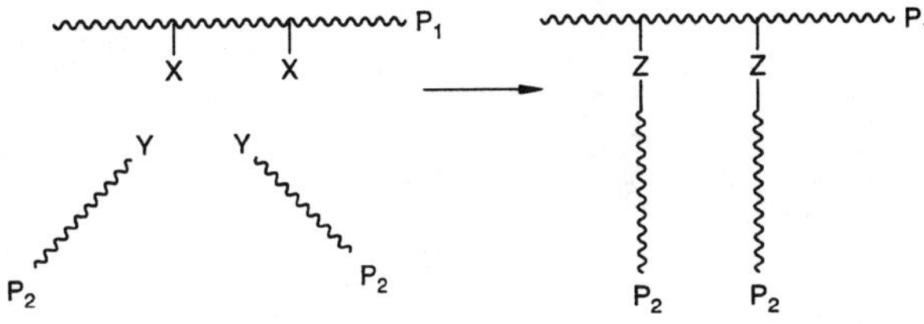


grafting from (PW3):

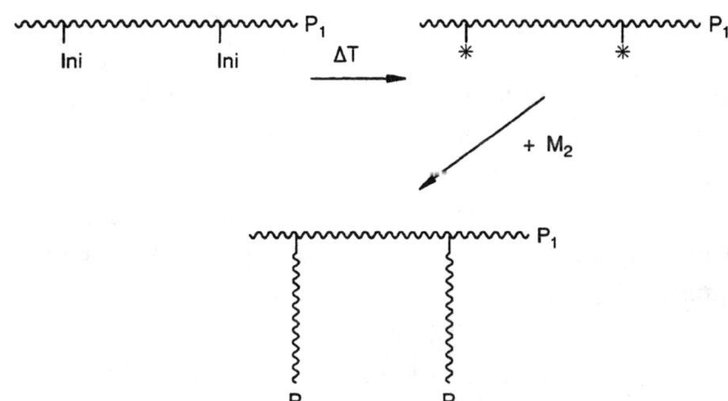


**Scheme 4.3**

ΔT

$+ M_2$

structure A

$R^{\bullet} + M_1 \rightarrow$ $P_2$

homopolymer

**Scheme 4.4**

The nature of the free radical R· is very important. High initiating ability of R· leads to formation of large quantities of homopolymer $P_2$ and the grafting will be only partly successful. Therefore, the incorporation of symmetrical azo compounds which form radicals of equal reactivitiy should be avoided. However, in the case of azo bridged networks, with difunctional symmetrical azo compounds incorporated [54,55,60], all radicals should be located at the polymer backbone and no homopolymerization should take place according to **scheme 4.5**.

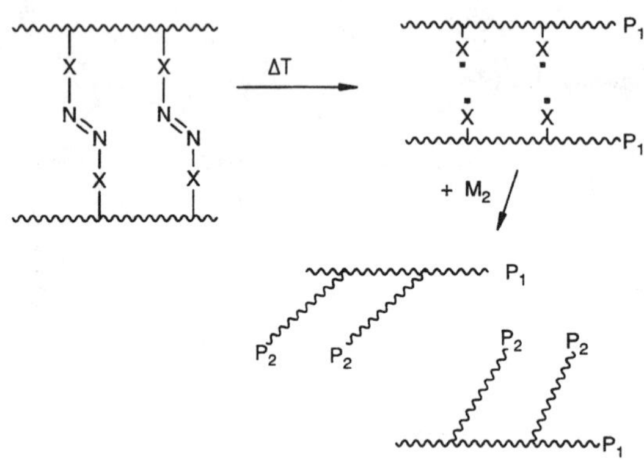


**Scheme 4.5**

Mostly, these polymers will still be crosslinked after grafting due to incomplete cleavage of the azo linkage or fast radical combination in the network. Moreover, the crosslinked azo polymer can only be swollen in the monomer 2 and the grafting process is not well-defined. Using these structures, for example Naarmann et al. [60] grafted MMA from a SCPAI based on ethylhexylacrylate (structure **54**) and Vollmert et al. [54] grafted styrene from azo linked poly(butylacrylate).

Some of the results obtained with the unsymmetrical, monofunctional azo monomers **59-62**, leading to the soluble structure A, will be discussed now in detail. For the successful detailed studies on the grafting reactions the following requirements apply:

- The boiling point of the monomer $M_2$ should not be lower than the decomposition temperature of the azo functions.
- Identification of the grafted polymer should be easy.
- Solubility differences in the products should allow separation of graft- and homopolymer.

Furthermore, grafting with monomers prefering combination termination can lead to combination of two growing chains and therefore to crosslinking or doubling of the molecular weight. Purification and analysis of the product has to be done carefully to allow statements on the characteristic values for grafting processes: grafting efficiency $\epsilon$, grafting success $U_G$, and degree of grafting $\pi$ [88].

$$\text{grafting efficiency in \%:} \quad \epsilon = \frac{P_2 \text{ in graft } P_1/P_2}{P_2 \text{ in graft } P_1/P_2 + \text{homo-}P_2} \times 100$$

$$\text{grafting success in \%:} \quad U_G = \frac{P_1 \text{ in graft } P_1/P_2}{P_1 \text{ in graft } P_1/P_2 + \text{homo-}P_1} \times 100$$

$$\text{degree of grafting in \%:} \quad \pi = \frac{P_2 \text{ in graft } P_1/P_2}{\text{graft } P_1/P_2} \times 100$$

Conclusions can be drawn from the $\epsilon$-values with respect to reactivity of the various radicals: High $\epsilon$ means low reactivity of the free radical $R^{\bullet}$ in homopolymerization of $M_2$ (scheme 4.4). $U_G = 100\%$ implies that each backbone contains at least one grafted side chain speaking at least one azo group was cleaved and effective in initiation of the radical polymerization of $M_2$. The degree of grafting $\pi$ (molar composition of the pure graftcopolymer) is controlled by the ratio of polymer $P_1$/monomer $M_2$, number of azo functions, and reaction conditions.

These characteristic values can only be determined after fractionation of the grafting product. Homopolymer $P_2$, not grafted prepolymer $P_1$ and the graftcopolymer $P_1/P_2$ have to be separated. The fractionation procedures are determined by the nature of $P_1$ and $P_2$ and vary from system to system. An example of the purification process is given in [5] and [65] for the graftcopolymer PS/PAN and in [81,82] for **72**/polyisoprene and **72**/PMMA. In many cases the fractionation is not fully successful and not all characteristics can be determined. For example, the graft product of polycarbonate **72** with styrene could be separated by extraction with ethyl acetate only in a polystyrene-rich and a polystyrene-poor fraction [80]. Therefore, ungrafted polycarbonate could be extracted, but separation of graftcopolymer and homopolystyrene was not possible. It was found that in all cases where grafting was carried out using SCPAI's, homopolymerization of $M_2$ could not be fully avoided. Important is the nature of the formed polymer radical and the free radical $R^{\bullet}$. Some results are given in Tab. 6.

**Tab.6** Grafting efficiency $\epsilon$ for different prepolymers of the structure (polymer)-$R_1$-N=N-$R_2$

| backbone poly-$R_1^{\bullet}$ | $R_2^{\bullet}$ | graft monomer | $\epsilon$ in % | ref. |
|---|---|---|---|---|
| PS-Ph$^{\bullet}$ | $^{\bullet}C(CN)_2CH_3$ | AN | 41-76 | 67,89,90,92 |
| PMMA-Ph$^{\bullet}$ (crosslinked) | $^{\bullet}C(CN)_2CH_3$ | MAN | 31-92 | 91 |
| p-carbonate-**72**-Ph$^{\bullet}$ | $^{\bullet}C(CN)_2CH_3$ | isoprene | 94 | 82 |
| p-carbonate-**72**-Ph$^{\bullet}$ | $^{\bullet}C(CN)_2CH_3$ | MMA | 85 | 81 |
| PS-Ph$^{\bullet}$ | $^{\bullet}$S-Ph | AN | 56 | 64 |
| PS-PhS$^{\bullet}$ | $^{\bullet}$Ph | AN | 27-34 | 64 |

Although several problems had to be overcome, "grafting from" using azo functions was successfully applied for the combination of several polymer structures. The structures from [56a], azo modified $\gamma$-poly(glutamic acid)s, were grafted with MMA up to 65wt% resulting in soluble material. Polymeric initiators based on polystyrene and about 0.05 mole% azo monomers **57** or **58** were grafted with 66 wt% vinyl acetate [62]. The products showed no phase separation as it was observed in blends of the homopolymers.

More informations are given on the graft products using the styrene derivatives **59** - **62**. These azo compounds were mainly copolymerized with styrene and grafting was carried out with styrene [63] or methacrylonitrile (MAN) [64,64a,65,91]. Generally $U_G$ values close to 100% can be achieved [65] and the degree of grafting can be easily controlled by the ratio monomer/$P_1$ and the number of azo groups ($\pi$ = 20-85%) [81,91]. However, the grafting efficiency $\epsilon$, was close to 50%, as seen in Tab.6, implying rather large amounts of homopolymer $P_2$. These values could be improved by increasing the $M_2$ concentration (from 41% to 76%).

The lowest results ($\epsilon$ < 30%) were observed using the azo monomer **61**[64]. This is easily explained by different reactivity of the PS-PhS$^\bullet$ and the Ph$^\bullet$ radicals towards monomer. Therefore exchange of radicals (PS-PhS$^\bullet$ by PS-Ph$^\bullet$ and Ph$^\bullet$ by PhS$^\bullet$) yields an increase of $\epsilon$ from 27% to 56%.

Copolymers of the MMA derivative **63** with MMA and styrene were successfully grafted with styrene, AN and MAN (67). Grafting efficiency is around 41% for the copolymer **63**/styrene and the graft monomer AN. Rather uncontrolled grafting efficiencies (31% to 92%) were observed in grafting experiments of MAN onto preformed networks of MMA, **60(R$_1$)** and BDDM [91], probably due to undefined formation of new grafting sites as a result of side reactions in the network.

Surprisingly high grafting efficiencies were obtained in the grafting of polycarbonates **72**. 85-94% of the graftcomonomer MMA or isoprene were actually connected to the polymer backbone [81,82]. This result can be partly explained by the termination of the growing chains via radical combination, which favors, from the statistically point of view, graftcopolymers over homopolymers. Combination of the homopolymer-radical with the growing chain end of a grafted chain yields a graftcopolymer. Assuming equal reactivitiy of the polymeric and the free radical, this leads to a 66% chance of graftcopolymers compared to a 33% chance of homopolymer. Therefore the grafting efficiency can be influenced by the nature of the monomer to be grafted.

Meanwhile increase of $\pi$ with [$M_2$] is expected, the increase of $\epsilon$ with [$M_2$], which was also found for the polycarbonates, is not yet well understood [81]. One possible explanation is the reduced danger of transfer to the solvent (leading to homopolymer formation) due to increased monomer concentration. In case of graftcopolymers with polycarbonate backbone it was also possible to determine the molecular weight of the sidearms after complete hydrolysis of the backbone (Tab.7).

**Tab.7** Grafting results of prepolymer **72** with MMA at 80°C in dioxane solution (last sample at 170°C) [81].

| SCPAI[a)c)] | $M_{GPC}$ graftcopolymer[b)] | sidearm[b)] | number of sidearms |
|---|---|---|---|
| 15,900 (F = 0.9) | 107,000 | 94,000 | 1 |
| 13,300 (F = 4.5) | 64,000 | 53,000 | 1 |
| 13,300 (F = 6) | 5,000 | 39,000 | 1 |
| 13,000 (F = 4.5) | 137,000[d)] | 45,000 | 2.7 |

*a) determined by GPC in THF, using polycarbonate standards*
*b) determined by GPC in THF, using polystyrene standards*
*c) in brackets the initiator functionality of the prepolymer*
*d) reaction temperature 170°C, carried out in an autoclave*

As it can be seen in Table 7, the average number of side chains is always close to 1 independent of the azo functionality of the SCPAI. Due to a glas effect (viscosity increase in the polymer solution) the conversion of monomer stops before the monomer is consumed completely which happens in general after less than one half life time of the initiator. Therefore, the number of sidearms could only be raised by increasing the decomposition rate of the initiators functions. As a result, the number of sidearms up to 2.7 could be obtained for a grafting reaction carried out at 170°C [81].

Azo groups allow very successful radical grafting of various monomers from a polymer backbone. The variability and the potential of this graft copolymerization technique is controlled by the nature of the polymer backbone, the reactivity of the formed polymeric radical, the number of azo functions per backbone, and the graft comonomer. The important characteristics are summarized in Table 8. Grafting can take place in bulk, solution, suspension, emulsion or in preformed networks.

Graftcopolymers which contain undecomposed azo functions can be used for two or multi-step grafting [82]. Various analytical methods allow exact characterization of the products.

## 4.3 Self-Crosslinking Networks

Polymer networks are formed by covalent, ionic (ionomers), and physical (thermoplastic elastomers) links between different polymer chains. One of the most important method for the synthesis of networks is the chemical crosslinking of prepolymers containing C,C-double bonds by adding a low-molar-mass crosslinking agent. This agent (for example sulfur) has to be mixed with the polymer to be crosslinked. Mixing in bulk can be difficult and is energy consuming.

**Tab.8** Important characteristics of the radical "grafting from" using SCPAI's

| grafting parameter | aim | what to do |
|---|---|---|
| $\epsilon$ | 100% | choice of monomer (termination by combination)<br>choice of polymeric radical (more reactive than free radical)<br>increase of $[M_2]$ |
| $U_G$ | 100% | increase of azo content |
| $\pi$ | variable | f($[M_2]$,[azo],t,conversion,T...) |

In order to avoid this additional mixing process, it is possible to use SCPAI's with unsaturated bonds in the main chain (structures **66** and **68** in section 3.2.4). Simple heating of the polymer samples to the temperature at which cleavage of the azo functions occurs, results in formation of polymeric and low molecular weight radicals. Both types of radicals can initiate polymerization reactions through the double bonds of the polymer backbone, yielding insoluble products [73,75,76,93].

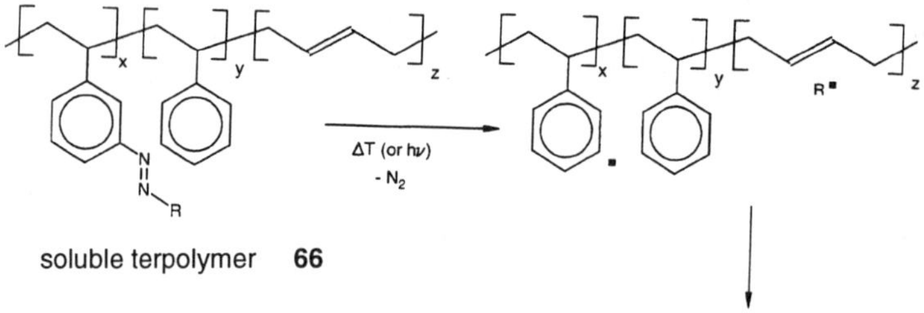

soluble terpolymer **66**

ΔT (or hν)
- $N_2$

crosslinked structure **73**

The network density is increased with reaction time and with the concentration of azo groups in the terpolymer **66** [76]. Network formation and density was analyzed by swelling experiments. The characteristic values 1/Q could be obtained:

$$1/Q = \frac{b}{a-b}$$

a = weight of the swollen polymer
b = dry weight of the polymer after swelling

Intensive studies were done with the terpolymers **66** containing the azo compound **60($R_3$)** yielding structure **73** after crosslinking.

**60(R₃)**

**60(R₃)a**: $R_x = R_y = CH_3$

**60(R₃)b**: $R_x = CH_3$, $R_y = Ph$

**60(R₃)c**: $R_x = R_y = Ph$

For this azo polymer, the crosslinking reaction was carried out at 134°C (**60(R$_3$)c**), 184°C (**60(R$_3$)b**) and at 264°C (**60(R$_3$)a**), depending on the thermal stability of the azo functions. 1/Q-values increase proportional with the crosslinking time as well as with the concentration of azo groups in the prepolymer. The highest crosslinking efficiency was observed for **66** containing **60(R$_3$)a** (1/Q up to 1.3), the lowest in the SCPAI with **60(R$_3$)c** (1/Q below 0.133) [76].

The crosslinked structures **73** were also studied by vulcametric and dynamic mechanical measurements and the results were compared with those obtained for conventionally with dicumylperoxide (DCP) crosslinked polybutadiene and SBR (styrene-butadiene-rubber) [93]. The vulcametric studies showed for **73** (with **60(R$_3$)c**) that the maximum network density was reached after about 10 min. at 170°C. Comparison of the mechanical properties (elongation at break for example) of the SCPAI network and those conventionally crosslinked led to the conclusion that the azo groups in the SCPAI are much less effective in crosslinking than the added DCP at equal concentration. Probably, only the polymeric radicals take part in the crosslinking process which decreases crosslinking efficiency by the factor of 2. Furthermore, the polymeric radicals are quite immobile compared to low molecular DCP and the cage effect can cause a high degree of radical combination before the start of any crosslinking reactions.

Similar observations [92] were made for the copolymer styrene/**60(R$_1$)** (structure of the azo compound given in section 3.2.1). The crosslinking effectiveness of this azo polymer was studied by mixing it with 1,4-cis-polybutadiene. Subsequent heating above the decomposition temperature of the azo functions should lead to crosslinking of the butadiene rubber. However, comparision of the 1/Q-values obtained through crosslinking by the SCPAI and crosslinking of polybutadiene by DCP showed that DCP was 10-20 times more effective [92].

Quite good results concerning crosslinking effectiveness of the SCPAI were observed for the terpolymer **73** containing the azo compound **60(R$_3$)a**. In this case polymeric radicals were not very effective in crosslinking but the formed low molar weight acetoxyradical $^{\bullet}C(CH_3)_2(OAc)$ with a similar crosslinking mechanism as found for DCP [93].

Grafting reactions in solution, as described earlier, seem to be successful since the accessibility and mobility of the polymeric radicals is given. However, the activity of the polymeric radicals in bulk is strongly reduced, compared to low molecular weight radicals. Therefore, the tempting idea of self-crosslinking systems based on polymeric azo initiators will probably not lead to any industrial application due to low crosslinking efficiency, the necessity of high azo content, and high costs of the azo polymers. An exception might be the terpolymer **73** containing the azo compound **60(R$_3$)a**.

## 4.4 Reactive Blending and Reactive Fillers

Similar to the idea of self-crosslinking systems, an application of the SCPAI s in reactive blending is new, interesting and a little bit extravagant. Homogeneous blends of two polymers with different structures are rare but intensively looked for. The combination of different polymers can lead to new properties or to major improvements of the properties of one of the blend components. However, in polymer mixtures with high phase separation as found for incompatible polymers, the mechanical properties drop drastically and no improvements can be obtained. Furthermore, due to the phase separation, large domains of the two polymers are formed and light scattering results in opaque products. Better results can be achieved by reducing the domaine size of the two phases for example with added compatibalizers or by crosslinking the two polymers before they can separate (interpenetrating networks).

Similar problems are found in the area of filler materials for example for rubbers. To improve their mechanical properties, especially long time deformation, abraison, and elongation break, filler particles, often silica gel, activated carbon, or $ZnO_2$, were added to rubbery polymers. Reinforcement is only observed, when the particles are small ($< 1 \mu m$), the surface is high, and at least physical binding exists between the filler material and the continuous phase, the polymer. However, the physical binding is not permanent and in such reinforced rubbers the polymers yields in the stress-strain experiments which decreases its mechanical strain before elongation break occurs. This effect can be explained by the simple scheme of the polymer chains slowly losing the contact to the filler particles. This problem can be overcome by chemical binding of the filler to the polymer instead of the physical binding.

SCPAI's were studied as reactive fillers and in reactive blends. Since the filler particle in these experiments are actually crosslinked polymer beads, one can not speak from reactive fillers but from reactive blending with the aim of reinforcement. In a procedure described in section 3.2.4, terpolymers of azo monomer **60**/AN/butadiene, some of them crosslinked with divinyl benzene (structures **68** and **69**), were synthesized in emulsion by redox initiation at low temperatures. These nitrile rubbers were coprecipitated from emulsion with natural rubber. The precipitate was then heated to 170°C in an aluminum tool to result in a rubber blend stabilized by covalent links between the different polymers [75]. The insoluble part of the natural rubber in these blends increased with increasing azo concentration and the immobilization of the natural rubber was further improved using SCPAI beads precrosslinked with divinyl benzene. By this method it was possible to obtain optical clear blends of nitrile rubber and natural rubber. The domain size of the polymer phases was kept below the dimensions of the wave length of the day light. For filler materials, the terpolymer **70** was synthesized in emulsion [79].

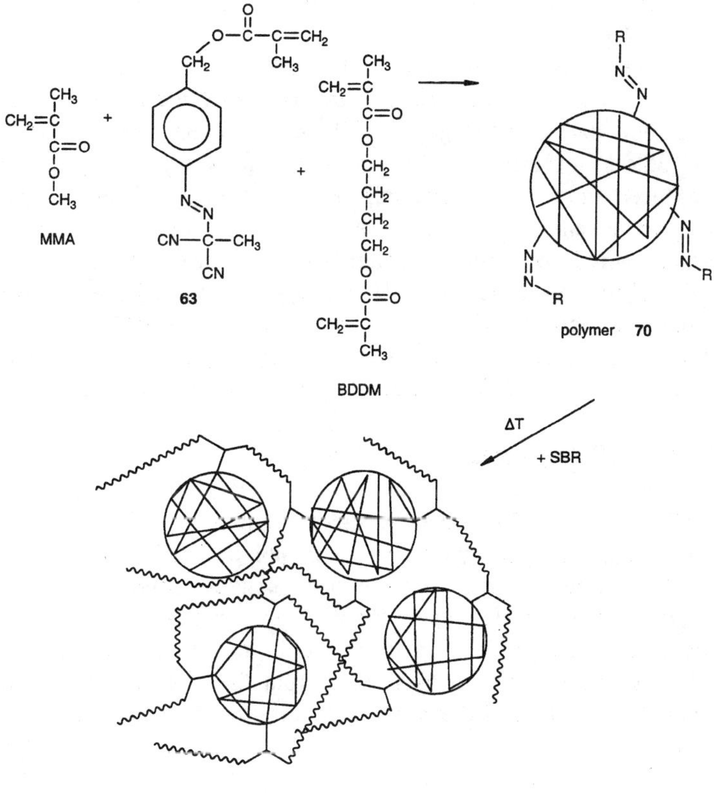

MMA + **63** + BDDM → polymer **70**

$\Delta T$, + SBR

reactive network **74**

The beads (diameter ~ 50 nm) were crosslinked with 0.5 mol% BDDM and the azo content was varied from 2 to 16 mole%. The polymerized emulsion was mixed with SBR or polybutadiene emulsions that contained 0.8 wt% dicumylperoxide and then precipitated. The preciptate was compressed and vulcanized for 25 min. at 150°C. The content of SCPAI in SBR was 10 to 50 phr (parts per hundred rubber). The reinforcing effect of the reactive latex particles on the rubber polymer was studied by stress-strain experiments using Mooney-Rivlin calculations [79]. The elongation at break could be increased by a factor of four compared to samples of SBR und polybutadiene filled with conventional filler particles. The crosslinking density, however, was almost unaffected by the azo filler and fully determined by the amount of DCP in the rubber emulsion which is in agreement with the observation of low crosslinking efficiency of SCPAI's (section 4.3). The crosslinked, reactive reinforced rubbers are optical clear in an azo rich system since agglomeration of the small PMMA latex particles to larger domaines can not occur due to the chemical binding. The domaine size of the PMMA phases was kept at about 50 nm. Blends of PMMA particles without azo functions and SBR are opaque and show phase separation. Disadvantage of these types of reactive fillers are the high costs of SCPAI compared to conventional filler materials. However, the strong reinforcing effect, the good control of the chemical bindings by the amount of azo functions, and the optical clear products may allow special applications.

There are other attempts on reactive fillers using azo compounds. However, not polymeric azo initiators were used but low molecular weight azo compounds linked to active carbon [94] or silica gel [95,96]. These azo functions could be cleaved by heat and radical grafting from $SiO_2$ and carbon was possible. The azo functions were introduced onto carbon by modifying the carbon with acid chloride, epoxy or carboxy groups and subsequent reaction of these functional groups with 2,2-azobis[2-(2-imidazolin-2-yl)propane]. Radical grafting with MMA, styrene and acrylonitrile could be carried out [94]. Microparticulate silica was modified with 4,4'-azobis(4-cyanopentanoic acid) **9** and high amount of PMMA was grafted from it [96]. Since these structures don't belong to the polymeric azo initiators, we would like to refer the reader to the literature for more details [94-96].

## 4.5 Core-Shell Polymers

The crosslinked, rather hard latex particles **70**, described for reactive fillers in 4.4, were also used in the synthesis of core-shell polymers. The principle is comparable to this of the reactive blends. Well-defined chemical bonds between the core and the shell polymer are created using the polymeric radicals formed from SCPAI and separation of the incompatible polymers can be therefore avoided. The synthesis of core-shell polymers is achieved by grafting from the precrosslinked beads **70** directly in emulsion. No isolation of the beads is necessary. The shell monomer $M_2$ (styrene, butylmethacrylate or isoprene) is added to the emulsion and, after some equilibration time, the mixture is heated to the decomposition temperature of the azo functions (70-80°C) for several hours.

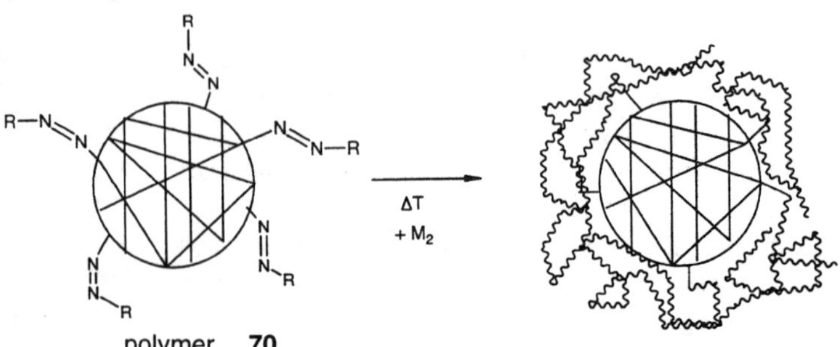

After analyzing the structures **75** by elemental analysis, DSC, REM and DCP (disc centrifuge photometer as particle sizer), it was found that the grafting was successful and yield an increase in diameter of the particles. Extraction of the core-shell polymers gave only a very small amount of homo-shell polymer not bound to the crosslinked particles (< 2%) [78].

Core-shell polymers open another pathway to combine the properties of incompatible polymers. Hard or soft cores of well-defined sizes can be imbedded in a soft or hard matrix without phase separation. Further studies on the effect of the SCPAI-core-shell polymers **75** in blends are in progress.

## REFERENCES


1. C. G. Overberger, M. T. O'Shaugnessy, H. Shalit, *J. Am. Chem. Soc.*, 71, 2661 (1949)
2. J. W. Hill (to E. I. du Pont de Nemours & Co., Inc.),U.S. Pat. 2,556,876 (June 12, 1951), C.A. 45, 9915 (1951)
3. O. Nuyken in "Encyclopedia of Polymer Science and Engineering", 2nd ed., New York, Wiley & Sons, Vol. 2, 158 (1985)
4. G. S. Kumar in " Azo functional polymers", Lancaster, Technomic Publications (1992)
5. O. Nuyken, R. Weidner, *Adv. Polym. Sci.*, 73/74, 145 (1986)
6. C. S. Sheppard in "Encyclopedia of Polymer Science and Engineering", 2nd ed., New York, Wiley & Sons, Vol. 2, 143 (1985)
7. M. Saffran, G. S. Kumar, D. C. Neckers, R. H. Jones, J. B. Field, *Biochem. Soc. Trans.*, 1990, 752 (1990)
8. H. Craubner, *J. Polym. Sci., Polym. Chem. Ed.*, 20, 1935 (1982)
9. C. S. Sheppard and R. E. McLeay (to Pennwalt Corp.), U.S. Pat. 3,937,761 (Feb 10, 1975), C.A. 84, 151243s (1976) and 4,101,522 (July 18, 1978), C.A. 90 39595f (1979),
10. K. B. Chandalia and F. J. Preston (to Olin Corp.), U.S. Pat 4,094,868 (Junc 13, 1978), C.A. 89,164347g(1978)
11. K. Matsukawa, A. Ueda, H. Inoue, *J. Polym. Sci., Part A, Polym. Chem.*, 28, 2107 (1990)
12. S. Nagai, Y. Hidaka, A. Ueda, Japan Kokai 74/17,895 (Feb. 17, 1974), C.A. 81, 92227c (1974)
13. M. H. George, J. R. Ward, *J. Polym. Sci., Polym. Chem. Ed.*, 11, 2909 (1973)
14. A. Ueda, S. Nagai, *J. Polym. Sci., Polym. Chem. Ed.*, 22, 1611 and 1783 (1984)
15. D. A. Smith, *Makromol. Chem.*, 103, 301 (1967)

15a. O. Nuyken, J. Dauth, W. Pekruhn, *Angew. Makromol. Chem.*, 190, 81 (1991); 187, 207 (1991) and J. Dauth, *PhD thesis, University of Bayreuth,* Germany, 1992

16. J. Laverty, Z. Gardlund, *Polym. Prep. Am. Chem. Soc., Div. Polym. Chem.*, 15, 306 (1974)

17. A. Ueda, S. Nagai, *J. Polym. Sci, Part A: Polym. Chem.*, 25, 3495 (1987)
18. A. Ueda, S. Nagai, *J. Polym. Sci, Part A: Polym. Chem.*, 24, 405 (1986)
19. Y. Yagci, Ü. Tunca, N. Bicak, *J. Polym. Sci., Part C: Polym. Lett.*, 24, 491 (1986)
20. Y. Yagci, Ü. Tunca, *J. Polym. Sci., Part A: Polym. Chem.*, 28, 1721 (1990)
21. H. Inoue, A. Ueda, S. Nagai, *J. Polym. Sci., Part A: Polym. Chem.*, 26, 1077 (1988)
21a. C. Simionescu, V. Harabagiu, E. Comanita, V. Hamciuc, D. Giurgiu, B. Simionescu, *Eur. Polym. J.*, 26, 565 (1990)
22. A. Akar, A. Acuner, G. Hizal, T. Öztürk, *Angew. Makromol. Chem.*, 168, 129 (1989)
23. H. Kinoshita, M. Ooka, N. Tanaka, T. Araki, *Makromol. Chem.*, 194, submitted (1993)
24. J. Furukawa, S. Takamori, S. Yamashita, *Angew. Makromol. Chem.*, 1, 92 (1967)
25. W. Heitz, H.-G. Stahl, R. Dicke (to Bayer A.-G.), Ger. Offen. 3,005,889 (Sep. 3, 1981), C.A. 95, 151474v (1982)
26. C. Oppenheimer, W. Heitz, *Angew. Makromol. Chem.*, 98, 167 (1981)
27. R. Walz, W. Heitz, *J. Polym. Sci., Polym. Chem. Ed.*, 16, 1897 (1978)
28. P. S. Anand, H.-G. Stahl, W. Heitz, G. Weber, L. Bottenbruch, *Makromol. Chem.*, 183, 1685 (1982)
29. R. Dicke, W. Heitz, *Makromol. Chem., Rapid Commun.*, 2, 83 (1981)
30. F. Tianru, X. Suxian,, L. Yan, *Polym. Bull.*, 22, 317 (1989)
31. S. Denizligil, Y. Yagci, *Polym. Bull.*, 22, 547 (1989)
32. Y. Yagci, S. Denizligil, *Eur. Polym. J.*, 27, 1401 (1991)
33. R. Walz, B. Bömer, W. Heitz, *Makromol. Chem.*, 178, 2527 (1977)
34. B. Hazer, *Makromol. Chem.*, 193, 1081 (1992)
35. G. Mino, S. Kaizerman, E. Rasmussen, *J. Polym. Sci.*, 38, 393 (1959)
36. Ü. Tunca, I. Serhatli, Y. Yagci, *Polym. Bull.*, 22, 483 (1989)
37. B. Hazer, I. Cakmak, S. Denizligil, Y. Yagci, *Angew. Makromol. Chem.*, 195, 121 (1992)
38. I. Cakmak, B. Hazer, Y. Yagci, *Eur. Polym. J.*, 27, 101 (1991)
39. T. Araki, H. Narita, H. Kinoshita, *J. Polym. Sci., Part A: Polym. Chem.*, 30, 333 (1992)
40. Y. Yagci, Y. Menceloglu, B. Baysal, A. Gungor, *Polym. Bull.*, 21, 259 (1989)
41. G. Hizal, Y. Yagci, *Polymer*, 30, 722 (1989)
42. A. Akar, A. Aydogan, N. Talinli, Y. Yagci, *Polym. Bull.*, 15, 293 (1986)
43. E. Franta, L. Reibel, J. Lehmann, S. Penczek, *J. Polym. Sci. Symp.*, 56, 139 (1976)
44. O. Nuyken, H. Kröner, S. Aechtner, *Makromol. Chem. Rapid Commun.*, 9, 671 (1988)
45. G. Vlasov, G. Rudkovskaya, L. Ovsyannikova, *Makromol. Chem.*, 183, 2635 (1982)
46. M. Abadie, D. Ourahmoune, *Eur. Polym. J.*, 26, 515 (1990)
47. Y. Vinchon, R. Reeb, G. Riess, *Eur. Polym. J.*, 12, 317 (1976)

48. G. Riess, R. Reeb, *J. Am. Chem. Soc. Symp. Ser.*, 166, 477 (1981)
49. V. Harabagiu, V. Hamciuc, D. Giurgiu, B. Simionescu, C. Simionescu, *Makromol. Chem. Rapid. Commun.*, 11, 433 (1990)
50. R. E. MacLeay, C. S. Sheppard (to Pennwalt Corp.), U.S. Pat. 4,075,286 (Feb. 21, 1978), C.A. 88, 170790s (1978)
50a. R. Nolte, *PhD thesis, TU München,* Germany (1989)
51. B. Hazer, Angew. *Makromol. Chem.*, 129, 31 (1985)
52. B. Hazer, A. Ayas, N. Besirli, N. Saltek, *Makromol. Chem.*, 190, 1987 (1989)
53. W. Hahn, A. Fischer, *Makromol. Chem.*, 21, 77 (1956)
54. B. Vollmert, H. Bolte, *Makromol. Chem.*, 36, 17 (1960)
55. S. N. Gupta, U. Nandi, *Makromol. Chem.*, 176, 3179 (1975)
56. H. Dekking (to Union Oil Co.), U.S. Pat. 3,511,898 (May 12, 1970), C.A. 73, 26062d (1970)
56a. N. Tsubokava, M. Inagaki, T. Endo, *J. Polym. Sci., Part A: Polym. Chem.*, 31, 563(1993)
57. C. S. Sheppard, R. E. MacLeay (to Pennwalt Corp.), U.S. Pat. 4,101,522 (Jul. 18, 1978), C.A. 90, 39595f (1979)
58. J. Dauth, O. Nuyken, P. Strohriegl, B. Voit, *Makromol. Chem.* 193, 723 (1992)
59. B. Voit, PhD thesis, *University of Bayreuth, Germany* (1990)
60. H. Naarmann, G. E. McKee, B. Schmitt (to BASF A.-G.), Ger. Offen. 2,749,579 (May 10, 1979), C.A. 91, 75252w (1979)
61. R. Kerber, O. Nuyken, R. Steinhausen, *Makromol. Chem.*, 177, 1357 (1976)
62. R. E. MacLeay, C. S. Sheppard (to Pennwalt Corp.), U.S. Pat. 4,218,370 (Aug. 19, 1980), C.A. 94, 140400w (1981)
63. R. Kerber, O. Nuyken, M. Dorn, *Makromol. Chem.*, 179, 1803 (1978)
64. O. Nuyken, M. Hofinger, R. Kerber, *Polym. Bull.*, 1, 679 (1979)
64a. O. Nuyken, M. Dorn, R. Kerber, *Makromol. Chem.*, 180, 1651 (1979)
65. O. Nuyken, H. Schuster, R. Kerber, *Makromol. Chem.*, 184, 2285 (1983)
66. O. Nuyken, B. Voit, *Makromol. Chem.*, 190, 1325 (1989)
67. O. Nuyken, R. Rengel, R. Kerber, *Makromol. Chem.*, 181, 1565 (1980)
68. O. Nuyken, J. Gerum, R. Steinhausen, *Makromol. Chem.*, 180, 1497 (1979)
69. R. Kerber, O. Nuyken, R. Steinhausen, *Makromol. Chem.*, 175, 3225 (1974)
70. O. Nuyken, T. Knepper, B. Voit, *Makromol. Chem.*, 190, 1015 (1989)
71. O. Nuyken, R. Weidner, B. Dörner, A. Grunow, *Makromol. Chem.*, 190, 469 (1989)
72. A. Stasko, K. Szaboova, V. Cholvad, O. Nuyken, J. Dauth, *J. Photochem. Photobiol. A: Chem.*, 69, 295 (1993)
73. O. Nuyken, R. Weidner, *Makromol. Chem.*, 189, 1331 (1988)
74. A. Stasko, O. Nuyken, B. Voit, S. Biskupic, *Tetrahedron Lett.*, 31, 5737 (1990)
75. O. Nuyken, A. Grunow, G. Pampus, *Makromol. Chem. Makromol. Symp.*, 30, 263 (1989)
76. O. Nuyken, R. Weidner, R. Schuster, *Makromol. Chem.*, 190, 459 (1989)
77. O. Nuyken, L. Dyckerhoff, R. Kerber, *Angew. Makromol. Chem.*, 143, 1 (1986)
78. W. Billig-Peters, *Ph.D. thesis, University of Bayreuth, Germany* (1991)
79. R. Bayer, *PhD thesis, University of Bayreuth, Germany* (1993)

80. R. Kerber, J. Gerum, O. Nuyken, *Makromol. Chem.*, 180, 609 (1979)
81. O. Nuyken, U. Presenz, *Polym. Bull.*, 21, 23 (1989)
82. O. Nuyken, J. Gerum, R. Kerber, *Angew. Makromol. Chem.*, 91, 143 (1980)
83. O. Nuyken, U. Presenz, *Polym. Bull*, 20, 335 (1988)
84. R. J. Ceresa in "Block and Graft Copolymers", Wiley, New York (1967); and J. P. Kennedy, B. Ivan "Designed Polymers by Carbocationic Macromolecular Engineering", Hanser Publishers, Munich, p.212 ff (1992)
85. J. P. Kennedy, *J. Appl. Polym. Sci., Appl. Polym. Symp.*, 30, 1 (1977)
86. A. Katai, C. Schuerch, *J. Polym. Sci., Part A-1* 4, 2683 (1966)
87. G. Mino, S. Kaizerman, *J. Polym. Sci.*, 31, 242 (1958)
88. J. P. Fischer, *Angew. Makromol. Chem.*, 33, 35 (1973)
89. O. Nuyken, *Habilitation thesis, TU München,* Germany (1979)
90. R. Steinhausen, *PhD thesis, TU München, Germany* (1976)
91. O. Nuyken, L. Dyckerhoff, R. Kerber, Angew. *Makromol. Chem.*, 143, 11 (1986)
92. R. Kerber, O. Nuyken, R. Steinhausen, *Makromol. Chem.*, 178, 1833 (1977)
93. O. Nuyken, R. Weidner, G. Pampus, *Kautschuk + Gummi, Kunststoffe*, 42, 11 (1989)
94. K. Fujiki, *J. Macromol. Sci., Chem.*, A28, 715 (1991)
95. R. Laible, K. Hamann, Angew. *Makromol. Chem.*, 48, 97 (1975)
96. G. Boven, M. Oosterling, G. Challa, A. Schouten, *Polymer*, 31, 2377 (1990)

# Macroinitiators Chapter 9

## BLOCK AND GRAFT COPOLYMERS BY RING-OPENING POLYMERIZATION USING VARIOUS MACROINITIATORS

**Masahiko Okada**

Faculty of Agriculture
Nagoya University
Furo-cho, Chikusa-ku
Nagoya 464-01, Japan





Macromolecular Design: Concept and Practice (Macromonomers, Macroinitiators, Macroiniferters, Macroinimers, Macroiniters, and Macroinifers) Pages 359 - 389

# 1. INTRODUCTION

Living polymerization is an essential technique for synthesizing block and graft copolymers having controlled molecular weights and narrow molecular weight distributions. Since the epochal discovery of cationic living polymerization of tetrahydrofuran [1], a variety of cyclic monomers including some cyclic ethers [2], cyclic amines [3,4], cyclic sulfides [5,6], lactones [7,8], lactams [9,10], amino acid N-carboxy anhydrides [11], oxazolines [12], and spiro orthocarbonates [13] have been found to polymerize without termination and chain transfer under properly selected polymerization conditions. Such marked development in the field of ring-opening polymerization has made it possible to design various heterochain block and graft copolymers of well defined structure.

Cyclic monomers generally undergo ring-opening polymerization by cationic and/or anionic mechanisms, although some cyclic monomers polymerize by a radical mechanism [14]. General consideration of macroinitiators is described in other chapters, and hence this chapter focuses on recent advance in the synthesis of block and graft copolymers by ionic ring-opening polymerization of cyclic monomers using a variety of macroinitiators acting as cationic or anionic initiating sites. Block and graft copolymers containing hetero-chains thus obtained often exhibit unique functions as well as excellent physical properties which would not be attained by homopolymers. Therefore properties and functions of some of these copolymers will be briefly described along with their syntheses.

Up to the present, there have been several review articles and books dealing with block and graft copolymerizations of cyclic monomers, in which use of macroinitiators is described [15-19]. In the present article, therefore, the author has attempted minimizing the overlap of these previous works.

# 2. BLOCK COPOLYMERIZATION USING MACROINITIATORS

## 2.1 Synthesis of Block Copolymers by Cationic Ring-Opening Polymerization

Polytetrahydrofuran possesses low melting point ($45^{0}$C) and glass transition temperature ($-86^{0}$C), and therefore it is often used as a constituent for the synthesis of block and graft copolymers in the modification of conventional polymers. Extensive investigations on the kinetics and mechanisms of ring-opening polymerization of tetrahydrofuran have established the conditions under which living polymerization occurs [20]. Polymers having a terminal group or groups which are transformed to cationic species initiating the polymerization of tetrahydrofuran by the direct addition mechanism can be used as macroinitiators for the synthesis of block copolymers containing polytetrahydrofuran as one of the components.

A soluble Ziegler-Natta catalyst composed of vanadium acetylacetonate and diethylaluminum chloride initiates living coordination polymerization of propylene at temperatures below -65$^0$ C to afford syndiotactic polypropylene with a narrow molecular weight distribution ($M_w/M_n$ = 1.05-1.20)[21]. The livingpolyprpylene chain end can be transformed to iodide by treatment with a solution of iodine in toluene. In combination with silver perchlorate, the polymeric iodide generates in situ a carbocation which initiates cationic polymerization of tetrahydrofuran [22].

$$\cdots CH_2\text{-}\underset{|}{\overset{CH_3}{C}}H\text{-}V^{3+} \xrightarrow{I_2} \cdots CH_2\text{-}\overset{CH_3}{CH}\text{-}I \xrightarrow{AgClO_4} \cdots CH_2\text{-}\overset{CH_3}{CH^+}\ ClO_4^- \xrightarrow{THF}$$

$$\cdots CH_2\text{-}\overset{CH_3}{CH}\text{-}\overset{+}{O}\bigcirc\ ClO_4^- \xrightarrow{THF} \text{-}(CH_2\text{-}\overset{CH_3}{CH})_m\text{-}[O(CH_2)_4]_n\text{-}$$

Block copolymers composed of polypropylene and polytetrahydrofuran were successfully synthesized at 0$^0$C, the block efficiency being practically 100 %. In contrast, the block copolymers produced at 20$^0$C contained a small, but not negligible, amount of polytetrahydrofuran. In addition, polypropylene was also detected in the reaction mixture. Probably a $\beta$-proton was eliminated from the cationic end of the macroinitiator in the initial stage to yield perchloric acid, which induced homopolymerization of tetrahydofuran.

A trimethylcellulose macroinitiator having a reactive chlorine atom at the reducing end was prepared by partially cleaving high molecular weight trimethylcellulose by 1N hydrogen chloride at 30$^0$ C in a mixture of dichloromethane and benzene (4:1, v/v) [23]. Polymerization of tetrahydrofuran was initiated by the macroinitiator in the presence of silver hexafluoroantimonate at temperatures between -10 and 23$^0$C. The polymerization was terminated either by methanolic potassium hydroxide to produce trimethylcellulose-polytetrahydrofuran diblock copolymer or by low molecular weight poly(4-vinylpyridine) (DP=ca.20) to yield trimethylcellulose-polytetrahydrofuran star-shaped block copolymers.

OR, O, RO, RO, O — HCl → OR, O, RO, RO, Cl — $AgSbF_6$ → OR, O, +, $SbF_6^-$, RO, RO

— THF → OR, O, RO, RO, +O, $SbF_6^-$ — THF → OR, O, RO, RO, $O(CH_2)_4$---

The solid state properties of these diblock and star-shaped block copolymers were examined by X-ray scattering, polarization microscopy, differential scanning calorimetry, and mechanical dynamic measurements. They were in agreement with a four phase morphology consisting of an amorphous and a crystalline phase of both trimethylcellulose and polytetrahydrofuran. The star-shaped block copolymers were effectively cross-linked, but they were not thermoplastically processable due to the decomposition of the samples [24].

A synthetic polysaccharide derivative ($M_n$ = 5000-28000) having a reactive chlorine atom at the reducing end was prepared by ring-opening polymerization of 1,6-anhydro-2,3,4-tri-O-benzyl-β-D-glucose(TBLG) with antimony pentachloride as an initiator in the presence of acetyl chloride in dichloromethane at -60°C [25]. The polysaccharide, when treated with silver hexafluoroantimonate in tetrahydrofuran, 2-methyltetrahydrofuran or tetrahydropyran at or below -60°C, acted as a macroinitiator to induce cationic polymerization of a bicyclic acetal, 6,8-dioxabicyclo[3.2.1]octane(DBO).

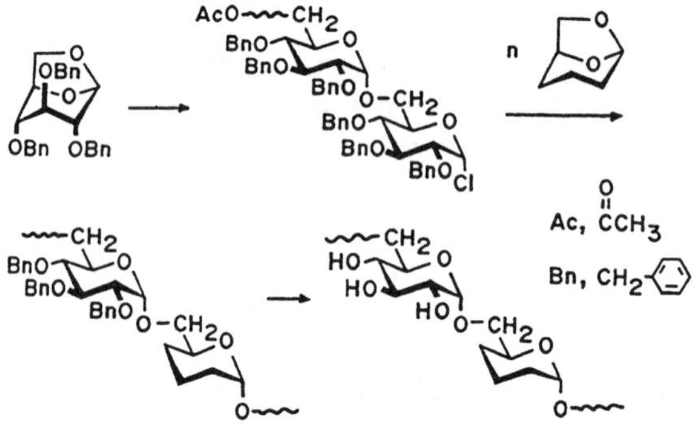

Fractionation of the reaction products gave a block copolymer composed of poly(TBLG) and poly(DBO) segments. Debenzylation of the block copolymer provided an amphiphilic block copolymer consisting of hydrophilic (1→6)-α-D-glucopyranan (dextran) and hydrophobic poly(DBO) [(1→6)-linked polysaccharide skeleton] segments. The amphiphilic block copolymer is soluble in dimethyl sulfoxide and swells in water and chloroform.

It is very uncommon to carry out cationic polymerization of DBO in cyclic ethers which are more basic than the monomer and even polymerizable. However, Hall et al. [26] reported that no incorporation of tetrahydrofuran into polymer occurred in the cationic polymerization of DBO in tetrahydrofuran at -78°C, in other words, tetrahydrofuran acts as solvent but not as monomer at low temperature. In the block copolymerization above, the basic cyclic ether solvents are needed to immediately stabilize a highly reactive, unstable oxacarbenium ion generated from the macroinitiator either by solvation or by nucleophilic addition of a solvent molecule.

Some cyclic amines such as 1-tert-butylaziridine [27], 1,3,3-trimethylazetidine [28], and conidine [29] polymerize cationically without termination and chain transfer. Polymerization of 1-tert-butylaziridine was attempted by using a poly(dimethylsiloxane) macroinitiator having a terminal tosylate group [30]. Although a low molecular weight model initiator, ethyl tosylate, smoothly induced the polymerization of the same monomer at $0^{0}C$, no polymerization occurred when the macroinitiator was used. Even at $100^{0}C$, both the conversion of the monomer and the initiator efficiency were low owing to the slow initiation and chain transfer to form side reaction products, cyclic oligomers.

Kinetics and mechanisms of ring-opening polymerization of cyclic imino ethers, especially 2-oxazolines, have been extensively investigated by Saegusa and Kobayashi [31]. These monomers are readily polymerized with alkyl halides and alkyl tosylates to yield living polymers. By utilizing the living character of the polymerization, a variety of block polymers have been prepared.

Polystyrene, poly(2-vinylnaphthalein), and poly(9-vinylphenanthrene) having haloalkyl groups at both chain ends were prepared by anionic polymerization followed by terminating the living anion with an excess of 1,2-dichloroethane or 1,4-dibromobutane [32]. These polymers acted as macroinitiators to polymerize 2-methyl-2-oxazoline to yield the corresponding block copolymers of 2-methyl-2-oxazoline and aromatic vinyl monomers.

$$Na^{+-}\underset{Ar}{CH}CH_2\text{-----}CH_2\underset{Ar}{CH}^-Na^+ \xrightarrow[\substack{n=2,\ X=Cl\\ n=4,\ X=Br}]{X(CH_2)_nX} X(CH_2)_n\underset{Ar}{CH}CH_2\text{-----}CH_2\underset{Ar}{CH}(CH_2)_nX$$

$$\xrightarrow{KI}\xrightarrow{MeOZO} \text{---}CH_2CH_2\underset{COCH_3}{N}(CH_2)_n\underset{Ar}{CH}CH_2\text{------}CH_2\underset{Ar}{CH}(CH_2)_n\underset{COCH_3}{N}CH_2CH_2\text{----}$$

Ar= phenyl, naphthyl, phenanthryl

This method has the following advantages: 1) all vinyl monomers which undergo anionic living polymerization can be utilized as the source of one of block seggments, 2) the control of the degree of polymerization of a block sequence can be achieved easily, and 3) since the selection of the initiator for the anionic polymerization of vinyl monomers gives polymers having haloalkyl groups either at one end or both ends, AB diblock or ABA triblock copolymer can be prepared at one's choice [32].

Higashimura et al. [33] developed hydrogen iodide/iodine or zinc halides binary initiator systems for cationic living polymerization of alkyl vinyl ethers and some styrene derivatives. On the basis of the living nature of the polymerization of vinyl ethers, a variety of block copolymers can be designed [34].

Polymerization of alkyl vinylethers (alkyl: n-butyl, i-octyl, and n-cetyl) was first carried out with hydrogen iodide as an initiator in hexane at -20$^0$C to produce the propagating species of poly(alkyl vinyl ether) having iodide at the chain end. Then, a cyclic imino ether (2-methyl- and 2-ethyl-2-oxazolines and 2-methyl-5,6-dihydro-4H-1,3-oxazine)was added to the mixture at -78$^0$C and allowed to react with the polymer to introduce an oxazolinium group at the terminal. On heating the reaction mixture, the polymerization of the cyclic imino ether was initiated from the active chain end to give hydrophobic poly(alkyl vinyl ether)-hydrophilic poly(N-acylalkyleneimine) block copolymers [35].

$$CH_2{=}CH(OR_1) \xrightarrow{HI} H{+}CH_2CH(OR_1){+}_m I \longrightarrow$$

$$H{+}CH_2CH(OR_1){+}_m N^{+}\text{(cyclic, } R_2\text{, } (CH_2)_p\text{, O)} \xrightarrow{\text{Monomer}} H{+}CH_2CH(OR_1){+}_m{+}N(COR_2)(CH_2)_p{+}_n I$$

These block copolymers exhibit good surface activities and hence are a family of nonionic polymer surfactants. The lowest $\gamma$ value reached 27.6 dyne/cm for the copolymer derived from n-butyl vinylether and 2-methyl-2-oxazoline. The C-N bond between the poly(alkyl vinyl ether) segment and poly(N-acylalkyleneimine) segment was readily cleaved under acidic conditions to give a mixture of both homopolymers having no surface activities.

Poly(alkyl vinyl ether)s (alkyl; methyl and butyl) having a chlorine atom at the chain end were prepared by the cationic living polymerization using the chloroethyl vinyl ether hydrogen iodide adduct together with zinc chloride in toluene at -20$^0$ C [36]. Polymerization of 2-ethyl-2-oxazoline using the functional poly(alkyl vinyl ethers)s as macroinitiators were carried out in the presence of sodium iodide in a mixed solvent of chlorobenzene and N-methyl-2-pyrrolidinone at 110$^0$C. Sodium iodide generates alkyl iodide from alkyl chloride in situ, and N-methyl-2-pyrrolidinone solvates sodium iodide to facilitate the exchange of iodide for chloride. The polymerization of 2-ethyl-2-oxazoline by the poly(alkyl vinyl ether) macroinitiators yielded diblock copolymers of these two components.

Siloxane containing block copolymers can be synthesized by the use of ring-opening polymerization techniques. Depending on the type and architecture of the block copolymer to be prepared, a mono- or difunctionally terminated siloxane oligomer is used as an macroinitiator [37].

$\alpha,\omega$-Benzyl chloride terminated poly(dimethylsiloxane) oligomers ($M_n$, 2000-20000) were prepared by the triflic acid catalyzed equilibration reactions of octamethylcyclotetrasiloxane with benzyl chloride terminated siloxane dimers in bulk at 65$^0$C [38]. Due to the equilibrium nature of these reactions, the final reaction mixture was composed of functional, linear oligomers and nonfunctional, cyclic siloxane species. These cyclic siloxanes were removed from the system by vacuum distillation, after the neutralization and/or removal of the equilibration catalyst.

$$X\text{-}R\text{-}Si\text{-}O\text{-}Si\text{-}R\text{-}X + \boxed{\text{-}(Si(Me)_2\text{-}O)_4\text{-}} \longrightarrow X\text{-}R(Si(Me)_2\text{-}O)_n Si\text{-}R\text{-}X + \boxed{\text{-}(Si(Me)_2\text{-}O)_m\text{-}}$$

$R\text{-}X = (CH_2)_2\text{-}C_6H_4\text{-}CH_2Cl$ $\quad m = 3, 4, 5 \cdots$

$\downarrow$ 2-ethyl-2-oxazoline

$$\text{-}[(CH_2)_2N(C{=}O)Et]_m\text{-}R\text{-}(Si(Me)_2\text{-}O)_n Si(Me)_2\text{-}R\text{-}[N(C{=}O)Et(CH_2)_2]_m\text{-}$$

ABA-type poly(2-ethyl-2-oxazoline)-poly(dimethylsiloxane) copolymers were obtained by the melt polymerization of the oxazoline monomer at $130^0$C using the benzyl halide terminated siloxane oligomers as macroinitiators in the presence of catalytic amounts of sodium iodide [38].

The molecular weights of the siloxane and oxazoline blocks were varied between 2000 and 6000 g/mol. All of the copolymers (siloxane content of about 30% by weight) were soluble in a water/ethanol (90 : 10, w/w) mixture and also in chlorinated hydrocarbons and tetrahydrofuran. The solubility behavior clearly indicates the amphipathic nature of these siloxane containing copolymers. When cast from solution, the high molecular weight copolymer yielded clear films with good mechanical integrity at room temperature as determined by TMA and penetration studies. DSC and TMA analysis also indicated the formation of two-phase morphologies, where polysiloxane and polyoxazoline transitions were observed at about -120 and +50$^0$C, respectively [38].

Poly(ethylene glycol) and its derivatives are finding a rapidly expanding range of chemical, biomedical, and industrial applications resulting from their low cost and useful properties, such as solubility in aqueous and organic solvents, metal complexing ability, biological compatibility, and ease of chemical modification. Improved and versatile methods for the synthesis and characterization of poly(ethylene glycol) derivatives have been developed [39].

Sulfonates of poy(oxyethylene) can be used as macroinitiators for cationic polymerization of cyclic imino ethers. The preparation of tosylates of poly(ethylene glycol) by the conventional procedure using tosyl chloride and pyridine is sometimes accompanied by chain cleavage. In order to avoid this, De Vos and Goethals [40] proposed two methods, one is a Shotten-Bauman type reaction between the diol and tosyl chloride using 4-(diethylamino)pyridine (DAP) as a base, and the other involves the conversion of the diol to its lithium alcholate with a stoichiometric amount of butyl lithium in benzene, followed by the addition of tosyl chloride. The advantage of the latter method is that the by-products (butane and lithium chloride) are readily removed. The tosylate with a functionality greater than 1.98 was obtained in 90 % yield.

$$HO\text{-}(CH_2CH_2O)_{\bar{n}}H \xrightarrow{TsCl/DAP} TsO\text{-}(CH_2CH_2O)_{\bar{n}}Ts$$

$$HO\text{-}(CH_2CH_2O)_{\bar{n}}H \xrightarrow{BuLi} LiO\text{-}(CH_2CH_2O)_{\bar{n}}Li \xrightarrow{TsCl} TsO\text{-}(CH_2CH_2O)_{\bar{n}}Ts$$

$Ts = p\text{-}CH_3C_6H_4SO_2\text{-}$

Block copolymers consisting of poly(N-acylethylenimine) and poly(oxyethylene) chains were prepared by initiating the polymerization of 2-methyl-2-oxazoline and 2-ethyl-2-oxazoline with $\alpha,\omega$-ditosylated or mesylated poly(oxyethylene) [41]. Since the polymerization proceeds with slow initiation and rapid propagation, it is postulated that block copolymers of ABA and AB types are formed depending on the polymerization conditions. Blends of the block copolymers (3 wt%) with Nylon 6 were spun to fiber which showed good anti-electrostatic property.

Block copolymers consisting of poly(oxypropylene) and poly(2-methyl-2-oxazoline) segments were prepared by poly(propylene glycol) p-nitrobenzenesulfonate [42]. The end groups of poly(propylene glycol) consist of primary and secondary hydroxyl groups. Since the sulfonation of the latter is generally difficult because of its reduced reactivity, p-nitrobenzenesulfonyl chloride was used in the presence of 2,6-lutidine for the esterification. However, the functionality of the sulfonate was 1.50-1.80 and hence, the polymerization of 2-methyl-2-oxazoline by the sulfonate macroinitiator led to a mixture of ABA and AB block copolymers [A, poly(2-methyl-2-oxazoline); B, poly(oxypropylene)]. The block copolymer, a class of nonionic surfactant, did not deposit from their aqueous solutions even at 100$^0$C, whereas commercially available poly(oxyethylene)-poly(oxypropylene) block copolymers "Pluronics" deposit from aqueous solution at 40-80$^0$C.

$$HO\text{-}(\underset{Me}{CH}CH_2O)_{\bar{n}}\text{-}\underset{Me}{CH}CH_2\text{-}(OCH_2\underset{Me}{CH})_{\bar{m}}OH + NO_2\text{-}C_6H_4\text{-}SO_2Cl \xrightarrow{2,6\text{-lutidine}}$$

$$NO_2\text{-}C_6H_4\text{-}SO_3\text{-}(\underset{Me}{CH}CH_2O)_{\bar{n}}\text{-}\underset{Me}{CH}CH_2\text{-}(O\underset{Me}{CH}CH_2)_{\bar{m}}\text{-}OSO_2\text{-}C_6H_4\text{-}NO_2 \xrightarrow{MeOZO}$$

$$HO\text{-}(CH_2CH_2\underset{COMe}{N})_{\bar{p}}\text{-}(\underset{Me}{CH}CH_2O)_{\bar{n}}\text{-}\underset{Me}{CH}CH_2\text{-}(OCH_2\underset{Me}{CH})_{\bar{m}}\text{-}(\underset{COMe}{N}CH_2CH_2)_{\bar{q}}OH$$

AB and ABA block copolymers from cyclic imino ethers (A) and tetrahydrofuran (B) were synthesized by utilizing the living nature of the cationic ring-opening polymerization of both monomers [43]. Ethyl triflate and triflic anhydride were used as initiators for the polymerization of tetrahydrofuran to produce mono- and difunctional living polymers, respectively. The cyclic trialkyl oxonium ions at the growing chain ends were transformed to the oxazolinium ions by the reaction with cyclic imino ether at -78°C. Then the polymerization of the cyclic imino ether using the polytetrahydrofuran as a macroinitiator was carried out at 60°C. The polymerization of cyclic imino ether took place exclusively even when unreacted tetrahydrofuran monomer was present because of the large difference in reactivity of propagating species between both monomers. These block copolymers are a group of nonionic polymer surfactants, exhibiting excellent surface activities. The lowest $\gamma$-value reached 28.2dyne/cm for a ABA-type triblock copolymer derived from tetrahydrofuran and 2-methyl-2-oxazoline.

An ABA-type block copolymer containing poly(N-acetylethyleneimine) (A) as a hard segment and poly(ethylene glycol adipate) (B) as a soft segment was obtained by bulk polymerization of 2-methyl-2-oxazoline using poly(ethylene glycol adipate) having tosylate groups at the both ends as a macroinitiator [44].

## 2.2 Synthesis of Block Copolymers by Anionic Ring-Opening Polymerization

It is well established that anionic ring-opening polymerization of lactams is effectively initiated in the presence of N-acyllactams or other acylating reagents such as isocyanates and chloroformates which readily react with a lactam to form an N-acyllactam [45,46]. These compounds are called differently in the literature: initiators, activators, cocatalysts, or promotors. In order to avoid confusion, these compounds are hereafter referred to as activators. Polymers having these functional groups at one or both of the chain ends can be used as macroactivators for synthesizing block copolymers containing polyamide segments.

Polyisobutene macroactivators having terminal isocyanate groups were prepared by the reaction of telechelic mono-, di-, and trihydroxyl-terminated polyisobutene with hexamethylene diisocyanate, toluene diisocyanate, N-chlorocarbonyl diisocyanate, and oxalyl chloride. The hydroxyl-terminated polyisobutenes were obtained by the "inifer" method. Nylon 6-polyisobutene diblock, triblock, and tristar radial block copolymers were synthesized by the anionic polymerization of $\epsilon$-caprolactam using the corresponding macroactivators [47]. The polymerization using the macroactivators obtained by the reaction of the hydroxyl-terminated polyisobutenes with N-chlorocarbonyl diisocyanate gave the highest yields and blocking efficiency.

$$CH_2{=}C(Me)_2 \xrightarrow[BCl_3]{C_6H_5\text{-}C(Me)_2\text{-}Cl} C_6H_5\text{-}C(Me)_2\text{+}CH_2\text{-}C(Me)_2\text{+}_n Cl \xrightarrow{t\text{-}BuOK}$$

$$\text{---}CH_2\text{-}C({=}CH_2)Me \xrightarrow[NaOH/H_2O_2]{9\text{-}BBN} \text{---}CH_2\text{-}CH(CH_2OH)Me \xrightarrow{ClCO\text{-}NCO}$$

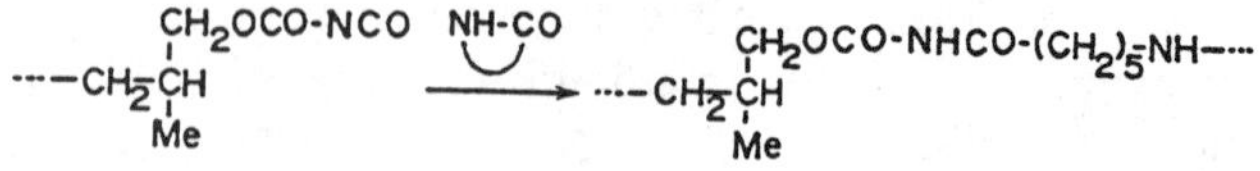

$$\text{---}CH_2\text{-}\underset{Me}{\overset{CH_2OCO\text{-}NCO}{CH}} \xrightarrow{\text{NH-CO}} \text{---}CH_2\text{-}\underset{Me}{\overset{CH_2OCO\text{-}NHCO\text{-}(CH_2)_5\text{-}NH\text{---}}{CH}}$$

In addition to the expected $T_g$ and $T_m$ characteristics of long polyisobutene and nylon 6 segments, DSC showed an intermediate glass transition at ca. -20°C. Transmission electron microscopy of the di-, tri-, and radial block copolymers revealed increasing segregation and orientation of rubbery/crystalline domains. Tensile strength and elongation of the block copolymers range from 16.5 to 41 MPa and 15 to 30 %, respectively, and stress-strain diagrams show the effect of block architecture on these properties.

Poly(oxyethylene) macroactivators containing N-carbonylcaprolactam end groups were prepared by the aminolysis of isophthaloylbis-, terephthaloylbis- and adipoylbiscaprolactam with $\alpha,\omega$-diamino-terminated poly(oxyethylene). The telechelic poly(oxyethylene) was obtained by the cyanoethylation of $\alpha,\omega$-dihydroxypoly-(oxyethylene) followed by hydrogenation. Anionic polymerization of $\epsilon$-caprolactam using these macroactivators provided ABA triblock copolymers consisting of poly($\epsilon$-caprolactam) (A) and poly(oxyethylene) (B) [48].

A bicyclic lactam, 8-oxa-6-azabicyclo[3.2.1]octan-7-one, readily undergoes anionic polymerization in dimethyl sulfoxide at room temperature to give a polyamide consisting of alternating tetrahydropyran ring and amide moieties. ABA block copolymers composed of the polyamide as the outer segment and poly(oxyethylene) as the inner segment were prepared by the anionic polymerization of the bicyclic lactam activated with poly(oxyethylene) having isocyanate groups at both chain ends [49]. The macroactivator was prepared by reaction of poly(oxyethylene glycol) ($M_n$ = 1030-19000) with a large excess of tolylene-2,4-diisocyanate in carbon tetrachloride at 60°C in a high vacuum line. A porous membrane was prepared from the triblock copolymer by casting a solution of a mixture of the block copolymer with poly(oxyethylene glycol) and drying up at room temperature, followed by immersing and rinsing in water. In a solute rejection test for some proteins, such as myoglobin and cytochrome c, the concentration of the permeate was found to be significantly higher than that in the feed, in other words, facilitated permeation of the solutes occurred [50].

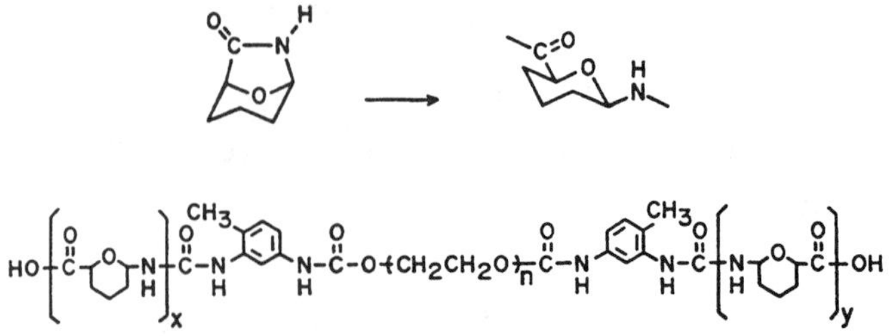

Multicomponent materials containing polysaccharide sequences are promising candidates for biodegradable or biomedical polymeric materials. A trimethylsilylated dextran derivative having an acyllactam group at the reducing end was prepared through a reaction sequence as shown in Scheme I. AB-type block copolymers composed of dextran and polyamide segments were prepared by the anionic polymerization of 8-oxa-6-azabicyclo[3.2.1]octan-7-one activated with the dextran derivative [51].

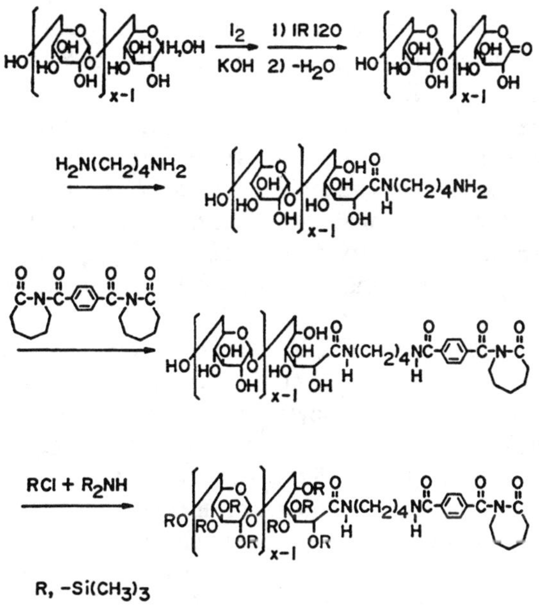

$I_2$, KOH; 1) IR120, 2) $-H_2O$; $H_2N(CH_2)_4NH_2$; RCl + $R_2NH$; R, $-Si(CH_3)_3$

**Scheme I.** Chemical modification of the reducing end of dextran and protection of its hydroxyl groups with trimethylsilyl groups

Anionic ring-opening polymerization of lactams proceeds by so-called activated monomer mechanism involving lactam anion. Since lactam anion is a strong base, macroactivators used for the synthesis of block copolymers should be inert under such basic conditions.

Owen et al. [52] reported that anionic polymerization of $\epsilon$-caprolactam using a poly(dimethylsiloxane)-based macroactivator gave poly(dimethylsiloxane)-poly($\epsilon$-caprolactam) block copolymer in nearly complete lactam conversion. However, Teyssie et al.[53] reinvestigated the same polymerization and found that the $\epsilon$-caprolactam conversion was limited to around 50%, and that only 8% of poly(dimethylsiloxane) were incorporated into the block copolymer instead of 61% as predicted from both the initial amount of poly(dimethylsiloxane) and the $\epsilon$-caprolactam conversion. On the basis of model reactions, they concluded that degradation of siloxane backbone occurs under the conditions required for the polymerization.

$$M^{+-}N\text{-}CO \quad R_1\text{-}Si\text{-}O\text{-}Si\text{-}R_2 \longrightarrow R_1\text{-}Si\text{-}O^-M^+ + O{=}C\text{-}N\text{-}Si\text{-}R_2$$

The lactam salt participates in two competitive processes, i.e., the polymerization of ϵ-caprolactam and the degradation of poly(dimethylsiloxane). The cleavage of a Si-O bond in poly(dimethylsiloxane) by the lactam salt generates a silanolate ion and silyllactam. The silanolate ion is unable to react with ϵ-caprolactam to form the corresponding lactam salt. As a consequence, the concentration of the lactam salt decreases in the course of the reaction, whereas the silyllactam does not act as an effective activator for polymerization of ϵ-caprolactam. Furthermore, the silanolate causes redistribution of poly(dimethylsiloxane) chain. This reaction pathway satisfactorily explains the observed degradation of poly(dimethylsiloxane), the low conversion of ϵ-caprolactam, and the poor yield of the block copolymer.

Poly(dimethylsiloxane) having an ω-N-acyllactam group was prepared by the reaction of poly(dimethylsiloxane) anion with chlorosilyl acyllactam. Anionic polymerization of ϵ-caprolactam using the poly(dimethylsiloxane) as a macroactivator was carried out in bulk with $LiAl(Ot\text{-}Bu)_2H_2$ as a catalyst instead of more conventional alkali caprolactamate or aluminum hydride [54]. This new catalyst with electron donating substituents lowers the nucleophilicity of the lactam anion by coordinating the aluminate with the lactam carbonyl to effectively delocalize the negative charge. This in turn reduces the detrimental reactivity of the lactam anion towards the siloxane but does not interfere with the activated monomer mechanism which requires the attack of lactam anion onto the terminal imide group.

$$Bu\text{-}(Si(Me)_2\text{-}O)_n\text{-}Si(Me)_2\text{-}O^-Li^+ \;-\; Cl\text{-}Si(Me)_2\text{-}(CH_2)_{10}\text{-}CO\text{-}N\text{-}CO \longrightarrow$$

$$Bu\text{-}(Si(Me)_2\text{-}O)_m\text{-}Si(Me)_2\text{-}(CH_2)_{10}\text{-}CO\text{-}N\text{-}CO \xrightarrow[LiAl(Ot\text{-}Bu)_2H_2]{HN\text{-}CO}$$

$$Bu\text{-}(Si(Me)_2\text{-}O)_m\text{-}Si(Me)_2\text{-}(CH_2)_{10}\text{-}CO\text{-}[NH\text{-}(CH_2)_5\text{-}CO]_x\text{-}N\text{-}CO$$

An anhydride-terminated poly(dimethylsiloxane) was prepared by equilibrating mixture of 1,2-bis(5-norbornyl-2,3-dicarboxylic anhydride-1,1,2,2-tetramethyldisiloxane) and ocatamethylcyclotetrasiloxane under acidic conditions. The oligomer was used as macroactivator to initiate the anionic polymerization of ϵ-caprolactam in the presence of a catalytic amount of sodium caprolactimide to yield block copolymer [55]. During the polymerization, the silicone backbone remained intact and no cleavage products such as silanolates or N-siloxylactams were formed. The nature of the linkage connecting the hard and soft blocks of the copolymer was presumed to be either an imide or mixed amide sodium carboxylate.

Block copolymers of poly(dimethylsiloxane) with poly(ϵ-caprolactam) were obtained by the anionic polymerization of ϵ-caprolactam using bifunctional macroactivators prepared by the partial aminolysis of N,N'-isophthaloylbis(ϵ-caprolactam) with $\alpha,\omega$-di(3-aminopropyl)poly(dimethylsiloxane) [56]. Copolymers containing 10-43 wt% poly(dimethylsiloxane) blocks were prepared by bulk polymerization at 150°C or in boiling toluene.

A variety of polypeptides have been synthesized by ring-opening polymerization of $\alpha$-amino acid N-carboxy anhydrides (NCAs). Mechanistic studies have established that polymerization of NCA by primary amines is initiated by direct nucleophilic addition of the amino group onto the carbonyl carbon (C-5) of the monomer and that the propagating species is primary amine [57,58]. In addition, the polymerization proceeds without termination and chain transfer under proper reaction conditions. Polymers having one or more terminal amino groups, therefore, can be used as macroinitiators for synthesizing block copolymers containing a polypeptide as one of the components.

Bifunctional initiators, bis(2-glycylaminophenyl) disulfide (1) and 2,2'-azo-2,2'-dimethyldipropionohydrazide (2), were used to obtain polymeric initiators of two types: 1) carbon chain initiators with amino end groups, which initiate the polymerization of amino acid N-carboxy anhydrides (NCAs), and 2) symmetric polypeptide initiators containing -S-S- or -N=N- groups, which decompose on exposure to UV light or elevated temperature forming free radicals. These polymeric initiators were used to synthesize block copolymers containing carbon chain and polypeptide blocks of ABA and AB structures [59]. These block copolymers form liquid-crystalline polymer structures and undergo controlled biodegradation by enzymes.

$H_2NCH_2CONH$ — $C_6H_4$ —S—S— $C_6H_4$ — $HNCOCH_2NH_2$ (1)

$H_2NNHCO-C(Me)_2-N=N-C(Me)_2-CONHNH_2$ (2)

$$CH_2=C(Me)(CO_2Me) \xrightarrow{1} H_2N-R-S-(CH_2-C(Me)(CO_2Me))_n-S-R-NH_2$$

$R = -CH_2CONH-C_6H_4-$

$$\xrightarrow{NCA} \text{(polypeptide)} - \text{(polyMMA)} - \text{(polypeptide)}$$

$$\text{R'CH-C(=O)-O-C(=O)-NH (NCA)} \xrightarrow{2} \text{H-(NH-CH(R')-CO)}_n\text{-NH-R-N=N-R-NH-(CO-CH(R')-NH)}_n\text{-H} \quad \text{R} = \text{-NHCOC(Me)}_2\text{-}$$

$$\xrightarrow{\text{MMA}} \text{(polypeptide)-(polyMMA)}$$

Poly(methyl methacrylate) and polystyrene having terminal amino groups were synthesized by the radical polymerization of these monomers in the presence of 2-mercaptoethylammonium chloride as a chain transfer agent [60]. Anionic polymerization of Glu(OBzl)NCA and Lys(z)NCA was effectively initiated by these macroinitiators to yield the corresponding AB-type block copolymers. By removing the protecting groups of the polypeptide segments, poly(methyl methacrylate)-poly(Glu), poly(methyl methacrylate)-poly(Lys), polystyrene-poly(Glu) and polystyrene-poly(Lys) block copolymers were synthesized with little side reactions. The side chain amino groups of poly(Lys) segment in the poly(methyl methacrylate)-poly(Lys) were sulfonated or stearoylated successfully.

$$CH_2=C(X)(Y) \xrightarrow[HS(CH_2)_2NH_3Cl]{AIBN} -(CH_2-C(X)(Y))_m-S(CH_2)_2NH_2$$

$$\xrightarrow{\text{R-CH-C(=O)-O-C(=O)-NH}} -(CH_2-C(X)(Y))_m-S(CH_2)_2NH-(C(=O)-CH(R)-NH)_n-H$$

Alternatively, similar ABA block copolymers consisting of poly($\alpha$-amino acid) (A) and polystyrene or poly(methyl methacrylate) (B) were synthesized from a mixture of terminally haloacetylated poly($\alpha$-amino acid) and styrene or methyl methacrylate by photoirradiation in the presence of $Mn_2(CO)_{10}$ or by heating with $Mo(CO)_6$ [61]. The ABA type block copolymers showed higher antithrombogenicity than the corresponding homopolymers. A film of the ABA type block copolymer of poly[Glu(OBzl)] and polystyrene possessed a microphase-separated structure and did not induce a conformational change of fibrinogen adsorbed.

ABA-type triblock copolymers composed of poly($\gamma$-benzyl or $\gamma$-ethyl L-glutamate) as the A component and polybutadiene as the B component were prepared by the polymerization of the corresponding NCAs using an amine terminated polybutadiene macroinitiator. The macroinitiator was prepared by bifunctional anionic living polymerization of butadiene, followed by carboxylation and amidation with N-aminoethylpiperadine [62,63]. Spectroscopic measurements along with X-ray diffraction and electron microscopy showed that the polybutadiene chains are in random coil conformation and form domains embedded in the matrix phase formed by the polypeptide chains of the $\alpha$-helical conformation.

Gervais et al. [64] synthesized triblock ABA copolymers consisting of polybutadiene as the inner segment and four different polypeptides as the outer segments. A comercially available dicarboxyl-terminated polybutadiene was converted into diamine-terminated polybutadiene by a coupling reaction with a large excess of diaminohexane in the presence of dicyclohexylcarbodiimide as coupling agent and N-hydroxysuccinimide as nucleophilic ant. ABA triblock copolymers containing polyγ-benzyl L-glutamate), poly-(N$^{\epsilon}$-trifluoroacetyl-L-lysine) and polysarcosine were obtained in tetrahydrofuran solution by the polymerization of the respectiveNCAs using the diamine terminated polybutadiene as a macroinitiator. Aminolysis of the copolymer having poly(γ-benzyl L-glutamate) blocks by a large excess of ethanolamine in tetrahydrofuran gave a copolymer containing poly(N$^5$-hydroxyethylL-glutamine) blocks.

X-ray photoelectron spectroscopy showed that the composition of the surface was the same as the overall bulk composition for the copolymers with hydrophobic polypeptide blocks: poly(γ-benzyl L-glutamate) and poly(N$^{\epsilon}$-trifluoroacetyl L-lysine). In contrast, the surface was rich in polybutadiene for the copolymers with hydrophilic polypeptide blocks: poly(N$^5$-hydroxyethylL-glutamine) and polysarcosine. The copolymers displayed lamellar structures perpendicular to the air-polymer interface [64].

Amino-terminated poly(oxyethylene)s can be obtained by several different methods [39, 65]: from the tosylate of poly(ethylene glycol) via phthalimide and subsequent hydrazinolysis or via direct reaction with potassium 2-aminoethoxide, and from the sodium salt of poly(ethylene glycol) by the reaction with 2-bromo-ethylphthalimide as a chain extender.

A water soluble ABA-type block copolymer of poly(L-proline) (A) and poly(oxyethylene) (B) was prepared by the anionic polymerization of L-proline N-carboxy anhydride using an amino terminated poly(oxyethylene) as a macroinitiator [66]. The poly(L-proline) component in the block copolymer exists nearly as Form II poly(L-proline) with trans-conformation, and interferes the crystal growth of poly(oxyethylene) component in the solid state. The addition of poly(methacrylic acid) into an aqueous triblock copolymer solution caused a 1 : 1 interpolymer complex formation between poly(methacrylic acid) and the block copolymer through hydrogen bonding.

$$\text{(L-proline NCA)} + H_2NCH_2CH_2O(CH_2CH_2O)_nCH_2CH_2NH_2 \longrightarrow$$

$$-(\text{Pro}-N-\overset{O}{\overset{\|}{C}})_m-NHCH_2CH_2O(CH_2CH_2O)_nCH_2CH_2NH-(\overset{O}{\overset{\|}{C}}-N-\text{Pro})_m-$$

A polysaccharide macroinitiator having an amino group at the reducing end ($M_n$ = 4800-11600) was prepared by cationic ring-opening polymerization of 1,6-anhydro-2,3,4-tri-O-benzyl-β-D-glucose followed by termination with ammonia [67]. Anionic polymerization of γ-benzyl L-glutamate N-carboxy anhydride initiated with this macroinitiator gave (1→6)-α-linked polysaccharide-polypeptide diblock copolymers.

Primary amine-terminated poly(dimethylsiloxane)s have been used as macroinitiators for the synthesis of block copolymers containing polypeptide. Block copolymers composed of poly(dimethylsiloxane) and poly(DL-phenylalanine) or poly(γ-benzyl L-glutamate) segments were synthesized by polymerizing the corresponding NCAs with primary amine-terminated poly(dimethylsiloxane) prepared by the reaction of "living" polysilanolate with γ-aminopropyltriethoxysilane [68].

ABA-type poly(ε-caprolactone) (A)-poly(dimethylsiloxane) (B) block copolymers were prepared by ring-opening polymerization of ε-caprolactone in the presence of hydroxyalkyl-terminated siloxane oligomers and stannous octoate [69]. Block copolymerization reactions were completed in two steps: The first step (initiation) was performed at 70-80°C and involved the capping of poly(dimethylsiloxane) with caprolactone. The second step (propagation) was conducted at 140°C for 3-4 h. The products were obtained in quantitative yields.

DSC analysis of the copolymers indicated the formation of two phase morphologies with the glass transition temperature of poly(dimethylsiloxane) at ca. -120° and the melting point of poly(ε-caprolactone) at ca. +60 to 65°C. These products can be used as reactive oligomers for the synthesis of segmented siloxane-urethane or siloxane-ester copolymers and in the siloxane modification of epoxy or phenoxy networks [69]. The presence of terminal poly(ε-caprolactone) blocks improves solubility of these copolymers in polar solvents such as dimethylformamide, dimethylacetamide, and N-methylpyrrolidone, which are widely used in the preparation of polyurethanes.

Poly(ε-caprolactone) gives miscible blends with a wide variety of organic polymers including poly(vinyl chloride), ABS and SAN resins, urethanes, esters, epoxy resins, and cellulose esters. Poly(dimethylsiloxane), on the other hand, is totally immiscible with these base resins. The triblock copolymers consisting of poly(ε-caprolactone) and poly(dimethylsiloxane) show good miscibility with the above resins. When blended, they provide permanent surface modification to the base system [69].

AB-type poly(butyl vinylether)-poly(ε-caprolactone) block copolymers were synthesized from hydroxyl-terminated vinyl ether oligomers by initiating ring-opening polymerization of ε-caprolactone [70]. Monofunctional poly(butyl vinyl ether)s were prepared by adding a hexane solution of hydrogen iodide to a solution of butyl vinyl ether in toluene at -20°C. Subsequently, zinc iodide was added into the reaction mixture to initiate the polymerization. The reaction was terminated with aqueous potassium carbonate to afford aldehyde-terminated poly(butyl vinyl ether) oligomers. Sodium hydride reduction of the aldehyde end group gave a primary hydroxyl group at the chain end, from which polymerization of ε-caprolactone was initiated in refluxing toluene with stannous octoate as a catalyst.

$$CH_2{=}\underset{OR}{CH} \xrightarrow{HI} CH_3{-}\underset{OR}{CH}{-}I \xrightarrow{ZnI_2} H{-}(CH_2{-}\underset{OR}{CH})_m I \cdots ZnI_2$$

$$\xrightarrow{K_2CO_3} H{-}(CH_2{-}\underset{OR}{CH})_m CH_2CHO \xrightarrow{NaBH_4} H{-}(CH_2{-}\underset{OR}{CH})_m (CH_2)_2OH$$

$$\xrightarrow[(C_7H_{15}CO_2)_2Sn]{\varepsilon\text{-CL}} H{-}(CH_2{-}\underset{OR}{CH})_m (CH_2)_2O{-}[\underset{O}{\overset{\|}{C}}{-}(CH_2)_5O]_n{-}H$$

The resulting diblock copolymers showed two transitions in the range from -70 to -51°C and a melting endotherm at 63°C in their DSC diagrams. The thermal properties suggest that the diblock copolymers phase-separate into two amorphous phases and one crystalline poly(ε-caprolactone) phase.

Teyssie et al. [71] established the perfectly "living"character of the polymerization of ε-caprolactone initiated by bimetallic μ-oxo alkoxides under mild conditions. They obtained poly(ε-caprolactone)-based block copolymers with these anionic coordination type catalysts [72].

The μ-oxo alkoxides have the general formula $M_d$-[O-$M_p(OR)_{p-1}]_d$, where $M_d$ and $M_p$ are di- and tri- or tetravalent metals, respectively, and OR is an alcoholate group with at least three carbon atoms. The aluminium zinc μ-oxo isopropoxide is, among others, a very efficient catalyst for polymerization of ε-caprolactone. The propagation has living character and proceeds through a selective acyl-oxygen cleavage of the lactone with insertion into the $M_p$-OR bond of the catalyst.

The alcoholate group can be exchanged for another alcohol R'OH group according to eq.1. R'OH must be a more acidic and less volatile alcohol than ROH. It is possible to substitute the alkoxy groups of the bimetallic μ-oxo alkoxides by a hydroxyl-terminated polymer and to use the modified bimetallic compounds as initiators for the polymerization of ε-caprolactone.

$$M_d{-}[O{-}M_p(OR)_{p-1}]_d + d(p-1)R'OH \longrightarrow M_d[O\,M_p(OR')_{p-1}]_d + d(p-1)ROH\uparrow \quad (1)$$

Three of the four isopropoxy groups of $Al_2Zn$ μ-oxo isopropoxide were first substituted by slow addition of a stoichiometric amount of 2-ethylhexanoic acid (eq. 2). 2-Ethylhexanoic acid (pKa = 5.2) was selected because of its low reactivity toward the Al-O-Zn bonds and its hydrocarbon backbone which enhances the solubility of the final bimetallic compound in organic media. The modified bimetallic catalyst (I) was then added to $\alpha,\omega$-dihydroxypolybutadiene (PBD) in a 1:2 [PBD]:[catalyst] ratio (eq.3). PBD was end-capped by two bimetallic entities, each of them carrying one potential insertion site for the lactone.

$$Zn[OAl(OR)_2]_2 + 3\,R'COOH \longrightarrow Zn[OAl(OCOR')_{1.5}(OR)_{0.5}]_2 + 3\,ROH \quad (2)$$

I

$$2\,I + HO(PBD)OH \longrightarrow (R'COO)_2AlOZnO\text{-}\underset{O(PBD)O}{\overset{RCOO}{Al}}\quad\underset{}{\overset{OCOR}{Al}}\text{-}OZnOAl(OCOR')_2 + 2\,ROH \quad (3)$$

II

$R = CH(CH_3)_2$; $R' = CH(C_2H_5)(CH_2)_3CH_3$

$\epsilon$-Caprolactone was added to a toluene solution of the macrocatalyst II at 25$^0$C and polymerized. The extraction of the crude polymer with hexane at -20$^0$C indicated the absence of homopolybutadiene.

$Al_2Zn$ $\mu$-oxo isopropoxide was modified by substituting either two or three of its isopropoxy groups by 2-ethylhexanoate ones; the remaining isopropoxy groups were then substituted by $\omega$-hydroxy polystyrene (PST) ($M_n$ = 6000) and two new catalytic species, $(PST)_2$-(2-ethylhexanoate)$_2Al_2O_2Zn$ (III) and PST-(2-ethylhexanoate)$_3Al_2O_2Zn$ (IV) were obtained. Catalysts III and IV polymerized $\epsilon$-caprolactone at 25$^0$C. The extraction of the crude products with selective solvents of polycaprolactone (acetic acid) and polystyrene (a 70/23/7 mixture of cyclohexane/hexane/toluene) respectively proved the absence of any trace of homopolycaprolactone, whereas small amounts of homopolystyrene were present [72].

ABA block copolymers consisting of poly($\epsilon$-caprolactone) (PCL) and poly(L-lactide) (PLA) were prepared by using Teyssie's catalyst. Since the growing AB block copolymer intermediate (PCL-PLA-OAl<) did not reinitiate the polymerization of $\epsilon$-caprolactone, it was extended with ethylene oxide at room temperature to form a hydroxyl-terminated oligo(oxyethylene) end-capped prepolymer. The prepolymer was further allowed to react with a modified Teyssie catalyst and replaced its isopropoxy group to form an active prepolymer-supported catalyst, PCL-PLA-$(OCH_2CH_2)_{2\text{-}4}$-OAl<. This catalyst initiated the polymerization of $\epsilon$-caprolactone in toluene at 90$^0$C to give PCL-PLA-PCL triblock copolymers with different segment lengths [73].

The synthesis and use of functional diethylaluminum alkoxides ($Et_2AlORX$) as initiators for the ring-opening polymerization of $\epsilon$-caprolactone were undertaken as a possible strategy to prepare ω-primary amino poly(ε-caprolactone) [74]. An aluminum alkoxide containing ethylphthalimide as the functional -RX group was synthesized and used for the ring-opening polymerization of ε-caprolactone. The protected amino group was successfully introduced at the end of poly(ε-caprolactone) chains. However, the treatment of the phthalimide end group with aqueous methylamine gave mainly α,ω-dihydroxy poly(ε-caprolactone) containing an amide function inside the chain.

In an alternative approach, an alkyl bromide was introduced at the polymer chain end by using diethylaluminum 12-bromo-1-dodecyl oxide as an initiator for the synthesis of α-hydroxy-ω-amino poly(ε-caprolactone). The end functionalization was quantitative and the bromo end group was converted into an azide group. It was reduced to a primary amino group by catalytic hydrogenation in the presence of ammonium formate in dry dimethylformamide.

$$Br(CH_2)_{12}OAlEt_2 \xrightarrow{\varepsilon\text{-CL}} Br(CH_2)_{12}O\!\left[\!\underset{\underset{O}{\|}}{C}(CH_2)_5O\right]_n\!H \xrightarrow{NaN_3}$$

$$N_3(CH_2)_{12}O\!\left[\!\underset{\underset{O}{\|}}{C}(CH_2)_5O\right]_n\!H \xrightarrow{HCO_2^-NH_4^+/Pd} H_2N(CH_2)_{12}O\!\left[\!\underset{\underset{O}{\|}}{C}(CH_2)_5O\right]_n\!H$$

Under similar hydrogenation conditions, α-hydroxy-ω-azido poly(ε-caprolactone) derived from diethylaluminum 2-bromoethoxide as an initiator did not give a terminal amino group but an amide group inside the polymer chain. In this case, a nucleophilic attack by the nascent amino group occurs on the carbonyl carbon of the ester linkage in the β-position of the azido end group of a second chain to produce an amide linkage.

The hydroxyl group of α-hydroxy-ω-amino poly(ε-caprolactone) must be protected before it is used as a macroinitiator for the polymerization of NCAs. Actually, the hydroxyl group of α-hydroxy-ω-bromo poly(ε-caprolactone) was protected by a quantitative reaction with acetic anhydride under basic conditions, since the resulting acetate end group is not expected to perturb the derivatization of the amine from the alkyl bromide end group. The ω-primary amino poly(ε-caprolactone) was proved to be an effective macroinitiator for the ring-opening polymerization of γ-benzyl L-glutamate N-carboxy anhydride [74].

# 3. GRAFT COPOLYMERIZATION USING MACROINITIATORS

## 3.1 Synthesis of Graft Copolymers by Cationic Ring-Opening Polymerization

Synthesis of graft copolymers can be achieved by three different methods, a) graft from, b) graft on, and c) graft through. The first method requires a stock polymer containing pendant functional groups which are capable of acting as initiator for the polymerization of second monomer. This section deals with cationic graft copolymerization of cyclic monomers by macroinitiators.

A 1,3-dioxolan-2-yliumion initiates cationic polymerization of tetrahydrofuran by direct addition on the oxygen atom of the monomer to form trialkyloxonium ion, from which propagation occurs to give polymer with a formate terminal group. Okada et al. [75] prepared polystyrene carrying 1,3-dioxolan-2-ylium moieties on some of its phenyl rings. They used the modified polystyrene as a macroinitiator to obtain polystyrene-polytetrahydrofuran graft copolymers with high graft efficiency. However, the chemical modification of polystyrene is somewhat cumbersome.

Burgess et al. [76] demonstrated that a combination of allyl bromide or benzyl bromide and silver perchlorate initiates living polymerization of tetrahydrofuran. The initiation reaction can be applied to the synthesis of polystyrene-polytetrahydrofuran and polydiene-polytetrahydrofuran graft copolymers.

Cai et al. [77] prepared partially $\alpha$-brominated polystyrene by the Wohl-Ziegler reaction of polystyrene with N-bromosuccinimide, and used it in the presence of silver perchlorate to initiate cationic graft polymerization of tetrahydrofuran. $\alpha$-Brominated polystyrene with 5 to 15 bromine atoms in every 100 monomeric units of polystyrene chains were appropriate to be used for the synthesis of graft copolymers.

The graft efficiency in the above graft copolymerization was 79% or higher at temperatures below $0^0$C. However, it decreased dramatically at higher temperatures, e.g.,21.8% at $15^0$C and 12.8% at $30^0$C. The marked decrease in graft efficiency can be interpreted in terms of $\beta$-proton elimination from the in situ generated carbenium ion as well as chain transfer to monomer at higher temperatures. Perchloric acid formed by the elimination reaction initiates the homopolymerization of tetrahydrofuran. Furthermore, at high temperature and particularly at high monomer conversion, gelation occurs due to cross-linking caused by chain transfer to polymer during polymerization. Therefore the cationic graft copolymerization of tetrahydrofuran should be carried out at a temperature below $0^0$C, and the initiation should be conducted at a still lower temperature, in order to prevent the formation of gel and homopolytetrahydrofuran.

$$\text{---}CH_2\text{-}\underset{C_6H_5}{\underset{|}{CH}}\text{---} \xrightarrow{NBS} \text{---}CH_2\text{-}\underset{C_6H_5}{\underset{|}{\overset{Br}{\overset{|}{C}}}}\text{---} \xrightarrow{AgClO_4} \text{---}CH_2\text{-}\underset{C_6H_5}{\underset{|}{\overset{+}{C}}}\text{---}\ ClO_4^-$$

$$\xrightarrow{THF} \text{---}CH_2\text{-}\underset{C_6H_5}{\underset{|}{\overset{\overset{\bigcirc}{O^+}}{\overset{|}{C}}}}\text{---}\ ClO_4^- \xrightarrow{THF} \text{---}CH_2\text{-}\underset{C_6H_5}{\underset{|}{\overset{O\text{-}(CH_2)_4\text{---}}{\overset{|}{C}}}}\text{---}$$

Bromobutyl rubber containing a small amount of brominated isoprene units, most of which carried a bromine atom in the active allylic position, was employed as a macroinitiator to polymerize tetrahydrofuran [78]. Reaction of the macroinitiator with silver perchlorate or silver triflate generated in situ carbocations capable of initiating the polymerization of tetrahydrofuran at these sites.

Initiation at -20$^0$C, followed by polymerization at 0$^0$C with a [$AgClO_4$]/[Br] ratio of about unity, gave the graft copolymers with grafting efficiency of higher than 90%. Copolymers containing up to 40% of grafted polytetrahydrofuran were obtained by this method. The butyl rubber-g-polytetrahydrofuran exhibited a two-phase morphology comprising amorphous phases of butyl rubber with crystalline domains of polytetrahydrofuran which function as points of reinforcement in the rubber up to about 300K.

Trimethylsilyl trifluoromethanesulfonate (TMSOTf) acts as a useful initiator for cationic polymerization of heterocyclics and alkenes [79]. However TMSOTf alone is not sufficiently active for initiating polymerization of tetrahydrofuran. Addition of a small amount of propylene oxide as a promoter makes the initiation faster and more efficient. Partially triflated poly(methylphenylsilylene) in which about 20 % of the phenyl groups were displaced by the reaction with trifluroromethanesulfonic acid was prepared and used as a macroinitiator to polymerize tetrahydrofuran in the presence of propylene oxide [80]. The resulting graft copolymer was irradiated by strong UV light to degrade the polysilylene chain. The number average molecular weight of the liberated polytetrahydrofuran chains was higher than the theoretically expected value. This fact suggests incomplete triflation, inefficient activation, and/or some side reactions.

Since alkyl halides as well as alkyl tosylates initiate cationic ring-opening polymerization of 2-oxazolines, polymers carrying these groups along their chains act as macroinitiators for the graft polymerization of 2-oxazolines.

2-Methyl-2-oxazoline was polymerized in benzonitrile or in bulk at 100$^0$ C by using 1-chloro-1,3-butadiene-butadiene copolymer with reactive allylic chlorine atoms as a macroinitiator [81]. Under the reaction conditions, the active species of the graft polymerization of 2-methyl-2-oxazoline is covalent-bonded alkyl chloride. Alkaline hydrolysis of the resulting graft copolymer gave poly(butadiene-g-ethyleneimine). The graft copolymerization was accompanied by the formation of poly(2-methyl-2-oxazoline). Probably, hydrogen chloride generated by the thermal degradation of the butadiene copolymer initiates the homopolymerization of 2-methyl-2-oxazoline, although a possibility of chain transfer to monomer cannot be excluded. The butadiene copolymer degrades at 147$^0$C, and in the presence of amines or acids, the degradation occurs at lower temperatures. Since the oxazoline monomer is basic, the degradation of the stock polymer possibly takes place during the graft copolymerization under the experimental conditions.

Poly(ethyleneimine) chelates heavy metal ions. However, the adsorption capacity for $Cu^{2+}$ ion of the polybutadiene-poly(ethyleneimine) graft copolymer was relatively small, 0.08 mmol/g, probably due to the lower content of ethyleneimine unit and the lower swelling capacity of the graft copolymer [81].

Graft copolymers having a polyethylene main chain and poly(N-acylethyleneimine) graft chain were prepared from ethylene-vinyl acetate copolymers [82]. The acetate moieties of the copolymer were converted to hydroxyl groups by saponification, which were subsequently tosylated or bromoacetylated. The tosylated or bromoacetylated copolymers were used as macroinitiators for the cationic ring-opening polymerization of 2-methyl-2-oxazoline or 2-ethyl-2-oxazoline.

$$\text{---}CH_2\underset{OAc}{CH}\text{---}CH_2CH_2\text{---} \xrightarrow{OH^-} \text{---}CH_2\underset{OH}{CH}\text{---}CH_2\underset{OAc}{CH}\text{---}CH_2CH_2\text{---}$$

$$\xrightarrow[\text{or } BrCH_2COCl]{TsCl} \text{---}CH_2\underset{OX}{CH}\text{---}CH_2\underset{OH}{CH}\text{---}CH_2\underset{OAc}{CH}\text{---}CH_2CH_2\text{---}$$

$X = SO_2C_6H_4CH_3$ or $COCH_2Br$

$$\xrightarrow{\text{2-R-2-oxazoline}} \text{---}CH_2\underset{Y\text{-}(\underset{RC=O}{N}CH_2CH_2)_n\text{-}}{CH}\text{---}CH_2\underset{OH}{CH}\text{---}CH_2\underset{OAc}{CH}\text{---}CH_2CH_2\text{---}$$

Y = none or $OCOCH_2$
R = $CH_3$ or $C_2H_5$

Graft efficiency was not high, i.e.,up to 15.5% for the starting tosylated polymer and 50.0% for the starting bromoacetylated polymer. The rate constants of initiation of 2-ethyl- or 2-methyl-2-oxazoline by a secondary alkyl tosylate and by an alkyl bromide species are at least 10 times and about 50 times smaller than those of propagation, respectively [83]. This means that a small number of relatively long chains may have been grafted, especially in the case of tosylates.

Homopolymers of 2-oxazolines such as 2-methyl-2-oxazoline and 2-n-propyl-2-oxazoline are miscible with a synthetic polymer such as poly(vinyl chloride), polystyrene, or poly(vinylidene fluoride) [84]. Introduction of the poly(2-oxazoline) chain into cellulose and its derivatives therefore makes these natural polymers compatible with a synthetic polymer in which the poly(2-oxazoline) chain act as a "compatibilizer".

Mercerized cellulose and cellulose diacetate were p-toluenesulfonylated or methanesulfonylated, and they were used as macroinitiators to induce the cationic polymerization of 2-methyl-2-oxazoline and 2-ethyl-2-oxazoline [85]. The resulting graft copolymers, cellulose-g-poly(N-acylethylenimine) and cellulose diacetate-g-poly

(N-acylethylenimine) were isolated by fractionation using chloroform and methanol. Sodium periodate oxidation of one of the graft copolymers showed that there was one N-acetylethyleneimine branch with an average chain length of 22 units per approximately 33 glucose units along the backbone chain.

Styrene-p-acetoxystyrene random copolymer was prepared by radical copolymerization, and its acetoxy groups were partially converted by alkaline hydrolysis to sodium phenoxide groups. Anionic polymerization of propylene oxide was initiated from the phenoxide groups. The resulting hydroxylend group of each polypropylene oxide branch was transformed to a tosylate group from which cationic polymerization of 2-methyl-2-oxazoline was initiated to yield poly[styrene-g-(propylene oxide-b-2-methyl-2-oxazoline)]s [86]. A film of the graft-block copolymer was tougher than that of polystyrene-poly(2-methyl-2-oxazoline) block copolymer. This feature probably arises from the stress relaxation of polypropylene oxide domain having a lower $T_g$ (-78°C). Alkaline hydrolysis of these graft-block copolymers gave poly[styrene-g-(propylene oxide-b-ethyleneimine)]s.

Catalysts that are inert under ordinary conditions and function as initiators solely by heating or photo-irradiation are useful as latent catalysts for cross-linking reactions such as curing of epoxy resins. Endo et al. [87,88] have explored several thermally latent catalysts such as benzyl dialkyl sulfonium salts, benzyl trialkyl ammonium salts, and benzyl pyridinium salts. All these onium salts act as effective initiators for the cationic polymerization of spiro orthocarbonates, bicyclic orthoesters, and epoxy resins. These onium salts generate a benzyl cation via an $S_N1$ process or they are attacked by monomer via an $S_N2$ process to form active species for cationic polymerization. Therefore graft copolymerization takes place when the polymer supported onium salts are used.

Polystyrene having pendant sulfonium moieties was prepared by radical polymerization of p-vinylbenzyl tetramethylene sulfonium hexafluoroantimonate [89]. It was used as a thermally latent cationic initiator for the polymerization of a bicyclic orthoester, 1-phenyl-4-ethyl-2,6,7-trioxabicyclo[2.2.2]octane. The cationic ring-opening isomerization

CH2=CH 1) S 2) NaSbF6 CH2Cl → CH2=CH CH2-S+ SbF6- AIBN → ---CH2-CH--- CH2-S+ SbF6-

Ph–(O,O,O)–Et → ---CH2-CH--- CH2-(O-CH2-C(Et)(O-C(=O)-Ph)-CH2)n polymerization of the bicyclic orthoester was effectively initiated by the benzyl carbocations generated by the thermal decomposition of the sulfonium salt at 120°C to give a graft copolymer.

Homopolymers and copolymers with styrene were prepared by radical polymerization from two vinyl monomers with pendant ammonium salt moieties, N-(p-vinylbenzyl)-N-methylpyrrolidinium hexafluoroantimonate and N-(p-vinylbenzyl)-p-cyanopyridinium hexafluoroantimonate. These polymeric onium salts initiated the cationic polymerization of a bicyclic orthoester, 1-phenyl-4-ethyl-2,6,7-trioxabicyclo[2.2.2]octane at 120°C to yield the corresponding graft polymers [90]. The initiation activity of the copolymer in the polymerization of the bicyclic orthoester was higher than that of the homopolymer but lower than that of the monomeric one, N-benzyl-p-cyanopyridinium hexafluoroantimonate.

## 3.2 Synthesis of Graft Copolymers by Anionic Ring-Opening Polymerization

Alkali metal carboxylates in combination with suitable crown ethers as well as tetraalkylammonium carboxylates induce anionic livingpolymerization of some lactones such as $\beta$-propiolactone, $\alpha,\alpha,$-dimethyl-$\beta$-propiolactone commonly termed pivalolactone, and $\epsilon$-caprolactone [91,92].

$$\text{---}CH_2\text{-}\underset{Me}{\overset{Me}{C}}\text{-}CH_2\text{-}\underset{C_6H_4CH_3}{CH}\text{---} \xrightarrow[\text{TMEDA}]{\text{BuLi}} \text{---}CH_2\text{-}\underset{Me}{\overset{Me}{C}}\text{-}CH_2\text{-}\underset{C_6H_4CH_2Li}{CH}\text{---} \xrightarrow[\text{2. }H^+]{\text{1. }CO_2}$$

$$\text{---}CH_2\text{-}\underset{Me}{\overset{Me}{C}}\text{-}CH_2\text{-}\underset{C_6H_4CH_2CO_2H}{CH}\text{---} \xrightarrow{Bu_4NOH} \text{---}CH_2\text{-}\underset{Me}{\overset{Me}{C}}\text{-}CH_2\text{-}\underset{C_6H_4CH_2CO_2NBu_4}{CH}\text{---}$$

$$\xrightarrow{\text{PVL}} \text{---}CH_2\text{-}\underset{Me}{\overset{Me}{C}}\text{-}CH_2\text{-}\underset{C_6H_4CH_2CO_2\text{-}(CH_2\text{-}\underset{Me}{\overset{Me}{C}}\text{-}CO_2)_n\text{-}}{CH}\text{---}$$

A polyisobutene macroinitiator having tetramethylammonium carboxylate groups was prepared from copolymers of isobutene with methylstyrenes by lithiation of the benzylic sites in the copolymer, subsequent carboxylation with carbon dioxide, and finally conversion of the carboxyl groups to tetramethylammonium salts [93]. Anionic ring-opening polymerization of pivalolactone initiated with the macroinitiator gave poly[(isobutene-co-methylstyrene)-g-pivalolactone] ofvaried composition and molecular weight. The graft copolymer could be melt-spun to elastic fibers that were characterized by unusually low stress decay. The outstanding property is attributable to resistance of the polyisobutene backbone to autoxidation.

Well phase-separated copolymer systems can display elasticity without chemical cross-linking if the hard phase is capable of a thermally reversible change of state. Several polysiloxane copolymers have the advantage that they can be used as elastomers at very low temperatures.

Poly(dimethylsiloxane) carrying pendant carboxyl group was prepared by copolymerization ofdichlorodimethylsilane with(dichloromethyl)(3-cyanopropyl)silane. The cyano group was hydrolyzed during the copolymerization procedures. The potassium salt of the carboxylated polysiloxane was used as a macroinitiator for anionic ring-opening polymerization of pivalolactone [94]. Various lengths of poly(pivalolactone) were grafted in the presence of 18-crown-6, and the grafting reaction proceeded even in the absence of the heterogeneous state without formation of homopoly(pivalolactone). However, no grafting occurred in the absence of the crown ether.

$$Cl\text{-}Si(Me)_2\text{-}Cl + Cl\text{-}Si(Me)((CH_2)_3CN)\text{-}Cl \xrightarrow{H_2O,\ H^+} \cdots\text{-}Si(Me)_2\text{-}O\text{-}Si(Me)((CH_2)_3CO_2H)\text{-}\cdots \xrightarrow{KOH}$$

$$\cdots\text{-}Si(Me)_2\text{-}O\text{-}Si(Me)((CH_2)_3CO_2K)\text{-}\cdots \xrightarrow[18\text{-}C\text{-}6]{\text{pivalolactone (Me, Me; O–C=O)}} \cdots\text{-}Si(Me)_2\text{-}O\text{-}Si(Me)((CH_2)_3CO_2\text{-}(CH_2\text{-}C(Me)_2\text{-}CO_2)_n\text{-}H)\text{-}\cdots$$

The copolymers were high in molecular weight, exhibited excellent phase separation, and possessed high melting points. The glass transition temperature of the polysiloxane phase was essentially the same as that of homopolysiloxane and remained unchanged regardless of the mole percent of the polysiloxane phase in the copolymer. This fact indicates excellent phase separation of the two segments.

Primary amines initiate living polymerization of $\alpha$-amino acid N-carboxy anhydrides (NCAs) by the direct nucleophilic addition mechanism. Polymers having pendant primary amino groups along the chains can be used as macroinitiators for graft polymerization of NCAs. Graft copolymers carrying polypeptide branches may find application as biocompatible polymeric materials.

$$\cdots\text{-}CH_2\text{-}CH(C_6H_4\text{-}(CH_2)_2\text{-}N(Me)\text{-}(CH_2)_2NH_2)\text{-}CH_2\text{-}C(Me)(CO_2Me)\text{-}\cdots \xrightarrow{\text{R-CH-C(=O)-O-C(=O)-NH (NCA)}} \cdots\text{-}CH_2\text{-}CH(C_6H_4\text{-}(CH_2)_2\text{-}N(Me)\text{-}(CH_2)_2NH\text{-}(CO\text{-}CH(R)\text{-}NH)_n\text{-}H)\text{-}CH_2\text{-}C(Me)(CO_2Me)\text{-}\cdots$$

$R = (CH_2)_n CO_2CH_2C_6H_5$ $(n = 1, 2)$

Poly(butyl methacrylate) having pendant primary amino groups were prepared by radical copolymerization of butyl methacrylate with a small amount of N-methyl-N-(4-vinylphenylethyl)ethylenediamine. Polymerization of $\beta$-benzyl L-aspartate N-carboxy anhydride and $\gamma$-benzyl L-glutamate N-carboxy anhydride was initiated by the amino groups of the poly(butyl methacrylate) to yield the corresponding poly(butyl methacrylate)-polypeptide graft copolymers [95].

A polymer membrane having a transmembrane permeating pathway ("channel") was prepared by casting a solution of the graft copolymer followed by debenzylation. The transmembrane continuous phases of the hydrophilic polypeptide are formed in the stable matrix from vinyl polymer and function as a permeating pathway for polar substances. The permeability of solutes through the membrane can be controlled by adjusting pH on which the conformation of the polypeptide segment depends.

The conformational change of the hydrophilic polypeptide is also induced by bivalent cations, urea, quaternary ammonium salts, etc. For example, the permeability of phenyl-1,2-ethanediol through the membrane was remarkably reduced in the presence of divalent cations such as $Ca^{2+}$ and recovered by addition of ethylenediaminetetraacetic acid (EDTA) [96]. A reversible change of the structure of the pathway-composing polypeptide chains was induced alternatively by the addition of $Ca^{2+}$ and EDTA.

Efficient utilization of renewable natural resources has become increasingly important from the standpoints of preserving mineral resources on one hand and producing "environmentally friendly" polymeric materials on the other hand. There have been a number of papers and patents concerning polysaccharide-based graft copolymers, particularly those based on cellulose and vinyl monomers by radical polymerization [97].

Block and graft copolymers composed of polysaccharide and polypeptide are potentially biocompatible composite materials. Above all, those composed of cellulose or its derivatives and polypeptides are of practical interest as promising candidates for biomedical materials. Avny et al. [98] were the first that prepared graft copolymers by the polymerization of an amino acid N-carboxy anhydrides using cellulose alkoxide derivatives as macroinitiators.

Miyamoto et al. [99] prepared cellulose-g-poly($\gamma$-benzyl L-glutamate) with various composition by the polymerization of the corresponding NCA in dimethyl sulfoxide at 20°C using aminoethyl cellulose with the degree of substitution of 0.05 as a macroinitiator. The solubility tests of the reaction products showed that all of the aminoethyl cellulose molecules were grafted with polypeptide chains and that no homopolypeptide was formed.

$$\text{Cell-O(CH}_2)_2\text{NH}_2 \xrightarrow{\text{GluNCA}} \text{Cell-O(CH}_2)_2\text{NH-(}\overset{\text{O}}{\overset{\|}{\text{C}}}\text{-}\underset{}{\overset{(\text{CH}_2)_2\text{CO}_2\text{CH}_2\text{C}_6\text{H}_5}{\overset{|}{\text{CH}}}}\text{-NH)}_n\text{H}$$

$\xrightarrow{HO(CH_2)_2NH_2}$ Cell-$O(CH_2)_2NH$-(C(=O)-CH($(CH_2)_2CO_2NH(CH_2)OH$)-NH$)_n$-H

Reaction of cellulose-g-poly($\gamma$-benzyl L-glutamate) with 2-aminoethanol gave cellulose-g-poly($N^5$-2-hydroxyethyl L-glutamate). The blood compatibilities of these graft copolymers were assessed in vivo by a method of peripheral vein indwelling sutures. The polyester sutures were coated with graft copolymers by casting either from formic acid or LiCl/dimethylacetamide solutions using water as the regenerating medium, and the polymer-coated sutures were implanted into a jugular and femoral vein of a dog. The graft copolymers exhibited excellent antithrombogenic properties.

Chitin is the most abundant natural amino polysaccharide and estimated to be produced annually almost as much as cellulose. However, its utilization has been very much limited because of difficulties in isolation, purification, and solubilization. Development of efficient utilization processes of this underutilized natural resource is therefore highly desired.

Chemical modification of chitin have been carried out mostly on solid chitin because of the lack of solubility in ordinary solvents. Kurita et al. [100] developed water-soluble chitin, a partially deacetylated chitin with about 50 % deacetylation prepared by homogeneous hydrolysis. The water-soluble chitin has a great advantage as a starting material for chemical modifications, since its water solubility allows smooth reactions under mild conitions.

A chitin derivative having polypeptide side chains was prepared by the anionic graft copolymerization of $\gamma$-methyl L-glutamate N-carboxy anhydride from water-soluble chitin in a mixture of water and ethyl acetate [101]. Both the conversion and grafting efficiency were high as expected from the homogeneous polymerization. The graft copolymers with the side chains of various lengths were obtained reproducibly by regulating the amount of the NCA monomer. The solubilities, infrared spectra, and X-ray diffraction patterns of the copolymers were largely dependent on the side-chain length. These observations clearly demonstrate the influence of $\alpha$-helix formation by the side chains. Alkaline hydrolysis of the ester groups of the side chains made the resulting copolymers water-soluble.

OH, O, HO, O, $NH_2$ $\xrightarrow{\gamma\text{-methyl GluNCA}}$ OH, O, HO, O, NH-(CO-CH($(CH_2)_2CO_2CH_3$)-NH$)_n$-H

$\xrightarrow{OH^-}$ OH, O, HO, O, NH-(CO-CH($(CH_2)_2CO_2H$)-NH$)_n$-H

In a similar manner, L-alanine N-carboxy anhydride was graft-polymerized from the amino group of water-soluble chitin in an aqueous ethyl acetate solution [102]. The resulting graft copolymer having a terminal free amino group in each branch was used for the immobilization of dihydronicotinamide groups, the active site of coenzyme NADH.

## 4. CONCLUDING REMARKS

As we have seen in the present review article, diverse molecular design of block and graft copolymers has become feasible by means of living ring-opening polymerization of cyclic monomers using various macroinitiators. However, the groups of monomers which undergo living polymerization and hence can be used for the synthesis of block and graft copolymers of well-defined structure are still limited at present. In order to broaden the scope of sophisticated polymer synthesis, further fundamental and critical research, in particular, development of new initiators or catalysts which enable living polymerization of a wider range of monomers, will be indispensable.

## REFERENCES

1. M. P. Dreyfuss and P. Dreyfuss, *Polymer*, **6**, 93 (1965).
2. T. Aida, R. Mizuta, Y. Yoshida, and S. Inoue, *Makromol. Chem.*, **182**, 1073 (1981).
3. A. Munir and E. J. Goethals, *J. Polym. Sci., Polym. Chem. Ed.*, **19,** 1985 (1981).
4. E. H. Schacht and E. J. Goethals, *Makromol. Chem.*, **167**, 155 (1973).
5. P. Guerin, S. Boileau, and P. Sigwalt, *Eur. Polym. J.*, **10**, 13 (1974).
6. T. Aida, K. Kawaguchi, and S. Inoue, *Macromolecules*, **23**, 3887 (1990).
7. A. Hamitou, T. Ouhadi, R. Jerome, and Ph. Teyssie, *J. Polym. Sci., Polym. Chem. Ed.*, **15**, 865 (1977).
8. K. Shimasaki, T. Aida, and S. Inoue, *Macromolecules*, **20**, 3076 (1987).
9. J. Sebenda and J. Hauer, *Polym. Bull.*, **5,** 529 (1981).
10. Hashimoto, H. Sumitomo, and A. Washio, *J. Polym. Sci.: Pat A: Polym. Chem.,* 27, 1915 (1989).
11. H. R. Kricheldorf, "*α-Amino Acid N-Carboxy Anhydrides and Related Heterocycles*",Springer Verlag, Berlin, 1987, p.59.
12. S. Kobayashi, *Prog. Polym. Sci.*, **15**, 751 (1990).
13. T. Endo, T. Sato, and T. Tanaka, *Macromolecules*, **21**, 1187 (1988).
14. W. J. Bailey, P. Y. Chen, S.-C. Chen, W.-B. Chao, T. Endo, B. Gapud, V. Kurganti, Y.-N. Lin, Z. Ni, C.-Y. Pan, S. E. Shaffer, L. Sidney, S.-R. Wu, N. Yamamoto, N. Yamazaki, K. Yonezawa, and L. L. Zhou, *Makromol. Chem., Macromol. Symp.*, **6**, 81 (1986).
15. A. Noshay and J. E. McGrath, "*Block Copolymers, Overview and Critical Survey*", Academic Press, New York, 1977. p.138 and 242.

16. Y. Yamashita, *Adv. Polym. Sci.,* **28,** 1 (1978).
17. G. Riess, G. Hurtrez, and P. Bahadur, "*Encyclopedia of Polymer Science and Engineering*", John Wiley & Sons, New York, 1985, Vol.2, p.324.
18. P. Dreyfuss and R. P. Quirk, "*Encyclopedia of Polymer Science and Engineering*", John Wiley & Sons, New York, 1987, Vol.7, p.551.
19. I. Goodman, "*Developments in Block Copolymers-1*", I. Goodman Ed., Applied Science, London, 1982, p.127.
20. S. Penczek, P.Kubisa, and K. Matyjaszewski, *Adv. Polym. Sci.*, **68/69**, 1 (1985).
21. T. Keii and Y. Doi, *Adv. Polym. Sci.*, **73/74**, 201 (1986).
22. Y. Doi, Y. Watanabe, S. Ueki, and K. Soga, *Makromol. Chem., Rapid Commun.*, **4,** 533 (1983).
23. C. Feger and H.-J. Cantow, *Polym. Bull.*, **3**, 407 (1980).
24. C. Feger and H.-J. Cantow, *Polym. Bull.*, **6**, 321 and 583 (1982).
25. M. Okada, H. Sumitomo, and T. Ito, *Makromol. Chem.*, **189**, 263 (1988).
26. H. K. Hall, Jr. and M. J. Steuck, *J. Polym. Sci., Polym. Chem. Ed.*, **11**, 1035 (1973).
27. A. Mumir and E. J. Goethals, *J. Polym. Sci., Polym. Chem. Ed.,* **19**, 1985 (1981).
28. E. H. Sacht and E. J. Goethals, *Makromol. Chem.*, **167,** 155 (1973).
29. K. Muhlbach and R.C. Schulz, *Makromol. Chem.*, **189,** 1267 (1988).
30. H. Kazama, Y. Tezuka, K. Imai, and E. J. Goethals, *Makromol. Chem.*, **189,** 985 (1988).
31. S. Kobayashi and T. Saegusa, "*Ring-Opening Polymerization*", K. J. Ivin and T. Saegusa, Eds., Vol. 2, Elsevier, London, 1984, p.761.
32. Y. Morishima, T. Tanaka, and S. Nozakura, *Polym. Bull.*, *5,* 19 (1981).
33. M. Sawamoto and T. Higashimura, Makromol. Chem., *Macromol. Symp.*, **54/55**, 41 (1992).
34. M. Minoda, M. Sawamoto, and T. Higashimura, *Macromolecules,* **23,** 1897 (1990); 25, 2796 (1992).
35. S. Kobayashi, H. Uyama, De R. Liu, and T. Saegusa, *Macomolecules,* **23**, 5075 (1990).
36. O. Liu, J. Bronk, A. Verma, J. Wilson, R. M. Davis, and J. S. Riffle, *Polym. Preprints,* **33**, No.1, 215 (1992).
37a. I. Yilgor and J. E. McGrath, *Adv. Polym. Sci.*, **86,** 1 (1988). b)Y. Tezuka, *Prog. Polym. Sci.*, **17,** 471 (1992).
38. I. Yilgor and W. P. Steckle, Jr., *J. Polym. Sci.:PartA: Polym. Chem.*, **27**, 3673 (1989).
39. J. M. Harris, E. C. Struck, M. G. Case, M. S. Paley, M. Yalpani, J. M. Van Alstine, and E. D. Brooks, *J. Polym. Sci., Polym. Chem. Ed.,* **22**, 341 (1984).
40. R. J. De Vos and E. J. Goethals, Makromol. Chem., *Rapid Commun.*, **6**, 53 (1985).
41. M. Miyamoto, Y. Sano, T. Saegusa, and S. Kobayashi, *Eur. Polym. J.*, **19**, 955 (1983).
42. M. Miyamoto, K. Aoi, H. Yamanaka, and T. Saegusa, *Polym. J.*, **24**, 405 (1992).

43. S. Kobayashi, H. Uyama, E. Ihara, and T. Saegusa, *Macromolecules*, **23**, 1586 (1990).
44. C. I. Simonescue, I. Rabia, and Z. Crisan, *Polym. Bull.*, **7**, 217 (1982).
45. H. Sekiguchi, "*Ring-Opening Polymerization*", J. K. Ivin and T. Saegusa Ed., Elsevier Applied Science, London and New York, 1984, Vol.2, p.809.
46. J. Sebenda, "*Comprehensive Polymer Science*", G. C. Eastmond, A. Ledwith, S. Russo, and P. Sigwalt Ed., Pergamon Press, Oxford, 1989, Vol.3, p.511.
47. R. H. Wondraczek and J. P. Kennedy, *J. Polym. Sci.: Polym. Chem. Ed.*, **20,** 173 (1982).
48. J. Stehlicek and J. Sebenda, *Eur. Polym. J.* **18**, 535 (1982)
49. K. Hashimoto, H. Sumitomo, and H. Yamamori, Polym. J., 19, 249 (1987).
50. K. Hashimoto, H. Sumitomo, and H. Yamamori, *Polym. J.*, **19**, 1139 (1987).
51. K. Hashimoto., S. Imanishi, M.Okada, and H. Sumitomo, *J. Polym. Sci.: Part A:Polym. Chem.*, **30**, 211 (1992).
52. M. J. Owen and T. Thompson, *Br. Polym. J.*, **4**, 297 (1972).
53. P. M. Lefebvre, R. Jerome, and P. Teyssie, *Makromol. Chem.*, **183,** 2453 (1982).
54. C. A. Veith and R. E. Cohen, *Makromol. Chem., Macromol. Symp.*, **42/43**, 241 (1991).
55. P. P. Policastro and P. K. Hernandez, *Polym. Bull.*, **16,** 43 (1986).
56. J. Stehlicek, Z. Tuzar, K., Kazmierski, J. Sebenda, and J. Chojnowski, *Eur. Polym. J.*, **26**, 509 (1990).
57. Y. Imanishi, "*Ring-Opening Polymerization*", J. K.Ivin and T. Saegusa Ed., Elsevier Applied Science, London and New York, 1984, Vol.2, p.523.
58. R. Kricheldorf, "*Comprehensive Polymer Science*", G. C. Eastmond, A. Ledwith, S. Russo, and P. Sigwalt Ed., Pergamon Press, Oxford, 1989, Vol.3, p.531.
59. G. P. Vlasov, G. D. Rudkovskaya, and L. A. Ovsyannikova, *Makromol. Chem.*, **183**, 2635 (1982).
60. M. Tanaka, A. Mori, Y. Imanishi, and C. H. Bamford, *Int. J. Biol. Macromol.*, **7**, 173 (1985).
61. Y. Imanishi, M. Tanaka, and C. H. Bamford, *Int. J. Biol. Macromol.*, **7**, 89 (1985).
62. A. Nakajima, T. Hayashi, K. Kugo, and K. Shinoda, *Macromolecules*, **12,** 840 (1979).
63. G.-W. Chen, T. Hayashi, and A. Nakajima, *Polym. J.*, **13**, 433 (1981).
64. M. Gervais, A. Douy, B. Gallot, and R. Erre, *Polymer*, **29**, 1779 (1988).
65. K. E. Geckeler, "*Telechelic Polymers: Synthesis and Application*", E. J. Goethals Ed., CRC Press, 1989, p. 229.
66. S. H. Jeon, S. M. Park, and T. Ree, *J. Polym. Sci.: Part A: Polym. Chem.*, **27,** 1721 (1989).
67. Y. Ikeda, M. Murakami, S. Yamashita, M. Okada, and H. Sumitomo, *Polym. Preprints,* Japan, **38**, 1590 (1989).
68. C. M. Kania, H. Nabizadeh, D. G. McPhillimy, and R. A. Patsiga, *J. Appl. Polym. Sci.*, **27**, 139 (1982).

69. I. Yilgor, W. P. Steckle, Jr., *J. Polym. Sci.: Part A: Polym. Chem.*, **27**, 3673 (1989).
70. A. Verma, M. Glagola, A. Prasad, H. Marand, and J.S. Riffle, *Makromol. Chem., Macromol. Symp.*, **54/55,** 95 (1992).
71. A. Hamiltou, T. Ouhadi, R. Jerome, and Ph. Teyssie, *J. Polym. Sci., Polym. Chem. Ed.,* **15**, 865 (1977).
72. J. Heuschen, R. Jerome, and Ph. Teyssie, *Macromolecules,* **14,** 242 (1981).
73. C. X. Sang and X. D. Feng, *Macromolecules*, **17**, 2764 (1984).
74. Ph. Degee, Ph. Dubois, R. Jerome, and Ph. Teyssie, *Macromolecules,* **25,** 4242 (1992).
75. M. Okada, H. Sumitomo, and T. Kakezawa, *Makromol. Chem.*, **162,** 285 (1972).
76. F. J. Burgess, A. V. Cunliffe, D. H. Richards, and D. Thompson, *Polymer*, **19,** 334 (1978).
77. G. Cai and D. Yan, *Makromol. Chem.*, **188,** 1005 (1987).
78. G. Gordon Cameron and K. Sarmouk, *Makromol. Chem.,* **191,** 17 (1990).
79. M. S. Gong and H. K. Hall, Jr., *Macromolecules,* **19**, 3011 (1986).
80. J. S. Hrkach and K. Matyjaszewski, *Polym. Preprints*, **33,** No. 1, 229 (1992).
81. T. Saegusa, A. Yamada, and S. Kobayashi, *Polym. J.,* **11,** 53 (1979).
82. S. Kobayashi, Y. Shimano, and T. Saegusa, *Polym. J.,* **23**, 1307 (1991).
83. S. Kobayashi, H. Uyama, and Y. Narita, *Macromolecules*, **22,** 353 (1990).
84. S. Kobayashi, M. Kaku, and T. Saegusa, *Macromolecules*, **21**, 334 (1988).
85. S. Kobayashi, M. Kaku, and T. Saegusa, *Macromolecules,* **21,** 1921 (1988).
86. S. Ishikawa, K. Ishizu, and T. Fukutomi, *Polym. Bull.*, **16**, 223 (1986).
87. T. Takata and T. Endo, 2*Macromolecules,* **21,** 900 (1988).
88. H. Uno and T. Endo, *J. Polym. Sci., Part C: Polym. Lett.*, **26**, 453 (1988).
89. H. Uno and T. Endo, *Chem. Lett.,* 1869 (1986).
90. H. Uno, T. Takata, and T. Endo, *Macromolecules,* **22,** 2502 (1989).
91. S. Slomkowski and S. Penczek, *Macromolecules,* **13**, 229 (1980).
92. R. W. Lenz, M. Dror, and R. Jorgensen, *Polym. Eng. Sci.,* **18,** 937 (1978).
93. J. F. Harris, Jr. and W. H. Sharkey, *Macromolecules,* **19,** 2903 (1986).
94. S. Wanigatunga and K. B.Wegener, *Macromolecules,* **22,** 4156 (1989).
95. S. Higuchi, T. Mozawa, M. Maeda, and S. Inoue, *Macromolecules*, **19**, 2263 (1986).
96. M. Maeda, M. Aoyama, and S. Inoue, *Makromol. Chem.,* **187,** 2137 (1986).
97. B. S. Morin, I. P. Breusova, and Z. A. Rogovin, *Adv. Polym. Sci.*, **42,** 139 (1982).
98. Y. Avny, S. Migdal, and A. Zilkha, *Eur, Polym. J.*, **2,** 355 (1966).
99. T. Miyamoto, S. Takahashi, S. Tsuji, H. Ito, and H. Inagaki, *J. Appl. Polym. Sci.*, **31,** 2303 (1986).
100. K. Kurita, M. Kanari, and Y. Koyama, *Polym. Bull.*, **14,** 511 (1985).
101. K. Kurita, A. Yoshida, and Y. Koyama, *Macromolecules*, **21,** 1579 (1988).
102. K. Kurita, S. Iwasaki, S. Ishii, and S. Nishimura, *J. Polym. Sci.: Part A: Polym. Chem.*, **30,** 1123 (1992).

# Macroinitiators Chapter 10

## MACROINITIATORS IN MULTI-MODE POLYMERIZATION

**Yusuf Yagci** and **Munmaya K. Mishra**

Department of Chemistry
Istanbul Technical University
Maslak, Istanbul - 80626
TURKEY

Polymers Research
Texaco Research Center
P.O.Box 509
Beacon, NY 12508, USA





Macromolecular Design: Concepts and Practice (Macromonomers, Macroinitiators, Macroiniferters, Macroinimers, Macroiniters, and Macroinifers) Pages 391 - 429

# 1. INTRODUCTION

In this chapter we have provided an overview of various "transformation reaction," with emphasis on the synthetic aspects/mechanism of the process.

Various methods have been proposed and used for the synthesis of block copolymers. These methods are generally based on the use of either telechelic oligomers in polycondensation reactions or living polymerization techniques. The end groups of a telechelic oligomer can react with the end groups of another telechelic oligomer or with a difunctional small molecule to give block copolymers **Scheme 1**.

$$n\,x\sim\sim\sim y + n\,x'\text{-}\square\text{-}\square\text{-}\square\text{-}y' \longrightarrow \left[x\sim\sim yx'\text{-}\square\text{-}\square\text{-}\square\text{-}y'\right]_n$$

$$n\,x\sim\sim\sim y + n\,x'\text{—}y' \longrightarrow \left[x\sim\sim yx'\text{—}y'\right]_n$$

**Scheme 1**

Although telechelic oligomers with hydroxyl, thiol, halogen, carboxyl and amine end groups can now be made by wide variety of polymerizations and processes, their use in block copolymerization suffer from a number of disadvantages. In particular, block lengths are limited to short chains, since when the average chain length is increased, the probability of end reactions leading to the block copolymer formation is reduced. As generally accepted, sequential monomer addition(SMA) in living polymerization is the best way to prepare block copolymers possessing well-defined and predetermined structures.

In addition to well established living anionic polymerization, several other living polymerization systems have been developed during the past years. Higashimura et al. [1-7] established that polymerization of alkyl vinylethers using $HI/I_2$ or $ZnI_2$ proceeds via a living mechanism. Another recent advance has been the development of the living carbocationic polymerization of isobutylene by Mishra et al. [8,9] and Kennedy et al.[10-12]. Well defined block copolymers are prepared by these living systems following the common strategy of SMA. A limitation to the SMA technique is that it excludes monomers that polymerize by other mechanisms.

Because of practical limitations on synthesizing different copolymers using a single type of propagating species, Richards [13a, 13b] developed the transformation approach. "Transformation reactions" involve synthesis in which two (or more) mutually exclusive polymerization mechanisms are sequentially combined [13, 14]. The transformation approach, involves different propagating species and allows multiple combination of monomers [15-18]. Thus, monomer 1 is polymerized by a mechanism 'A' to produce a polymer with a functional group F (by initiation or termination) that is capable of initiating polymerization by a different mode/ mechanism 'B'. The overall process is shown in **Scheme 2**. The polymer formed by mechanism A is used as a macroinitiator for the polymerization of other monomers using mechanism B. Indirectly one Polymerization mechanism is transformed into another in order to incorporate into the copolymer combinations of blocks or grafts that can not be prepared by a single mechanism.

$$I + nM_1 \xrightarrow{\text{mechanism A}} I\text{-}\left[M_1\right]_{n-1} M_1^* \xrightarrow{\text{termination}} I\text{-}\left[M_1\right]_n \mathbf{F}$$

$$I\text{-}\left[M_1\right]_n \mathbf{F} + mM_2 \xrightarrow{\text{mechanism B}} I\text{-}\left[M_1\right]_n\text{-}block\text{-}\left[M_2\right]_m$$

**Scheme 2**

The application of multi-mode polymerization i.e., the transformation reactions depends critically on the feasibility of the polymerization modes to prepare polymers with functional groups to act as macroinitiators for the second monomer. Functionalization may be achieved by initiation, termination and transfer processes. However, most work concentrated on transformations between anionic, cationic and free radical polymerization. The range of transformation reactions has been expanded in recent years as a result of the development of new polymerization systems. Many attempts have been made to increase the number of combinations of monomers which can be incorporated into block and graft copolymers by transforming one mode of chain propagation into another and simultaneously changing the polymerizable monomer.

Although a few reviews related to polymerizations involving transformation approach have already been published [13, 19, 20] there is a growing interest in the area. Many new studies have been reported recently regarding the synthesis of block and graft copolymers using the transformation approach. In this chapter, transformation reactions that involve primarily macroinitiators will be discussed. Such reactions are classified on the basis of interconversion between propagation mechanisms: metathesis, anionic, cationic, free radical, group transfer, activated monomer and Ziegler-Natta. Coupling reactions, which also enable the formation of block copolymers through the reaction between the two different propagating species (usually anion and cation, and recently cation [21] and group transfer [22] but do not involve macroinitiator concept, will not be discussed.

## 2. ANIONIC TO RADICAL TRANSFORMATION

This approach is the most widely applied process of the transformation techonolgy. In principle, a wide range of block copolymers of anionically polymerizable monomers (styrenes and dienes etc.) and free radically polymerizable monomers (acrylates, methacrylates, acrylonitriles and other vinyl monomers) are accessible with this route. An important characteristic of this process is the fact that the overall structure of the resultant block copolymer is determined by both the functionality of the anionic initiator and termination mode of the particular monomer involved in radical process. By suitably selected combinations, di-, tri- and multi-block copolymers may be prepared.

In early studies, two methods were applied to change the propagating site from anionic to free radical; (1) functionalization with oxygen [23,24], or (2) termination by azo [25-27] or peroxy [28] compounds. For example, living polystyrene chains were reacted with oxygen to generate hydroperoxide end groups. Coupling reactions resulting in electron transfer to oxygen, which leads to the formation of non-functionalized polymers may be prevented by the use of excess oxygen (**Scheme 3**).

$$\sim CH_2\text{-}\underset{Ph}{\underset{|}{CH^-}}\,Li^+ + O_2 \longrightarrow \sim CH_2\text{-}\underset{Ph}{\underset{|}{CH}}\text{-}O\text{-}O^-Li^+ \xrightarrow{H^+} \sim CH_2\text{-}\underset{Ph}{\underset{|}{CH}}\text{-}O\text{-}OH + Li$$

$$\sim CH_2\text{-}\underset{Ph}{\underset{|}{CH^-}}\,Li^+ + O_2 \longrightarrow \sim CH_2\text{-}\underset{Ph}{\underset{|}{CH^\cdot}} + O_2^- \longrightarrow \sim CH_2\text{-}\underset{Ph}{\underset{|}{CH}}\text{-}\underset{Ph}{\underset{|}{CH}}\text{-}CH_2 \sim$$

**Scheme 3**

Heating or reacting polymeric hydroperoxide with ferrous sulphate (redox system) in the presence of a second monomer afforded block copolymers. Apart from block copolymers, homopolymers were also formed, especially in the former case where a hydroxyl radical was produced (**Scheme 4**).

$$\sim CH_2\text{-}\underset{Ph}{\underset{|}{CH}}\text{-}O\text{-}OH \xrightarrow{150\,^\circ C} \sim CH_2\text{-}\underset{Ph}{\underset{|}{CH}}\text{-}O\cdot + OH\cdot$$

$$\sim CH_2\text{-}\underset{Ph}{\underset{|}{CH}}\text{-}O\text{-}OH \xrightarrow{Fe^{+2}} \sim CH_2\text{-}\underset{Ph}{\underset{|}{CH}}\text{-}O\cdot + Fe^{+3} + OH^-$$

**Scheme 4**

The other technique [28] to introduce peroxy groups, involves, deactivation of living polystyrene and polyisoprene with 4,4´-bromomethyl-dibenzoylperoxide. Thermal decomposition was then used to initiate the polymerization of methyl methacrylate and vinylchloride (**Scheme 5**).

In this connection, it should be pointed out that the nucleophilicity of the polystyryl anion must be reduced [18], e.g., by adding diphenyl ethylene, to prevent significant nucleophilic attack on the perester carbonyl during the anionic termination step which leads to the formation of nonfunctionalized polymer. It has been demonstrated [29] that multiblock copolymers may also be prepared utilizing bifunctional anionic initiators in the first stage.

Azoinitiators having terminal electrophilic groups such as nitrile and carboxyl groups are commercially available and applicable to anion to radical transformation reactions. In analogy to peroxy compounds discussed above, 2,2´-azobisisobutyronitrile (AIBN), was used [25,26] to deactivate living polystyrene according to the following reactions (**Scheme 6**).

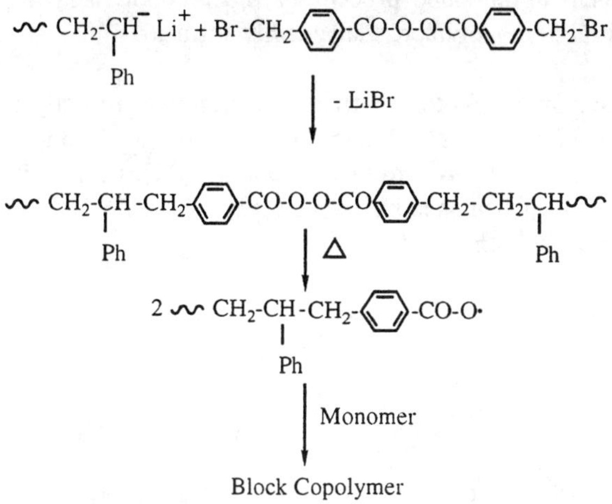

$$\sim CH_2\text{-}\underset{Ph}{CH^-}\ Li^+ + Br\text{-}CH_2\text{-}C_6H_4\text{-}CO\text{-}O\text{-}O\text{-}CO\text{-}C_6H_4\text{-}CH_2\text{-}Br$$

$$\downarrow -LiBr$$

$$\sim CH_2\text{-}\underset{Ph}{CH}\text{-}CH_2\text{-}C_6H_4\text{-}CO\text{-}O\text{-}O\text{-}CO\text{-}C_6H_4\text{-}CH_2\text{-}CH_2\text{-}\underset{Ph}{CH}\sim$$

$$\downarrow \Delta$$

$$2 \sim CH_2\text{-}\underset{Ph}{CH}\text{-}CH_2\text{-}C_6H_4\text{-}CO\text{-}O\cdot$$

$$\downarrow \text{Monomer}$$

Block Copolymer

**Scheme 5**

$$2\ Mn^-Mz^+ + CH_3\text{-}\underset{CN}{\overset{CH_3}{C}}\text{-}N{=}N\text{-}\underset{CN}{\overset{CH_3}{C}}\text{-}CH_3$$

(a)

$$CH_3\text{-}\overset{CH_3}{C}\text{-}N{=}N\text{-}\overset{CH_3}{C}\text{-}CH_3 \quad \text{with } Mz^{+-}N{=}\underset{Mn}{C} \text{ and } \underset{Mn}{C}{=}N^-Mz^+$$

$$\downarrow H^+/H_2O$$

$$CH_3\text{-}\underset{\underset{Mn}{C=O}}{\overset{CH_3}{C}}\text{-}N{=}N\text{-}\underset{\underset{Mn}{C=O}}{\overset{CH_3}{C}}\text{-}CH_3$$

(b)

$$CH_3\text{-}\underset{Mn}{\overset{CH_3}{C}}\text{-}N{=}N\text{-}\underset{Mn}{\overset{CH_3}{C}}\text{-}CH_3 + 2\ CH_3CN$$

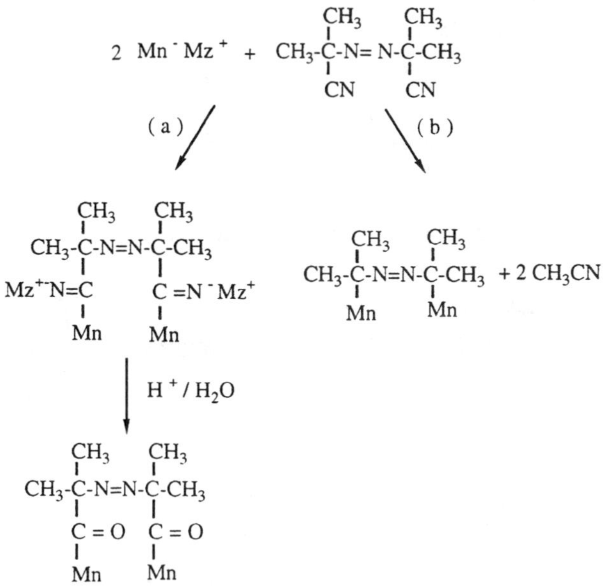

**Scheme 6**

Although both modes in **Scheme 6** eventually lead to the incorporation of azo group into the polymer, investigation [27] of this termination by using AIBN labelled with $^{14}C$ on the nitrile groups revealed that mode (b) was the predominant one. The resultant product generates polymeric radical species to polymerize vinylchloride and methylmethacrylate. However, no details on the efficiency of block copolymerization were given. In another work, Abadie et al. [30] utilized COCl derivative of AIBN as terminating agent in the same procedure. Block copolymers together with significant amount of initial homopolymer were readily formed.

The other major approach to this transformation was recently developed by Yagci et al. [31] and involves the introduction of azo groups in anionic segments using an organometallic initiator. According to Jenkins [32], polymerization of acrylonitrile proceeds smoothly, when it is mixed with tetrakis (dimethyl amino) titanium (TDT) as depicted below in **Scheme 7**.

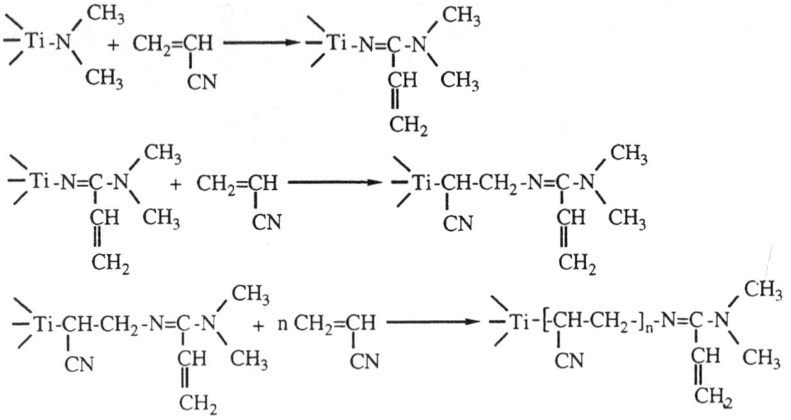

$$\gt Ti\text{-}N(CH_3)_2 + CH_2{=}CH(CN) \longrightarrow \gt Ti\text{-}N{=}C(CH{=}CH_2)\text{-}N(CH_3)_2$$

$$\gt Ti\text{-}N{=}C(CH{=}CH_2)\text{-}N(CH_3)_2 + CH_2{=}CH(CN) \longrightarrow \gt Ti\text{-}CH(CN)\text{-}CH_2\text{-}N{=}C(CH{=}CH_2)\text{-}N(CH_3)_2$$

$$\gt Ti\text{-}CH(CN)\text{-}CH_2\text{-}N{=}C(CH{=}CH_2)\text{-}N(CH_3)_2 + n\,CH_2{=}CH(CN) \longrightarrow \gt Ti\text{-}[\text{-}CH(CN)\text{-}CH_2\text{-}]_n\text{-}N{=}C(CH{=}CH_2)\text{-}N(CH_3)_2$$

**Scheme 7**

Similarly, coordination type reaction between TDT and AIBN under anhydrous conditions at ambient temperatures leads to the formation of an adduct (**Scheme 8**).

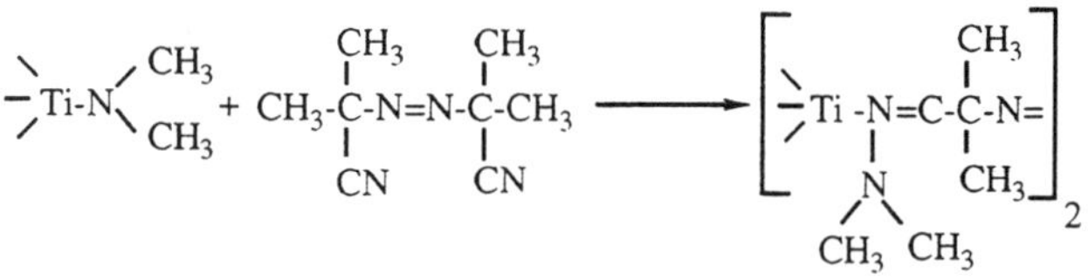

$$\gt Ti\text{-}N(CH_3)_2 + CH_3\text{-}C(CH_3)(CN)\text{-}N{=}N\text{-}C(CH_3)(CN)\text{-}CH_3 \longrightarrow \left[\gt Ti(N(CH_3)_2)\text{-}N{=}C\text{-}C(CH_3)_2\text{-}N{=}\right]_2$$

**Scheme 8**

The polymerization of acrylonitrile initiated by the preformed adduct is an efficient process that yields a polymer with -N=N- groups in the main chain. This can then be used to initiate radical polymerization to yield acrylonitrile-vinyl block copolymers.

Another approach [33] is based on the incorporation of unstable carbon-metal bonds by living anionic polymerization. Richards and co-workers [34] polymerized isoprene and styrene by an anionic living process and reacted the resulting polymer with a trialkyl lead halide. The efficiency of the reaction was quantitative. Lead containing polymers may decompose either by reacting with a soluble silver salt such as $AgNO_3$, AgCN and $AgClO_4$ or thermally to yield initiating radicals (**Scheme 9**).

$$\sim M_1^- Li^+ + ClPbR_3 \longrightarrow \sim M_1PbR_3 + LiCl$$

$$\sim M_1PbR_3 + AgX \longrightarrow \sim M_1\text{-}Ag + PbR_3X$$

$$\sim M_1\text{-}Ag \longrightarrow \sim M_1\bullet + Ag^+ \xrightarrow{n\,M_2} \text{Block copolymer}$$

$$\sim M_1PbR_3 \xrightarrow{\Delta} \sim M_1\bullet + R_3Pb\bullet \xrightarrow{n\,M_2} \text{Block copolymer}$$

**Scheme 9**

It was found that the former processes resulted in lower block efficiency due to the side reactions such as $\beta$-elimination, especially in the case of polystyrene which deactivates the polymeric radical (**Scheme 10**).

$$2 \sim CH_2\text{-}CH(C_6H_5)\text{-}Ag \longrightarrow \sim CH{=}CH(C_6H_5) + \sim CH_2\text{-}CH_2(C_6H_5) + Ag$$

**Scheme 10**

One way [33] to improve transformation efficiency is by using complexing agents such as crown ethers which retard the reaction between the silver salt and the polystyrene lead adduct, enhancing polymeric radical formation. However, thermally induced homolytic scission of the carbon-lead bond from trialkyl lead functionalized polyisoprene has been shown [34] to initiate the block copolymerization with higher efficiency. Homopolymer formation from reaction of trialky lead radical is claimed to be negligible due to the rapid combination of these radicals to give inert product.

$$4R_3Pb\bullet \quad \Rightarrow \quad 3R_4P + Pb\downarrow$$

In connection with lead-polymer adduct described above, deactivation of living polystyrene with mercuric salts has also been conveniently applied [35,36] to the synthesis of block copolymer by anion to radical transformation reaction. Actually, preparation procedure as well as the radical generation by thermolysis are closely relevant in both systems. But overall transformation efficiency was relatively low in this case (**Scheme 11**).

$$\sim M_1^- Li^+ + HgCl_2 \longrightarrow \sim M_1HgCl + LiCl$$

$$\sim M_1HgCl \xrightarrow{\Delta} \sim M_1\bullet + HgX \xrightarrow{n\,M_2} \text{Block copolymer}$$

**Scheme 11**

Another system involves the termination of living anionic polymerization with cyclic thioether in order to generate and isolate a polymeric mercaptan radical transfer agent. However, contamination of homopolymer is unavoidable in this type of reactions (**Scheme 12**).

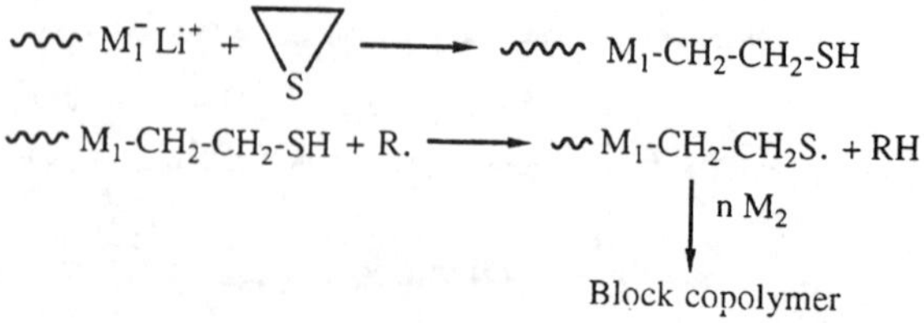

$$\sim\sim M_1^- Li^+ + \underset{S}{\bigtriangledown} \longrightarrow \sim\sim M_1\text{-}CH_2\text{-}CH_2\text{-}SH$$

$$\sim\sim M_1\text{-}CH_2\text{-}CH_2\text{-}SH + R. \longrightarrow \sim M_1\text{-}CH_2\text{-}CH_2S. + RH \xrightarrow{n\,M_2} \text{Block copolymer}$$

**Scheme 12**

Valsova et al. [37] have reported the block copolymer of polypeptide and polymethyl methacrylate. The azo groups are introduced into the polypeptide chain by the anionic polymerization of amino acid N-carboxy anhydrides (AA-NCAs) with 2,2'-azo-2,2'-dimethyldipropionohydrazide [37]. The resultant azo containing polymer was used to initiate radical polymerization of MMA to yield the block copolymer as depicted in **Scheme 13**.

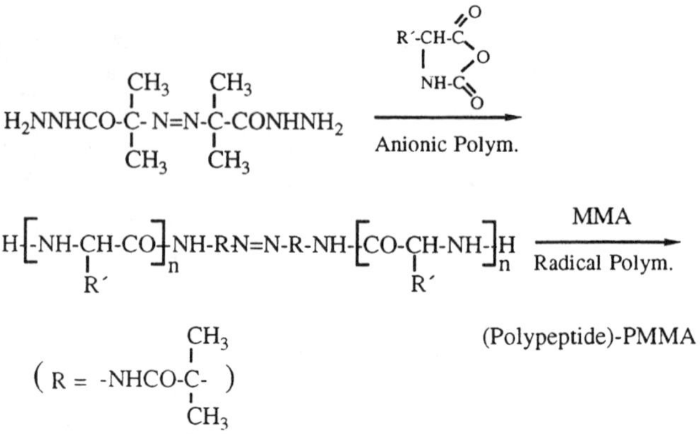

$$H_2NNHCO\text{-}C(CH_3)_2\text{-}N{=}N\text{-}C(CH_3)_2\text{-}CONHNH_2 \xrightarrow[\text{Anionic Polym.}]{R'\text{-CH-C(=O)-O-C(=O)-NH (NCA)}}$$

$$H\text{-}[NH\text{-}CH(R')\text{-}CO]_n\text{-}NH\text{-}RN{=}N\text{-}R\text{-}NH\text{-}[CO\text{-}CH(R')\text{-}NH]_n\text{-}H \xrightarrow[\text{Radical Polym.}]{MMA} \text{(Polypeptide)-PMMA}$$

$$(R = \text{-NHCO-}C(CH_3)_2\text{-})$$

**Scheme 13**

Polybutadiene chains prepared by anionic polymerization were terminated with bis (2-chloroethyl) 2, 2'- azodisobutyrate. This macroinitiator was used for the radical polymerization of MMA [38] (**Scheme 14**).

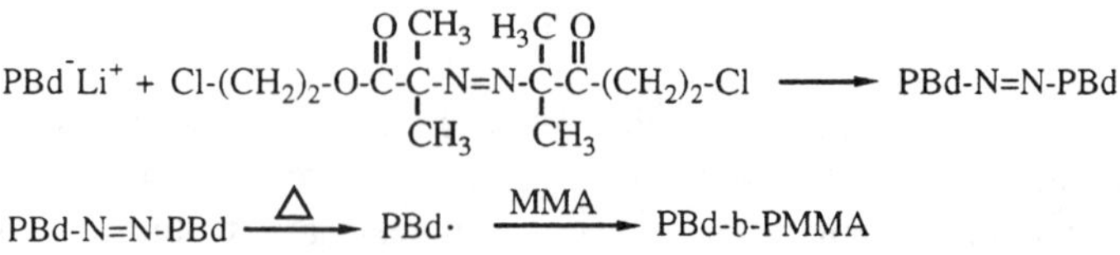

$$PBd^-Li^+ + Cl\text{-}(CH_2)_2\text{-}O\text{-}C(=O)\text{-}C(CH_3)_2\text{-}N{=}N\text{-}C(CH_3)_2\text{-}C(=O)\text{-}(CH_2)_2\text{-}Cl \longrightarrow PBd\text{-}N{=}N\text{-}PBd$$

$$PBd\text{-}N{=}N\text{-}PBd \xrightarrow{\Delta} PBd\cdot \xrightarrow{MMA} PBd\text{-}b\text{-}PMMA$$

**Scheme 14**

Eastmond et al. [39] have prepared bromine terminated PST by anionic polymerization and then used as a macroinitiator with $Mn_2CO_{10}$ for the radical polymerization of vinyl monomers. Bromine terminated PST was prepared by terminating the anionic polymerization of styrene, via a Grignard intermediate, with bromine to give terminal group.

Also, they have prepared the bromine terminated PST by a different reaction sequence as shown in **Scheme 15**. The free radical polymerization using these macroinitiators were carried out in conjunction with dimanganese decacarbonyl. The radical generation from the macroinitiator may be represented by **Scheme 16**. The generated radical initiates the polymerization of the monomer present.

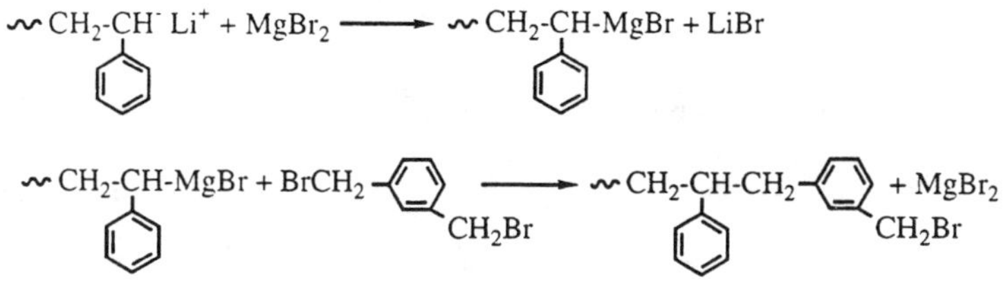

$$\sim CH_2\text{-}CH(C_6H_5)^- Li^+ + MgBr_2 \longrightarrow \sim CH_2\text{-}CH(C_6H_5)\text{-}MgBr + LiBr$$

$$\sim CH_2\text{-}CH(C_6H_5)\text{-}MgBr + BrCH_2\text{-}C_6H_4\text{-}CH_2Br \longrightarrow \sim CH_2\text{-}CH(C_6H_5)\text{-}CH_2\text{-}C_6H_4\text{-}CH_2Br + MgBr_2$$

**Scheme 15**

$$\sim CH_2\text{-}CH(C_6H_5)\text{-}CH_2\text{-}C_6H_4\text{-}CH_2Br + M(O) \longrightarrow \sim CH_2\text{-}CH(C_6H_5)\text{-}CH_2\text{-}C_6H_4\text{-}\dot{C}H_2 + M(I)$$

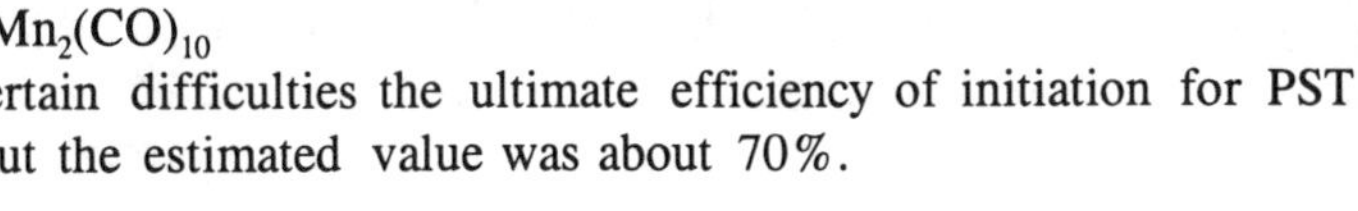

**Scheme 16**

M(o) = active manganese species in its zeroth oxidation state and produced on photolysis of $Mn_2(CO)_{10}$
Because of certain difficulties the ultimate efficiency of initiation for PST was not determined, but the estimated value was about 70%.

Anionic-cationic-radical transformation [40] was achieved by anionic functionalizaion of styrene (bromine terminated) followed by cationic polymerization of THF (bromine functionalization) and then used for the radical polymerization of MMA.

## 3. CATIONIC TO RADICAL TRANSFORMATION

In addition to anionic to radical transformations, corresponding cationic to radical transformation reactions has become attractive because of the recent discovery of living cationic polymerization systems involving vinylethers, styrene and isobutylene. They are based on the rapid equilibration between small amounts of active carbenium ions and large amount of either dormant covalent species or dormant onium ions. Until very recently, only tetrahydrofuran polymerization fulfilled the living conditions which is critical requirement in the preparation of macroinitiators for a quantitative introduction of functional groups. Therefore, most of the reported cation to radical transformation involved polytetrahydrofuran (PolyTHF) as cationic segment.

The first example of this type of transformation was the introduction of a common radical initiator group, such as peroxide into the center of polyTHF. Expectedly, subsequent thermolysis would yield two polymeric radicals. Burgess [41] has reported that termination of living polyTHF with the sodium salt of succinic acid peroxide forms polyTHF with peroxy end-group. These polymers were used to initiate polymerization of suitable monomers to produce AB and ABA type block copolymers. However, the overall transformation efficiency was not reported (**Scheme 17**).

$$\sim M_1^+PF_6^- + Na^+ \, ^-O\text{-}\overset{\overset{O}{\|}}{C}\text{-}(CH_2)_2\text{-}\overset{\overset{O}{\|}}{C}\text{-}O\text{-}O\text{-}\overset{\overset{O}{\|}}{C}\text{-}(CH_2)_2\text{-}\overset{\overset{O}{\|}}{C}\text{-}O^- \, Na^+$$

$$\downarrow - NaPF_6$$

$$\sim\sim M_1\text{-}O\text{-}\overset{\overset{O}{\|}}{C}\text{-}(CH_2)_2\text{-}\overset{\overset{O}{\|}}{C}\text{-}O\text{-}O\text{-}\overset{\overset{O}{\|}}{C}\text{-}(CH_2)_2\text{-}\overset{\overset{O}{\|}}{C}\text{-}O\text{-}M_1\sim\sim$$

$$\Delta \downarrow M_2$$

Block copolymer

**Scheme 17**

However, more recent development in this theme has been associated with the use of functional azo initiator. Yagci et al. [42,43] have synthesized polyTHF containing a central azo-linkage by cationic ring opening polymerization of THF initiated by 4,4´azobis(cyanopentanoyl chloride) in the presence of a silver salt with non-nucleophilic counter ion. Spectroscopic and degradation studies [44] showed that polymers obtained via this initiation method contain one azo linkage per chain. The functionalized polyTHF was used [45] as an initiator for copolymerization of styrene and divinyl benzene to prepare macroporous beads with good swelling properties as polyTHF segments stay out of the crosslinked domain (**Scheme 18**).

Recent application of this technique demonstrated a good example of the flexibility of the functional initiator approach. Using the above type of transformation, it has been possible to obtain new block copolymers composed of crystalline and liquid crystalline blocks [46] of the following structure (**Scheme 19**).

While a substantial inhibition of the crystallization propensity of polyTHF block is obtained depending on the copolymer composition, the thermal behavior of the liquid crystalline block is practically unaffected by the presence of polyTHF block. It is clear that this reaction can be extended to other mesogenic comonomer pairs for developing a great variety of macromolecular architectures with liquid crystalline properties.

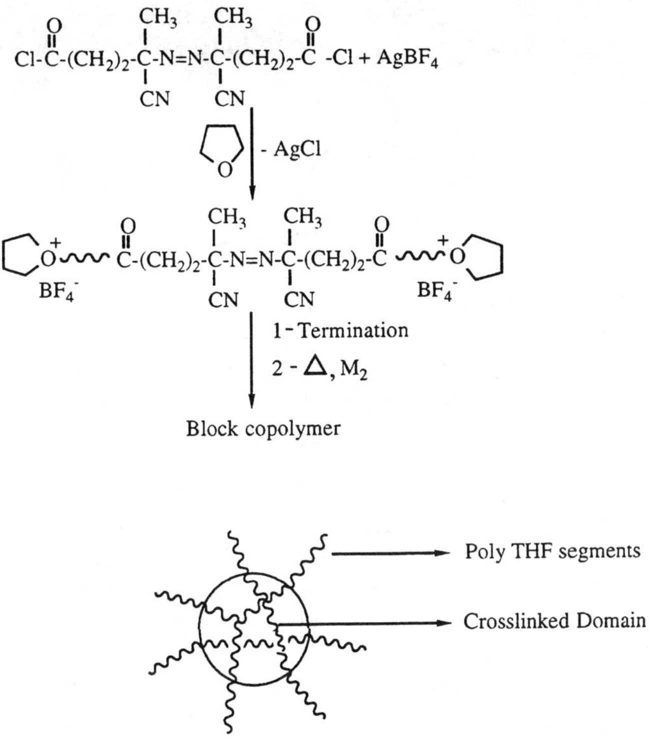

$$Cl\text{-}\overset{O}{\overset{\|}{C}}\text{-}(CH_2)_2\text{-}\underset{CN}{\overset{CH_3}{C}}\text{-}N{=}N\text{-}\underset{CN}{\overset{CH_3}{C}}\text{-}(CH_2)_2\text{-}\overset{O}{\overset{\|}{C}}\text{-}Cl + AgBF_4$$

THF, - AgCl

$$\text{THF}^{+}O\sim\overset{O}{\overset{\|}{C}}\text{-}(CH_2)_2\text{-}\underset{CN}{\overset{CH_3}{C}}\text{-}N{=}N\text{-}\underset{CN}{\overset{CH_3}{C}}\text{-}(CH_2)_2\text{-}\overset{O}{\overset{\|}{C}}\sim O^{+}\text{THF} \quad BF_4^- \quad BF_4^-$$

1 - Termination

2 - $\Delta$, $M_2$

Block copolymer

Poly THF segments

Crosslinked Domain

**Scheme 18**

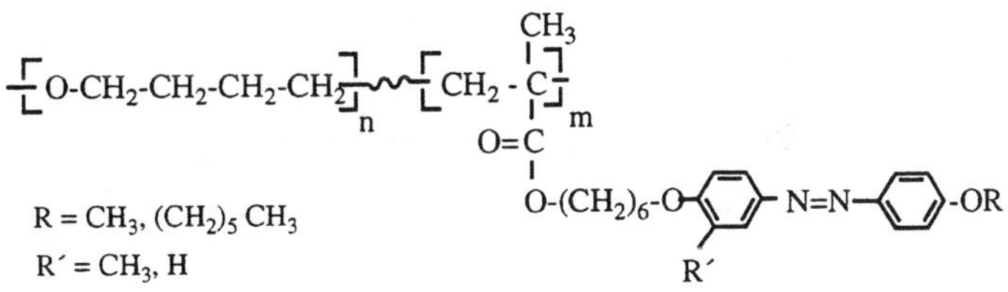

$$-[O\text{-}CH_2\text{-}CH_2\text{-}CH_2\text{-}CH_2]_n\sim[CH_2\text{-}C(CH_3)(C{=}O\text{-}O\text{-}(CH_2)_6\text{-}O\text{-}C_6H_3R'\text{-}N{=}N\text{-}C_6H_4\text{-}OR)]_m-$$

$R = CH_3, (CH_2)_5CH_3$

$R' = CH_3, H$

**Scheme 19**

Azo functionality may also be introduced to polyTHF segments in the termination step [47]. Thus, a dicarboxylate disodium 4,4′-azobis(4-cyanopentanoate) was reacted with living PolyTHF. For practical reasons, e.g for controlling the living end groups and easy handling, living oxonium end groups were converted to 4-membered cyclic ammonium ion prior to termination (**Scheme 20**). The azo group thus introduced could generate macromolecular radical species upon heating at 60°C in the presence of styrene and ethyl methacrylate to produce the corresponding block copolymers consisting of ionically obtained polyTHF and radically obtained styrene and methyl methacrylate segments.

$$\text{Poly(THF)}\sim\sim Z^{+}\!\!\supset + 2\ \text{NaOOC-(CH}_2)_2\text{-}\underset{\text{CN}}{\overset{\text{CH}_3}{\text{C}}}\text{-N=N-}\underset{\text{CN}}{\overset{\text{CH}_3}{\text{C}}}\text{-(CH}_2)_2\text{-COONa}$$

$$\downarrow$$

$$\text{Poly THF}\sim\sim\sim \text{N = N} \sim\sim\sim \text{Poly THF}$$

**Scheme 20**

Another radical generating groups, peroxides, may also be introduced [48-50] to polymers by using functional initiator approach. The cationic polymerization of styrene, lactones such as $\epsilon$-caprolactone and THF was initiated by chloromethyl derivative of benzoyl peroxide in conjunction with silver salts to produce peroxide containing polymers.

It was shown in the previous chapter "Macroinitiators for Chain Polymerization" that terminal photoactive groups such as $CBr_3$ groups can function as free radical sources in the synthesis of linear block copolymers. This approach dealt with systems that a metal carbonyl compound acts as chromophore for visible light such as $Mn_2(CO)_{10}$ or $Re_2(CO)_{10}$. An interesting variation of the manganese carbonyl method applies to active center transformation from cation to radical. In this method [51] the polyTHF is terminated by the sodium salt of bromoacetic acid in quantitative manner and subsequently irradiated in the presence of the manganese carbonyl, to generate macroradical followed by the polymerization of vinyl monomers to generate block copolymers (**Scheme 21**). The great advantage of this particular transformation is based on the fact that homopolymer contamination is avoided because the first step involves living polymerization and no small radicals but only macroradicals are produced in the second step.

$$\sim\sim\sim\overset{+}{O}\!\!\bigcirc + Li^{+}\,{}^{-}O\text{-}\overset{O}{\overset{\|}{C}}\text{-}CH_2\text{-}Br \longrightarrow \sim\sim\left[O\text{-}(CH_2)_4\right]_n O\text{-}\overset{O}{\overset{\|}{C}}\text{-}CH_2\text{-}Br$$

$$\sim\sim\left[O\text{-}(CH_2)_4\right]_n O\text{-}\overset{O}{\overset{\|}{C}}\text{-}CH_2\text{-}Br \xrightarrow[h\nu]{Mn_2(CO)_{10}} \sim\sim\left[O\text{-}(CH_2)_4\right]_n O\text{-}\overset{O}{\overset{\|}{C}}\text{-}CH_2\cdot$$

$$\downarrow n\,M_2$$

$$\text{Block copolymer}$$

**Scheme 21**

The other technique which was developed by Nuyken et al. [52] to incorporate azo initiators to polymers during the cationic initiation and termination step is conceptually similar to that discussed above and involves living carbocationic polymerization of vinyl ethers. Many initiating systems of this type of living cationic polymerization [53] of vinyl ethers have been reported. The main advantage of these systems is the absence of side reactions such as chain transfer dominating vinylether polymerization observed with classical initiators. Nuyken et al. [52] prepared polyisobutyl vinylether containing azo end groups by nucleophilic substitution of the labile end groups with the alcoholic azo compound (**Scheme 22**).

$$\sim CH_2\text{-}\underset{OR}{CH}\text{-}I + HO\text{-}\underset{CH_3}{CH}\text{-}C_6H_4\text{-}N{=}N\text{-}\overset{CH_3}{\underset{CH_3}{C}}\text{-}CN \longrightarrow \sim CH_2\text{-}\underset{OR}{CH}\text{-}O\text{-}\underset{CH_3}{CH}\text{-}C_6H_4\text{-}N{=}N\text{-}\overset{CH_3}{\underset{CH_3}{C}}\text{-}CN$$

**Scheme 22**

Although thermolysis of the resulting polyether in the presence of benzylmethacrylate has been shown to yield the required block copolymer, conversion of the second monomer and the overall transformation were very low. This was attributed to the limited interaction of polymeric radicals with the monomer. In order to avoid the formation of low molar mass radical $\{CH_3\text{-}C{\bullet}(CN)_2\}$ during thermolysis, an alternative technique involving the living polymerization system has been developed by the same group [54]. Thus, polyvinyl ether possessing central azo group are obtained by using azo-divinylether and adding sequentially HI and coinitiator and isobutyl vinylether as monomer (**Scheme 23**).

$$CH_2{=}CH\quad CH{=}CH_2 + 2HI \longrightarrow CH_3\text{-}\overset{I}{CH}\quad \overset{I}{CH}\text{-}CH_3 \qquad (\text{both CH bonded to } O\text{-}R'\text{-}O)$$

$$\xrightarrow[\text{co-initiator}]{CH_2 = CH(OR'')}$$

$$I\text{-}[\text{-}\underset{OR''}{CH}\text{-}CH2\text{-}]_n\text{-}\overset{CH_3}{CH}\text{-}O\text{-}R'\text{-}\overset{CH_3}{CH}\text{-}[\text{-}CH2\text{-}\underset{OR''}{CH}\text{-}]_n\text{-}I$$

$$R' = \text{-}[\text{-}CH_2\text{-}CH_2\text{-}O\text{-}\overset{O}{\overset{\|}{C}}\text{-}CH_2\text{-}CH_2\text{-}\overset{CH_3}{\underset{CN}{C}}\text{-}N{=}]_2 \quad R'' = \text{-}CH_2\text{-}\overset{CH_3}{CH}\text{-}CH_3$$

**Scheme 23**

The resulting azo linked polyvinylether showed similar initiation efficiency in the polymerization of methacrylonitrile under identical reaction conditions. AB type block copolymers were formed in this process since propagating methacrylonitrile radicals terminates mainly by disproportionation. However, no details on blocking and overall transformation efficiency were given.

Cationic to radical transformation [40] was also carried out by preparing polyTHF via cationic polymerization followed by functionalization (with terminal bromine atom) and used for the radical polymerization of MMA and/or styrene. To incorporate terminal bromine functionality into the cationically prepared living polytetrahydrofuran chain, the polymerization was terminated by addition of lithium bromoacetate (**Scheme 24**).

$$\sim O(CH_2)_4\text{-}\overset{+}{O}\bigcirc \; PF_6^- + BrCH_2\text{-}COOLi \longrightarrow \sim O\text{-}(CH_2)_3\text{-}CH_2\text{-}O\text{-}\overset{O}{\overset{\|}{C}}\text{-}CH_2Br + LiPF_6$$

**Scheme 24**

Bromomethyl groups activated by proximity to a carbonyl group are reactive in free-radical formation using metal carbonyl plus halide systems and were used to effect a transformation from cation to free-radical propagation. PTHF-Br was shown to be an effective macroinitiator for the radical polymerization of styrene and MMA in conjuction with photolyzed $Mn_2(CO)_{10}$. It was demonstrated that terminally brominated PTHF (PTHF-Br), of molecular weight atleast 20,000 g/mole can act as an effective macroinitiator in conjuction with $Mn_2(CO)_{10}$(photolyzed at 436 nm), and that the kinetics of polymerization for STY and MMA are closely similar to those for initiation from low molecular weight initiators.

At the end of this section, which concerns cation to radical transformation via photoactive terminal groups, a method [55] which has been applied so far only to polyTHF is dealt with. However, it may be useful in other cases.

When the cationic polymerization of THF is performed in the presence of 2,2-dimethoxy-2-phenylacetophenone (DMPA), degradative chain transfer occurs and polyTHF both with and without photoactive end groups is formed as shown in **Scheme 25**.

UV irradiation of the polymer resulted in the generation of a benzoyl radical and a macroradical according to the following reaction **Scheme 26**. Both radicals are capable of initiating the polymerization of vinyl monomers such as styrene and methyl methacrylate. By this means, a mixture of homopolymer (initiated by benzoyl radicals) and block copolymer are produced.

## 4. RADICAL TO CATIONIC TRANSFORMATIONS

So far, all applied radical to ionic transformation polymerizations are based on mainly radical to cationic process using redox or transfer agent to functionalize the radicals. For example, the addition of redox compounds such as ferric bromide to a macro radical yields a polymeric bromine adduct [56]. This polymer was employed to generate a carbocation which is capable of initiating copolymerization by reacting with silver salts such as $AgClO_4$, $AgPF_6$ or $AgSbF_6$ (**Scheme 27**).

Alternatively, a bromo functionalized polymer may be obtained by the use of carbon tetrabromide as chain transfer agent in free radical polymerization of styrene. This

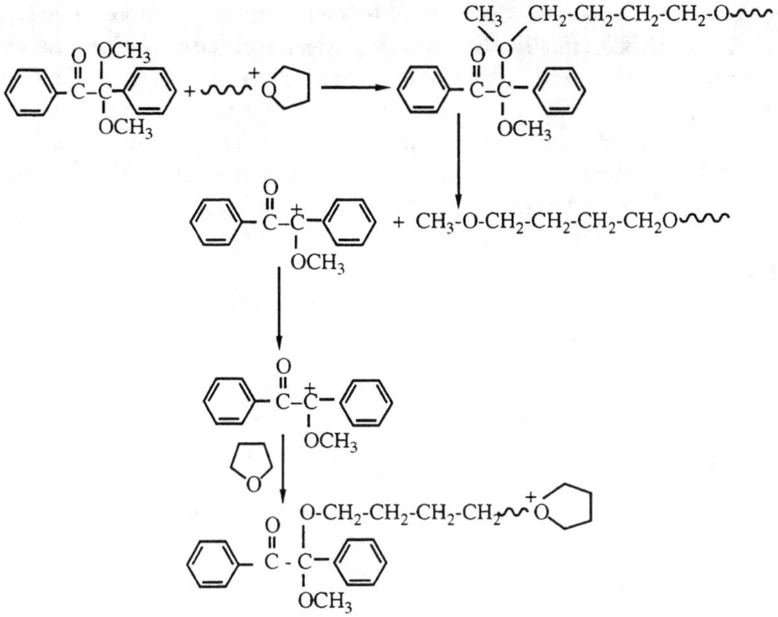

Scheme 25

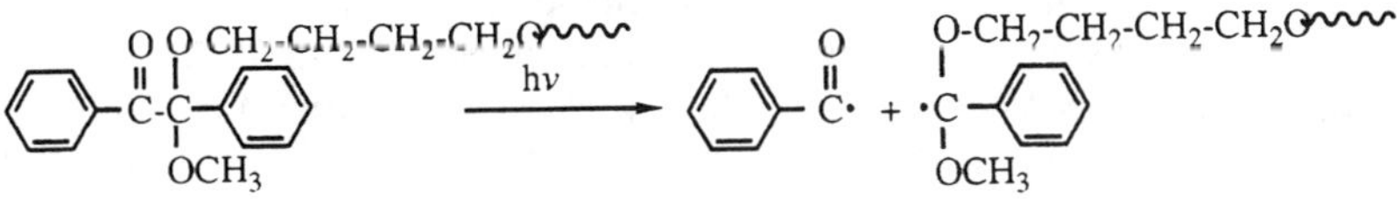

Scheme 26

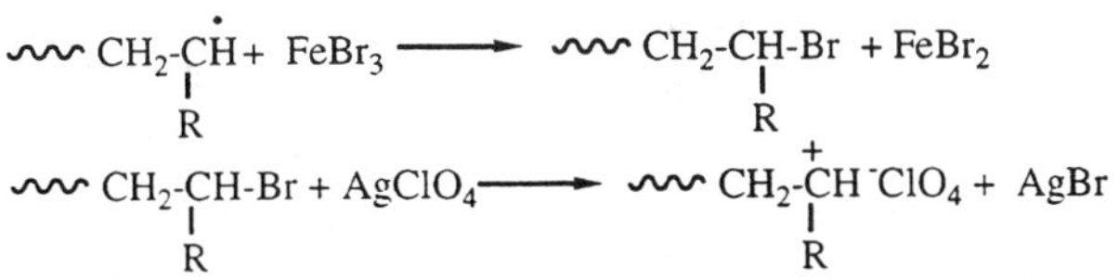

$$\sim CH_2\text{-}\dot{C}H(R) + FeBr_3 \longrightarrow \sim CH_2\text{-}CH(R)\text{-}Br + FeBr_2$$

$$\sim CH_2\text{-}CH(R)\text{-}Br + AgClO_4 \longrightarrow \sim CH_2\text{-}\overset{+}{C}H(R)\,^{-}ClO_4 + AgBr$$

Scheme 27

polymer in conjunction with a silver salt has been used [31] to give block copolymers with THF, *N*-vinylcarbazol and *N*-vinylpyrrolidone. However, although block copolymer formation was evidenced by investigation of polymers with NMR spectroscopy, no details of transformation efficiencies were reported.

Recently, Yagci [57] reported the radical to cation transformation which involves functional azo initiator. Actually, this approach was the reverse mode of the cation to radical transformation. Free radical polymerization of styrene initiated by 4,4′-azobis(4-cyanopentanoic acid) (ACPA) and subsequent treatment of these polymers with $PCl_5$ yielded two acylchloride end groups. This method suffers from the difficulty in treating a small amount of functional groups present in the long chain with $PCl_5$ and successive leaching for the removal of phosphorous oxychloride from the polymer. To parry this problem an alternate route was proposed [58] to obtain directly, acyl chloride terminated polymers having appropriate functionality for subsequent cationic polymerization leading to the formation of block copolymers. This involves polymerization of vinyl monomers using a chlorinated initiator (**Scheme 28**).

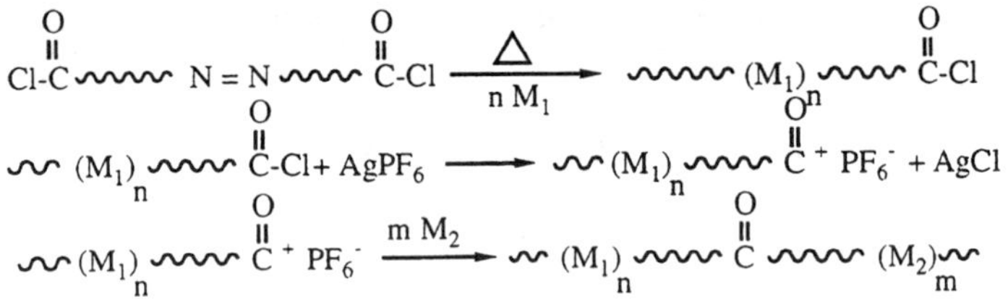

$$\text{Cl-}\overset{O}{\overset{\|}{C}}\sim\sim N=N\sim\sim\overset{O}{\overset{\|}{C}}\text{-Cl} \xrightarrow[n\,M_1]{\Delta} \sim\sim(M_1)_n\sim\sim\overset{O}{\overset{\|}{C}}\text{-Cl}$$

$$\sim(M_1)_n\sim\sim\overset{O}{\overset{\|}{C}}\text{-Cl} + AgPF_6 \longrightarrow \sim(M_1)_n\sim\sim\overset{O}{\overset{\|}{C}}{}^+\,PF_6^- + AgCl$$

$$\sim(M_1)_n\sim\sim\overset{O}{\overset{\|}{C}}{}^+\,PF_6^- \xrightarrow{m\,M_2} \sim(M_1)_n\sim\sim\overset{O}{\overset{\|}{C}}\sim\sim(M_2)_m\sim$$

**Scheme 28**

An interesting example of this theme is accomplished with photoactive polymers obtained by using azo-benzoin initiators [59-61]. It is well known that oxidation of electron donor radicals to corresponding cations may be used to initiate cationic polymerization of compounds such as epoxides., cyclic ethers and alkyl vinylethers. The overall process may be presented by the following reaction and useful oxidants include onium salts [62] and pyridium salts [63].

$$R\bullet \xrightarrow{-e} R+$$

Irradiation of benzoin terminated polymers in conjunction with pyridium salt in the presence of cyclohexene oxide as a cationically polymerizable monomer leads to the formation of block copolymers [64] of different chemical nature (**Scheme 29**). The resulting polymer mixture contained up to about 50% by weight block copolymer (concluded from fractionation experiments). Successful blocking was confirmed by a

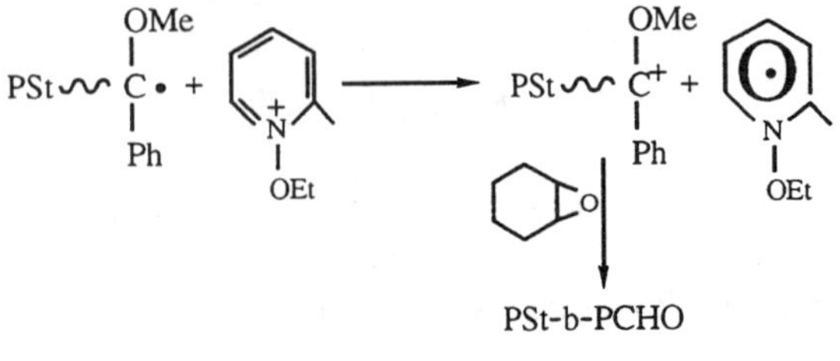

PSt∼ C(OMe)(Ph)• + 2-methyl-N-ethoxypyridinium ⟶ PSt∼ C⁺(OMe)(Ph) + 2-methylpyridine + OEt

(cyclohexene oxide) ⟶ PSt-b-PCHO

**Scheme 29**

strong change in the molecular weight of the prepolymer and the block copolymer as well as IR and NMR studies.

Okada et al. [65] prepared polystyrene (PST) carrying 1,3-dioxolan-2-ylium (DY) moieties on some of its phenyl rings. DY ion initiates cationic polymerization of THF by direct addition on the oxygen atom of the monomer to form trialkyloxonium ion, from which propagation occurs to give polymer with a formate terminal group. By this method polystyrene-polytetrahydrofuran graft copolymers were obtained. The polymeric initiator was synthesized from PST by the following route (**Scheme 30**). The grafting reaction was accompanied by the homopolymerization of THF and cross-linked polymer was obtained when THF was polymerized at a relatively high conversion.

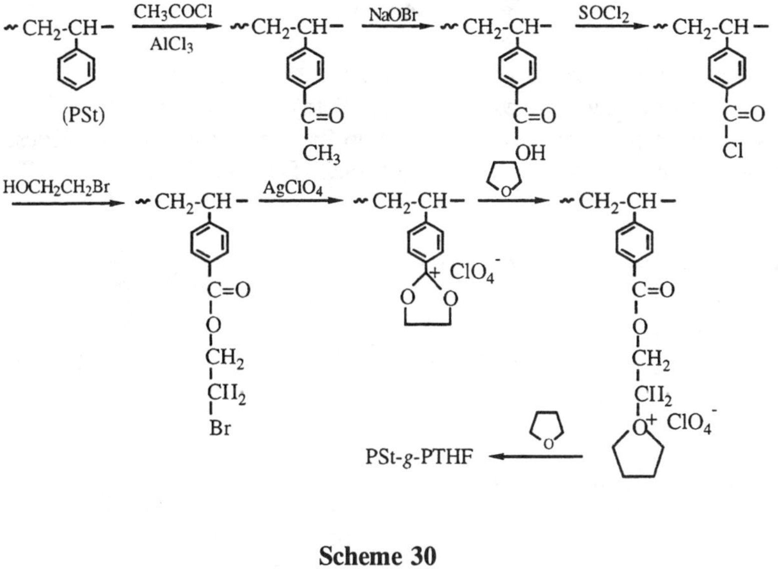

$CH_3COCl$ / $AlCl_3$; NaOBr; $SOCl_2$; $HOCH_2CH_2Br$; $AgClO_4$; $ClO_4^-$; (PSt); PSt-*g*-PTHF

**Scheme 30**

Cai and Yan [66] prepared partially α-brominated PST by bromination with N-bromosuccinimide. The so called macroinitiator was used in the presence of $AgClO_4$ to initiate cationic graft copolymerization of THF as shown in the reaction **Scheme 31**.

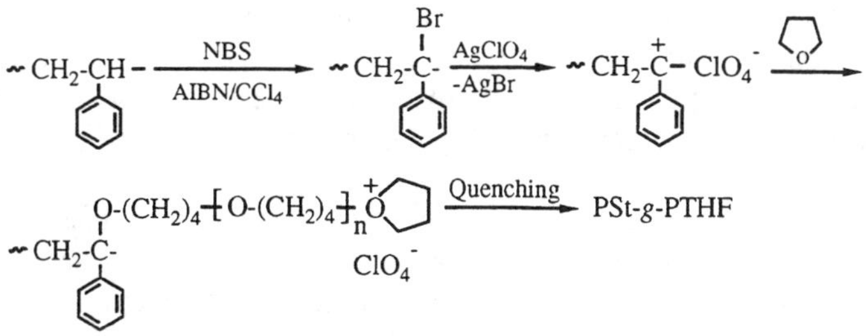

NBS / AIBN/$CCl_4$; $AgClO_4$ / -AgBr; $ClO_4^-$; $O\text{-}(CH_2)_4\text{-}[O\text{-}(CH_2)_4]_n\text{-}O^+$; Quenching; PSt-*g*-PTHF

**Scheme 31**

Increasing reaction temperature enhances grafting efficiency at reaction temperatures below $0^0$ C, however, above $0^0$C grafting efficiency deteriorates. The reason for this is that a higher reaction temperature is more likely to bring about the $\beta$-proton elimination reaction (**Scheme 32**).

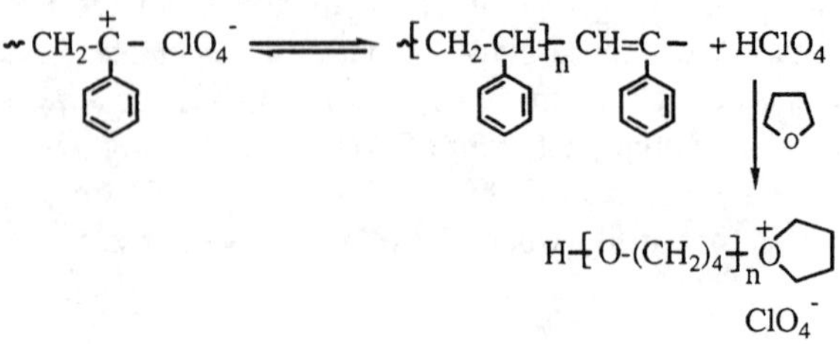

$$\sim CH_2\text{-}\overset{+}{C}(C_6H_5)\text{-}\ ClO_4^- \rightleftharpoons \sim\!\!\left[CH_2\text{-}CH(C_6H_5)\right]_n CH{=}C(C_6H_5)\text{-} + HClO_4$$

$$\xrightarrow{\text{THF}} H\!\left[O\text{-}(CH_2)_4\right]_n\overset{+}{O}(CH_2)_4\ \ ClO_4^-$$

**Scheme 32**

Kobayashi et al. [ 67] have polymerized 2-methyl-2-oxazoline or 2-ethyl-2-oxazoline by using tosylated or bromoacylated copolymer of ethylene and vinylacetate. The acetate moieties of the copolymer were converted to hydroxyl group by saponification, that were subsequently tosylated or bromoacetylated (**Scheme 33**).

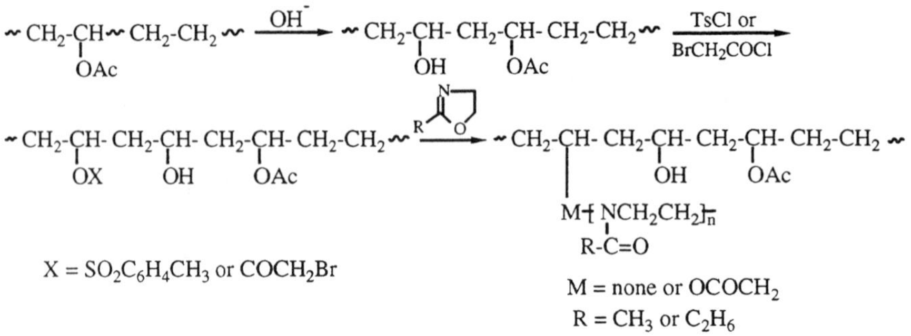

$$\sim CH_2\text{-}CH(OAc)\sim CH_2\text{-}CH_2\sim \xrightarrow{OH^-} \sim CH_2\text{-}CH(OH)\text{-}CH_2\text{-}CH(OAc)\text{-}CH_2\text{-}CH_2\sim \xrightarrow[BrCH_2COCl]{TsCl\ or}$$

$$\sim CH_2\text{-}CH(OX)\text{-}CH_2\text{-}CH(OH)\text{-}CH_2\text{-}CH(OAc)\text{-}CH_2\text{-}CH_2\sim \xrightarrow{\text{2-R-2-oxazoline}} \sim CH_2\text{-}CH(M\!\left[N(R\text{-}C{=}O)CH_2CH_2\right]_n)\text{-}CH_2\text{-}CH(OH)\text{-}CH_2\text{-}CH(OAc)\text{-}CH_2\text{-}CH_2\sim$$

$X = SO_2C_6H_4CH_3$ or $COCH_2Br$

$M$ = none or $OCOCH_2$

$R = CH_3$ or $C_2H_6$

**Scheme 33**

The random copolymer of styrene and p-acetoxy styrene was prepared by radical copolymerization and its acetoxy groups were partially converted to sodium phenoxide groups by alkaline hydrolysis. The sodium phenoxide groups of the copolymer intiated the polymerization of propylene oxide to generate hydroxyl end group at each propylene oxide branch. The hydroxyl groups were converted to tosylate groups from which cationic polymerization of 2-methyl-2-oxazoline monomer was initiated [68] to yield poly {styrene-g-[polypropylene oxide -b-poly(2 methyl-2-oxazoline)]}(**Scheme 34**).

The latent catalysts are very useful because they are inert under ordinary conditions and function as initiators by heating or photo-irradiation. Polystyrene having pendant sulfonium moieties behave as a latent cationic initiator in some cases. The radical polymerization of p-vinylbenzyltetramethylene sulfonium hexafluoroantimonate yielded PST with pendant sulfonium moieties. These latent cationic initiators were used for

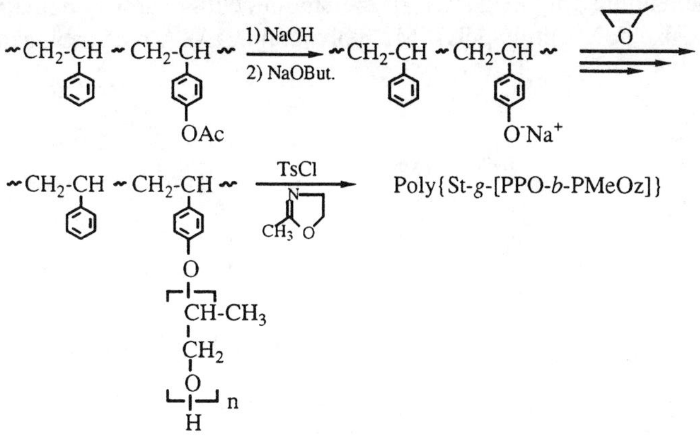

Scheme 34

the cationic ring opening isomerization-polymerization of a bicyclic orthoester, 1-phenyl-4-ethyl-2,6,7,-trioxabicyclo[2.2.2]octane[69]. The polymerization was initiated by the benzyl carbocation generated by the thermal decomposition of the sulfonium salt above >100$^0$C, produces a graft copolymer (**Scheme 35**).

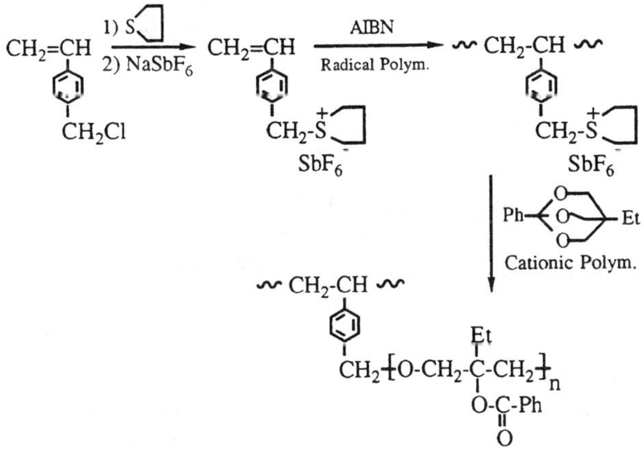

Scheme 35

# 5. RADICAL TO ANIONIC TRANSFORMATION

Valsova et al. [37] have used the bifunctional intiator such as bis(2-glycylaminophenol) disulphide as a radical initiator for the polymerization of MMA to yield polyMMA with amino end groups. The telechelic amino end groups were obtained during termination of the propagating chains by mutual combination. These amino terminated polyMMA was used for the anionic polymerization of the amino acid N-carboxy anhydrides (AA-NCAs) to synthesize polypeptide-polyMMA-polypeptide, ABA type triblock copolymer (**Scheme 36**).

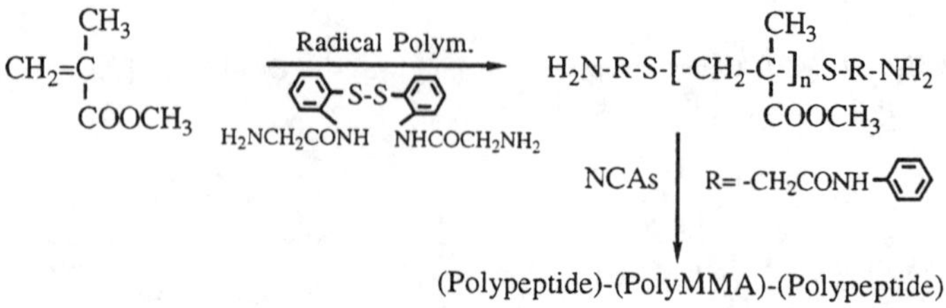

$$CH_2{=}C(CH_3)COOCH_3 \xrightarrow[\text{H}_2\text{NCH}_2\text{CONH-C}_6\text{H}_4\text{-S-S-C}_6\text{H}_4\text{-NHCOCH}_2\text{NH}_2]{\text{Radical Polym.}} H_2N\text{-}R\text{-}S\text{-}[\text{-}CH_2\text{-}C(CH_3)(COOCH_3)\text{-}]_n\text{-}S\text{-}R\text{-}NH_2$$

NCAs; $R = -CH_2CONH-C_6H_4-$

(Polypeptide)-(PolyMMA)-(Polypeptide)

**Scheme 36**

Similarly, the radical polymerization of MMA and styrene in presence of 2-mercaptoethylammonium chloride as a chain transfer agent [70] yielded polymers having terminal amino group. Anionic polymerization of amino acid NCAs was effectively initiated by the macoinitiator to yield corresponding AB type block copolymers **(Scheme 37)**.

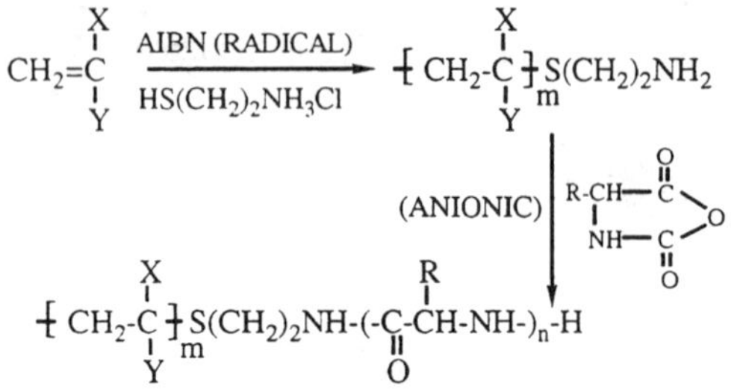

$$CH_2{=}C(X)(Y) \xrightarrow[\text{HS(CH}_2)_2\text{NH}_3\text{Cl}]{\text{AIBN (RADICAL)}} -[CH_2\text{-}C(X)(Y)]_m\text{-}S(CH_2)_2NH_2$$

(ANIONIC) R-CH—C(=O)—O—C(=O)—NH (NCA ring)

$$-[CH_2\text{-}C(X)(Y)]_m\text{-}S(CH_2)_2NH\text{-}(\text{-}C(=O)\text{-}CH(R)\text{-}NH\text{-})_n\text{-}H$$

**Scheme 37**

Inoue et al. [71] have used similar concept to initiate the polymerization of AA-NCAs. They prepared poly(butyl methacrylate) (PolyBMA) with pendant amino groups by radical copolymerization of butyl methacrylate with a small amount of N-methyl-N-(4-vinylphenyl ethyl)ethylenediamine. These primary amine containing polyBMA was used as an anionic macroinitiator for the polymerization of the AA-NCAs to yield the corresponding polyBMA-g-polypeptide copolymer (**Scheme 38**). The efficiency of the reaction was unknown. These graft copolymer was used as membranes.

Another example is the synthesis of a styrene graft polypropylene oxide copolymer [68] is represented in (**Scheme 39**).

Radical to anionic transformation [72] was also carried out by the radical copolymerization of $\alpha$-acetoxy styrene and styrene and subsequently the anionic polymerization of MMA by trating the copolymer with BuLi. The conversion of the acetate group to the hydroxyl group in the copolymer was accomplished in 90% yield. The graftinf efficiency of the reaction was not stated.

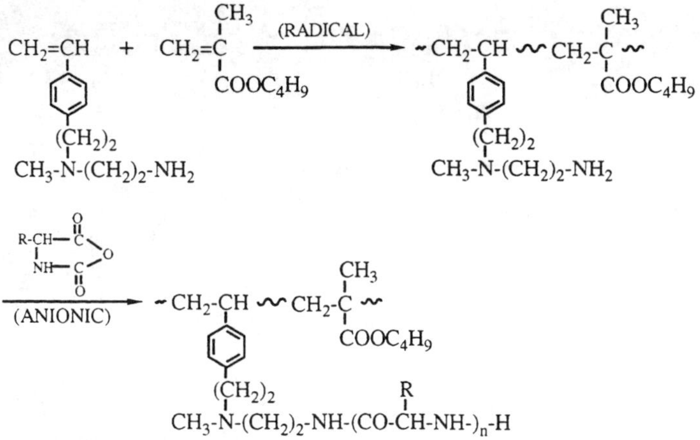

$$CH_2{=}CH(C_6H_4(CH_2)_2N(CH_3)(CH_2)_2NH_2) + CH_2{=}C(CH_3)COOC_4H_9 \xrightarrow{\text{(RADICAL)}} \sim CH_2\text{-}CH(C_6H_4(CH_2)_2N(CH_3)(CH_2)_2NH_2) \sim CH_2\text{-}C(CH_3)(COOC_4H_9) \sim$$

$$\xrightarrow[\text{(ANIONIC)}]{R\text{-}CH\text{-}C(=O)\text{-}O\text{-}C(=O)\text{-}NH} \sim CH_2\text{-}CH(C_6H_4(CH_2)_2N(CH_3)(CH_2)_2NH\text{-}(CO\text{-}CH(R)\text{-}NH\text{-})_n\text{-}H) \sim CH_2\text{-}C(CH_3)(COOC_4H_9) \sim$$

**Scheme 38**

$$CH_2{=}CH(C_6H_5) + CH_2{=}CH(C_6H_4OAc) \xrightarrow[\text{Polym.}]{\text{Radical}} \sim CH_2\text{-}CH(C_6H_5) \sim CH_2\text{-}CH(C_6H_4OAc) \sim \xrightarrow[\text{hydrolysis}]{\text{Alkaline}}$$

$$\sim CH_2\text{-}CH(C_6H_5) \sim CH_2\text{-}CH(C_6H_4O^-Na^+) \sim \xrightarrow{\text{Propylene oxide}} \text{Graft copolymer}$$

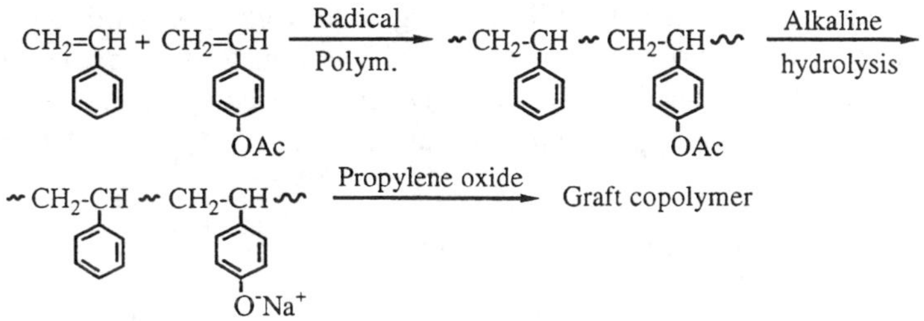

**Scheme 39**

The synthesis of poly(styrene-graft-ethylene oxide) has been demonstrated by Rempp et al. [73] and Ito et al. [74] using macromonomer techniques. However, recently Jannasch and Wesslen [75] have demonstrated the synthesis of graft copolymers containing poly(ethylene oxide) side chains on a polystyrene backbone using macroinitiator concept. The styrene copolymers containing between 5 and 15 mole % acrylamide and methacrylamide were synthesized by free radical mechanism using AIBN initiator. The amide groups in the copolymers were ionized by using potassium tert.-butoxide or potassium naphthalene, and grafting was achieved by utilizing the amide anion as initiating sites for the polymerization of ethylene oxide in 2-ethoxyethyl ether at $65^0$ C. The general route for the preparation of the graft copolymers is shown in **Scheme 40.**

## 6. ANIONIC TO CATIONIC TRANSFORMATION

Anion to cation or reverse transformation reactions were successfully employed to prepare block copolymers. The particular advantage of these transformations is that both anionic and cationic block can be prepared under livingpolymerization conditions.

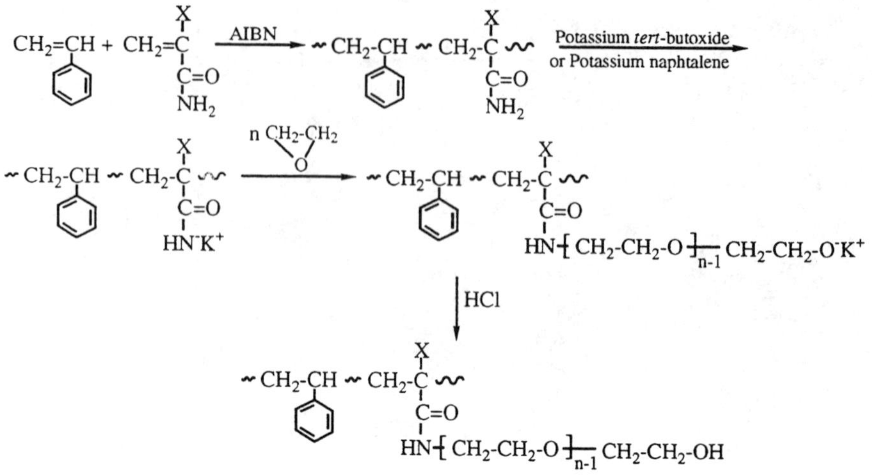

$$CH_2{=}CH(C_6H_5) + CH_2{=}C(X)(C{=}O)NH_2 \xrightarrow{AIBN} \sim CH_2\text{-}CH(C_6H_5) \sim CH_2\text{-}C(X)(C{=}O)NH_2 \sim \xrightarrow[\text{or Potassium naphtalene}]{\text{Potassium } tert\text{-butoxide}}$$

$$\sim CH_2\text{-}CH(C_6H_5) \sim CH_2\text{-}C(X)(C{=}O)HN^-K^+ \sim \xrightarrow{n\ CH_2\text{-}CH_2\text{(-O-)}} \sim CH_2\text{-}CH(C_6H_5) \sim CH_2\text{-}C(X)(C{=}O)HN\text{-}[CH_2\text{-}CH_2\text{-}O]_{n-1}CH_2\text{-}CH_2\text{-}O^-K^+ \sim$$

$$\downarrow HCl$$

$$\sim CH_2\text{-}CH(C_6H_5) \sim CH_2\text{-}C(X)(C{=}O)HN\text{-}[CH_2\text{-}CH_2\text{-}O]_{n-1}CH_2\text{-}CH_2\text{-}OH \sim$$

**Scheme 40**

In this connection, prominence must be given again to work of Richards and coworkers [76-78], demonstrating the great versatility of the transformation reactions. They prepared bromine terminated polystyrene by direct reaction of excess bromine or xylene dibromide with living polystyrene (**Scheme 41**).

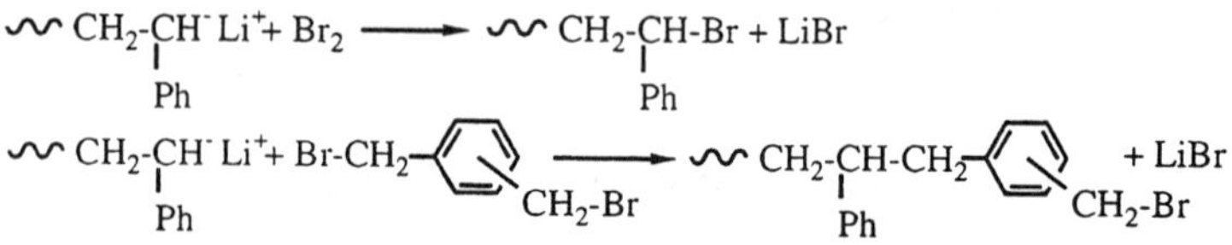

$$\sim CH_2\text{-}CH(Ph)^- Li^+ + Br_2 \longrightarrow \sim CH_2\text{-}CH(Ph)\text{-}Br + LiBr$$

$$\sim CH_2\text{-}CH(Ph)^- Li^+ + Br\text{-}CH_2\text{-}C_6H_4\text{-}CH_2\text{-}Br \longrightarrow \sim CH_2\text{-}CH(Ph)\text{-}CH_2\text{-}C_6H_4\text{-}CH_2\text{-}Br + LiBr$$

**Scheme 41**

It was found that competing Wurtz coupling reactions may be prevented [79] using a Grignard intermediate. The desired bromo functionalized polystyrene was obtained at up to 95% efficiency (**Scheme 42**).

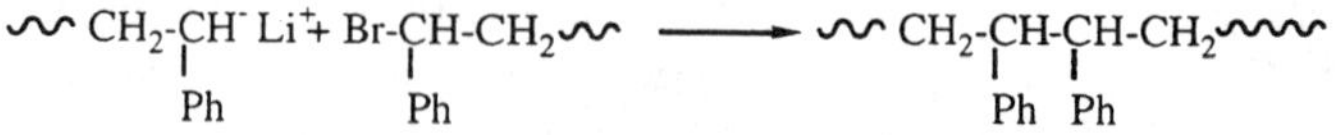

$$\sim CH_2\text{-}CH(Ph)^- Li^+ + Br\text{-}CH(Ph)\text{-}CH_2 \sim \longrightarrow \sim CH_2\text{-}CH(Ph)\text{-}CH(Ph)\text{-}CH_2 \sim$$

**Scheme 42**

Bromo functionalized polymer was employed to prepare block copolymer upon generating carbocations by reacting suitable silver salts (**Scheme 43**).

Although, block copolymers with narrow polydispersity were obtained, quantitative transformation efficiency was not achieved even at low temperatures due to the

$$\sim CH_2\text{-}\underset{\mathrm{Ph}}{CH}\text{-}Br + AgClO_4 \longrightarrow \sim CH_2\text{-}\underset{\mathrm{Ph}}{\overset{+}{CH}}\ {}^{-}ClO_4 + AgBr$$

**Scheme 43**

$\beta$-proton elimination reactions. Termination by $\beta$-proton elimination may be avoided by using xylene dibromide in halogenation process. However, it did not improve the overall transformation efficiency (**Scheme 44**).

$$\sim CH_2\text{-}\underset{\mathrm{Ph}}{CH}\text{-}Br + AgClO_4 \longrightarrow \sim CH_2\text{-}\underset{\mathrm{Ph}}{\overset{+}{CH}}\ {}^{-}ClO_4 + AgBr$$

$$\downarrow$$

$$\sim CH_2\text{=}\underset{\mathrm{Ph}}{CH} + HClO_4$$

**Scheme 44**

Recently, a conceptually similar approach was used by Mulbach and Schulz [80] to prepare styrene and 1-azabicyclo[4.2.0]octane block copolymer. By endcapping living polystyrene with ethylene oxide and subsequent reaction with bromoacetyl bromide, a polymer with bromoacetyl groups were obtained. This polymer together with $AgClO_4$ acted as a macroinitiator for living polymerization of the cyclic monomer. Very little homopolymer was formed. Two distinct glass transitions at 10 and 94$^0$C were observed with the block copolymer corresponding to poly(1-azabicyclo[4.2.0]octane) and polystyrene sequences respectively, which indicates that both blocks are incompatible and phase separated (**Scheme 45**).

$$\sim[\text{-}CH_2\text{-}\underset{\mathrm{Ph}}{CH}\text{-}]_n\text{-}CH_2\text{-}CH_2\text{-}O\text{-}\overset{O}{\overset{\|}{C}}\text{-}CH_2\text{-}Br$$

$$\downarrow AgClO_4,\ m\ \text{(1-azabicyclo[4.2.0]octane)}$$

$$\sim[\text{-}CH_2\text{-}\underset{\mathrm{Ph}}{CH}\text{-}]_n\text{-}CH_2\text{-}CH_2\text{-}O\text{-}\overset{O}{\overset{\|}{C}}\text{-}CH_2\text{-}[N(\text{piperidine ring})\text{-}CH_2\text{-}CH_2]_m$$

**Scheme 45**

Cationic polymerization of cyclic amines are well known [81-83]. Low molecular weight initiators such as ethyl tosylate induce the polymerization of cyclic amines such as 1-tert.-butylaziridine. The idea of using a macronitiator having a tosylate end group to polymerize cyclic amines, prompted Kazama et al. [84] to polymerize 1-tert.-butyl aziridine using poly(dimethylsiloxane) having a terminal tosylate group. However, no polymerization occured when the macroinitiator was used. But this clearly indicates the initiative behind the transformation reaction between anionic and cationic polymerizations.

Anionic polymerization is one of the best methods to prepare end-functionalized polymers. The halo alkyl groups at one end or both ends of the vinyl polymers were prepared by the anionic polymerization followed by termination of the living anion with an excess of 1,2-dichloroethane or 1,4-dibromobutane [85]. These polymers served as macroinitiators for the polymerization of 2-methyl-2-oxazoline to yield the corresponding block copolymers of 2-methyl-2-oxazoline (MeOZO) and aromatic vinyl monomers (**Scheme 46**).

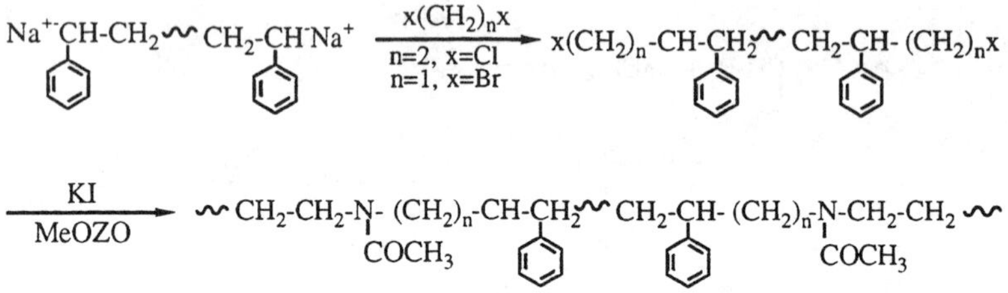

$Na^{+-}CH(Ph)\text{-}CH_2\sim CH_2\text{-}CH(Ph)^{-}Na^{+} \xrightarrow[n=2,\ x=Cl;\ n=1,\ x=Br]{x(CH_2)_nx} x(CH_2)_n\text{-}CH(Ph)\text{-}CH_2\sim CH_2\text{-}CH(Ph)\text{-}(CH_2)_nx$

$\xrightarrow[MeOZO]{KI} \sim CH_2\text{-}CH_2\text{-}N(COCH_3)\text{-}(CH_2)_n\text{-}CH(Ph)\text{-}CH_2\sim CH_2\text{-}CH(Ph)\text{-}(CH_2)_n\text{-}N(COCH_3)\text{-}CH_2\text{-}CH_2\sim$

**Scheme 46**

Block copolymers consisting of poly(N-acylethyleneimine) and poly(oxyethylene) chains were prepared [86] by initiating the polymerization of 2-methyl-2-oxazoline and 2-ethyl-2-oxazoline with $\alpha,\omega$-ditosylated or mesylated poly(oxyethylene) (**Scheme 47**). The blocking efficiency was close to $\sim 100\%$.

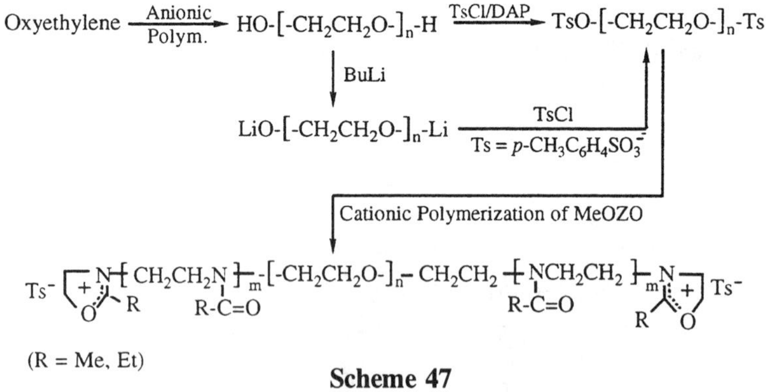

Oxyethylene $\xrightarrow[Polym.]{Anionic}$ $HO\text{-}[\text{-}CH_2CH_2O\text{-}]_n\text{-}H$ $\xrightarrow{TsCl/DAP}$ $TsO\text{-}[\text{-}CH_2CH_2O\text{-}]_n\text{-}Ts$

$\downarrow$ BuLi

$LiO\text{-}[\text{-}CH_2CH_2O\text{-}]_n\text{-}Li$ $\xrightarrow[Ts = p\text{-}CH_3C_6H_4SO_3^-]{TsCl}$ (to $TsO\text{-}[\text{-}CH_2CH_2O\text{-}]_n\text{-}Ts$)

$\downarrow$ Cationic Polymerization of MeOZO

$Ts^-$ (oxazolinium, R) $N\text{-}[CH_2CH_2N(R\text{-}C{=}O)]_m\text{-}[\text{-}CH_2CH_2O\text{-}]_n\text{-}CH_2CH_2\text{-}[N(R\text{-}C{=}O)CH_2CH_2]_m\text{-}N$ (oxazolinium, R) $Ts^-$

(R = Me, Et)

**Scheme 47**

Simonescue et al. [87] have used polyethyleneglycol adipate having tosylate groups at both ends as macroinitiators for the cationic polymerization of 2-methyl-2-oxazoline to produce ABA type block copolymers. The concept was further explored and the block copolymers [88] consisting of poly(oxypropylene) and poly(2-methyl-2-oxazoline) were prepared using polypropyleneglycol p-nitrobenzene sulfonate as a macroinitiator for the cationic polymerization of 2-methyl-2-oxazoline. As the conversion to the sulfonate functionality was not quantitative, the polymerization of 2-methyl-2-oxazoline by the macroinitiator produced a mixture of AB and ABA type block copolymer (**Scheme 48**).

$$HO\text{-}[\text{-}\underset{|}{\overset{Me}{C}}HCH_2O\text{-}]_n\text{-}\overset{Me}{C}HCH_2\text{-}[\text{-}OCH_2\overset{Me}{C}H\text{-}]_m\text{-}OH + NO_2\text{-}C_6H_4\text{-}SO_2Cl \xrightarrow{\text{2,6-Lutidine}}$$

$$NO_2\text{-}C_6H_4\text{-}SO_2\text{-}O\text{-}[\text{-}\overset{Me}{C}HCH_2O\text{-}]_n\text{-}\overset{Me}{C}HCH_2\text{-}[\text{-}OCH_2\overset{Me}{C}H\text{-}]_m\text{-}O\text{-}SO_2\text{-}C_6H_4\text{-}NO_2 \xrightarrow{\text{MeOZO}}$$

$$HO\text{-}[\text{-}CH_2CH_2\underset{COMe}{N}\text{-}]_p\text{-}[\text{-}\overset{Me}{C}HCH_2O\text{-}]_n\text{-}\overset{Me}{C}HCH_2\text{-}[\text{-}OCH_2\overset{Me}{C}H\text{-}]_m\text{-}[\text{-}\underset{COMe}{N}CH_2CH_2\text{-}]_q\text{-}OH$$

**Scheme 48**

## 7. CATIONIC TO ANIONIC TRANSFORMATION

The reverse mode cation to anion transformations have been reported by several research groups. In 1982, Richards and co-workers [89] had reported the termination of living polyTHF with lithium salt of an alcohol possessing a styryl unit or with primary amine [90] to yield polymers with a terminal group capable of reacting with a strong nucleophile (**Scheme 49**).

$$\sim\overset{+}{O}(\text{THF ring}) \xrightarrow{LiOCH_2CH=CH(Ph)} \sim O\text{-}(CH_2)_4\text{-}OCH_2CH=CH(Ph)$$

$$\sim\overset{+}{O}(\text{THF ring}) \xrightarrow[\text{2) } OH^-]{\text{1) } R\text{-}NH_2} \sim O(CH_2)_4\text{-}NHR$$

**Scheme 49**

Upon reaction with butyllithium, these end groups could be converted to a secondary anion and alkali metal salt, respectively, which in turn could be used as macroinitiators to polymerize a second monomer anionically to create block copolymers. However, low blocking efficiencies were attained in both cases (**Scheme 50**).

$$\sim O\text{-}(CH_2)_4\text{-}OCH_2CH=CH(Ph) + BuLi \longrightarrow \sim O\text{-}(CH_2)_4\text{-}OCH_2\overset{Bu}{C}H\text{-}\overset{-}{C}H(Ph)\ Li^+$$

$$\sim O(CH_2)_4\text{-}NHR + BuLi \longrightarrow \sim O(CH_2)_4\text{-}\overset{-}{N}R\ Li + Bu\text{-}H$$

**Scheme 50**

With the ready availability of functional polymers by new polymerization techniques, efforts have been made to prepare block copolymers by cation to anion transformation. Such transformation can be realized, for example, by first preparing a tert-chlorine terminated polyisobutylene (PIB), dehydrohalogenating and reacting with butyllithium. This reactive carbanion can then be used [91] for example, to induce anionic polymerization of butadiene (Bd) (**Scheme 51**). The reactivity of the $PIB^-$ increases significantly if the dehydrochlorination-metalation-polymerization step is carried out with n-BuLi/t-ROK (t-BuOK) combinations. The PIB-b-PBd products contain no detectable homo-PBd at low Bd dose level. However, at higher Bd concentrations the homo PBd content was increased.

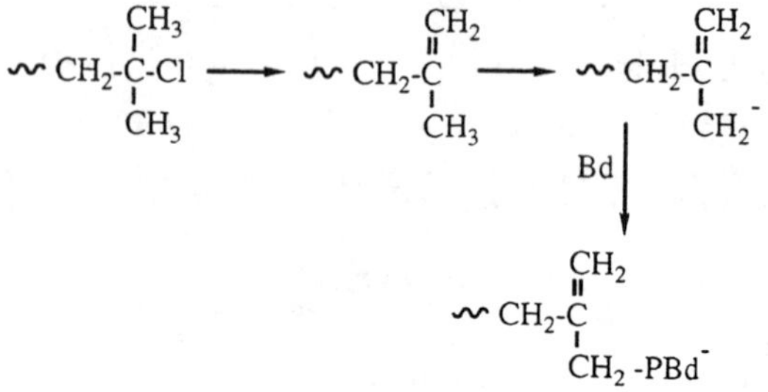

$$\sim CH_2\text{-}\underset{CH_3}{\overset{CH_3}{C}}\text{-}Cl \longrightarrow \sim CH_2\text{-}\underset{CH_3}{\overset{CH_2}{\overset{\|}{C}}} \longrightarrow \sim CH_2\text{-}\underset{CH_2^-}{\overset{CH_2}{\overset{\|}{C}}} \xrightarrow{Bd} \sim CH_2\text{-}\underset{CH_2\text{-}PBd^-}{\overset{CH_2}{\overset{\|}{C}}}$$

**Scheme 51**

Another cation to anion transformation is associated with benzyl anion-telechelic polyisobutylenes. Poly(methylmethacrylate-b-isobutylene-b-methylmethacrylate) (PMMA-PIB-PMMA) triblock copolymer has been readily prepared [92] by the lithiation of p-tolyl polyisobutylene, followed by the quantitative addition of 1,1-diphenyl ethylene to the polyisobutylene dianions which induce the living polymerization of methyl methacrylate. The block copolymer exhibits two glass transition points resulting from the two different blocks. This demonstrates that the product exhibits a two phase morphology (**Scheme 52**).

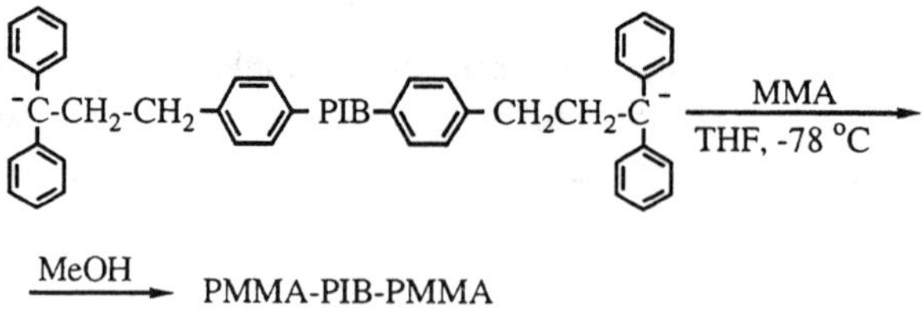

$$^-C(Ph)_2\text{-}CH_2\text{-}CH_2\text{-}C_6H_4\text{-}PIB\text{-}C_6H_4\text{-}CH_2CH_2\text{-}C(Ph)_2^- \xrightarrow[THF,\ -78\ ^\circ C]{MMA} \xrightarrow{MeOH} \text{PMMA-PIB-PMMA}$$

**Scheme 52**

Wilczek, Mishra and Kennedy [93] had used the concept of transfomation reactions using macroinitiator techniques to synthesize PIB-siloxane di- and multi-block copolymers. The macroinitiator was prepared by converting the hydroxyl terminated telechelic PIB to the alkoxide with use of butyl lithium. The macroinitiator was used for the anionic ring opening polymerization of hexamethylcyclotrisiloxane ($D_3$) monomer (**Scheme 53**). Subsequent deactivation of living $D_3$ polymerization charges with $Me_3SiCl$ yielded the target triblock whereas stoichiometric amounts of $Me_3SiCl_2$ gave the multiblock copolymer. Initiation (i.e., "blocking")of $D_3$ polymerization from Li phenolate-ended PIB was also studied. However, the phenolate exhibited a much

$HOCH_2 \sim PIB \sim CH_2OH$

$\downarrow$ n - BuLi

$LiOCH_2 \sim PIB \sim CH_2OLi$

$\downarrow D_3$

$LiOSi(Me)_2 - (O - Si(Me)_2 -)_m OCH_2 \sim PIB \sim CH_2O(- Si(Me)_2 - O)_m - Si(Me)_2OLi$

Excess $(CH_3)_3SiCl$( - LiCl) $\downarrow$ ; $(CH_3)_2SiCl_2$ $\rightarrow$

$(CH_3)_3SiOSi(Me)_2 - (O - Si(Me)_2 -)_m OCH_2 \sim PIB \sim CH_2O(- Si(Me)_2 - O)_m - Si(Me)_2OSi(CH_3)_3$

PDMS - b - PIB - b - PDMS

$[ - OSi(Me)_2 - (- O - Si(Me)_2 -)_m OCH_2 \sim PIB \sim CH_2O( - Si(Me)_2 - O)_m Si(Me)_2 - O - Si(Me)_2 - ]_n$

- ( - PDMS - b - PIB - b - PDMS - ) n

**Scheme 53**

lower initiating efficiency from $D_3$ polymerization than the alcoholate under the same conditions. The triblock copolymer exhibited satisfactory hydrolytic resistance under certain conditions. The blocking efficiency was about 100 %.

Hatada and co-workers [94] also prepared triblock copolymer composed of PMMA and PIB. Using $\alpha,\omega$-dihydroxy PIBs as starting material, they prepared the isobutyrate derivative followed by deprotonation to prepare the $\alpha,\omega$-dilithiated PIB. Low temperature polymerization of MMA afforded the desired triblock copolymer (**Scheme 54**). This method is similar to that described by Kennedy et al. [92], except the starting material is being hydroxyl terminated polymer instead of chlorine terminated one.

Nylon 6-PIB diblock, triblock and tristar radial block copolymers were synthesized [95] by the anionic polymerization of $\epsilon$-caprolactone using the functional PIB (macroactivator/macroinitiator). PIB macroactivators/macroinitiators having terminal isocyanate groups were prepared by the reaction of telechelic hydroxy functional PIB with hexamethylene diisocyanate, toluene diisocyanate, N-chlorocarbonyl diisocyanate, and oxalyl chloride (**Scheme 55**). Prepolymers with molecular weights from 6000 to 38,000 were employed. Derivatization with N-chlorocarbonyl diisocyanate and subsequent anionic caprolactam polymerization gave highest yield and blocking efficiencies. Transmission electron microscopy of di-, tri- and radial blocks showed increasing segregation and orientation of rubbery/crystalline domains. The blocking efficiencies were usually 40 ~ 80 %, however, very low values (15 ~ 20 %) were obtained with hexamethylene diisocyanate.

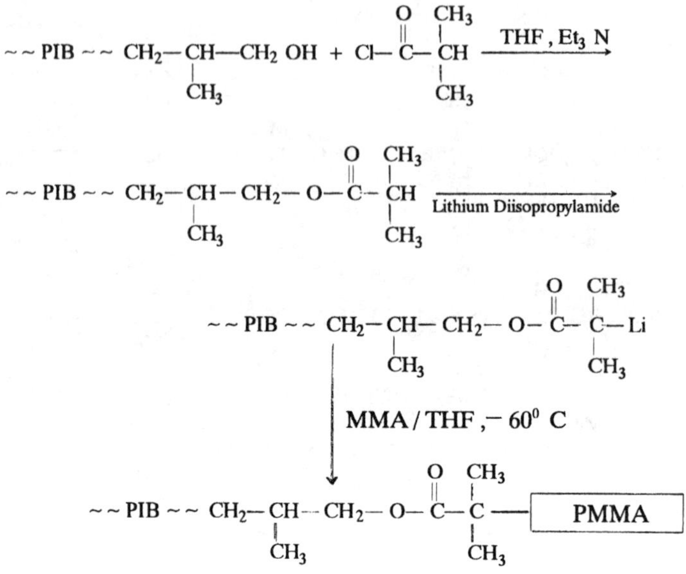

$$\sim\sim PIB \sim\sim CH_2{-}\underset{\displaystyle CH_3}{\underset{|}{CH}}{-}CH_2\,OH + Cl{-}\overset{\displaystyle O}{\overset{||}{C}}{-}\underset{\displaystyle CH_3}{\underset{|}{\overset{\displaystyle CH_3}{\overset{|}{CH}}}} \xrightarrow{THF,\ Et_3\ N}$$

$$\sim\sim PIB \sim\sim CH_2{-}\underset{\displaystyle CH_3}{\underset{|}{CH}}{-}CH_2{-}O{-}\overset{\displaystyle O}{\overset{||}{C}}{-}\underset{\displaystyle CH_3}{\underset{|}{\overset{\displaystyle CH_3}{\overset{|}{CH}}}} \xrightarrow{\text{Lithium Diisopropylamide}}$$

$$\sim\sim PIB \sim\sim CH_2{-}\underset{\displaystyle CH_3}{\underset{|}{CH}}{-}CH_2{-}O{-}\overset{\displaystyle O}{\overset{||}{C}}{-}\underset{\displaystyle CH_3}{\underset{|}{\overset{\displaystyle CH_3}{\overset{|}{C}}}}{-}Li$$

$$\downarrow \text{MMA / THF}, -60^0\ C$$

$$\sim\sim PIB \sim\sim CH_2{-}\underset{\displaystyle CH_3}{\underset{|}{CH}}{-}CH_2{-}O{-}\overset{\displaystyle O}{\overset{||}{C}}{-}\underset{\displaystyle CH_3}{\underset{|}{\overset{\displaystyle CH_3}{\overset{|}{C}}}}{-}\boxed{\text{PMMA}}$$

**Scheme 54**

$$\sim\sim\sim\sim CH_2 - \underset{\displaystyle CH_3}{\underset{|}{\overset{\displaystyle CH_2\,OH}{\overset{|}{CH}}}} \xrightarrow{Cl\,CO - NCO} \sim\sim\sim\sim CH_2 - \underset{\displaystyle CH_3}{\underset{|}{\overset{\displaystyle CH_2\,OCO - NCO}{\overset{|}{CH}}}} \xrightarrow{\text{NH - CO (cyclic)}}$$

$$\sim\sim\sim\sim CH_2 - \underset{\displaystyle CH_3}{\underset{|}{\overset{\displaystyle CH_2\,OCO - NHCO - (CH_2)_3 - NH ---}{\overset{|}{CH}}}}$$

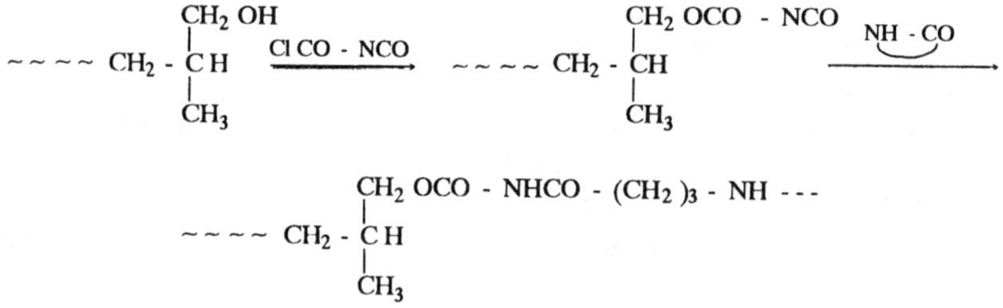

**Scheme 55**

Harris et al. [96] have prepared PIB macroinitiator having tetrabutylammonium carboxylate groups from copolymers of isobutylene with methylstyrenes by lithiation of the benzylic sites in the copolymer, subsequent carboxylation with carbon dioxide followed by the conversion of carboxyl groups to tetrabutylammonium salts. Anionic polymerization of pivolactone by the macroinitiator produced poly[(isobutylene-co-methylstyrene)-graft-pivolactone)] of varied composition (**Scheme 56**). With this method isobutylene polymers have been converted to thermoplastic elastomers by the growing of pivolactone segments at initiating sites positioned along the polyisobutylene backbone. The effeciency of the reaction was not stated.

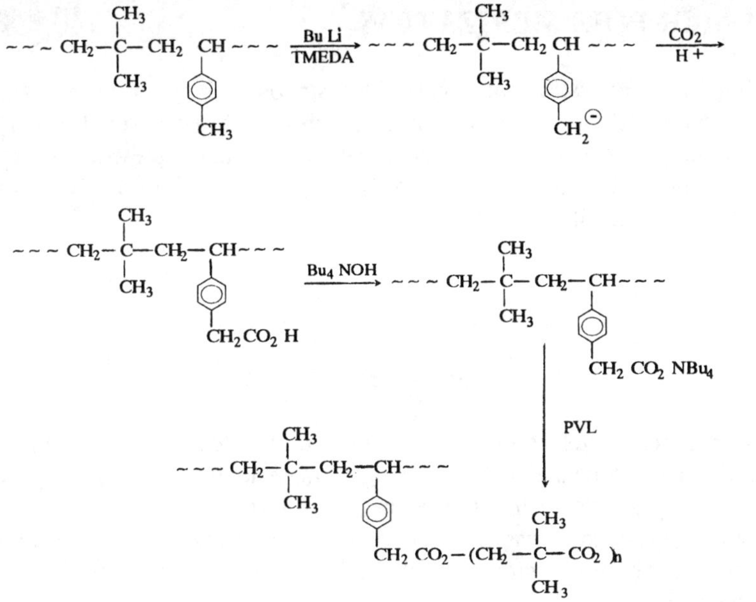

$$\sim\sim CH_2-C(CH_3)_2-CH_2-CH(C_6H_4CH_3)\sim\sim \xrightarrow[TMEDA]{Bu\,Li} \sim\sim CH_2-C(CH_3)_2-CH_2-CH(C_6H_4CH_2^{\ominus})\sim\sim \xrightarrow[H+]{CO_2}$$

$$\sim\sim CH_2-C(CH_3)_2-CH_2-CH(C_6H_4CH_2CO_2H)\sim\sim \xrightarrow{Bu_4\,NOH} \sim\sim CH_2-C(CH_3)_2-CH_2-CH(C_6H_4CH_2CO_2NBu_4)\sim\sim$$

$$\xrightarrow{PVL} \sim\sim CH_2-C(CH_3)_2-CH_2-CH(C_6H_4CH_2CO_2-(CH_2-C(CH_3)_2-CO_2)_n)\sim\sim$$

**Scheme 56**

Verma et al. [97] have demonstrated the preparation of poly(butyl vinyl ether)-poly($\epsilon$-caprolactone) block copolymers from hydroxyl terminated vinyl ether oligomers by initiating ring opening polymerization of $\epsilon$-caprolactone in presence of stannous octoate as a catalyst **(Scheme 57)**. The relative good agreement between the molecular weight values obtained using different methods, together with agreement with the targeted molecular weights based on monomer to initiator feed ratio, further confirms both the living nature of the process involved, as well as the quantitative aspects of the functionalization reactions.

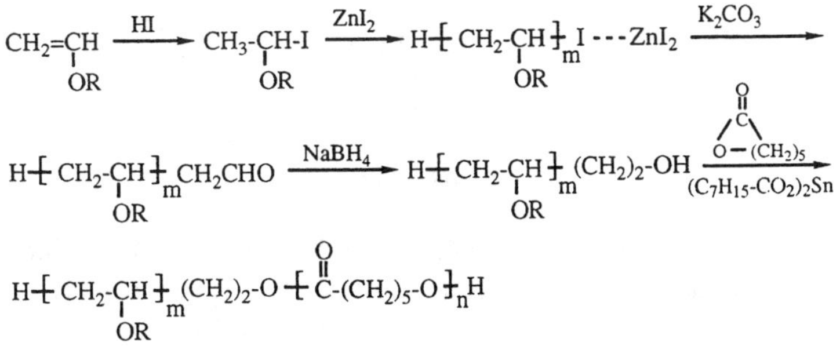

$$CH_2=CH(OR) \xrightarrow{HI} CH_3-CH(OR)-I \xrightarrow{ZnI_2} H\!+\!CH_2-CH(OR)\!+_m I\cdots ZnI_2 \xrightarrow{K_2CO_3}$$

$$H\!+\!CH_2-CH(OR)\!+_m CH_2CHO \xrightarrow{NaBH_4} H\!+\!CH_2-CH(OR)\!+_m (CH_2)_2-OH \xrightarrow[(C_7H_{15}-CO_2)_2Sn]{\epsilon\text{-caprolactone } (O-(CH_2)_5-C=O)}$$

$$H\!+\!CH_2-CH(OR)\!+_m (CH_2)_2-O\!+\!C(=O)-(CH_2)_5-O\!+_n H$$

**Scheme 57**

## 8. TRANSFORMATIONS INVOLVING ACTIVATED MONOMER POLYMERIZATION

Polymerization by the activated monomer mechanism opened up promising pathways for transformation reactions. Recently, Penczek, Kubisa and co-workers [98-101] have found that cationic polymerization of oxiranes may proceed in the presence of hxdroxyl containing compounds by the activated monomer (AM) mechanism which may be represented generally as follows (**Scheme 58**):

$$H\text{-}\overset{+}{O}\!\triangleleft + HO\text{-}R \longrightarrow H\text{-}O\frown O\text{-}R + H^+$$

**Scheme 58**

Thus, the propagation involves the reaction of a protonated (activated) monomer molecule with a nucleophilic site in the neutral growing macromolecule. More recently, Yagci et al. [102] adapted this procedure to transformation reactions. Thus, 4,4´-azobis(4-cyanopentanol) was used in AM polymerization of epichlorohydrin (ECH) to produce polymers with azo linkage in the main chain according to the following **Scheme 59**.

$$H^+ + O\!\triangleleft\text{-}CH_2Cl \longrightarrow H\text{-}\overset{+}{O}\!\triangleleft\text{-}CH_2Cl$$

$$H\text{-}\overset{+}{O}\!\triangleleft\text{-}CH_2Cl + HO\text{-}CH_2\text{-}CH_2\text{-}CH_2\text{-}C(CH_3)(CN)\text{-}N{=}N\text{-}C(CH_3)(CN)\text{-}CH_2\text{-}CH_2\text{-}CH_2\text{-}OH \longrightarrow$$

$$HO\text{-}CH_2\text{-}CH(CH_2Cl)\text{-}O\text{-}CH_2\text{-}CH_2\text{-}CH_2\text{-}C(CH_3)(CN)\text{-}N{=}N\text{-}C(CH_3)(CN)\text{-}CH_2\text{-}CH_2\text{-}CH_2\text{-}O\text{-}CH(CH_2Cl)\text{-}CH_2\text{-}OH + 2H^+$$

$$\Big\downarrow\ O\!\triangleleft\text{-}CH_2Cl$$

$$H\text{-}[\text{-}O\text{-}CH_2\text{-}CH(CH_2Cl)\text{-}]_n\text{-}O\text{-}CH_2\text{-}CH_2\text{-}CH_2\text{-}C(CH_3)(CN)\text{-}N{=}N\text{-}C(CH_3)(CN)\text{-}CH_2\text{-}CH_2\text{-}CH_2\text{-}O\text{-}[\text{-}CH(CH_2Cl)\text{-}CH_2\text{-}O\text{-}]_n\text{-}H$$

**Scheme 59**

Polymerization was conducted under typical conditions, i.e. by slow addition of ECH to the solution of initiator containing catalyst. Reaction was considerably slower than in the presence of simple diols (e.g. ethylene glycol) and only 28 % conversion was obtained under conditions sufficient to reach complete conversion in the polymerization initiated by ethylene glycol. Poly(epichlorohydrin) PECH prepared this way, was used in the polymerization of styrene to produce block copolymer (**Scheme 60**) .

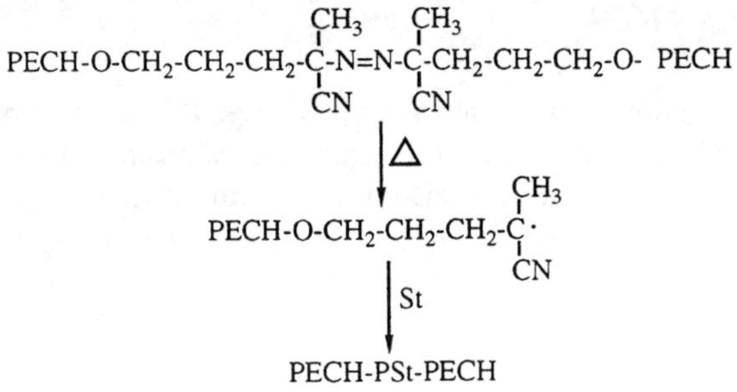

$$\text{PECH-O-CH}_2\text{-CH}_2\text{-CH}_2\text{-}\underset{\text{CN}}{\overset{\text{CH}_3}{\text{C}}}\text{-N=N-}\underset{\text{CN}}{\overset{\text{CH}_3}{\text{C}}}\text{-CH}_2\text{-CH}_2\text{-CH}_2\text{-O- PECH}$$

$$\downarrow \Delta$$

$$\text{PECH-O-CH}_2\text{-CH}_2\text{-CH}_2\text{-}\underset{\text{CN}}{\overset{\text{CH}_3}{\text{C}}}\cdot$$

$$\downarrow \text{St}$$

$$\text{PECH-PSt-PECH}$$

**Scheme 60**

The reverse mode of this transformation is also accessible. For instance, Stewart [20] has shown that by using a free radically prepared hydroxyl terminated polybutadiene as a macro-catalyst in AM polymerization of ECH, block copolymers are produced with a quantitative yield (**Scheme 61**). AM polymerization offers other transformation possibilities, since hydroxy telechelics may be prepared by various polymerization modes.

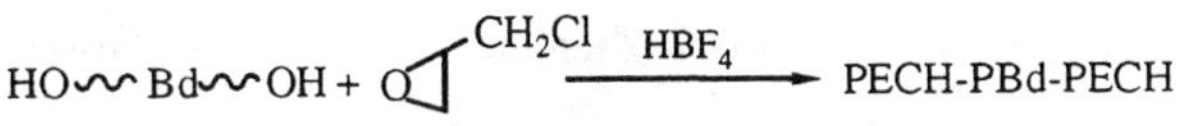

$$\text{HO}\sim\text{Bd}\sim\text{OH} + \text{(epichlorohydrin, CH}_2\text{Cl)} \xrightarrow{\text{HBF}_4} \text{PECH-PBd-PECH}$$

**Scheme 61**

# 9. ANIONIC TO METATHESIS TRANSFORMATION

The transformation route for the synthesis of block copolymers of styrene and cyclopentene has been reported by Amass et al. [103]. In this work, the authors main aim was to elucidate the mechanism of initiation of the conventional metathesis catalyst. Using polystyryl lithium as co initiator together with tungsten hexachloride, the formation of block copolymer suggested that the active center was essentially covalent and the propagating center might have bridge structure as shown in **Scheme 62**.

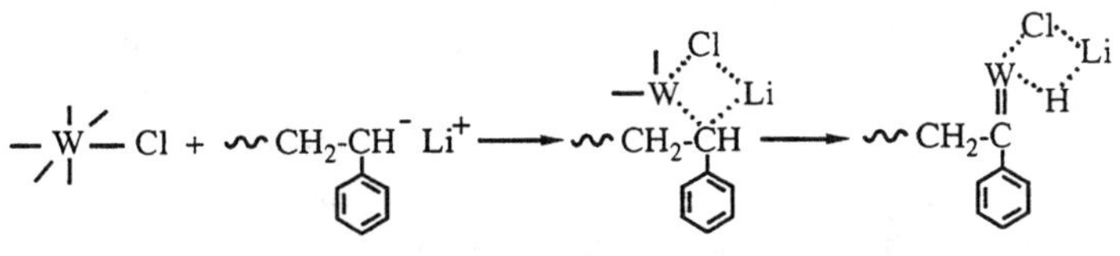

$$-\text{W}-\text{Cl} + \sim\text{CH}_2\text{-}\underset{\text{Ph}}{\text{CH}^-}\ \text{Li}^+ \longrightarrow \sim\text{CH}_2\text{-}\underset{\text{Ph}}{\text{CH}}\cdots\text{W(Cl)}\cdots\text{Li} \longrightarrow \sim\text{CH}_2\text{-}\underset{\text{Ph}}{\text{C}}=\text{W}\cdots\text{Cl}\cdots\text{Li}\cdots\text{H}$$

**Scheme 62**

# 10. TRANSFORMATIONS INVOLVING ZIEGLER-NATTA POLYMERIZATION

Ziegler-Natta polymerization is well known and involves a two stage process. In the first stage an aluminum alkyl such as trialkyl aluminum is reacted with $TiCl_4$ in order to give the active $\beta$-$TiCl_3$. Alkyl radicals, also produced in this reaction, terminated by coupling and give inert products.Subsequent alkylation of $\beta$-$TiCl_3$ then occurs to generate the titanium species capable of initiating polymerization of olefines such as ethylene **(Scheme 63)**.

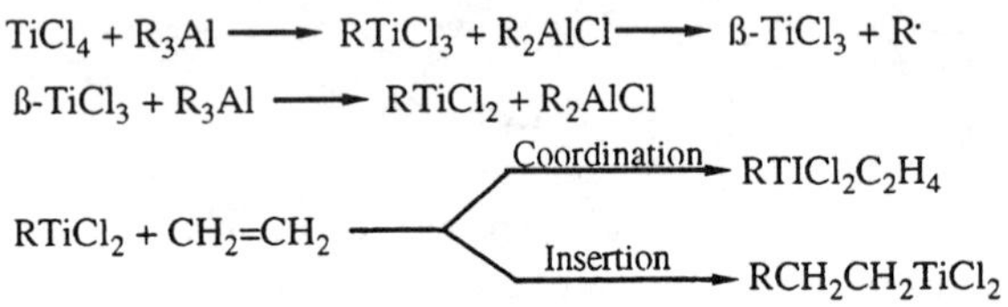

$$TiCl_4 + R_3Al \longrightarrow RTiCl_3 + R_2AlCl \longrightarrow \beta\text{-}TiCl_3 + R^{\cdot}$$

$$\beta\text{-}TiCl_3 + R_3Al \longrightarrow RTiCl_2 + R_2AlCl$$

$$RTiCl_2 + CH_2{=}CH_2 \xrightarrow{\text{Coordination}} RTICl_2C_2H_4$$

$$RTiCl_2 + CH_2{=}CH_2 \xrightarrow{\text{Insertion}} RCH_2CH_2TiCl_2$$

**Scheme 63**

As the polymerization results the incorporation of alkyl ligand in the final product, polymeric aluminum compounds may conveniently be employed instead of small molecule analogous to effect an anion to Ziegler-Natta transformation process yielding novel block copolymers [104-107].Indeed, polymeric aluminum compounds have been successfully synthesized by successive alkylation of aluminum halide with living anionic polymers as shown in **Scheme 64**. The method has been used to prepare block copolymers of styrene with ethylene and acetylene. Notably, extremely low transformation efficiencies were obtained. However, purified block copolymer of the later has potential use as electroactive polymer [108].

$$AlCl_3 + 3\sim\!\!\sim M^-Li^+ \rightleftharpoons (\sim\!\!\sim M)_3Al$$

**Scheme 64**

Another useful active site transformation from Ziegler-Natta to radical has been achieved by Agouri and co-workers [109]. Alkenic-vinylic type block copolymers containing crystalline and amorphous sequences were prepared by using this transformation. Ziegler-Natta polymerization was induced with diethylzinc as the transition complex and peroxy groups were incorporated to crystalline polymer through oxidation of the carbon zinc bond. Thermolysis of the macroinitiator formed this way resulted in the generation of a pair of radicals. Both radicals are capable of initiating the polymerization of vinyl monomers such as methyl methacrylate. By this means, a mixture of homopolymer (initiated by ethoxy radicals) and block copolymer is produced (**Scheme 65**).

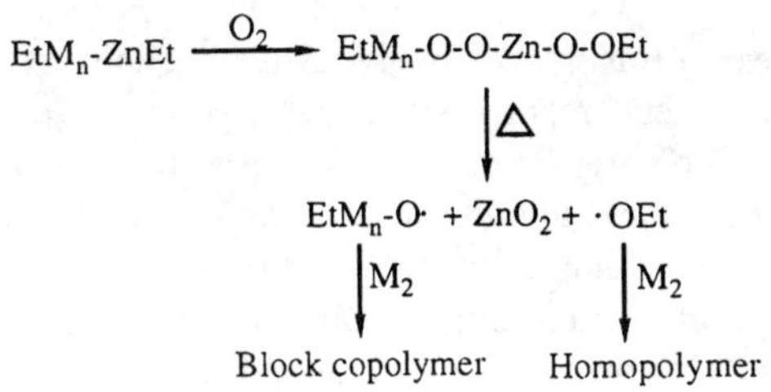

$$EtM_n\text{-}ZnEt \xrightarrow{O_2} EtM_n\text{-}O\text{-}O\text{-}Zn\text{-}O\text{-}OEt \xrightarrow{\Delta} EtM_n\text{-}O\cdot + ZnO_2 + \cdot OEt$$

$EtM_n\text{-}O\cdot \xrightarrow{M_2}$ Block copolymer; $\cdot OEt \xrightarrow{M_2}$ Homopolymer

**Scheme 65**

Doi et al. [110, 111] have demonstrated the living coordination polymerization of propylene by a soluble Z-N catalyst composed of vanadium acetylacetonate and diethylaluminum chloride. The living polypropylene chain end can be transformed to iodide by treatment with a solution of iodine in toluene. In combination with $AgClO_4$ the polymer containing iodide generates a carbocation which initiates cationic polymerization of THF at 0° C [110] as shown in **Scheme 66**. However, the blocking efficiency was negligible at $20^0$C .

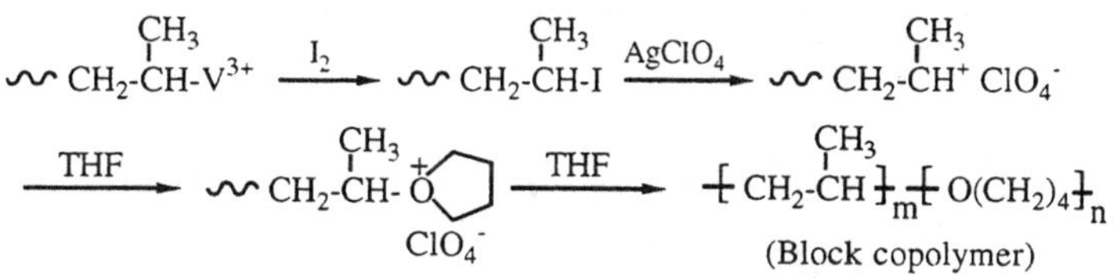

$$\sim CH_2\text{-}CH(CH_3)\text{-}V^{3+} \xrightarrow{I_2} \sim CH_2\text{-}CH(CH_3)\text{-}I \xrightarrow{AgClO_4} \sim CH_2\text{-}CH(CH_3)^+ ClO_4^-$$

$$\xrightarrow{THF} \sim CH_2\text{-}CH(CH_3)\text{-}\overset{+}{O}(CH_2)_4\ ClO_4^- \xrightarrow{THF} \left[CH_2\text{-}CH(CH_3)\right]_m\left[O(CH_2)_4\right]_n$$

(Block copolymer)

**Scheme 66**

## 11. TRANSFORMATIONS INVOLVING GROUP TRANSFER POLYMERIZATION

The mechanism of group transfer polymerization [112, 113] implies that the initiating silyl keten acetal moiety remains attached to the propagating chain.

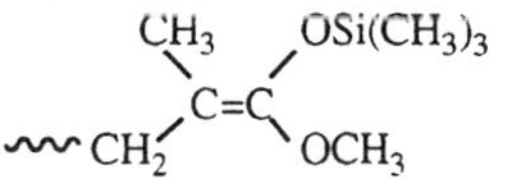

$$\sim CH_2(CH_3)C{=}C(OSi(CH_3)_3)(OCH_3)$$

Eastmond et al. [114] have reported the transformation of group transfer to radical polymerization using the metal carbonyl photoinitiating system. They synthesized block copolymers of methyl methacrylate with other monomers starting with methyl methacrylate macroinitiator of narrow molecular weight distribution (obtained by means of group transfer polymerization). Termination with bromine leads to the desired bromine containing terminal group in the dead polymer. This polymer was employed to prepare block copolymers with styrene. The product was essentially a triblock copolymer of the structure ABA with $M_w$ of the polystyrene blocks ranging from 6 x $10^4$ to 3 x $10^5$ g/mole.

The other approaches of this theme involve switching active centers to group transfer polymerization. One recent report is the publication of preliminary findings on a cationic to group transfer polymerization by Ruth et al. [115]. Thus, $\alpha,\omega$-dihydroxyPIB prepared by living cationic polymerization was converted to the diester which is sequentially treated with lithium diisopropylamide (LDA) and chlorotrimethylsilane to afford the silylketen acetal macroinitiator. Reaction of the polymeric silylketen acetal with methacrylate in the presence of tetrabutyl ammonium dibenzoate (TBABB) produced a block copolymer (**Scheme 67**). Although methylmethacrylate conversion was quantitative, significant amount of unreacted polyisobutylene was reported by the authors, so contamination problems exist. However, this work confirms the potential use of the transformation reactions for the preparation of thermoplastic elastomers.

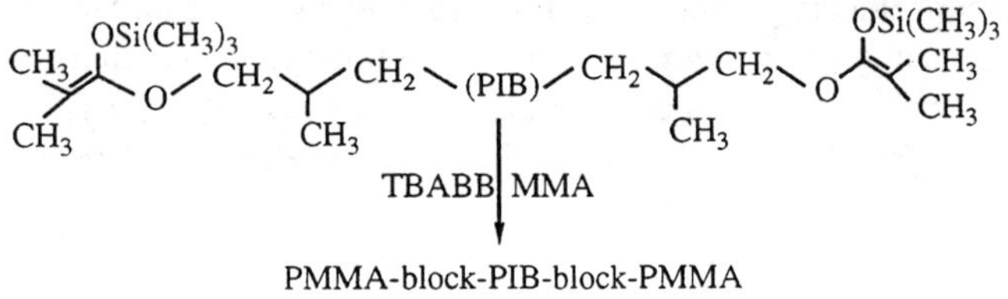

$OSi(CH_3)_3$ $CH_3$ $CH_3$ O $CH_2$ $CH_2$ $CH_3$ (PIB) $CH_2$ $CH_2$ $CH_3$ O $OSi(CH_3)_3$ $CH_3$ $CH_3$

TBABB | MMA

PMMA-block-PIB-block-PMMA

**Scheme 67**

Attempts to extend this technology to the other living polymerization were also, reported. For this purpose, using $HI/ZnI_2$ initiating systems, monofunctional silylketen acetal terminated polyalkylvinylether with precise functionality was prepared. Since the keten acetal structure is also end group in the group transfer polymerization of acrylates, block copolymers of related monomers can be obtained [23] by quenching the living polymerization of alkyl vinyl ether with polyacrylates prepared by group transfer polymerization **(Scheme 68)**.

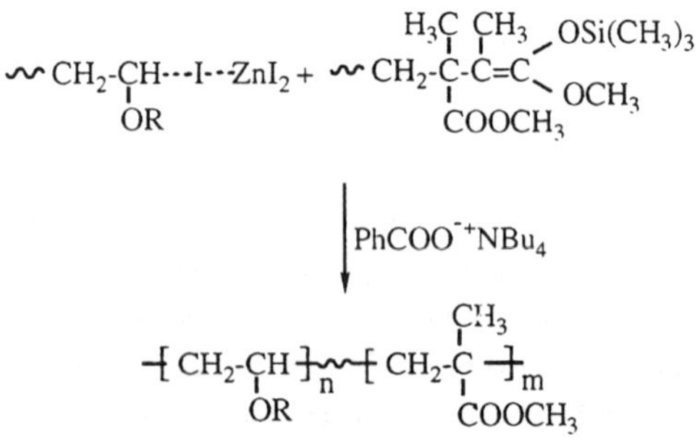

$\sim CH_2\text{-}CH(OR)\cdots I\cdots ZnI_2 + \sim CH_2\text{-}C(CH_3)(COOCH_3)\text{-}C(CH_3){=}C(OSi(CH_3)_3)(OCH_3)$

$PhCOO^{-+}NBu_4$

$\text{-[}CH_2\text{-}CH(OR)\text{]}_n \sim \text{[}CH_2\text{-}C(CH_3)(COOCH_3)\text{]}_m$

**Scheme 68**

It is obvious that macroinitiator concept involving indirect transformation may also be useful to prepare these materials, requiring only the initiation of group transfer polymerization using polymeric silyl keten acetal and a co-catalyst in a manner described before [115]. In this connection it is interesting to note the work of Jenkins et al. [116] who basically applied this approach to prepare styrene-*graft*-methylmethacrylate copolymers. In this case, polystyrene, containing a minor proportion of residues with functional groups, served as macroinitiator for the subsequent group transfer polymerization.

## 12. REACTION EFFICIENCY

It may be worthwhile to briefly discuss the efficiency of the transformation process. The question may arise that what is "transformation efficiency," "blocking/grafting efficiency" and how they are related? Most of the cited papers assume that the blocking or grafting efficiency is a measure of the transformation efficiency. Blocking/grafting efficiency may be defined as the moles of copolymer formed per mole of prepolymer. It may be helpful to consider the following scenario as depicted in **Scheme 69**.

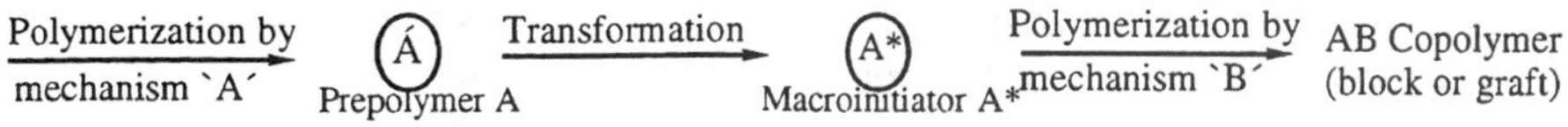

Polymerization by mechanism \`A´ → (Á) Prepolymer A —Transformation→ (A*) Macroinitiator A* —Polymerization by mechanism \`B´→ AB Copolymer (block or graft)

**Scheme 69**

In **Scheme 69**, the true blocking/grafting efficiency is the moles of copolymer (AB) formed per mole of macroinitiator A*. However, it may be difficult to determine the number of moles of macroinitiator A* and the efficiency of the transformation from prepolymer A to the macroinitiator A*. The <u>overall blocking/grafting efficiency</u> i.e., "transformation efficiency" may be defined as the moles of copolymer (AB) formed per mole of prepolymer A prepared by the first mode of polymerization.

## 13. CONCLUSION

Although we have tried our best to include all possible references related to the transformation process, some references may have been left out unintentionally. We hope this chapter will provide a practical guide for the scientist in designing new or modified macromolecular structures. The application of the materials obtained by transformation reactions are of interest. As shown in this chapter, a full range of possible block and graft copolymers built from monomers with different chemical nature (block copolymers of hydrophilic and hydrophobic segments, of liquid crystalline and crystalline segments and thermoplastic elastomers) are accessible through transformation reactions.

There are various polymeric products on the market basically consisting of random copolymers containing two or three different monomer units. Novel polymeric materials may be synthesized from new monomers, i.e. from monomers different in chemical nature from the classical ones. However since classical monomers are in most cases easily and cheaply available, their utilization as the basis of novel materials deserves intense consideration from practical viewpoints. This explains why transformation reactions have recently attracted a revitalized world wide interest.

## REFERENCES

1. M. Miyamoto, M. Sawamoto, T. Higashimura, *Macromolecules*, **17**, 265 (1984)
2. T. Enoki, M. Sawamoto and T. Higashimura, *J. Polym. Sci., Polym. Chem. Ed.*, **24**, 2261 (1986)
3. M. Sawamoto, T. Enoki and T. Higashimura, *Macromolecules*, **20**, 1 (1987)
4. M. Sawamoto, T. Enoki and T. Higashimura, *Polym. Bull.*, **19**, 359 (1988)
5. M. Sawamoto, T. Enoki and T. Higashimura, *Polym. Bull.*, **16**, 117 (1986)
6. M. Sawamoto, C. Okamoto and T. Higashimura, *Macromolecules*, **20**, 2693 (1987)
7. For a review, see: T. Higashimura, S. Aoshima and M. Sawamoto, *Makromol. Chem., Macromol. Symp.*, **13/14** 457 (1988)
8. M. K. Mishra and A. Osman, US Pat., 4,908,421(1990)
9. M. K. Mishra and A. Osman, US Pat., 4,943,616(1990)
10. J. P. Kennedy and M. K. Mishra, US Pat., 4,929,683(1990)
11. J. P. Kennedy and R. Faust, US Pat., 4,910,321(1990)
12. J. P. Kennedy and B. Ivan, Designed Polymers by Carbocationic Macromolecular Engineering: Theory and Practice (Hanser Publisher, Munich) 1991
13. a) D. H. Richards. *Dev. Polym.*, **1**, 1 (1979); b) *Br. Polym. J.*, **12**, 89 (1980)
14. a) R. P. Quirk, D. J. Kinning, and L. J. Fetters, in Comprehensive Polymer Sci., V 7 (G. Allen and J. C. Bevington, eds.), Pergamon, Oxford, 1989, Chapter 1.; b) P. Rempp, E. Franta and J.-E. Herz, *Adv. Polym. Sci.*, **86**, 145 (1988) c) F. Schue, in Comprehensive Polymer Sci., vol 7 (G. Allen and J. C. Bevington, eds.), Pergamon, Oxford, 1989, Chapter 10
15. M. J. M. Abadie and D. H. Richards, *Int. Chim.*, **208**, 135 (1980)
16. D. H. Richards, *ACS. Symp. Ser.*, **166**, 343 (1981)
17. D. H. Richards, *ACS. Symp. Ser.*, **286**, 87 (1985)
18. M. J. M. Abadie and D. Qurahmoune, *Br. Polym. J.*, **19**, 247 (1987)
19. F. Schue in "Encyclopedia of Polymer Science and Engineering," J. Wiley & Sons, New York, Vol 11 p 359 (1985)
20. M. J. Stewart in "New Methods of Polymer Synthesis," (Ed. J. R. Ebdon) Blackie & Sons, New York, p. 107 (1991)
21. D. H. Richards, S. B. Kingston and Th. Souel, *Polymer*, **19**, 68, 806 (1978)
22. A. Verma, A. Nielsen, J. E. McGrath and J. S. Riffle, *Polym. Bull.*, **23**, 563 (1990)
23. R. Reeb, Y. Vinchon, G. Riess, J. M. Catula and J. Brossas, *J. Bull. Soc. Chim.* France, **11-12**, 2717 (1975)
24. J. M. Catula, G. Riess and J. Brossas, *Makromol. Chem*, **178**, 1249 (1979)
25. Sumitomo Chemical co., Osaka, Japan, Ger. offen., 200, 9066 (1971)

26. Y. Vinchon, R. Reeb and G. Riess, *Eur. Polym. J.*, **12**, 317 (1976)
27. G. Riess and R. Reeb, *Polym. Prep.* **21**, 55 (1980)
28. Z. Nicolova-Nankova, F. Palacin, F. Raviola and G. Kiess, *Eur. Polym. J.*, **11**, 301 (1975)
29. B. Hazer, I. Cakmak, S. Kucukyavuz and T. Nugay, *Eur. Polym. J.*, **28**, 1295 (1992)
30. M. J. M. Abadie, D. Qurahmoune and H. Mendjel, *Eur. Polym. J.*, **26**, 515 (1990)
31. Y. Yagci, Y. Z. Menceloglu, B. Baysal and A. Gungo, *Polym. Bull.*, **21**, 259 (1989)
32. a) A. D. jenkins, M. F. Lapper and R. C. Srivastava, *Eur. Polym. J.*, **7**, 289 (1971) b) N. C. Billingham, L. M. Boxall and A. D. Jenkins, *Eur. Polym. J.*, **8**, 1045 (1972) c) N. C. Billingham, L. M. Boxall, A. D. Jenkins and P. D. Leeds, *Eur. Polym. J.*, **8**, 1045 (1972)
33. M. J. M. Abadie, F. Schue, T. Souel and D. H. Richards, *Polymer,* **22,** 1076 (1981)
34. T. Souel, F. Schue, M. J. M. Abadie and D. H. Richards, *Polymer*, **14**, 1292 (1977)
35. A. V. Cunliffe, G. F. Hayes and D. H. Richards, *J. Polym. Sci., Polym. Lett. Ed.* **14**, 483 (1976)
36. W. E. Lindbell, D. M. Service, I. Soutar and D. H. Richards, *Brit. Polym. J.* ,**19**, 255 (1987)
37. G. P. Vlasov, G. D. Rudkovskaya, and L. A. Ovsyannikova, *Makromol. Chem.*, **183**, 2635 (1982)
38. Q. Ren, H. Zhang, X. Zhang, B. Huang, *J. Polym. Sci., Polym. Chem.,* **31**(4), 847(1993)
39 a) G. C. Eastmond, K.J.Parr and J. Woo, *Polymer,* **29**(5), 950 (1988)
40. G. C. Eastmond and J. Woo, *Polymer*, **31**(2), 358 (1990)
41. F. Burgess, M.S. Thesis, University of St. Andrews (1976)
42. Y. Yagci, *Polymer Commun.*, **27**, 25 (1985)
43. I. E. Serhatli, G. Hizal and Y. Yagci, *Turk. J. Chem.*, **16**, 91 (1992)
44. G. Hizal and Y. Yagci, *Polymer*, **30,** 722 (1989)
45. A. Akar, A. C. Aydogan, N. Talinli, Y. Yagci, *Polym. Bull.*, **15**, 293 (1986)
46. G. Galli, E. Chiellini, Y. Yagci, E. T. Serhatll, M. Laus, M. C. Bignozzi and A. S. Angeloni, *Makromol. Chem. Rapid. Commun.*, **14**, 185 (1993)
47. F. D. Haese, E. J. Goethals, Y. Tezuka and K. Imai, *Makromol. Chem., Rapid. Commun.*, **7**, 165 (1986)
48. S. Handa and N. Tsubotawa, *Polym. Preprints*, **39**, 184 (1990)
49. S. Handa and N. Tsubotawa, *Polym. Preprints*, **40**, 2567 (1991)
50. B. Hazer, *Eur. Polym. J.*, **26**, 1167 (1990)
51. G. C. Eastmond and J. Woo, Proc. 28th. I. U. P. A. C. Makromol. Symp. 194 (1982)
52. O. Nuyken, H. Kroner and S. Aechtner, *Polym. Bull.*, **24**, 513 (1990)
53. M. Sawamoto, *Prog. Polym. Sci.,* **16**, 111 (1991)

54. O. Nuyken, H. Kroner and S. Aechtner, *Makromol. Chem. Rapid Commun.*, **9**, 671 (1988)
55. Y. Yagci and A. Ledwith, *J. Polym. Sci., Polym. Chem. Ed.*, **26**, 1911 (1988)
56. C. Nicora, G. Borsini and L. Ratti, *J. Polym. Sci., Polym. Lett. Ed.*, **4,** 151 (1966)
57. Y. Yagci, *Polym. Commun.*, **27**, 21 (1986)
58. G. Hizal, H. Tasdemir and Y. Yagci, *Polymer,* **7,** 31 (1990)
59. A. Onen, Y. Yagci, *J. Macromol. Sci. Chem.*, A **27,** 755 (1990)
60. A. Onen, Y. Yagci, *Angew. Makromol. Chem.* **181**, 191 (1990)
61. Y. Yagci, A. Onen, *J. Macromol. Sci. Chem.*, A **28**, 129 (1991)
62. Y. Yagci, W. Schnabel, *Makromol. Chem., Macromol. Symp.*, **13 - 14,** 161 (1988)
63. Y. Yagci, W. Schnabel, *Makromol. Chem., Macromol. Symp.* **60**, 133 (1992).
64. Y. Yagci, A. Onen, *Macromolecules*, **24,** 4620 (1991)
65. M. Okada, H. Sumitomo, and T. Kakezawa, *Makromol. Chem.*, **162,** 285 (1972)
66. G. Cai and D. Yan, *Makromol. Chem.*, **188**, 1005 (1987)
67. S. Kobayashi, Y. Shimano, and T. Saegusa, *Polym. J.*,**23**, 1307 (1991)
68. S. Ishikawa, K. Ishizu, and T. Fukutomi, *Polym. Bull.*, **16**, 223 (1986)
69. H. Uno and T. Endo, *Chem. Lett.,* 1869 (1986)
70. M. Tanaka, A. Mori, Y. Imanishi, and C. H. Bamford, *Int. J. Biol. Macromol.*, **7**, 173 (1985)
71. S. Higuchi, T. Mozawa, M. Maeda, and S. Inoue, *Macromolecules*, **19**, 2263 (1986)
72. J. E. Mulvaney and R. O. Ottaviani, *J. Polym. Sci., Polym. Chem. Ed.,* **20**(7),1941(1982)
73. P. Rempp, P. Lutz, P. Masson, P. Chaumont and E. Franta, *Makromol. Chem., Suppl.* **13**, 47 (1985)
74. K. Ito, K. Hashimura, S. Itsuno and E. Yamada, *Macromolecules*, **24**, 3977 (1991)
75. P. Jannasch and B. Wesslen, *J. Polym. Sci., Polym. Chem.*, V**31**, 1519 (1993)
76. F. J. Burgess, A. V. Cunliffe, J. R. MacCallum and D. H. Richards, *Polymer,* **18**, 718 (1977)
77. F. J. Burgess, A. V. Cunliffe, J. R. MacCallum and D. H. Richards, *Polymer,* **18**, 726 (1977)
78. F. J. Burgess, A. V. Cunliffe, J. V. Dawkins and D. H. Richards, *Polymer*, **18**, 733 (1977)
79. F. J. Burgess and D. H. Richards, *Polymer,* **17**, 1020 (1976)
80. K. Muhlbach and R. C. Schulz, *Makromol. Chem.*, **190**, 2551 (1989)
81. A. Mumir and E. J. Goethals, *J. Polym. Sci., Polym. Chem. Ed.,* **19**, 1985 (1981)
82. E. H. Sacht and E. J. Goethals, *Makromol. Chem.,* **167,** 155 (1973)
83. K. Muhlbach and R. C. Schulz, *Makromol. Chem.*, **189**, 1267 (1988)
84. H. Kazama, Y. Tezuka, K. Imai, and E. J. Goethals, *Makromol. Chem.*, **189,** 985 (1988)
85. Y. Morishima, T. Tanaka, and S. Nozakura, *Polym. Bull.*, **5**, 19 (1981)
86. M. Miyamoto, Y. Sano, T. Saegusa, and S. Kobayashi, *Eur. Polym. J.*,**19**, 955 (1983)
87. C. I. Simonescue, I. Rabia, and Z. Crisan, *Polym. Bull.*,**7**, 217 (1982)
88. M. Miyamoto, K. Aoi, H. Yamanaka, and T. Saegusa, *Polym. J.*,**24**, 405 (1992)

89. M. J. M. Abadie, F. Schue, Th. Souel and D. H. Richards, *Polymer,* **23**, 445 (1982)
90. P. Cohen, M. J. M. Abadie, F. Schue and D. H. Richards, *Polymer*, **23**, 1105 (1982)
91. S. Nemes and J. P. Kennedy, *J. Macromol. Sci. Chem.,* A **28** , 311-328 (1991)
92. J. P. Kennedy and J. Price, *Polym. Mat. Sci, Eng.*, **64**, 40 (1991)
93. L. Wilczek, M. K. Mishra and J. P. Kennedy, *J. Macromol. Sci. Chem.* A**24**(9), 1033 (1987)
94. T. Kitayama, T. Nishiura and K. Hatada, Polym. Bull., **26,** 513 (1991)
95. R. H. Wondraczek and J. P. Kennedy, *J. Polym. Sci. Polym. Chem. Ed.*, **20,** 173 (1982)
96. J. F. Harris, Jr. and W. H. Sharkey, *Macromolecules,* **19,** 2903 (1986)
97. A. Verma, M. Glagola, A. Prasad, H. Maranda and J. S. Riffle, *Makromol. Chem., Macromol Symp.,* **54/55,** 95 (1992)
98. S. Penczek, P. Cubisa, K. Matysasewski and R. Szymanski in Cationic Polymerization and Related Process, Ed. Goethals, Academic Press, New York, 1984 p. 140.
99. K. Brzezinska, R. Szymanski, P. Kubisa and S. Penzek, *Makromol. Chem., Rapid Commun.*, **7**, 1 (1986)
100. S. Penczek, P. Kubisa and R. Szymanski, *Makromol. Chem., Macromol. Symp.,* **3**, 203 (1986)
101. P. Kubisa, Makromol. Chem., *Makromol. Symp.*, **13/14**, 203 (1988)
102. Y. Yagci, I. E. Serhatli, P. Kubisa and T. Biedron, *Macromolecules*, **26**, 2397 (1993)
103. A. J. Amass and D. Eregory, *Br. Polym. J.*, **19,** 263 (1987)
104. M. Aldissi, *J. Chem. Soc. Chem. Commun.,* **20**, 1347 (1984)
105. M. Aldissi, A. K. Bishop, *Polymer,* **26**, 622 (1985)
106. M. Aldissi, *Synthetic Metals,* **13**, 87 (1986)
107. J. A. Stowell, A. J. Amass, M. S. Beevers and T. R. Farren, *Polymer,* **30**, 195 (1989)
108. M. Aldissi, M. Hou and J. Farrell, *Synthetic Metals*, **17,** 229 (1984).
109. E. Agnuri, G. Favier, R. Laputte, T. Philardeau and J. Rideau, "Symposium on Block and Graft Copolymerization: Preprints; GFP, Mulhouse (Fra.), 1972, p55.
110. Y. Doi, Y. Watanabe, S. Ueki, and K. Soga, *Makromol. Chem., Rapid Commun.*, **4,** 533 (1983)
111. T. Keii and Y. Doi, *Adv. Polym. Sci.*, **73/74**, 201(1986)
112. O. W. Webster, W. R. Hertler, D. Y. Sogah, W. B. Farnham and T. V. Rajanbabu., *J. Am. Chem. Soc.,* **105**, 5706 (1983)
113. For a Recent Review See. G. C. Eastmond, O. W. Webster in " New Methods of Polymer Synthesis", (Ed. J. R. Ebdon) Blackie & Sons, New York, p. 22 (1991)
114. G. C. Eastmond and J. Grigor, *Makromol. Chem., Rapid Commun,* **7**, 375 (1986)
115. W. G. Ruth, C. G. Moore, W. J. Brittain, J. Si and J. P. Kennedy, *Polymer Preprints*, ACS, **34** (1), 479 (1993)
116. A. D. Jenkins, E. Tsartolia and, D. R. M. Walton, *Makromol. Chem.*, **191**, 2511 (1990)

# Macroiniferters Chapter 11

## APPLICATION OF THERMAL INIFERTERS IN FREE RADICAL POLYMERIZATION: A NEW TREND IN MACROMOLECULAR ENGINEERING

**C.P. Reghunadhan Nair*, P. Chaumont and G. Clouet†**

Institut Charles Sadron (C.R.M. - E.A.H.P.)
6 rue Boussingault
67083 Strasbourg Cedex
FRANCE

---

* Author for correspondance, permanent address, Polymers & Special Chemical Division, Vikram Sarabhai Space Centre, Trivandrum - 695022, India

† *Deceased on April 26, 1993. The first two authors dedicate this article to his fond memory. His enthusiasm and passion for research has been responsible for most of the results described here.*



Macromolecular Design: Concepts and Practice (Macromonomers, Macroinitiators, Macroiniferters, Macroinimers, Macroinifers, and Macroiniters) Pages 431 - 470

## 1. INTRODUCTION

Polymers play a vital role in shaping modern man's life as sophisticated and comfortable as it is today. The boom in science and technology in the recent decades owes its share to the development in the area of polymer science as well. Moreover the technological challenges for the near future in all areas of science and technology ranging from domestic appliances to microelectronics, from biomedical science, to telecommunication, and finally to aerospace forcast a parallel boom in polymer technology to cope up with the exponential growth in those fields, since, polymers do play a crucial role in one phase or other of their development.

Depending on the particular need, the polymeric materials need to satisfy certain requirements, in terms of their processability, durability, resistance to environment, cost-effectiveness, mechanical performance,etc. A thorough understanding of the structure-property relationships has helped to design the polymers tailored for specific applications.

For many industrial applications, a single polymer system either thermoplastic or thermoset as such often do not serve the particular purpose. In many occasions, they need to be physically or sometimes structurally modified by a process often called functionalization. The functional groups range from simple organic groups like OH, $NH_2$-SH, Si-H, etc., to polymeric groups in which case the resultant polymers are either block or graft copolymers. Needless to say that for majority of industrial applications, these functionalized polymers are the material of choice rather than the corresponding homopolymers. Typically, terminal functional polymers (or telechelics) are required as chain extenders in propellant formulations, as components in adhesives, as reactive toughening agents in matrix resins in composites, and in adhesive formulations, as components in paints, etc.

As it is often the practice, a compromise in physico-mechanical properties is achieved by way of using blends of polymers. Blends usually pose the problems of incompatibility and rapid aging due to the diffusion of the components into their microdomains. This can be obviated to a great extent if the components of the blends constitute an integral part of one and the same molecule by way of realizing block or graft copolymers.

Block copolymers also find unique applications arising from the incompatibility of the blocks. They are used as thermoplastic elastomers [1], as adhesives and sealants with high strength and recovery characteristics [2], and in polymer blends, as surface modifiers for fillers and fibres for improving their surface characteristics [3-5]. They are used as binders for laminates [6,7] and as constituents in coating materials [8-11]. They are particulary suitable in approaching problems of interfacial adhesion and modulus profile design in fibre reinforced composites.

Their surfactant characteristics project block copolymers as useful dispersants, emulsifiers, wetting agents, foam stabilizers, solubilizers, fluocculants, demulsifiers, etc., in many industrial and pharmaceutical preparations [12,13].

An important use lies in the ability of block copolymers to stabilize latices in dispersion polymerization in both aqueous and non aqueous media [14-17].They are increasingly used to form membranes for applications like gas and liquid separation, ultrafiltration, desalination, pollution control, etc. [18,19].Biomedical area needs block copolymers to form surgical dressings, blood transfusion tubes, and blood storage containers [20] and in controlled drug delivery therapy [21]. Other applications include viscosity improvers for base lubricating oils, in cosmetics, etc.

Telechelics and block copolymers are conventionally realized through the living anionic polymerization technique [22-26].This technique however is loosing importance owing to the stringent operation conditions and its limitation to a few vinyl monomers. Same is the case with cationic technique too, although some break-throughs have been claimed by Kennedy et al by using their "inifer" technique [27,28].This technique addresses to the usually encountered problems of transfer in cationic polymerization.

Free radical polymerization is the technique of choice of the day, since it obviates many problems associated with other techniques. The method has the added advantages of the ease of processability and good kinetic control. Practically all vinyl monomers can be homo- or copolymerized by this method.

It is of relevance in the context of the objective of this article to cite here some of the common methods for producing telechelics and block copolymers by free radical route.

## 2. TELECHELICS BY FREE RADICAL ROUTE

### 2.1 Via Functional Initiators

Use of functional initiators of azo or peroxy type can lead to telechelics if the termination takes place by way of recombination. Bamford et al. [29] have prepared carboxy terminated PS and PMMA using azobiscyanovaleric acid. But the functionality in the case of PMMA is never two, because a large fraction of polymer radicals terminates by disproportionation. For several monomers, even if the termination occurs exclusively by recombination, inadvertent transfer like that to monomer, polymer, impurities, etc. renders the functionality less than two. A few other functional initiators have also been reported towards this end [30]. Among the functional initiators can be found peresters [31,32], bi-initiators like azo-peroxy, and azo-carbonates [33], and peroxy-peroxy [34,35].

Hydroxy terminated polybutadiene, an important telechelic polymer used in propellant formulation is made by the polymerization of butadiene initiated by $H_2O_2$ [36].

Tetraphenyl ethane derivatives constitute another class of interesting initiators with potentiality for telechelic formation [37-39].

## 2.2 Via Functional Chain Transfer Agents

Functional chain transfer agents of the disulfide type can give rise to telechelics if the transfer constant is good ( i.e. if it is close to the unity ). Since they warrant simulltaneous use of an initiator, the functionality will again depend upon the ratio of the initiator to the chain transfer agent.

$$M + [F]-S-S-[F] \xrightarrow{AIBN} -\underset{CN}{\overset{|}{C}}-[M]_n-S-[F] + [F]-S-[M]_n-S-[F]$$

( M : monomer )

Considering the Mayo's equation in case of chain transfer :

$$\frac{1}{DP} = \frac{1}{DP_0} + C_{tr}\frac{[T]}{[M]} = \frac{k_t R_p}{k_p^2[M]^2} + C_{tr}\frac{[T]}{[M]}$$

Where $C_{tr}$ is the chain transfer constant and [T], the concentration of the transfer agent. Rearranging ,

$$\frac{DP\ k_t R_p}{k_p^2[M]^2} + C_{tr}\frac{DP[T]}{[M]} = p_{nf} + p_f = 1$$

The first term on the left hand side represents the probability $P_{nf}$ of finding the non-functional chain ends and the second one $P_f$ that of finding the functional ones resulting from the chain transfer. It is evident that the possibility of all chain terminals getting functionalized is only a limiting case [42]. For obtaining oligomers it is essential to increase either $R_p$ (by the way of more initiators or by raising polymerization temperature) or [T], or use compounds having very high $C_{tr}$ values. If $C_{tr}$ is far from unity, this will result in a very broad molecular weight distribution even for small conversions [43]. The first method creates more non functional chain ends. Increasing [T] beyond a certain limit in some cases causes tremendous retardation, which tends to stagnate the molecular weight. In fact $R_p$ decreases inversely with the concentration of chain transfer agent [44] when degradative transfer is operative.

Aliphatic disulfides are poor transfer agents [45,46]. Various aromatic disulfides bearing functional groups like carboxy [47], isocyanate [48], amine, hydroxy, cyano, cyanate [49,50],etc. have been reported as possessing good transfer properties. Among the aliphatic disulfides, activated disulfides of the xanthogen disulfide and thiuram disulfides type only possess meaningful transfer properties. Xanthogen disulfides are very good chain transfer agents with $C_{tr}$ values greater than unity and have been used as chain modifier in emulsion polymerization of dienes [50a]. Functional xanthogen disulfides have been commercially used to generate telechelics for use in paints and coatings [51,52]. Thiuram disulfides will be dealt with in detail in this chapter.

It is also possible to use halogenated telogens to derive halogen telechelics, warranting subsequent chemical transformation which is not always easy [53]. Boutevin et al. [54, 55] have carried out immense studies on the use of halogenated telogens and two exellent reviews have been devoted to telomerisation.

The functionality problem can be solved by way of matched chain transfer i.e. the use of initiators and chain transfer agents both carrying same functional groups. Very recently Otsu et al. [56] have made some studies on these two-component **ini**tiator-trans**fer** (inifer) system. This way, a combination of 4,4'-azobiscyanovaleric acid (initiator) with dithiodiglycolic acid (transfer agent) gave carboxyl telechelic polystyrene. But the system was not efficient for monomers like MMA, methyl acrylate, and butadiene.

## 3. FREE RADICAL ROUTES FOR BLOCK COPOLYMERS

The free radical route for realizing block copolymers have been centred mainly on using polymeric azo or peroxy initiators, or using multifunctional initiator species having both azo and peroxy groups on the same molecule but with different decomposition temperatures. In the latter case one monomer is polymerized at the decomposition temperature of one type of initiator followed by the addition of the second monomer which is polymerized at the decomposition temperature of the second initiator [57,57a]. Another example cites utilization of polyazoester in the partial polymerization of one monomer resulting in polymeric species possessing reactive initiator groups which can be used in the polymerization of a second monomer [58]. Initiators born on polymer backbone give directly the block copolymers [59-60]. This technique however works only for monomers which terminate by recombination. The use of azo-benzoin type multi-initators demands a combination of thermal and photochemical polymerization [61].

The weak C-C bond in pinacolates have been exploited by Crivello et al. [62,63] for realizing silicone-vinyl block copolymers [62,63]and the technique extended by Santos et al. [64] for preparing polyoxyethylene containing vinyl blocks. The use of polymers with end groups that can be converted into radicals by organometallic compounds is another effective approach towards this goal [65]. Manganese and rhenium carbonyls in presence of suitable halide additives have been successfully used in the preparation of block copolymers including a series of completely alternating block copolymers [66,67].Other organometallics like aluminium alkyls have been used as chain transfer agents to derive Al-functional polymers, which in presence of a peroxide act as macro-initiators to polymerize a second monomer [68]. Utilization of polymeric chain transfer agent for preparation of block copolymers has been reported by Hallensleben [68a].

Although the above-discussed techniques permit realisation of telechelic and block copolymers, some of which are otherwise difficult to prepare, they result generally in polymers which are polydisperse in molecular weight, composition and structure and are often accompanied by a significant percentage of homopolymers. The reason is that, in these reactions, one exploits either only the initiation or the chain transfer properties of the species, which invariably leads to imperfect functionalization. This problem can be addressed to, if it is possible to envisage unique systems possessing both initiation and transfer properties. Thiuram disulides ideally fall into this category of compounds. In this chapter the discussion will be confined to the application of thiuram disulfides as thermal iniferters ( **ini**tiaton, trans**fer**, **ter**mination ) and their application in the free radical design of end-functional polymers and tri- or multiblock copolymers.

## 4. INIFERTERS

The free radical initiation, chain transfer and chain termination ( iniferter ) properties of thiuram disulfides had been identified by Tobolsky and Ferrington [69,70]and their kinetics of polymerization was studied by Staudner and Beniska [71,72].Otsu et al. [73] observed that the polymers formed in the presence of tetraalkyl thiuram disulfides are end-capped at both terminals with dithiocarbamate moieties. It was further observed that the dithiocarbamate moiety can undergo reversible photodissociation giving rise to an alkyl radical and a dithiocarbamate radical ( **Scheme 1** ).

$$\text{R-S-}\overset{\text{S}}{\overset{\|}{\text{C}}}\text{-N}\!\begin{smallmatrix}\text{R'}\\ \\ \text{R'}\end{smallmatrix} \;\overset{h\nu}{\rightleftharpoons}\; \text{R}^{\bullet} + {}^{\bullet}\text{S-}\overset{\text{S}}{\overset{\|}{\text{C}}}\text{-N}\!\begin{smallmatrix}\text{R'}\\ \\ \text{R'}\end{smallmatrix}$$

**Scheme 1**

This property of dithiocarbamates was exploited for the photoiniter ( initiation, termination ) method towards accomplishing multiblock, graft and star copolymers [74-76]. It was assumed that the dithiocarbamate radical does not initiate polymerization. Ideal reaction schemes leading to graft and block copolymers are depicted in **Scheme 2**.

Otsu's claim has been contested by Turner [77] and Sigwalt [78] who observed that the dithiocarbamate radical can really initiate vinyl polylmerization leading to the formation of homopolymers also in the system.

On a similar line Otsu et al. [79] showed that compounds of phenylazo-triphenyl methane type can act as thermal initer where the triphenylmethyl radical can act as a counter inactive radical; exhibiting a kind of living radical nature. One of the two polymer chain-ends is capped with the trityl group which undergoes reversible dissociation on heating. The molecular weight of the polymers obtained increased with conversion, claimed as an evidence for the living nature.

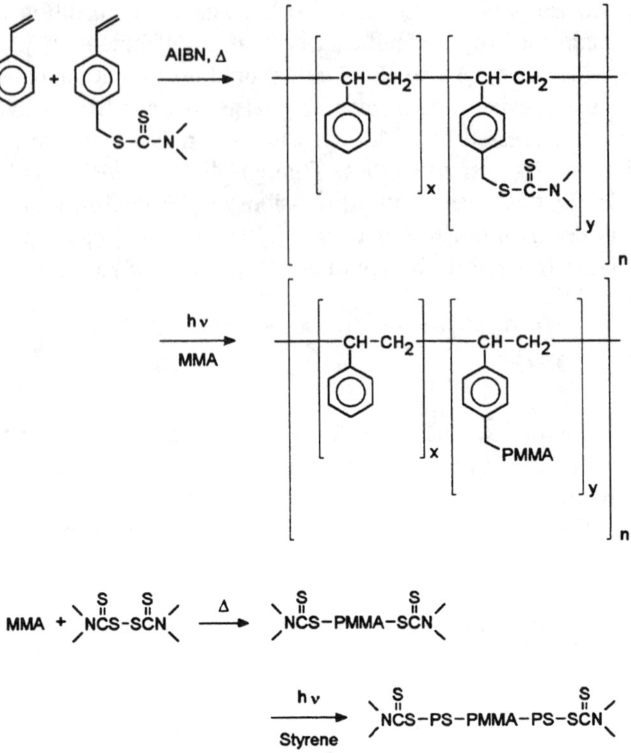

Scheme 2

Alternatively, the trityl end-capped polymer can serve as macro initer for the polymerization of a second monomer yielding block copolymers (**Scheme 3** ).

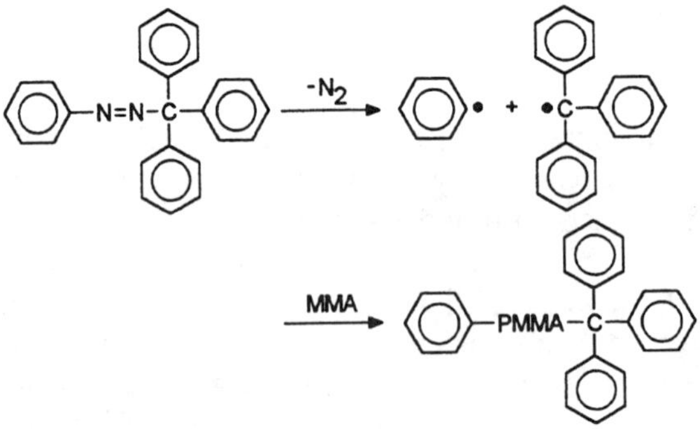

Scheme 3

Niwa et al. [79a] identified that xanthogen disulfides can act as photo iniferters in the polymerization of MMA, Styrene, etc., to give rise to xanthate end-capped polymers which can subsequently undergo photodissociation generating polymer radicals. This phenomenone opens the way for block coplymer synthesis in a manner similar to Otsu's method ( **Scheme 4** ).

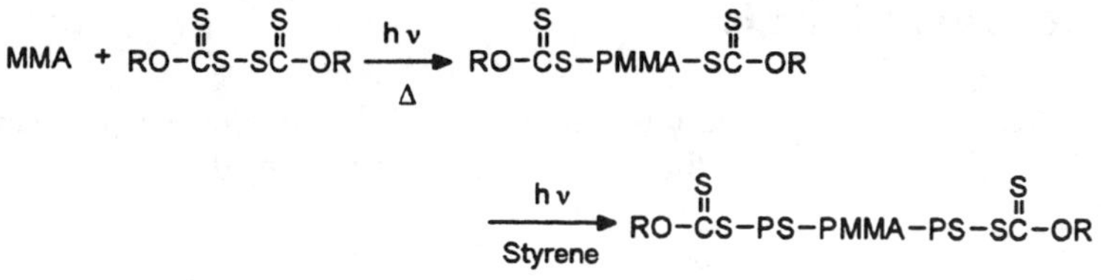

MMA + RO–C(=S)–S–S–C(=S)–OR —(hν, Δ)→ RO–C(=S)–S–PMMA–S–C(=S)–OR

—(hν, Styrene)→ RO–C(=S)–S–PS–PMMA–PS–S–C(=S)–OR

**Scheme 4**

# 5. THIURAM DISULFIDE AS THERMAL INIFERTERS

Otsu's observation of the iniferter characteristics of thiuram disulfides has been exploited to derive telechelics, tri- and multiblock copolymers. It is based on the concept of locating the required function on the alkyl group of the thiuram disulfide and using it in the thermal polymerization ( **Scheme 5** ).

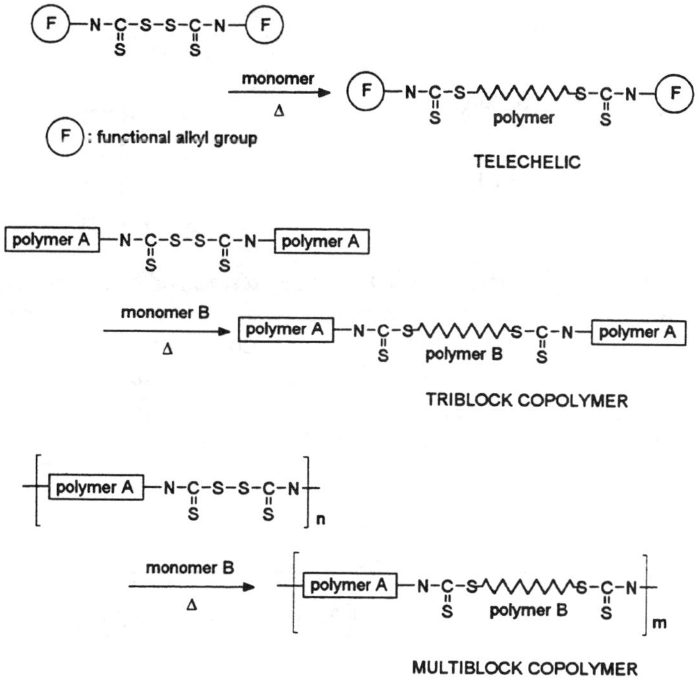

(F)–N–C(=S)–S–S–C(=S)–N–(F)

—(monomer, Δ)→ (F)–N–C(=S)–S~~~~polymer~~~~S–C(=S)–N–(F)

(F) : functional alkyl group

TELECHELIC

[polymer A]–N–C(=S)–S–S–C(=S)–N–[polymer A]

—(monomer B, Δ)→ [polymer A]–N–C(=S)–S~~~~polymer B~~~~S–C(=S)–N–[polymer A]

TRIBLOCK COPOLYMER

–[ [polymer A]–N–C(=S)–S–S–C(=S)–N ]$_n$–

—(monomer B, Δ)→ –[ [polymer A]–N–C(=S)–S~~~~polymer B~~~~S–C(=S)–N ]$_m$–

MULTIBLOCK COPOLYMER

**Scheme 5**

On the other hand, if one designs a polymeric thiuram disulfide and use it in thermal polymerization, it should lead to the formation of mid-chain functional polymer. When the functions are polymeric sequences, it results in block copolymers.

## 5.1 Polymerization Kinetics

As iniferter, the thiuram disulfide follows an unusual kinetics in the polymerization. The various reactions controlling the kinetics can be represented as follows :

| | | |
|---|---|---|
| $R-S-S-R \xrightarrow{\Delta, k_d} 2\,RS^{\bullet}$ | (a) | thermal dissociation |
| $RS^{\bullet} + M \xrightarrow{k_i} RSM^{\bullet}$ | (b) | primary radical initiation |
| $RSM^{\bullet} + M \xrightarrow{k_p} RSP^{\bullet}$ | (c) | propagation |
| $2\,RSP^{\bullet} \xrightarrow{k_t} RSPPSR + RSP$ | (d) | termination |
| $RSP^{\bullet} + RS^{\bullet} \xrightarrow{k_{tpr}} RSPSR$ | (e) | primary radical termination |
| $RSP^{\bullet} + RSSR \xrightarrow{k_{tr}} RSPSR + RS^{\bullet}$ | (f) | transfer to initiator |
| $RS^{\bullet} + RSSR \xleftrightarrow{k_{tr}} RSSR + RS^{\bullet}$ | (g) | primary radical transfer |
| $RS^{\bullet} + RS^{\bullet} \xrightarrow{k_{tmt}} \text{secondary products}$ | (h) | mutal combination of primary radicals |

The terms $k_p$, $k_t$ etc., are the rate constants for the respective reactions. The kinetic scheme resembles that of a free radical copolymerization between monomer M and the thiuram disulfide (RSSR) considered as a second monomer and a kinetic expression for a conventional radical copolymerization can be suitably applied to this system when the termination is chemically controlled ( as it is for low conversions ). This gives an expression for the rate of polymerization [80] as :

$$R_p = \frac{\dfrac{k_p}{k_{tr}}[M]R_i^{1/2}}{\left[\dfrac{k_t}{k_{tr}^2}[M]^2 + \dfrac{2k_{tpr}}{k_{tr}k_i}[M][RSSR] + \dfrac{k_{tmt}}{k_i^2}[RSSR]^2\right]^{1/2}}$$

which can be rearranged as :

$$R_p^2 = \frac{A^2[M]^2[RSSR]}{1+2B\frac{[RSSR]}{[M]}+C\frac{[RSSR]^2}{[M]^2}}$$

where A, B and C are complex constants :

$$A = \frac{k_p}{k_t^{1/2}}(2fk_d)^{1/2}$$

$$B = \frac{k_{tpr}k_{tr}}{k_i k_t} \qquad C = \frac{k_{tmt}}{k_i}\left(\frac{k_{tr}}{k_t}\right)^2$$

The complex form of the equation predicts that the rate of polymerization is no longer linear to the square root of the initiator concentration, but should show an initial increase, reaching a maximum and then a decrease. The decrease in $R_p$ is a consequence of the significance of the various retardation reactions in the kinetics and consequently the corresponding terms $k_{tpr}/k_i$, $k_{tmt}/k_i$, etc. appear in the kinetic expression for $R_p$. In other words the rate of initiation $k_i$ is inferior to $k_p$. In earlier studies it was shown that addition of the dithiocarbamate radical to MMA needs a higher $E_A$ ($\approx$ 21 kJ/mole) than for propagation ($\approx$ 15 kJ/mole) [81]. Participation of primary radicals in polymer radical termination and in mutual termination reactions decreases the effective rate of polymerization. Hence the deviation of the kinetics from the normal initiation behaviour would depend upon the extent to which $k_i$ and $k_{tpr}$ are influenced by the nature of the monomer. **Fig. 1** shows the typical kinetic behaviour in case of the iniferter for MMA, isoprene, and styrene. The $R_p$ - [I] ( I : iniferter ) curve follows the path predicted by the complex equation. The extent of deviation from normal kinetics is in the order MMA > isoprene > styrene which is the decreasing order of monomer reactivity towards a given radical. In the concentration range studied the constant C is not significant in the expression for $R_p$ for styrene because styrene is a very reactive monomer, easily scavanging the dithiocarbamate radicals in the system ( increase in $k_i$ ). The electron-rich polystyryl radical has lesser propensity to couple with a soft S-born radical. On the other hand MMA is electron-deficient as is the case for the resultant polymer radical. Hence initiation is slow but termination with primary radical is fast. Isoprene exhibits an intermediate behaviour strictly consistant with the anticipated monomer reactivity. In the case of acrylates the extreme situation arises so that thiuram disulfide is an inhibitor for acrylates and acrylonitriles. In other words the system works only for electron rich monomers like MMA, styrene, butadiene, vinyl acetate, vinyl ethers, etc.

## 5.2 Molecular Weight

It can be shown that molecular weight of the polymers formed depends upon the iniferter concentration as per the following relation [80] :

$$\frac{1}{DP}=\frac{k_tR_p}{k_p^2[M]^2}\left[1-\frac{\alpha}{2}+B\frac{[RSSR]}{[M]}\right]+C_{tr}\frac{[RSSR]}{[M]}+C_m$$

where $C_{tr}$ is chain transfer constant to the iniferter, $C_m$ that to the monomer, and $\alpha$ is the extent of termination by recombination. The modified equation takes into account the correction factor for the reduced initiation and primary radical termination. Let $\Delta$ be the quantity :

$$\Delta=\frac{1}{DP}-\frac{k_tR_p}{k_p^2[M]^2}\left[1-\frac{\alpha}{2}+B\frac{[RSSR]}{[M]}\right]$$

$C_{tr}$ is determined by plotting $\Delta$ versus [RSSR]/[M]. **Fig. 2** shows some typical plots and colinearity of the experimental points clearly demonstrates validity of the equation.

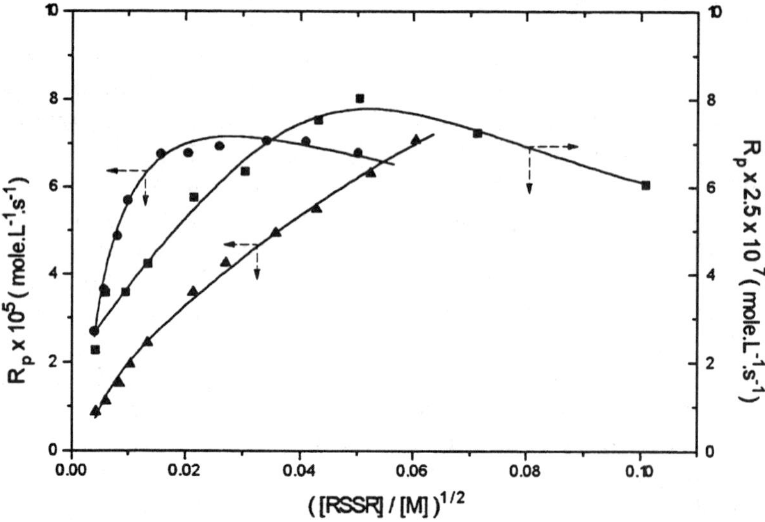


*Fig. 1. Dependance of polymerization rate on Iniferter concentration for various monomers* [ Ref. 82, 83 ] ( ● ): MMA, DPTD, ( ▲ ) : styrene, DPTD, ( ■ ) : isoprene, BPTD.

**Fig. 3** shows the theoretical variation of both the polymerization rate and the molecular weight with iniferter concentration for a representative case of polymerization of MMA at 70° C. It can be seen that molecular weight has a tendency to stagnate at high iniferter concentration. This is a consequence of the retardation observed at high iniferter loading. Thus very low molecular weight telechelics can be obtained only at the cost of the polymer yield since $R_p$ also decreases tremendously.

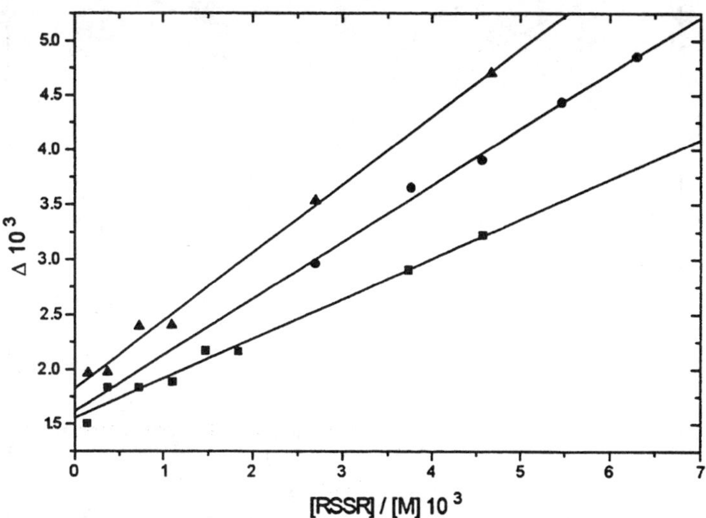


*Fig 2. Determination of chain transfer constant by modified Mayo's equation in the case of a PDMS based macroiniferter* [Ref. 95] (■) 80° C, (●) 90° C, (▲) 100° C.

In the above discussion it was pointed out that the kinetic parameters do depend on the monomer type. For a given monomer, the molecular environement of the iniferter can have an effect. **Table 1** lists the various kinetic parameters determined for MMA, isoprene, and styrene using different types of simple iniferters. **Table 2** gives the values for different macro and polymeric thiuram disulfides.

*Table 1 Kinetic parameters for various iniferters*

| iniferter (a) | monomer | temp °C | $2fk_d 10^6$ $s^{-1}$ | B | C $10^{-6}$ | $C_{tr}$ | $E_d$ kJ/mole | $E_{tr}$ kJ/mole |
|---|---|---|---|---|---|---|---|---|
| DHTD | MMA | 60 | 0.53 | 2770 | 3.2 | 0.23 | | |
| | | 70 | 4.41 | 3090 | 5.0 | 0.27 | 130 | 37.6 |
| ( ref 81 ) | | 85 | 25.90 | 4430 | 5.6 | 0.35 | | |
| | | 95 | 48.44 | 4650 | 6.2 | 0.46 | | |
| DPTD | MMA | 70 | 2.11 | 2950 | 1.1 | 0.41 | - | - |
| ( ref 82 ) | Styrene | 70 | 3.92 | 168 | - | 0.70 | - | - |
| DMPTD | MMA | 65 | - | - | - | 0.46 | - | |
| | | 75 | - | - | - | 0.50 | - | 42 |
| ( ref 83a ) | | 90 | - | - | - | 0.62 | - | |
| | | 100 | - | - | - | 0.76 | - | |
| TMTD | MMA | 60 | 0.64 | 3350 | 3.3 | - | - | - |
| ( ref 81 ) | | 70 | 2.51 | 3130 | 4.0 | - | - | - |
| BPTD | isoprene | 80 | 1.56 | 990 | 0.15 | - | - | - |
| | | 85 | 2.14 | 1040 | 0.18 | - | - | - |
| ( ref 83 ) | | 90 | 6.74 | 3500 | 0.50 | - | - | - |
| | | 100 | 5.36 | 2210 | 0.31 | - | - | - |

(a) for the description of the iniferters see discussions later

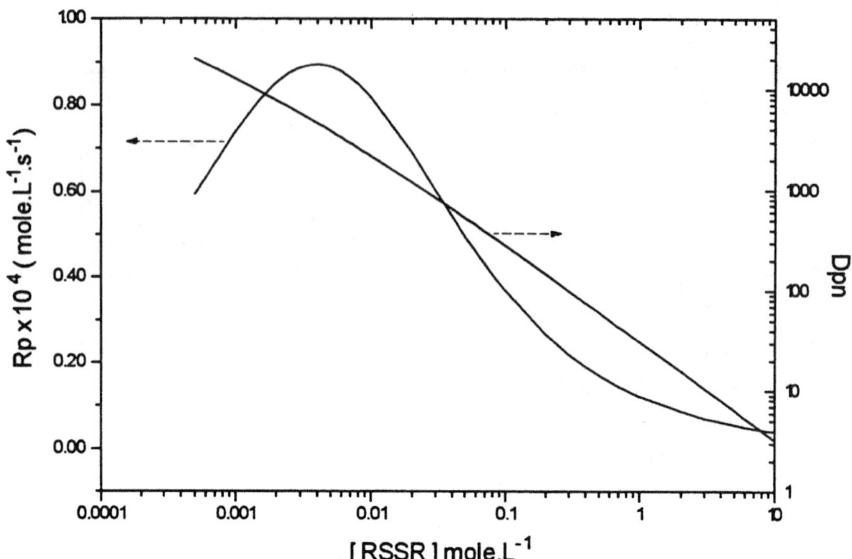


*Fig. 3 Variation of polymerization rate and polymer molecular weight in a typical case of MMA polymerization using an iniferter (70° C).* Note that the iniferter concentration and Dpn are both shown on a logarithmic scale

The overall dissociation constant (2 $fk_d$) varies slightly with the chemical structure. OH group in DHTD facilitates easy addition to MMA through possible H-bonding enhancing the efficiency factor f . The transfer constants vary from structure to structure but do not show any definite trend. The 2 $fk_d$ factor is always higher for styrene, implying that f is higher. The primary radicals are easily consumed by this monomer in the initiation step. Hence B is smaller and the term $C[RSSR]^2/[M]^2$ is negligible for this monomer. Extrapolated to the same temperature, the value of B, a measure of the degradative transfer, and of the deviation from normal kinetics is in the order of MMA > isoprene > styrene; which is the decreasing order of the polymer radical reactivity towards tne dithiocarbamate radical. The $2fk_d$ values show that, as initiators, the iniferters are only 1/7 th or 1/10 th time as efficient as AIBN. But their transfer constants are comparable to those of conventional thiols.

When the iniferter moieties are located in macromolecular chain, their properties can be definitely influenced by the physical environment. The results of the investigations carried out in respect of several macro- and polymeric iniferters have been compiled in Table 2.

An examination of the results as a whole indicates that :

- ☐ The activation energy for thermal dissociation of the disulfide is quite independant of the molecular environment.
- ☐ The overall initiation constant is reduced for the polymeric version *vis-à-vis* the microiniferter ( see **Table 1** )

Increase in chain length of the polymeric iniferter decreases the reactivity of the polymer radical with respect to transfer and termination. This is clear from a comparison of the $C_{tr}$ and B values of different PDMS based disulfides at a given temperature. Mutual reaction between macroiniferter radicals is significantly reduced. (The maximum in the $R_p$-[I] plot is shifted to higher concentrations of iniferter. In some cases it is not explicit in the concentration range studied ).

*Table 2 Kinetic parameters for various macro and polyiniferters*

| iniferter type | chain length g/mole | monomer | temp °C | $2 f k_d\ 10^6$ $s^{-1}$ | B | C $10^{-6}$ (*) | $E_d$ kJ/mole | $C_{tr}$ |
|---|---|---|---|---|---|---|---|---|
| polymeric PDMS | 3050 | MMA | 70 | 2.2 | 2730 | - | - | - |
| polymeric phosphorylamide | 3450 | MMA | 70 | 0.20 | 960 | - | 138 | - |
| | 3450 | styrene | 70 | 0.71 | - | - | - | - |
| macro PEA | 15 000 | MMA | 70 | 1.96 | 6 600 | - | - | - |
| | 10 000 | MMA | 70 | 2.12 | 5 200 | - | - | - |
| | 7 000 | Styrene | 70 | 4.61 | - | - | - | - |
| macro PDMS | 2 800 | MMA | 80 | 3.4 | 1560 | 1.9 | | 0.46 |
| | | | 90 | 10.5 | 980 | 3.5 | 112 | 0.61 |
| | | | 100 | 26.1 | 1350 | 3.2 | | 0.72 |
| macro PDMS | 4200 | MMA | 80 | 1.9 | 130 | 6.5 | | 0.38 |
| | | | 90 | 8.3 | 450 | 6.7 | 139 | 0.52 |
| | | | 100 | 23.7 | 490 | 9.8 | | 0.62 |
| macro PDMS | 4200 | Styrene | 80 | 7.0 | 1340 | - | | 0.72 |
| | | | 90 | 16.9 | 595 | - | 153 | 0.79 |
| | | | 95 | 37.5 | 910 | - | | - |
| | | | 100 | 130.5 | 2720 | - | | |

(*) - : indicates that C was not explicit in the kinetics in the concentration range studied

The reduced rates of transfer and primary radical termination of the polymeric iniferters vis-a-vis the microiniferter arise from the incompatibility of the two different types of polymer chains involved in the reaction. It is also likely that these reaction becomes more diffusion controlled than the initiation. Same is true for the mutual termination of the primary radicals.

Although for simple thiuram disulfides, a generalized equation with averaged values of constants A, B and C can help predict the instantaneous kinetics, for a macromolecular species, it is essential to determine the individual parameters before any reliable prediction can be done. It is also imperative to take into account the relevant diffusion controlled phenomena operative in such kinetics for a precise prediction.

## 5.3 Variation of Molecular Weight With Conversion

In conventional free radical polymerization, the molecular weight and molecular weight distribution do not evolve considerably with conversion unless the reaction is carried out to very high conversion or the Trommsdorff's effect comes into picture. The principal factor determinig the molecular weight is the rapport [initiator]$^{½}$/[monomer] $(=[I]^{½}/[M])$ in the absence of transfer to initiator. The relative rate of change of these concentration terms in case of simple initiation, under steady state condition is given as :

$$\frac{d[I]^{1/2}}{d[M]} = \frac{k_d[I]^{1/2}}{2R_p}$$

For standard experimental conditions :

$$[I] = 10^{-2}\ \text{mole.L}^{-1},\ Rp = 10^{-5}\ \text{mole.L}^{-1}.\text{s}^{-1},\ k_d = 10^{-5}\ \text{s}^{-1}$$

Hence this becomes :

$$\frac{d[I]^{1/2}}{d[M]} = 0.05$$

In other words the change is not significant. Whereas when transfer to initiator takes place, the relative rate of change of the corresponding terms as a consequence of transfer is given as :

$$\frac{d[I]^{1/2}}{d[M]} = \frac{C_{tr}}{2[I]^{1/2}}$$

For the same conditions and $C_{tr} = 0.5$, this becomes $\dfrac{d[I]^{1/2}}{d[M]} = 2.5$

The relative rate of change is about 50 times faster than when initiation alone is operative. The rapid shift in $[I]^{½}/[M$ ratio causes abrupt change in molecular weight ( increase in this case ) and its distribution as conversion increases. The result is shown in **Fig. 4** for a typical case of the polymerization of MMA. The increase in is in conformation to the above theory assuming simple initiation and transfer behaviours.

## 5.4 Polymerization of Acrylates With Thiuram Disulfide

One apparent difficulty observed with these disulfides was their inability to polymerize electron-deficient monomers like acrylonitrile and acrylates. This arises from two facts. The dithiocarbamate radical is stable and soft and hence tends to react with electron rich-monomers like styrene, isoprene, MMA, etc. For monomers like acrylates, the initiation is not favored. If at all it initiates, the resulting radical being very reactive and electron deficient, gets rapidly terminated by another primary radical. The ratio $k_{tpr}\backslash k_i$ in the complex constant B becomes so large that apparently the propagation does not proceed at all. This can be overcome if reactive monomer like MMA or styrene is added to the system. It addresses to the initiation problem, but the termination is still favored with the result that the conversion and DP become kinetically limited.

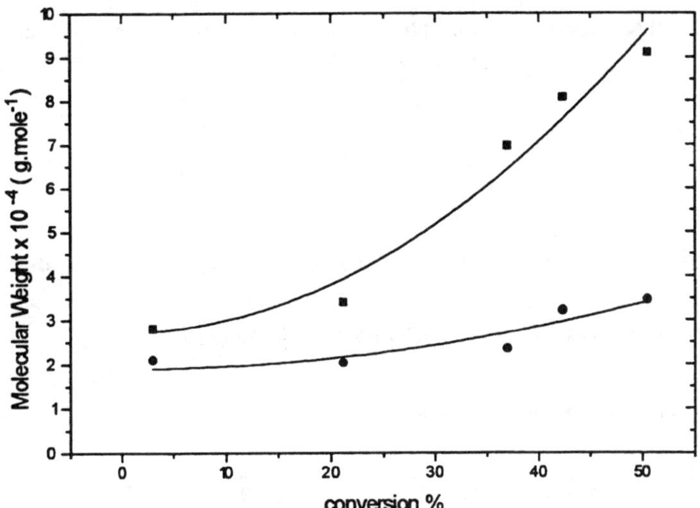


*Fig.4* *Molecular weight variation with conversion for the polymerization of MMA using DHTD.* (●) $M_n$ , (■) $M_w$ [ Ref. 81 ]

The more the "reactive" monomer, better the yield and higher the DP; depending also on the reactivity ratio of the two monomers. The results obtained for the copolymerization of MMA and butylacrylate in the presence of an iniferter is in conformation to this theoritical expectation[1] It was seen that as the concentration of MMA increases, both the polymerization rate and DP increase. This is just the reverse of the trend expected if the initiation was done in presence of a conventional initiator for this particular monomer pair. One of the ends of such polymer is invariably terminated with the dithiocarbamate moiety linked to the electron rich monomer.

## 6. FUNCTIONAL INIFERTERS FOR TELECHELICS

Following are some of the functional iniferters used for deriving telechelics.

$$\left[\begin{matrix} HO-CH_2-CH_2 \\ CH_3-CH_2 \end{matrix} \!\!>\! N-\overset{S}{\overset{\|}{C}}-S- \right]_2$$

**Scheme 6** : N,N'-Diethyl-N,N'Bis[2-Hydroxyethyl] Thiuram Disulfilde DHTD [81]

---

[1] *G. Clouet and M. C. Richou (unpublished results)*

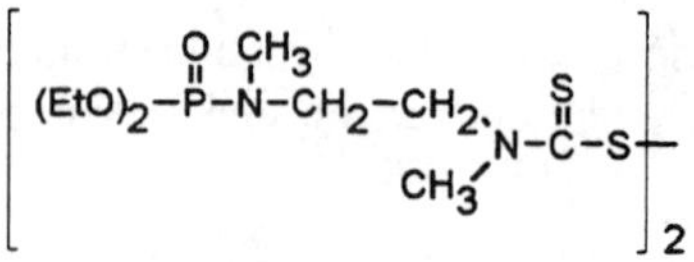

$$\left[ (EtO)_2-\overset{O}{\overset{\|}{P}}-\overset{CH_3}{\overset{|}{N}}-CH_2-CH_2\underset{CH_3}{\diagdown} N-\overset{S}{\overset{\|}{C}}-S- \right]_2$$

Scheme 7 : N,N'-Dimethyl -N,N'-Bis[2-(N,diethoxy phosphinoyl- N-methylamino) ethyl] thiuram disulfide DPTD[82]

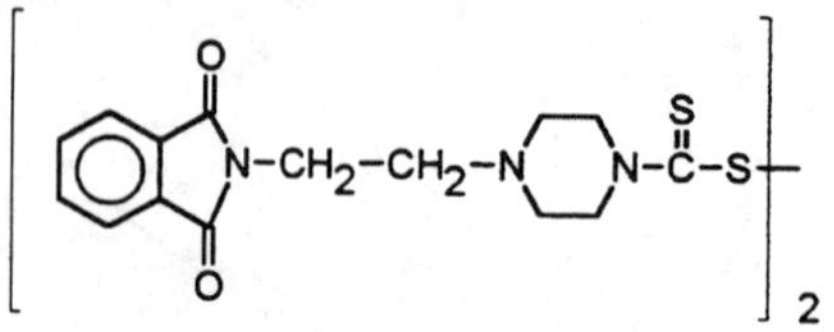

$$\left[ \text{(phthalimido)}N-CH_2-CH_2-N\langle\rangle N-\overset{S}{\overset{\|}{C}}-S- \right]_2$$

Scheme 8 : N,N' -Bis[N-(-2 phtalimido ethyl) piperazine] thiuram disulfide BPTD[83]

The Scheme 9 represents the typical telomerisation reaction.

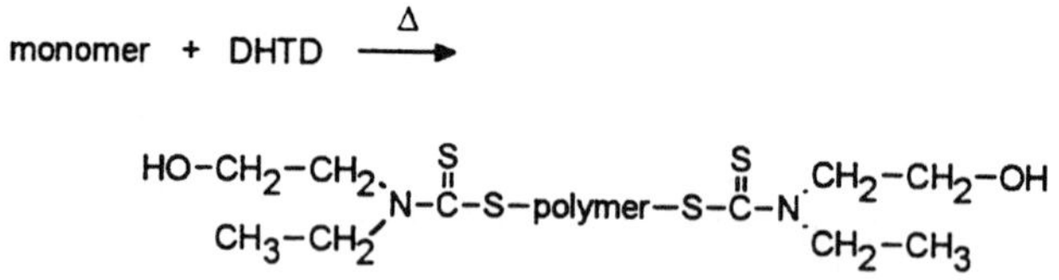

$$\text{monomer} + \text{DHTD} \xrightarrow{\Delta}$$

$$\begin{matrix} HO-CH_2-CH_2 \\ CH_3-CH_2 \end{matrix} N-\overset{S}{\overset{\|}{C}}-S-\text{polymer}-S-\overset{S}{\overset{\|}{C}}-N \begin{matrix} CH_2-CH_2-OH \\ CH_2-CH_3 \end{matrix}$$

Scheme 9

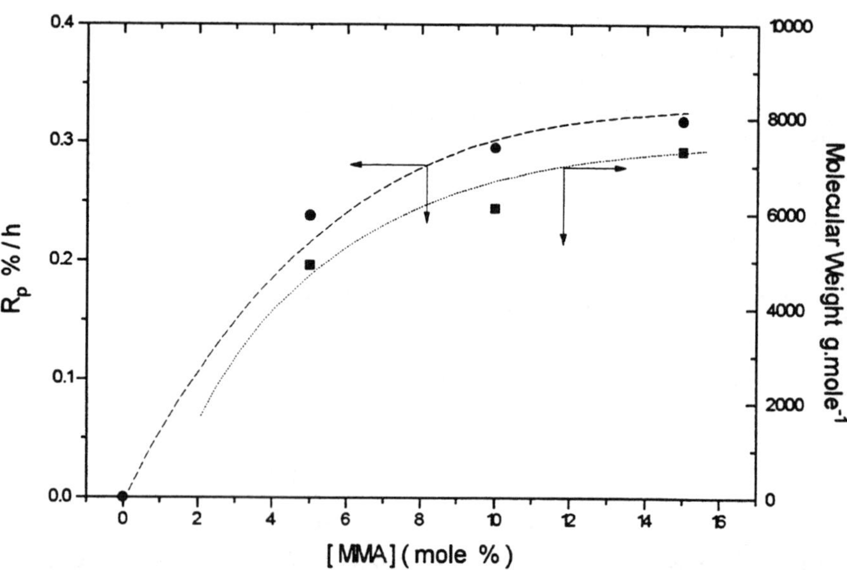


*Fig.5 Dependance of polymerization rate and polymer molecular weight on the feed composition for the copolymerization of Butyl acrylate with MMA in presence of an iniferter at fixed monomer and iniferter concentrations.* [Ref. 90]

DHTD and DPTD were used to derive OH and phosphorylamide telechelics rspectively. The latter gave polymers which exhibited enhanced flame resistance thanks to the terminal phosphorous functional groups.

BPTD was used to derive amine-telechelics of isoprene. The imide groups were cleaved to liberate the amine in the final polymer. With a view to end-cap polymer chain with functional aromatic group, N,N' dimethyl bis(phenetyl) thiuram disulfide (DMPTD) was examined as a model iniferter [83a] (**Scheme 10**).

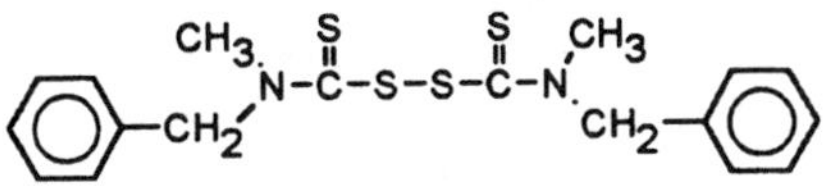

**Scheme 10**

The phenyl group apparently showed no particular influence on its iniferter properties. It is noteworthy that the phenyl group can serve as a template for carrying a variety of useful functional groups.Generally the disulfides are derived from the corresponding secondary amines by reaction with $CS_2$ and $I_2$ (**Scheme 11**). The amine is synthesised by appropriate method.e.g. synthesis of the amine precursor for DPTD is done by the reaction of diethoxy phosphonyl chloride on N,N'dimethyl 1,6-hexanediamine (**Scheme 12** ).

***Table 3*** *Functionalities of the End Functional Polymers*

| reaction system | [RSSR] mole.L$^{-1}$ | Temp. °C | Mn $10^{-4}$ (GPC) | Method of End Group estimation | Functionality |
|---|---|---|---|---|---|
| MMA, DHTD | 0.061 | 95 | 2.10 | S content (*) | 2.0 |
| | 0.061 | 85 | 2.73 | " | 2.3 |
| | 0.092 | 70 | 2.96 | " | 2.5 |
| | 0.031 | 95 | 3.24 | " | 2.2 |
| | 0.061 | 70 | 3.50 | " | 2.2 |
| MMA, DPTD | 0.061 | 90 | 2.50 | " | 2.3 |
| | 0.123 | 90 | 1.70 | " | 2.3 |
| | 0.246 | 90 | 1.10 | " | 1.8 |
| Styrene, DPTD | 0.032 | 70 | 2.20 | " | 2.3 |
| | 0.160 | 70 | 0.67 | " | 2.0 |
| | 0.250 | 70 | 0.47 | " | 1.8 |
| | 0.320 | 70 | 0.42 | " | 1.9 |
| Isoprene, BPTD, $CH_2Cl_2$ | 0.0015 | 80 | 3.5 | GPC / UV (**) | 1.7 |
| | 0.0015 | 85 | 3.1 | " | 1.5 |
| | 0.0015 | 90 | 2.6 | " | 1.7 |
| | 0.0015 | 100 | 2.2 | " | 2.1 |

(*) Elemental analysis for Sulfur

(**) Estimation of dithiocarbamate group by UV absorption at $\lambda$max = 280 nm, $\varepsilon$ = 2300 L/mole.cm

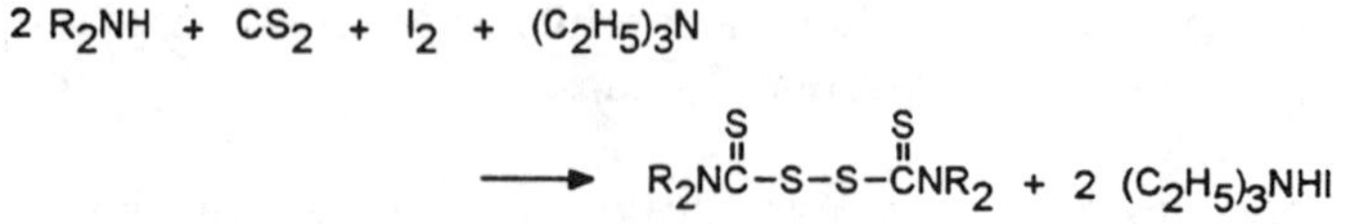

$$2\,R_2NH + CS_2 + I_2 + (C_2H_5)_3N \longrightarrow R_2N\overset{\overset{S}{\|}}{C}{-}S{-}S{-}\overset{\overset{S}{\|}}{C}NR_2 + 2\,(C_2H_5)_3NHI$$

**Scheme 11**

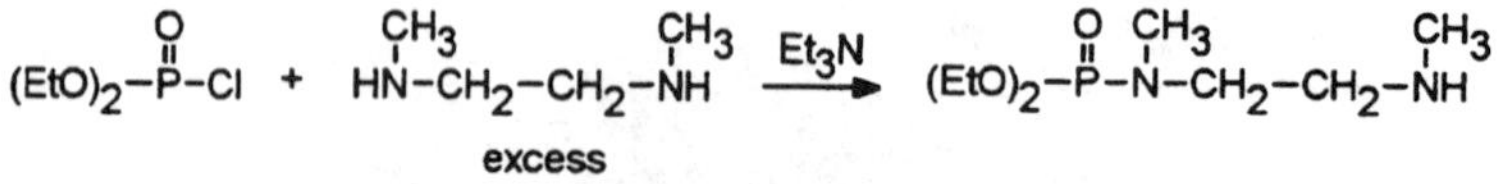

$$(EtO)_2{-}\overset{\overset{O}{\|}}{P}{-}Cl + \underset{\text{excess}}{H\overset{\overset{CH_3}{|}}{N}{-}CH_2{-}CH_2{-}\overset{\overset{CH_3}{|}}{N}H} \xrightarrow{Et_3N} (EtO)_2{-}\overset{\overset{O}{\|}}{P}{-}\overset{\overset{CH_3}{|}}{N}{-}CH_2{-}CH_2{-}\overset{\overset{CH_3}{|}}{N}H$$

**Scheme 12**

**Table 3** cites typical reaction conditions for obtaining the telechelics. Their functionalities are close to two within experimental error.

In these systems the possibility for the formation of non functional polymers is negligible because of the triple function of the iniferter. The length of the polymer sequence inserted between the disulfide is governed by the modified expression for DP. The above examples cite a few typical functions only. The technique can be successfuly extended to other functions as well.

## 7. THEORETICAL PREDICTION OF FUNCTIONALITIES

If all chains are initiated by dithiocarbamate radicals, then the polymer functionality will be determined exclusively by the mode of termination and their relative probabilities. In the case of monomers like styrene, the functionality is always two since recombination is the only mode of termination apart from transfer and primary radical terminations. In the case of monomers like MMA, non functional chain ends can result from the disproportionation reaction. Their proportion is given by the equation :

$$p_{nf} = \frac{\lambda k_t [P^\bullet]^2}{\lambda k_t [P^\bullet]^2 + k_{tr}[P^\bullet][RSSR] + k_{tpr}[P^\bullet][RS^\bullet]}$$

where $P_{nf}$ represents the probability for formation of a non functional polymer end, $\lambda$ the fraction of termination by disproportionation and $[P\bullet]$ is the concentration of macroradicals.

By appropriate substitution :

$$p_{nf} = \frac{\lambda k_t R_p/[M]k_p}{\lambda k_t R_p/[M]k_p + k_{tr}[RSSR] + R_p \dfrac{k_{tpr}}{k_i} C_{tr} \dfrac{[RSSR]}{[M]}}$$

The ratio $k_{tpr} \backslash k_i$ can be calculated from the constant B.

Fig. 6 shows the theoritical weight fraction of non functional chain ends at two different temperatures for typical polymerization of MMA at various concentration of the iniferter. It can be seen that the proportion of these chain ends becomes negligible (less than 0.5% ) even at very low iniferter concentrations. Since the telechelics are made with a significant concentration of the iniferter, it can be assumed that practically all chain-ends are functionalized. It is also possible to get an estimation of the nonfunctional terminals arising from thermal initiation. The iniferters are about 1/5$^{th}$ time as efficient as AIBN. The proportion of thermal initiation and hence of the nonfunctional terminals is given as

$$R_{i,T} = \frac{R_{pT}^2}{R_{pT}^2 + R_{pi}^2} = p_{nf}$$

where $R_{pT}$ is the rate of thermal polymerization and $R_{pi}$ that initiated by the iniferter. Even if the thermal polymerization amounts to 1/10$^{th}$ of the iniferter induced polymerization, this fraction worksout to be less than 0.01 only.

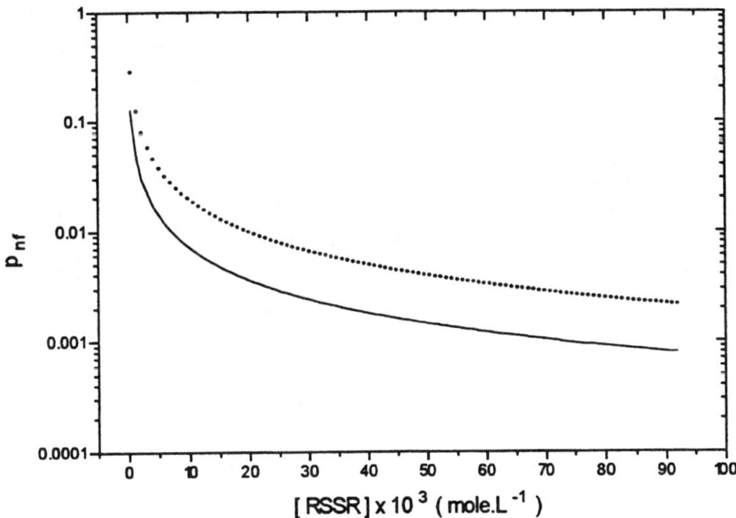


*Fig. 6 Fraction of nonfunctional chain-ends resultingfrom disproportionation reaction in the case of MMA at different concentration of iniferter*.....95° C, - 75° C.

The dithiocarbamate group does not undergo easy thermal degradation as was evident from the thermogravimetric analysis. In fact, they undergo degradation only along with the main chain.

# 8. TRIBLOCK COPOLYMERS THROUGH MACROINIFERTER

Replacement of the simple functions in the thiuram disulfide by polymer sequences gives rise to macrothiuramdisulfides and their utilization should lead to triblock copolymers (**Scheme 13**). Molecules bearing a single thiuramdisulfide group between two polymer segments are referred to as macrothiuram disulfides or macroiniferters to distinguish them from polymeric iniferters which contain repeating thiuramdisulfides and polymer segment units in the backbone.

$$▒-\overset{S}{\overset{\|}{C}}-S-S-\overset{S}{\overset{\|}{C}}-▒ \xrightarrow{CH_2=C(R)(R'),\ \Delta} ▒-\overset{S}{\overset{\|}{C}}-S\left[CH_2-\underset{R'}{\overset{R}{C}}\right]_n S-\overset{S}{\overset{\|}{C}}-▒$$

Scheme 13 ▒ represents the polymer sequence

The vinyl block gets inserted across the disulfide group. The salient features of this technique are :

- ☐ It permits a choice of the type of the iniferter-forming block (e.g. condensation or addition), having different nature ( hard or soft, hydrophobic or hydrophilic, etc.)
- ☐ Depending upon the choice of the monomer, and the iniferter, tailor made block copolymer can be obtained e.g. hard-soft, amphiphilic, etc.
- ☐ The sequence length of the terminal blocks are prefixed based on the iniferter of choice.
- ☐ The length of the vinyl sequence inserted depends upon the monomer, the concentration of the disulfide, its kinetic parameters and the conditions of operation.. Generally as the disulfide concentration or the temperature is increased, the length of the vinyl block is kinetically reduced. Knowledge of A, B C, and $C_{tr}$ can predict this length.
- ☐ The macro disulfide can be synthesised by an appropriate strategy.

## 8.1 Synthesis of Soft-Hard-Soft Triblock Copolymers

The following are some of the examples of triblock copolymers of this type synthesised by adoption of this technique.

### 8.1.1 Poly(ethyl acrylate)-Polystyrene or Poly(ethyl acrylate)-Poly(methyl methacrylate) [84]

The secondary amine terminated polyethylacrylate (PEA) is prepared by way of chain transfer ( **Scheme 14** ). In this context we have noticed that the aminomercaptan as such has very poor chain transfer constant and is hence not good for deriving amine terminated PEA. Upon transformation of the amine to their salt, the $C_{tr}$ increased tremendously [85].

$$CH_2{=}\overset{H}{C}{-}COOEt + AIBN + Cl^-, {}^{+}H\overset{H}{\underset{Bu}{N}}{-}CH_2{-}CH_2{-}SH$$

$$\longrightarrow PEA{-}S{-}CH_2{-}CH_2{-}\overset{H}{\underset{Bu}{N}}H^+, Cl^-$$

$$\xrightarrow[CS_2]{Et_3N} \left[ PEA{-}S{-}CH_2{-}CH_2{-}\underset{Bu}{N}{-}\underset{S}{\overset{}{C}}{-}S{-} \right]_2 = PEA{-}MI$$

$$PEA{-}MI \xrightarrow[\Delta]{Styrene} PEA{-}PS{-}PEA$$

$$PEA{-}MI \xrightarrow[MMA]{\Delta} PEA{-}PMMA{-}PEA$$

**Scheme 14**

The **Table 4** gives some data pertaining to the synthesis of triblock copolymers. It can be seen that triblocks are easily formed in case of PS, although tendencies to form diblocks exist in the case of PMMA at low MI loading.

*Table 4 Characteristics of PEA / PS copolymers obtained using MI of PEA with chain length 8100 g/mole*

| Polymerization conditions | [MI] $g.L^{-1}$ | Molecular weight | | | EA Weight % | average number of PEA blocks per PMMA or PS block |
|---|---|---|---|---|---|---|
| | | Mw $10^4$ | Mn $10^4$ | I | | |
| [MMA] = 4.7 mole.$L^{-1}$ | 12.5 | 48.6 | 19.3 | 2.5 | 5.5 | 1.30 |
| Solvent : THF | 18.8 | 38.8 | 17.8 | 2.2 | 6.6 | 1.50 |
| Temp : 80°C | 31.3 | 31.6 | 16.1 | 2.0 | 7.9 | 1.60 |
| | 43.8 | 27.7 | 11.2 | 2.5 | 13.8 | 1.90 |
| | 66.3 | 23.3 | 10.6 | 2.2 | 17.7 | 1.70 |
| [St] = 4.53 mole.$L^{-1}$ | 12.5 | 16.0 | 9.33 | 1.7 | 28.1 | 2.40 |
| Sovent : THF | 18.8 | 12.9 | 7.7 | 1.7 | 30.0 | 2.10 |
| Temp : 80°C | 31.3 | 9.7 | 5.6 | 1.7 | 36.6 | 1.93 |
| | 43.8 | 7.6 | 4.4 | 1.7 | 52.8 | 2.18 |
| | 66.3 | 7.0 | 4.3 | 1.6 | 58.7 | 2.23 |

In fact, the length of the vinyl sequence inserted across the disulfide for a given concentration of it was quite independant of the chain length of PEA. **Fig. 7** depicts variation of this vinyl sequence length as a function of the iniferter concentration for two macroiniferters of different molecular weights. The chain length dependance caused by the difference in reactivity arising from the incompatibility between the two reacting chains was not significant because the PEA sequences chosen here were shorter.

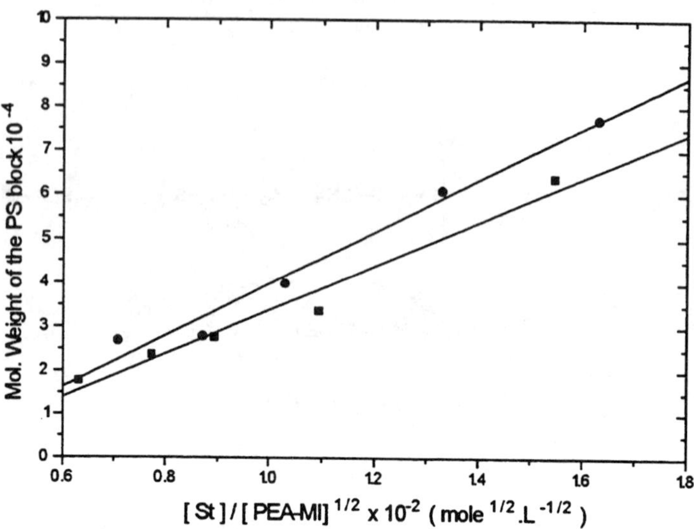


*Fig. 7 Dependance of PS sequence inserted between S-S linkage in the PEA-based macroiniferter on iniferter concentration and PEA chain length.* [ Ref. 84 ]
■ PEA chain length 3700 g/mole
● PEA chain length 11,600 g/mole

### 8.1.2 Polydimethylsiloxane-Polystyrene-Polydimethylsiloxane [86,87]

Macroiniferters based on polydimethylsiloxane (PDMS) when utilized for the polymerization of MMA or styrene led to triblock copolymers with PDMS terminal chains. In the kinetic studies it was shown that the chain transfer and primary radical terminations were clearly dependent on the chain length due to the formation of microphases or to the incompatibility between the reacting chains. The iniferter could be synthesised from the Si-H terminated PDMS by the hydrosilylation reaction using allyl N-methylamine. The macrodiamines were transformed to the thiuram disulfide by conventional reaction ( **Scheme 15** ).

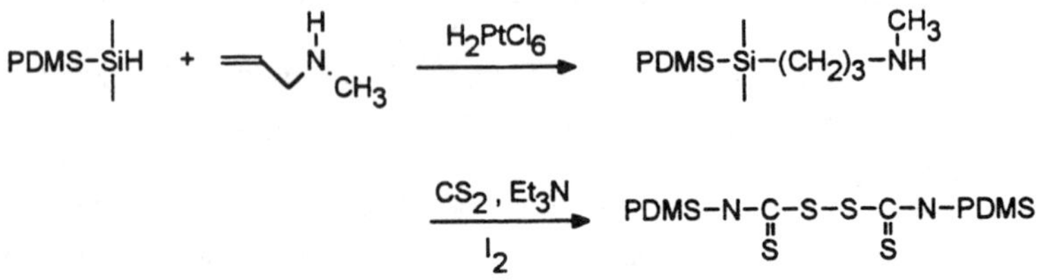

PDMS–SiH + $CH_2$=CH–$CH_2$–N(H)–$CH_3$ $\xrightarrow{H_2PtCl_6}$ PDMS–Si–$(CH_2)_3$–NH–$CH_3$

$\xrightarrow[I_2]{CS_2,\ Et_3N}$ PDMS–N–C(=S)–S–S–C(=S)–N–PDMS

**Scheme 15**

The same macroiniferters have been used to synthesise triblock copolymers with poly perfluoro methacrylate forming the central blocks[2] Bulk polymerization were found to be ineffective owing to the strong incompatibility of the two polymer sequences, but addition of a miscibilising phase like toluene led to the expected product. The **Table 5** furnishes some results pertaining to some of these polymers. PDMS content is apparently low, since the terminal blocks are chosen to be of small size.

*Table 5 Polymerization of perfluoromethacrylate in presence of PDMS based macroiniferter (PDMS-MI). temperature : 70 C, polymerization time 17 h, Mn PDMS 2500 g/mole*

| concentration of PDMS-MI mol/L $10^3$ | conversion % | $\overline{M}_n\ 10^{-3}$ g/mole | I (polydispersity index) | weight fraction of PDMS in triblock copolymer |
|---|---|---|---|---|
| 6.1 | 27.5 | 180 | 3.9 | 1.5 |
| 9.5 | 27.4 | 110 | 3.7 | 2.1 |
| 12.4 | 24.4 | 110 | 3.2 | 2.2 |
| 15.8 | 22.7 | 90 | 3.3 | 3.6 |

## 8.2 Hard-Soft-Hard Triblocks

In these systems, the central soft blocks can be comprised only of vinyl sequences, polymerizable by iniferters.

The macroiniferter technique looks effective for realizing triblocks of this type, by a proper choice of the central and terminal blocks. When it is required to choose a soft central block based on systems like the acrylates, homo-polyacrylates cannot be envisaged since thiuram disulfide does not homopolymerize them. Adding small quantities of a second monomer like MMA as already discussed in the preceeding section, can help overcome this difficulty. By this way triblocks of type PS-PBuA-PS and PMMA-PBuA-PMMA using PS- and PMMA- based macroiniferters have been synthesised. Some details regarding their synthesis can be found in **Table 6**.

DSC analyses of such block copolymers revealed microphase separation of the two sequences. The glass transitions due to the two sequences were separately seen. The PBuA sequence showed a $T_g$ of 251 K about 10 K more than that of the homopolymer, due to the incorporation of PMMA. An alternative possibility is to choose low $T_g$-polymethacrylates or polydienes as central blocks.

## 8.3 Amphiphilic Triblock Copolymers

The macroiniferter technique provides unique ways to derive amphiphilic triblock copolymers, in which either the central or terminal blocks can be chosen as hydrophilic.

[2] *D. Kayser and G. Clouet, unpublished results*

*Table 6 Polymerization of Butyl acrylate ( BuA ) in presence of PS and PMMA-based macroiniferters, in THF ( 50% by volume ), containing 10% / mole of MMA*

| macroiniferter | concentration of MI mole/L | duration of polymerization h | conversion % | molecular weight (peak) g/mole |
|---|---|---|---|---|
| PS Mn : 1 300 g/mole | $10^{-2}$ | 10 | 16 | 17 300 |
| PMMA Mn : 25 500 g/mole | $10^{-3}$ | 20 | 15 | 100 500 |
| | $5.10^{-3}$ | 20 | 9 | 101 900 |
| | $10^{-2}$ | 20 | 10 | 87 800 |

### 8.3.1 Terminal Hydrophilic Triblocks [88]

Typically, polyoxyethylene based macroiniferters were synthesised from the OH-terminated PEO by suitable chemical modification as shown in **Scheme 16.**

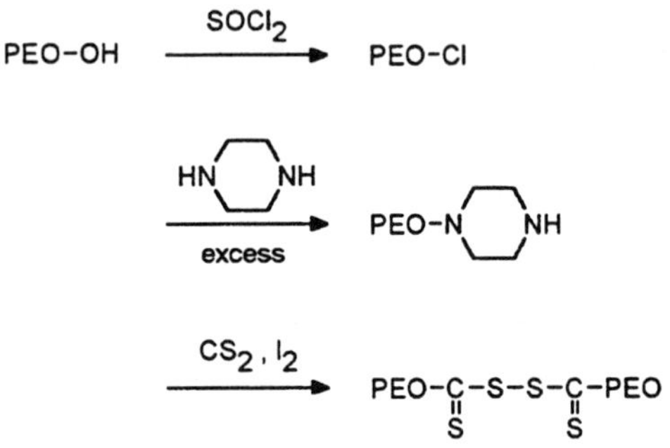

$$\text{PEO-OH} \xrightarrow{SOCl_2} \text{PEO-Cl}$$

$$\xrightarrow[\text{excess}]{\text{HN}\langle\rangle\text{NH}} \text{PEO-N}\langle\rangle\text{NH}$$

$$\xrightarrow{CS_2,\ I_2} \text{PEO-C(=S)-S-S-C(=S)-PEO}$$

**Scheme 16**

This when used as thermal iniferter for polymerization of styrene gave PEO-PS-PEO triblocks. The central block can be changed in nature and in chain length by an appropriate choice of the monomer and iniferter concentrations. **Table 7** gives representative data for the preparation of PEO-PS-PEO triblocks in this way.

Another combination of polymer sequences of this type is poly(acrylic acid)-PMMA-(polyacrylic acid) [89] (**Scheme 17** ).

### 8.3.2 Terminal Hydrophobic Blocks

A typical example is a sequence of the type PS-PMAA-PS which has been prepared as per the reaction **Scheme 18** [90].

*Table 7 Synthesis conditions and characteristics of PEO-PS-PEO triblock copolymers*

| Molecular weight of macroiniferter | [MI] $g.L^{-1}$ | Yield % | Mn of copolymer (GPC $10^4$) | I polyd. index | %POE Calc. | %POE Found | Vinyl block length $10^{-4}$ |
|---|---|---|---|---|---|---|---|
| 4 000 | 50 | 55 | 8.80 | 2.2 | 5.3 | 8.3 | 8.34 |
| | 100 | 50 | 4.30 | 2.6 | 9.3 | 10.0 | 3.90 |
| | 150 | 55 | 3.40 | 2.3 | 13.5 | 13.6 | 2.94 |
| | 200 | 57 | 2.86 | 2.6 | 16.1 | 16.7 | 2.40 |
| 10 000 | 100 | 45 | 8.5 | 1.9 | 16.5 | 13.5 | 7.1 |
| | 167 | 48 | 7.2 | 1.8 | 19.4 | 18.7 | 5.8 |
| | 233 | 48 | 6.5 | 1.8 | 21.5 | 21.6 | 5.1 |
| | 300 | 63 | 5.7 | 2.0 | 24.6 | 29.0 | 4.3 |
| | 500 (*) | 22 | 4.1 | 1.5 | 34.2 | 36.5 | 2.7 |
| | 700 (*) | 22 | 3.9 | 1.5 | 35.9 | 36.6 | 2.5 |

(*) in presence of Toluene ( 100 % )

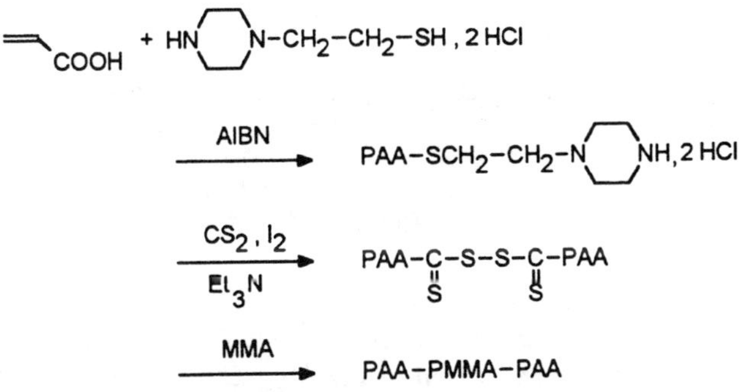

$CH_2{=}CH{-}COOH$ + HN(piperazine)N-$CH_2$-$CH_2$-SH , 2 HCl

$\xrightarrow{AIBN}$ PAA-$SCH_2$-$CH_2$-N(piperazine)NH, 2 HCl

$\xrightarrow[Et_3N]{CS_2,\ I_2}$ PAA-C(=S)-S-S-C(=S)-PAA

$\xrightarrow{MMA}$ PAA-PMMA-PAA

**Scheme 17** ( PAA : polyacrylic acid )

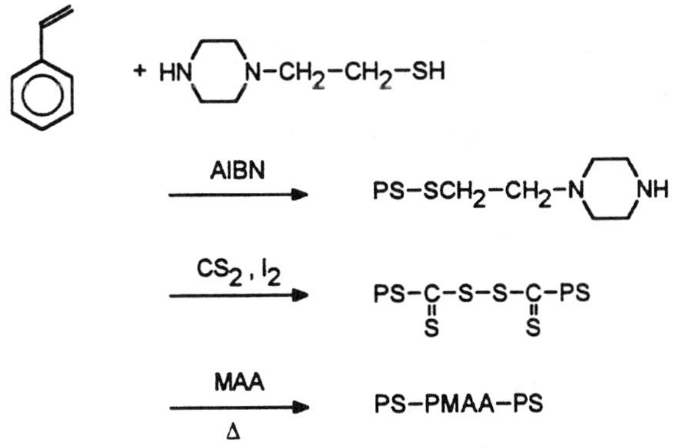

Styrene + HN(piperazine)N-$CH_2$-$CH_2$-SH

$\xrightarrow{AIBN}$ PS-$SCH_2$-$CH_2$-N(piperazine)NH

$\xrightarrow{CS_2,\ I_2}$ PS-C(=S)-S-S-C(=S)-PS

$\xrightarrow[\Delta]{MAA}$ PS-PMAA-PS

( MAA : methacrylic acid )

**Scheme 18**

The copolymer possessed molecular weights ranging from 14,000 to 170,000 g/mole with polystyrene sequence having molecular weight of the order of 1000 g/mole. These copolymers exhibited interesting behavior in aqueous solution arising from the association of the COOH groups and from their repulsions on saponification, projecting their potentiality for application as thickeners in aqueous emulsion systems.

## 9. POLYMERIC INIFERTERS AND THEIR APPLICATION IN THE SYNTHESIS OF MULTIBLOCK COPOLYMERS

An extension of the technique of using functional iniferters for deriving telechelics, to polymeric iniferters should be an ideal way to obtain mid-chain functional polymers with the functions linking the polymer sequences ( **Scheme 19** )

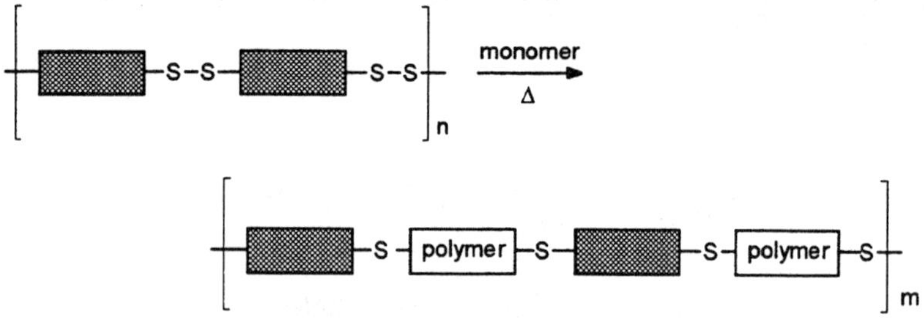

$$\left[ \blacksquare-S-S-\blacksquare-S-S \right]_n \xrightarrow[\Delta]{\text{monomer}} \left[ \blacksquare-S-\boxed{\text{polymer}}-S-\blacksquare-S-\boxed{\text{polymer}}-S \right]_m$$

**Scheme 19**

When simple organic functions were chosen, it was found that the resulting polymeric iniferters were not soluble in any organic solvent. To render them soluble, disulfides groups are to be sufficiently spaced using long chain sequences. When polymeric sequences are chosen as the spacer, the resulting functional polymers are multiblock copolymers ( **Scheme 20** ).

Evidently in such situation, length of the new sequences formed will be kinetically controlled. The number of A-B blocks per chain will also depend upon experimental conditions especially the monomer conversion.

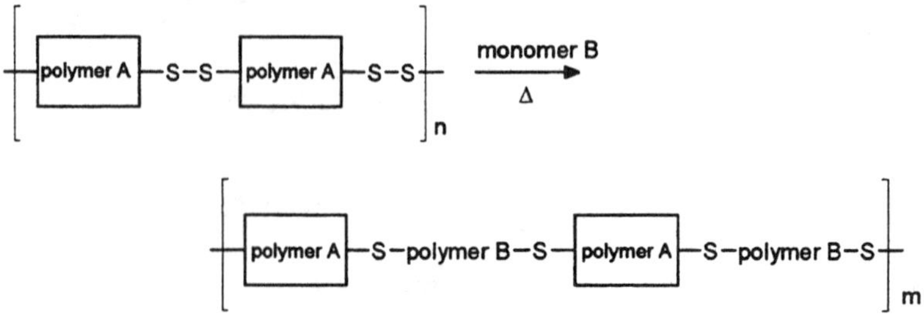

$$\left[ \boxed{\text{polymer A}}-S-S-\boxed{\text{polymer A}}-S-S \right]_n \xrightarrow[\Delta]{\text{monomer B}} \left[ \boxed{\text{polymer A}}-S-\text{polymer B}-S-\boxed{\text{polymer A}}-S-\text{polymer B}-S \right]_m$$

**Scheme 20**

## 9.1 Silicone-Vinyl Multiblock Copolymers

Using the iniferter technique it was possible to realize this type of multiblock copolymer. The polymeric iniferters based on PDMS were synthesised in a manner similar to their macroiniferter starting from $\alpha, \omega$ Si-H terminated PDMS ( **Scheme 21**).

$$\text{H-Si-PDMS-Si-H} \xrightarrow[H_2PtCl_6]{CH_2=CH-CH_2-NH-CH_3} CH_3\text{-}NH(CH_2)_3\text{-Si-PDMS-Si-}(CH_2)_3NH\text{-}CH_3$$

$$\xrightarrow{CS_2,\ I_2} \left[\text{-PDMS-}N(CH_3)\text{-}C(=S)\text{-S-S-}C(=S)\text{-}N(CH_3)\text{-}\right]_n$$

**Scheme 21**

n varies from 22 to 26 indicating near perfect functionalization of Si-H terminals. In fact, the formation of the polythiuramdisulfide resembles a polycondensation reaction of PDMS macrodiamine with $CS_2$. Block copolymers of PDMS with MMA or styrene were formed by heating them with the corresponding monomer. **Table 8** presents some details regarding their synthesis in a typical case. Since the isolated polymer could contain reactive disulfide groups, they were aged at high temperature to decompose them. The data show that under a given set of conditions, the molecular weight of the polymer formed increases with conversion so does also the number of A-B blocks per chain. This behaviour is typical for this kind of polymer preparation since the chain grows by the successive intercalation of the vinyl blocks across the disulfide linkage. Block copolymers containing as many as 8 A-B blocks per chain could be synthesised and these block copolymers presented narrow molecular weight distribution despite high conversion. The characteristics of some of these polymers (polymerization was carrried out to 100%) can be found in **Table 9.**

It is seen that the number of A-B blocks increases slightly with increase in the concentration of iniferter, at the cost of the vinyl block length which decreases systematically as expected according to the kinetics. The number of A-B blocks and hence the efficiency increases for small PDMS sequences, due probably to the better compatibility between the propagating PS radical and the PDMS during the transfer and termination processes as already noticed during the examination of the chain length dependance of the kinetic parameters. The studies have been limited to small PDMS segment-containing polymers. The multiblock copolymers so obtained exhibited the usual physico-chemical characteristics of PDMS-Vinyl block copolymers obtained by other ways.

*Table 8 Synthesis conditions and characteristics of PDMS-PS or PDMS-PMMA multiblock copolymers*

| monomer | experim. cond. | conversion % | PDMS weight % | copolymer $\overline{M}_n.10^{-4}$ | polydispersity index | A-B blocks per chain |
|---|---|---|---|---|---|---|
| MMA | bulk, 85°C | 7.5 | 10.9 | 3.9 | 2 | 2 |
| | [PI](*) = | 20 | 8.1 | 4.1 | 2.2 | 2.4 |
| | 100 g/L | 40 | 6.6 | 4.7 | 2 | 2.5 |
| | | 64 | 4.3 | 5.2 | 2.1 | 2.8 |
| Styrene | bulk, 90°C | 17.8 | 8.6 | 7.0 | 2.6 | 5.6 |
| | [PI](*) = | 33 | 8.4 | 7.8 | 2.5 | 7.1 |
| | 80 g/L | 49.6 | 7.4 | 8.5 | 2.5 | 7.1 |
| | | 71 | 6.3 | 10.5 | 2.7 | 8.6 |

(*) PI with PDMS sequence length 1 550, containing 26 units per chain

*Table 9 Characteristicsof PDMS-PS multiblock copolymers preparedat 100% conversion*

| experimental conditions | conc. of PI g/L | PDMS content weight % | copolymer $\overline{M}_n.10^{-4}$ | I polyd. index | A-B block | mol. weight of styrene sequence |
|---|---|---|---|---|---|---|
| styrene | 83 | 5.3 | 14.4 | 1.9 | 2.5 | 82 000 |
| bulk, 85 °C | 133 | 8.5 | 10.8 | 1.8 | 2.8 | 49 000 |
| PI-4 500 (*) g/mole | 200 | 10.7 | 8.4 | 1.8 | 2.9 | 38 000 |
| | 300 | 15.1 | 7.3 | 1.8 | 3.5 | 25 500 |
| | 400 | 17.7 | 7.2 | 1.6 | 3.8 | 21 000 |
| styrene | 67 | 5.0 | 13.3 | 2.5 | 8.3 | 29 500 |
| bulk, 85 °C | 117 | 7.6 | 8.7 | 2.4 | 8.2 | 19 000 |
| PI-1 550 | 167 | 12.0 | 7.6 | 2.4 | 11.2 | 11 300 |
| | 202 | 14.2 | 4.4 | 2.8 | 11.1 | 9 400 |

* PDMS chain length 4 500, contains 22 units per chain.

## 9.2 Block Copolymers of Phenylphosphonamide ( PPA ) with MMA or Styrene [92]

This was another case where it was demonstrated that the PI-technique can be successfully applied to the synthesis of condensation-addition block copolymers. The copolymers like phenyl phosphonamide-MMA and PPA-PS were synthesised with a view to improve the flame resistance characteristics of the thermoplastics ( PMMA, and PS ). The scheme for synthesis is despicted as per **Scheme 22**.

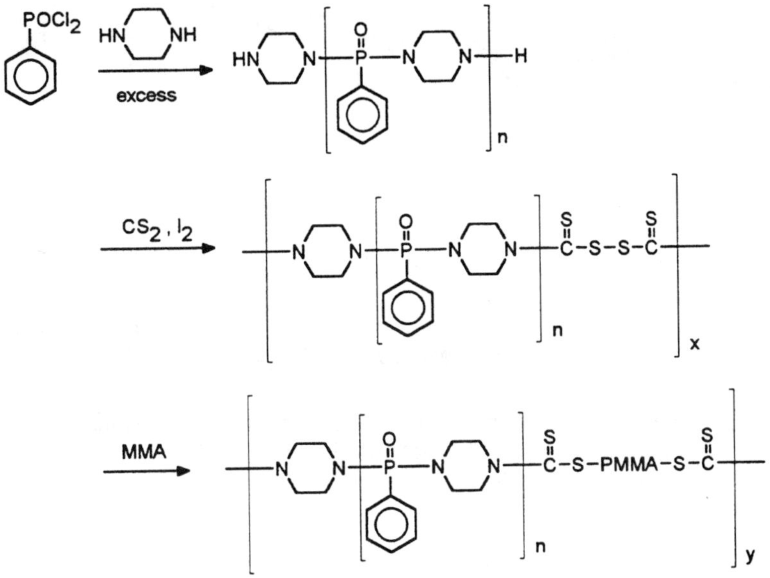

Scheme 22

The diamine precursor of desired molecular weight was obtained through a polycondensation reaction between phenyl phosphonic dichloride and piperazine. The polymeric iniferter on reacting with either MMA or styrene gave their corresponding multiblock copolymers. These copolymers exhibited enhanced thermal stability and glass transition temperature and showed strong tendancy for microphase seggregation. Association among the phosphonamide block is so strong that the polymer only swells in solvents like THF due to physical crosslinking. Thermogravimetric analyses of these polymers revealed that phosphonamide blocks enhanced the thermal stability of the copolymer. The Limiting Oxygen Index ( L.O.I. ) value of the block copolymers as a function of phosphorous content is shown in **Fig. 8**. The phosphonamide moiety was found to impart flame resistance to the base polymers.

## 9.3 PEO-PS Multiblock Copolymers[88]

In a way similar to their triblock copolymers, the multiblock copolymers of PEO and PS were prepared starting from a PI based on PEO, whose structure is shown in **Scheme 23**. Copolymers containing as many as 6 A-B blocks could be realized. The quantity of PEO could be enhanced to 45%.

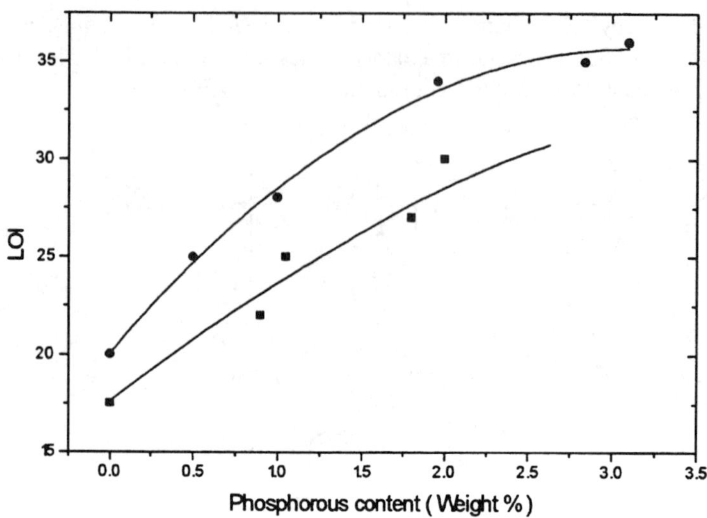


*Fig. 8 L.O.I values of block copolymers as a function of the P-content* [ Ref. 93 ]
■ *phenylphosphonamide-MMA,* ● *PPA-Styrene*

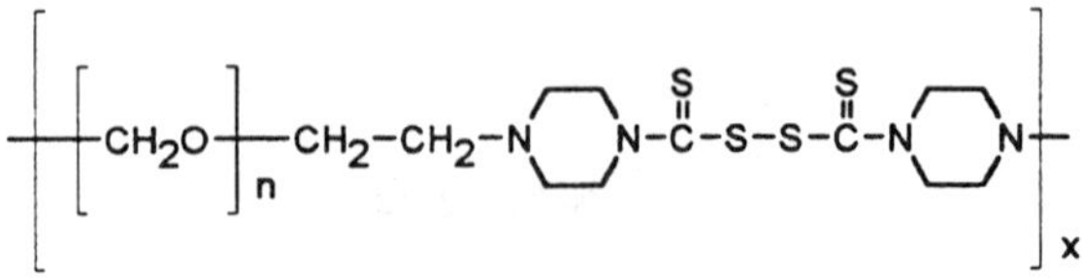

$$\left[ \left[ CH_2O \right]_n - CH_2-CH_2-N\bigcirc N-\overset{S}{\overset{\|}{C}}-S-S-\overset{S}{\overset{\|}{C}}-N\bigcirc N \right]_x$$

**Scheme 23**

The block copolymers exhibited the usual microphase separation as evidenced in their DSC analysis, each segment manifesting its physical properties independent of the other (see **Fig. 9** )

**Table 10** compiles some experimental conditions for their synthesis and the characteristics of the block copolymer formed.

## 10. INIFERTER PROPERTIES OF MOLECULES SIMILAR TO DITHIOCARBAMATES

The molecules having close structural resemblance to the dithiocarbamates are thiocarbamates, xanthates, and their oxygen version (**Scheme 24**). Of these (1) and (3) are known to possess initiation properties. (4) is a known chain transfer agent. (2), and (5) have not been reported previously for their behavior in free radical polymerization.

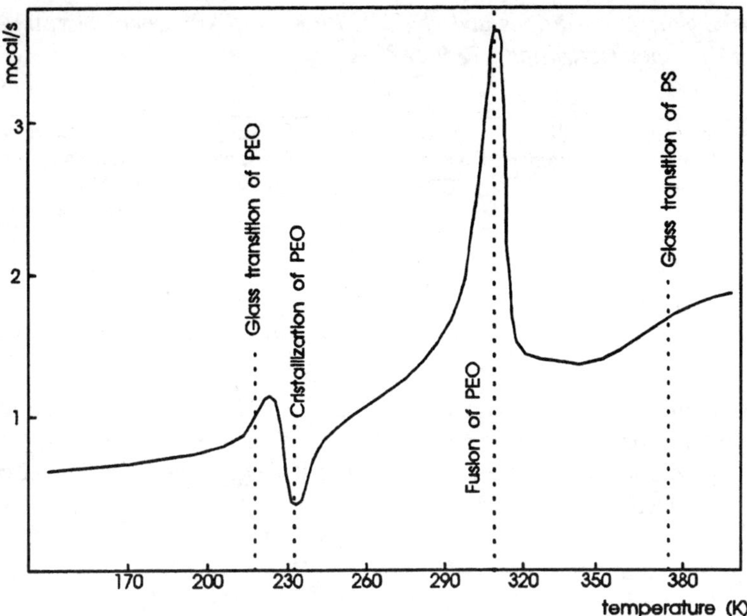


*Fig. 9* *DSC thermogram for a multiblock copolymer PEO-PS containing 40% PEO. PEO chain length 3,400 g/mole, copolymer molecular weight 30,200 g/mole. Heating rate : 20⁰ C /min* [ Ref. 88 ]

| | | |
|---|---|---|
| $[\,>N-C(=S)-S-\,]_2$ | thiuram disulfide | (1) |
| $[\,>N-C(=O)-S-\,]_2$ | biscarbamoyl disulfide | (2) |
| $[\,>N-C(=O)-O-\,]_2$ | carbamoyl peroxide | (3) |
| $[\,RO-C(=S)-S-\,]_2$ | xanthogen disulfide | (4) |
| $[\,RO-C(=S)-O-\,]_2$ | bis(alkoxy thiocarbonyl) peroxide | (5) |

**Scheme 24**

*Table 10 Synthesis conditions and characteritics of PEO-PS copolymer, time of polymerization : 72 h at 85⁰ C*

| (PI) g.L-1 | Volume % of Toluene | % Monomer Conv. | Mn 10-4 Copolymer (GPC) | I | % PEO found by analysis | Average number of alternating blocks in the copolymer | Average length of the styryl sequence |
|---|---|---|---|---|---|---|---|
| 115 | 150 | 39 | 4.02 | 2.4 | 25.2 | 5.0 | 15 200 |
| 200 | 225 | 33 | 3.54 | 2.0 | 32.5 | 5.8 | 11 350 |
| 300 | 300 | 23 | 3.23 | 2.1 | 40.4 | 6.6 | 8 140 |
| 400 | 400 | 17 | 3.02 | 2.1 | 43.1 | 6.6 | 6 170 |

Upon studying the biscarbamoyl disulfide (2) as initiator and chain transfer agent in the polymerization of MMA, it was found that the small molecules do not possess any initiation properties, and that they are poor transfer agents [94].The model compounds studied are given in **Scheme 25**.

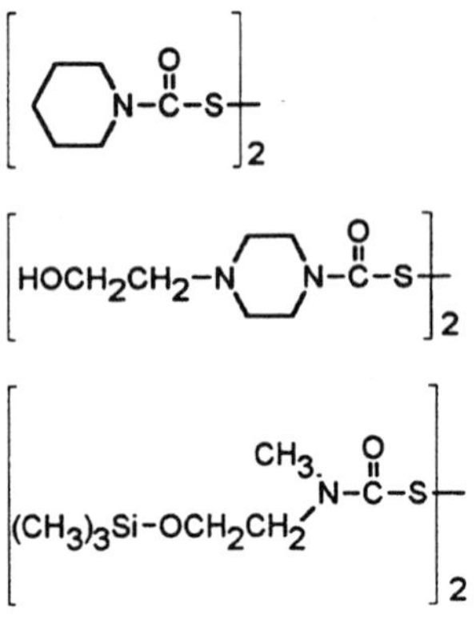

$$\left[ \text{C}_5\text{H}_{10}\text{N-C(=O)-S-} \right]_2 \qquad (1)$$

$$\left[ \text{HOCH}_2\text{CH}_2\text{-N}(\text{C}_4\text{H}_8)\text{N-C(=O)-S-} \right]_2 \qquad (2)$$

$$\left[ (\text{CH}_3)_3\text{Si-OCH}_2\text{CH}_2(\text{CH}_3)\text{N-C(=O)-S-} \right]_2 \qquad (3)$$

**Scheme 25**

The chain transfer constants of various biscarbamoyl disulfides for the methyl methacrylate polymerization has been represented in **Table 11**. They were synthesised by the reaction of carbonyl sulfide with the amine similar to the thiuram analogues (**Scheme 26** ).

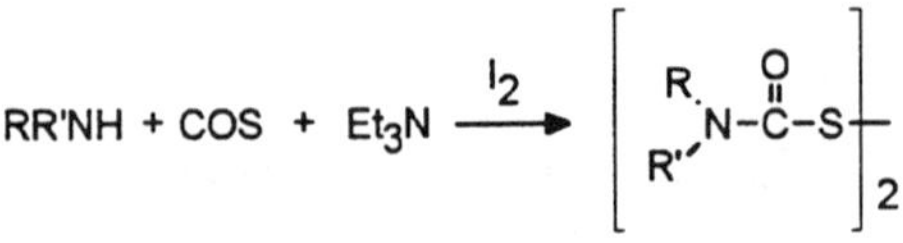

$$\text{RR'NH} + \text{COS} + \text{Et}_3\text{N} \xrightarrow{I_2} \left[ \text{R(R')N-C(=O)-S-} \right]_2$$

**Scheme 26**

*Table 11 Chain transfer constants of various biscarbamoyl disulfides for the polymerization of MMA*

| Transfer agent | Temperature °C | $10^3$.Ctr |
|---|---|---|
| (1) | 110 | 1.5 |
| (2) | 100 | 18 |
| (3) | 100 | 2.8 |

Surprisingly, when these groups were carried on a polymer (like PDMS) backbone, they were found to possess some iniferter properties like their thiuram analogues, although with lesser efficiency.

The macrobiscarbamoyl disulfides were synthesised in a way similar to their thiuram analogues [95].The kinetics of polymerization using them demonstrated that they were also characterized by a behavior similar to the thiuram analogues. **Fig 10** shows some typical Rp-[I] relationship for these compounds having defined PDMS chain length. The kinetic expression for $R_p$ was found to be the same as for the classical iniferters.

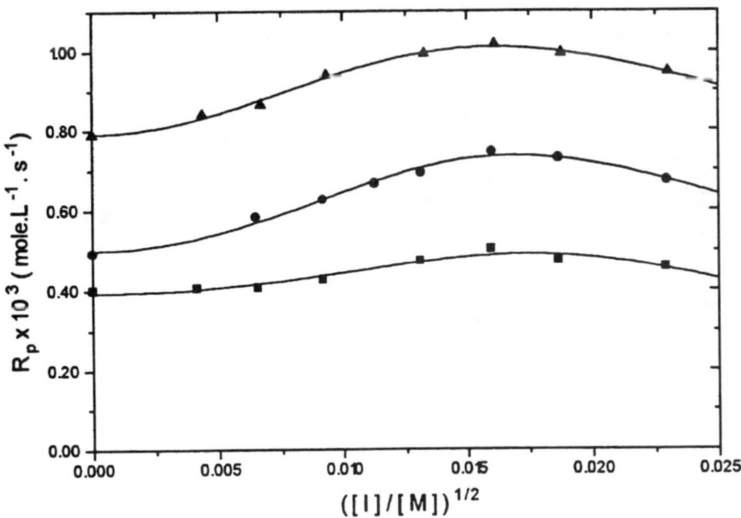


*Fig. 10 Dependance of Rp on iniferter concentration for the polymerization of MMA using macro biscarbamoyl disulfide based on PDMS.* [Ref. 95]
■ 96.5° C, ● 106.5°C, ▲ 120° C

The various kinetic parameters calculated by applying this model are given in **Table 12a** for the macro biscarbamoyldisulfide ( MBD ) along with the corresponding values for the thiuram analogues MTD ( in **Table 12b** ) for comparative purpose [96].

***Table 12a*** *Kinetic parameters for the polymerization of MMA using MBD's*

| | 96.5°C | | 106.5°C | | 120°C | |
|---|---|---|---|---|---|---|
| | MBD-1 (a) | MBD-2 (b) | MBD-1 | MBD-2 | MBD-1 | MBD-2 |
| $10^3.A$ ($mol/L.s^{1/2}$) | 7.1 | 7.23 | 11.3 | 14.7 | 29.2 | 22.1 |
| B | 2220 | 0 | 970 | 0 | 1350 | 760 |
| $10^{-6}.C$ | 0 | 3.6 | 2.2 | 6.7 | 4.3 | 5.8 |
| $10^6$ ($2fk_d$) $s^{-1}$ | 6.3 | 6.3 | 12 | 18 | 46 | 26 |
| $\tau_{1/2}$ (c) | 61 | 61 | 35 | 21 | 8.5 | 15 |
| $C_{tr}$ | - | 0.31 | 0.64 | 0.47 | 1.24 | 0.55 |
| $E_d$ | 102 KJ/mol (MTD-1) | | | 98 KJ/mol (MTD-2) | | |
| $E_{tr}$ | - | | | 60 KJ/mol (MTD-2) | | |

(a), (b) MBD : Macro biscarbamoyl sulfide of PDMS, molecular weight, (a) 8 500 , (b) 13 000

(c) apparent half-life in hour

***Table 12b*** *Kinetic parameters for the polymerization of MMA using MTD's*

| | 80°C | | 90°C | | 100°C | |
|---|---|---|---|---|---|---|
| | MTD-1 (a) | MTD-2 (b) | MTD-1 | MTD-2 | MTD-1 | MTD-2 |
| $10^3.A$ ($mol/L.s^{1/2}$) | 3.87 | 2.83 | 8.1 | 7.2 | 15.5 | 14.6 |
| B | 1560 | 129 | 980 | 448 | 1350 | 491 |
| $10^{-6}.C$ | 1.9 | 6.5 | 3.5 | 6.7 | 3.2 | 9.8 |
| $10^6$ ($2fk_d$) $s^{-1}$ | 3.4 | 1.9 | 10.5 | 8.3 | 26.1 | 23.7 |
| $\tau_{1/2}$ (c) | 113 | 203 | 37 | 46 | 15 | 16 |
| $C_{tr}$ | 0.46 | 0.38 | 0.61 | 0.52 | 0.72 | 0.62 |
| $E_d$ | 112 KJ/mol (MTD-1) | | | 139 KJ/mol (MTD-2) | | |
| $E_{tr}$ | 56 KJ/mol (MTD-1) | | | 58 KJ/mol (MTD-2) | | |

(a), (b) MTD : Macro thiuram disulfide of PDMS, molecular weight, (a) 4 150 , (b) 6 100

(c) apparent half-life in hour

It is clear from values of $2fk_d$ that biscarbamoyl disulfide needs sufficiently higher thermal activation than the thiuramdisulfide to show any meaningful initiation behaviour, despite that the energy of activation for the thermal dissociation is lower. This apparent contradiction comes probably from the fact that the formed radicals might be too stable to initiate the polymerization and that they need undergo secondary reactions to produce reactive radicals. At higher temperature, the radical generated by the secondary reaction might be causing the polymerization.

However they were found to be excellent chain transfer agents. The MBD as well as the MTD showed the usual chain length dependance for the transfer reactions. The reason for the differences in the properties of simple carbamoyl disulfides and their polymeric version is not fully understood.

A polymeric version of the biscarbamoyl disulfides, based on PDMS backbone was also synthesised in a way similar to the polythiuram disulfides [95]. Its iniferter capability was also found to be more or less the same as that of the macrodisulfides. Block copolymers of PDMS with MMA and styrene could be synthesized using them. These copolymers however, carried less of PDMS, owing to the concomitant formation of non negligible quantities of homopolymers induced by thermal polymerization.

## 11. CONCLUSION

The search for newer molecules and methodologies for the facile synthesis of functional polymers, and block and graft copolymers by free radical route is gaining importance in view of the increasing demand of these materials in various phases of modern technology. The investigation on thiuram disulfides as thermal iniferters gives some thrust to this area and it adds a new chapter to the free radical polymerization chemistry. The iniferter method of synthesis of telechelics and block copolymers has uniqueness in many respects, but it is not free from shortcomings. The beauty of the technique lies in the facility of realization of both iniferters and polymers, possibility to have different combinations of blocks, etc. Its incapability to initiate pure electron-deficient monomers like acrylates is a major drawback. The method works well for microiniferters and short chain containing-polymeric and macroiniferters but when the sequence becomes long, thermodynamic and diffusion controlled phenomena take over the chemically controlled kinetics which tend to reduce their efficiency.

## REFERENCES


1. G. Reiss, and P. Bahadur, "Block copolymers," in "Encyclopedia of Polymer Science and Engineering," Eds. H.F. Mark, N.M. Bikales, C.G. Overberger, and G. Menges, Wiley Interscience, Vol. 2, p. 324
2. D.H. Kaelble, *Polym. Sci. and Technol.* **9A**, 199 (1975)
3. D.R. Paul, in "Polymer Blends", Vol 2, Academic Press, Eds D.R. Paul and S. Newman, New-York, 35-60 (1978)
4. E.N. Kresge in ref. 3 p 293-317
5. G. Krauss in ref. 3 p 243-260
6. US Pat 4,027,072( May 31, 1977), R.E. Molari Jr. ( to General Electric Co )
7. Jpn. Kokai Tokkyo Koho 7,950,084( April 19, 1979), T. Yamamoto ( to Gunz Ltd)
8. Jpn. Kokai 7,611,823( Jan. 30, 1976), A. Aoki, K. Satake and K. Sakamoto, (to Asahi Chem. Industries Co Ltd.)
9. A.C. Levy and R. Sabia, Proc. Int. Wire Cable Symp. **31**, 313 (1982)

10. U.S.S.R. Pat 668,306( Oct. 7, 1980), I.P. Storozhuk, E. I. Levin, P.M. Valetskii, S. V. Vinogradova, V.V.Korshak, A.S. Kogan, E. V. Karpecheva, V.A. Sergeev, A.G. Sinaiskii, ( to Institute of Heterogenic Compounds, Academy of Science, USSR )
11. G. Riess, M. Bourdeaux, M. Brié, and G. Jouquet, International Conference on Carbon Fibres, London, Feb. 1, (1974). The Plastic Institute, London (1974)
12. I.R. Schmolka, *J. Am. Oil Chem. Soc.*, **54**, 110 (1977)
13. L.G. Lunsted, and I.R. Schmolka, "Block and Graft Copolymerization". Ed. R.J. Caresa, Vol. 2, John Wiley & Sons Inc., New-York (1976)
14. K.E.J. Barett, "Dispersion Polym. in Organic Media", John Wiley & Sons Inc., New York (1975)
15. D.H. Everett and J.F. Stageman, *J. Coll. Polym. Sci.*, **255**, 293 (1977)
16. J.V. Dawkins and G. Taylor, *Eur. Polym. J.*, **15**, 453 (1979)
17. J.V. Dawkins and G. Taylor, *J. Chem. Soc., Faraday Trans.* **1**, 76, 1263 (1980)
18. O.H. Lablanc, Jr. and W.T. Grubb, *Anal. Chem.*, **48**, 1658 (1976)
19. U.S. Pat. 4,177,222(Dec. 4, 1979), R.M. Thompson ( to Suntech Inc.)
20. U.S. Pat. 4,314,561( Feb. 9, 1982), D.S. Kaplan ( to American Cyanamid Co.)
21. U.S. Pat 4,215,691(Aug. 5, 1980), P.S. Wong ( to Alza Corp.)
22. M. Swarc, *Nature,* **178**, 1168 (1956)
23. R.P. Quirk, "Anionic synthesis of polymers with functional groups" in "Comprehensive Polymer Science", Eds S. Aggarwal and S. Russo, first supplement, Chapt. 5
24. S. Bywater, "Anionic Polymerization", Ref. 1, Vol. 2
25. S. Bywater, *Prog. Polym. Sci.*, **4**, 27 (1974)
26. J.E. McGrath, A.C.S. Symp. Ser., **166**, A.C.S. Washington D.C. (1981)
27. J.P. Kennedy "Macromolecular Engineering by Carbocationic Polymerization", In Proceeding of the Robert A. Welch Foundation Conference on Chemical Research, Houston, (1982), Vol XXVI, "Synthetic Polymers", Chap. III
28. J.P. Kennedy, and E. Marechal, "Carbocationic Polymerization", Wiley Interscience, New-York (1982)
29. C.H. Bamford and A.D. Jenkins, *Nature,* **176**, 78 (1955)
30. B.B. Idage, S.P. Vernekar and N.D. Ghatge, *J. Polym. Sci., Chem. Ed.*, **4**, 307 (1983)
31. Brit. Pat. 957,652( Sept. 17, 1962) (to Thiokol Corp.)
32. Y.N. Barentsevich and S.S. Ivanchev, *Polym. Sci. USSR*, **25**, 2341 (1983)
33. C.S. Sheppard, R.A. Bafford and R.E. Macleay, Fr. Pat. 1,905,915(1970), Chem. Abst. **72**, 13201
34. V.I. Galibei and S.S. Ivanchev, *Zh. Organ. Khim.*, **6**, 1585 (1970)
35. Ibid. **44**, 1630 (1974)
36. J.C. Brosse, D. Derouet, F. Epaillard, J.C. Soutif, G. Legay and K. Dusek, *Adv. in Polym. Sci.*, **289**, 167 (1987)
37. A. Bledski, *Kunststoffe,* **73**(3), 156 (1983)
38. D. Braun, *Makromol. Chem. Symp.*, **4**, 41 (1986)

39. A. Bledski and D. Braun, *Makromol. Chem.*, **187**, 2599 (1986)
40. A.J. Constanza, R.J. Coleman, R.M. Pierson, C.S. Marvec and C. King, *J. Polym. Sci.*, **17**, 319 (1955)
41. R.M. Pierson, A.J. Constanza and A.H. Weinstein, *J. Polym. Sci.*, **17**, 221 (1955)
42. C.P. R. Nair, *Eur. Polym. J.*, **28**(12), 1527 (1992)
43. T. Corner, *Adv. Polym. Sci.*, **62**, 97 (1984)
44. C.P.R. Nair and G. Clouet, unpublished results
45. L.A. Miller and V. Stannett, *J. Polym. Sci.* A1, **7**, 3159 (1969)
46. V.A. Dinaburg and A.A. Vansheidt, *Zh. Obshch. Khim.*, **24**, 840 (1954)
47. Brit. Pat. 945,713 (1964) (to Thiokol Corp.)
48. T. Okano, M. Katayama and I. Shinohala, *J. Appl. Polym. Sci.*, **22**, 369 (1978)
49. K. Tsuda and T. Otsu, *Bull. Chem. Soc. Jpn*, **39**, 2206 (1966)
50. T. Otsu, Y. Kinoshita and M. Imoto, *Makromol. Chem.*, **73**, 225 (1964)
50a. C.A. Uraneck, H.L. Hsich, and R.J. Sonnenfeld, *J. Appl. Polym. Sci.*, **13**, 149 (1969)
51. T. Okano, M. Katayama and I. Shinohala, *J. Appl. Polym. Sci.*, **22**, 369 (1978)
52. T.E. Bishop, G. Pasternack and J.M. Zimmerman, Eur. Pat. Appl No. 0,162,304 (20.04.85)
53. C.M. Starks, "Free Radical Telomerizations", Academic Press, New-York (1974)
54. B. Boutevin, *Adv. Polym. Sci.*, **94**, 70 (1990)
55. B. Boutevin and Y. Pietrasanta "Telomerization" in "Comprehensive Polymer Science" Vol. 3, Chain Polymn. Part 1, p. 185, Pergamon Press
56. T. Otsu, M. Yoshioka, and T. Tanaka, *Eur. Polym. J.*, **28**, 1325 (1992)
57. C. Simionescu, K.G. Sik, S. Dumitriu, E. Comanita, and M. Pastravanu, Polym. Bull., **15**, 503 (1986) and Prog. Polym. Sci., **12**, 1 (1986)
57a. B. Hazer, A. Ayas, N. Besirli and N. Saltek, *Makromol. Chem.*, **190**, 1987 (1989)
58. P.S. Anand, H.G. Stahl, W. Heitz , G. Weber and L. Bottenbruch, *Makromol. Chem.* **183**, 1685 (1982)
59. J.L. Laverty and Z.G. Gardlung, *J. Polym. Sci., Polym. Chem. Ed.*, **15**, 2001 (1977)
59a. B. Hazer and B.M. Baysal, *Polymer*, **27**, 961 (1986)
60. T.O. Ahn, J.H. Kim, J.C. Lee, H.M. Jeong and J.Y. Pable, *J. Polym. Sci., Chem. Ed.*, **31**, 435 (1993)
61. Y. Hepuzer, M. Bektas, D.S. Denizligil, A. Onen and Y. Yagci, *Macromol. Report,* **A30**, 111 (1993)
62. J.V. Crivello, D.A. Conlon and J.L. Lee, *J. Polym. Sci., Polym. Chem. Ed.*, **24**, 1197 and 1251 (1986)
63. J.V. Crivello, J.L. Lee and D.A. Conlon, *Polym. Bull.*, **16**, 905 (1986)
64. R.G. Santos, P. Chaumont, J.E. Herz and G. Beinert, *Eur. Polym. J.*, **28**, 1263 (1992)
65. G. Malsch, J. Barton, V. Horanska-Vaskova and P. Fritzsche, *Acta Polym.*, **31**, 118 (1980)

66. C.H. Bamford and S.U. Mullik, *Polymer*, **17**, 94 (1976)
67. C.H. Bamford and X.Z. Han, *Polymer*, **22**, 1299 (1981)
68. E.B. Milovskaya and L.V. Zamoiskaya, *Vysokomol. Soedin. Ser.*, **B18**, 300 (1976)
68a. M.L. Hallensleben, *Eur. Polym.* J., **13**, 437 (1977)
69. T.E. Ferington and A.V. Tobolsky, *J. Am. Chem. Soc.*, **77**, 4510 (1955)
70. T.E. Ferington and A.V. Tobolsky, *J. Am. Chem. Soc.*, **80**, 3215 (1958)
71. E. Staudner and J. Beniska, *Eur. Polym. J. Suppl.* 537 (1969)
72. E. Staudner, G. Kysela, J. Beniska and D. Mikilaj, *Eur. Polym. J.*, **14**, 1067 (1978)
73. T. Otsu and L.M. Yoshida, *Makromol. Chem. Rapid Commun.*, **3**, 127 (1982)
74. T. Otsu, T. Matsunaga, A. Kuriyama and M. Yoshioka, *Eur. Polym. J.*, **25**, 643 (1989)
75. A. Kuriyama and T. Otsu, *Polym. J.*, **16**, 511 (1984)
76. T. Otsu, K. Yamashita and K Tsuda, *Macromolecules*, **19**, 287 (1986)
77. S.R. Turner and R.W. Blevins, *Polym. Prepr.*, **29**, 6 (1988)
78. P. Lambrinos, M. Tardi, A. Polton and P. Sigwalt, *Eur. Polym. J.*, **26**, 1125 (1990)
79. T. Otsu and T. Tazaki, *Polym. Bull.*, 16, 277 (1986); a). M. Niwa, T. Matsumoto and H. Izumi, *J. Macromol. Sci., Chem. Ed.*, **24**, 567 (1987)
80. C.P.R. Nair and G. Clouet, *J. Macromol. Sci. Rev.*, **C31** (2&3), 311 (1991)
81. C.P.R. Nair, G. Clouet and P. Chaumont, *J. Polym. Sci., Polym. Chem. Ed.*, **27**, 1795 (1989)
82. C.P.R. Nair and G. Clouet, *Makromol. Chem.*, **190**, 1243 (1989)
83. G. Clouet and H.J. Juhl, *Makromol. Chem.* (in press)
83a. S.A. Haque and G. Clouet, *Makromol. Chem.* (in press)
84. C.P.R. Nair, M.C. Richou and G. Clouet, *Makromol. Chem.*, **192**, 579 (1991)
85. C.P.R. Nair, M.C. Richou, G. Clouet and P. Chaumont, *Eur. Polym. J.*, **26**, 811 (1990)
86. G. Clouet and D. Kayser, *J. Polym. Sci., Polym. Chem. Ed.* (in press)
87. G. Clouet and D. Kayser, *J. Polym. Sci., Polym. Chem. Ed.* (in press)
88. C.P.R. Nair, P. Chaumont and G. Clouet, *J. Macromol. Sci. Chem.* **A27**, 791 (1990)
89. S.A. Haque and G. Clouet, *J. Polym. Mater.* (in press)
90. G. Clouet and M.C. Richou (unpublished results)
91. C.P.R. Nair and G. Clouet, *Macromolecules*, **23**, 1361 (1990)
92. C.P.R. Nair and G. Clouet, *Polymer*, **29**, 1909 (1988)
93. C.P.R. Nair, G. Clouet and Y. Guilbert, *Polym. Degrad. Stab.*, **26**, 305 (1989)
94. R. Popielarz and G. Clouet, *Makromol. Chem.* (in press)
95. D. Kayser, Thèse de Doctorat, Université Louis Pasteur, Strasbourg, France (1992)
96. D. Kayser, P. Chaumont, C.P.R. Nair and G. Clouet, *Polym. Prepr.* (Am. Chem. Soc., Div. Polym. Chem.) August 1993

# Macroiniferters Chapter 12

## MACROINIFERTERS: CONTROLLED SYNTHESIS AND DESIGN THROUGH LIVING RADICAL POLYMERIZATION

**Takayuki Otsu*** **and Akikazu Matsumoto**

Department of Applied Chemistry
Faculty of Engineering
Osaka City University
Sugimoto, Sumiyoshi-ku
Osaka 558, JAPAN



---

* *Present address: Faculty of Biology-Oriented Science and Technology, Kinki University,* Uchida-cho, Naka-gun, Wakayama 649-64, Japan.



Macromolecular Design: Concepts and Practice (Macromonomers, Macroinitiators, Macroiniferters, Macroinimers, Macroinifers, and Macroiniters) Pages 471 - 486

# 1. INTRODUCTION

Radical polymerization is the most useful method for large-scale preparation of various vinyl polymers, and more than 50% of all plastics have been produced industrially through radical polymerization, because this method has a number of merits for the vinyl polymer synthesis beyond ionic or coordination polymerizations. However, the radical polymerization still has some unsolved problems; one of which is the structure control of the resulting polymer. The reactivity control of the monomers and their radicals is now the most important problem to design the polymer architechtures with well-controlled molecular weight, molecular weight distribution, monomer sequences, chain end groups, and branching, etc.

Some attempts have been done to find living radical polymerization systems [1] since Szwarc's first discovery of the living polymer in anionic polymerization [2]. However, because the propagating radical in radical polymerization is very short-lived in a homogeneous system, it is impossible to obtain a long-lived (living) propagating radical except the case in which its mobility decreases markedly. Actually, propagating radicals are detected in the polymerization systems in precipitation, microgel, microsphere, highly viscous medium, and frozen state. In these polymerizations, however, we cannot expect a living radical polymerization mechanism so that some ideas should be indispensable for the living radical polymerization, be different from living anionic polymerizations.

That is,

(i) short-lived radicals such as a propagating radical exist as covalent coupling product,

(ii) such covalent bonds produce a propagating radical by thermal or photo-dissociation, and

(iii) the reaction (termination) between radicals reproduce covalent bond which has an identical structure with dissociation ability.

In 1982, we have proposed a concept of iniferter (<u>ini</u>tiator-trans<u>fer</u> agent-<u>ter</u>minator) for designing of the polymer chain-end structure [3], and simultaneously a new model for living radical polymerization in a homogeneous system by using some iniferters [4]. These concept and model are summarized in the following section.

# 2. CONCEPT OF INIFERTER AND MODEL FOR LIVING RADICAL POLYMERIZATION

The polymer formation in radical polymerization of vinyl monomers (M) initiated by an initiator (B-B) is expressed by eq. (1), if mutual termination proceeds by recombination or disproportionation and no chain transfer reaction occurs.

$$\text{B–B} + 2n\,\text{M} \longrightarrow \text{B}\text{–}(\text{M})_{2n}\text{–B}$$

or

$$\text{B}\text{–}(\text{M})_{n-1}\text{–M(+H)} + \text{B}\text{–}(\text{M})_{n-1}\text{–M(–H)} \qquad (1)$$

When the termination involves only recombination, the polymerization gives a polymer with two initiator fragments at its both chain ends. However, in ordinary radical polymerizations, the termination by disproportionation and the chain transfer reactions occur importantly, and it is impossible to control the ratio of recombination and disproportionation, which is determined by the structure of polymer radicals (i.e., structure of monomers) and temperature. Therefore the number of initiator fragments per one polymer molecule is always less than 2.

If we use initiators which have very high reactivities for chain transfer to the initiator and/or primary radical termination, i.e., ordinary bimolecular termination (recombination and disproportionation) is negligible, it is expected to be obtained a polymer with two initiator fragments at the chain ends [eq. (2)].

$$\begin{aligned} &\text{B–B} + n\,\text{M} \longrightarrow \text{B}\!-\!(\text{M})_n\!-\!\text{B} \\ &\text{A–B} + n\,\text{M} \longrightarrow \text{A}\!-\!(\text{M})_n\!-\!\text{B} \end{aligned} \quad (2)$$

Therefore, these radical polymerizations may simply be considered as an insertion of monomer molecules into the B-B and A-B bonds of the B-B and A-B type iniferters, respectively, leading to the polymer with two initiator fragments. For the initiators with such functions we proposed to call iniferter [3].

Tetraethylthiuram disulfide (TD) was used first in 1956 for polymerization of styrene (St) or methyl methacrylate (MMA), and the resulting polymers bearing dithiocarbamate group (TD-polymer) were found to serve further as a polymeric iniferter, i.e. macroiniferter, under irradiation of light to give block copolymers [5]. Thereafter, tetraphenylethane derivatives for MMA polymerization and the resulting poly(MMA) were observed to act as an iniferter and a macroiniferter, respectively [6-8].

$$(C_2H_5)_2N\text{–}\overset{\overset{\displaystyle S}{\|}}{C}\text{–S–S–}\overset{\overset{\displaystyle S}{\|}}{C}\text{–}N(C_2H_5)_2 \xrightarrow{n\,\text{M}} (C_2H_5)_2N\text{–}\overset{\overset{\displaystyle S}{\|}}{C}\text{–S}\!-\!(\text{M})_n\!-\!\text{S–}\overset{\overset{\displaystyle S}{\|}}{C}\text{–}N(C_2H_5)_2 \quad (3)$$

TD (B-B type) TD-Polymer (A-B type)

From a viewpoint of the tailor-made polymer synthesis, the iniferters can be classified into several types; mono-, di-, tri- or poly-functional iniferters; thermal or photoiniferters; monomeric and macroiniferters; monomer- or macromonomer-iniferters, etc. For example, some sulfur compounds having *N,N*-diethyldithiocarbamate moiety were found to serve as excellent photoiniferters. When benzyl *N,N*-diethyldithiocarbamate (BDC) and *p*-xylylene bis(*N,N*-diethyldithiocarbamate) (XDC) were used as A-B type mono- and difunctional photoiniferters of the polymerization of St, the resulting polymers which can serve as mono- and difunctional macroiniferters, respectively, are produced.

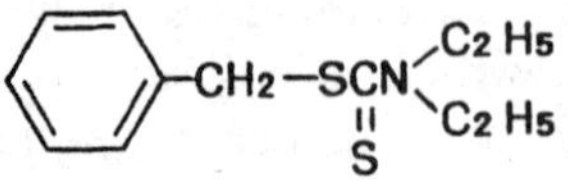

$$C_6H_5{-}CH_2{-}S{-}\overset{\overset{S}{\|}}{C}{-}N(C_2H_5)_2 \qquad (C_2H_5)_2N{-}\overset{\overset{S}{\|}}{C}{-}S{-}CH_2{-}C_6H_4{-}CH_2{-}S{-}\overset{\overset{S}{\|}}{C}{-}N(C_2H_5)_2$$

BDC 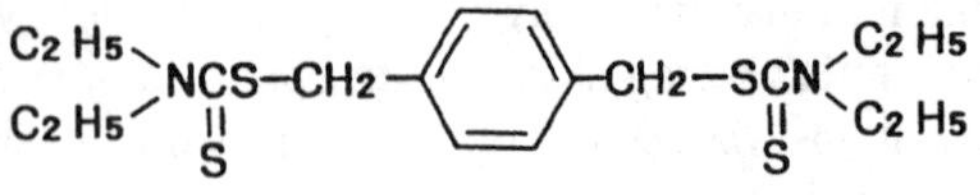 XDC

If the end groups of the polymers obtained by radical polymerization using certain iniferters still have an iniferter function, this radical polymerization is expected to proceed via a living radical mechanism even in a homogeneous system, i.e.,both yield and molecular weight ($M$) of the polymers produced increase with a reaction time. The proposed model is shown in eq. (4) [4].

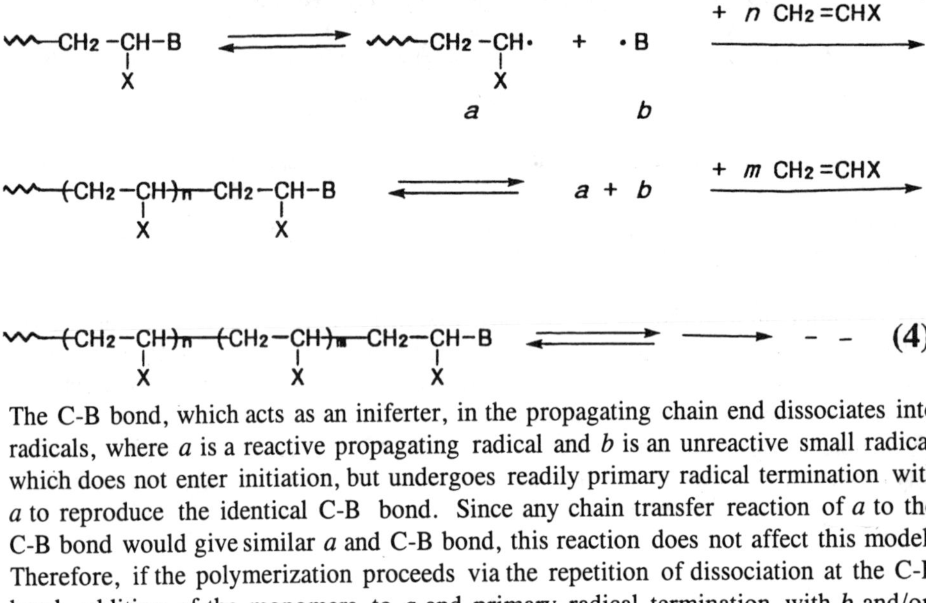

$$\sim\!CH_2{-}\underset{X}{CH}{-}B \rightleftharpoons \sim\!CH_2{-}\underset{X}{CH}\cdot\ (a) + \cdot B\ (b) \xrightarrow{+\,n\,CH_2=CHX}$$

$$\sim\!(CH_2{-}\underset{X}{CH})_n{-}CH_2{-}\underset{X}{CH}{-}B \rightleftharpoons a + b \xrightarrow{+\,m\,CH_2=CHX}$$

$$\sim\!(CH_2{-}\underset{X}{CH})_n{-}(CH_2{-}\underset{X}{CH})_m{-}CH_2{-}\underset{X}{CH}{-}B \rightleftharpoons \longrightarrow -\,- \qquad (4)$$

The C-B bond, which acts as an iniferter, in the propagating chain end dissociates into radicals, where $a$ is a reactive propagating radical and $b$ is an unreactive small radical which does not enter initiation, but undergoes readily primary radical termination with $a$ to reproduce the identical C-B bond. Since any chain transfer reaction of $a$ to the C-B bond would give similar $a$ and C-B bond, this reaction does not affect this model. Therefore, if the polymerization proceeds via the repetition of dissociation at the C-B bond, addition of the monomers to $a$ and primary radical termination with $b$ and/or chain transfer of $a$ to the C-B bond, such polymerization may proceed via a living radical mechanism. As an extreme case, if the polymerization proceeds via a stepwise insertion of one monomer molecule into the C-B bond (i.e., repeated cycles of dissociation, one monomer addition, and subsequent primary radical termination), it would result in a successive reaction, in which $a$ is represented by an intermediate radical paired with $b$.

The polymerizations of St and MMA with some organic compounds, especially containing dithiocarbamate groups as photoiniferters [9-14], and of MMA with phenylazotriphenylmethane [15] or tetraphenylethane derivatives [6,7] as thermal iniferters were found to proceed via a mechanism close to the model proposed in eq. (4). Therefore, the choice of such iniferters in vinyl polymerization seems to open a novel route for synthesizing of various types of polymers such as functional, telechelic, block, and graft polymers via a radical polymerization technique.

## 3. CONTROLLED SYNTHESIS AND DESIGN OF BLOCK COPOLYMERS

**Figures 1-4** show the time-conversion and time-*M* relationships in the photopolymerization of St and MMA with TD, BDC, and XDC. The yield and *M* of the resulting polymer increased with the polymerization time, i.e., conversion. The fact that *M* increases with the conversion is quite different from ordinary radical polymerization with common initiators, and it may suggest that these polymerizations proceed via a living radical polymerization mechanism according to eq. (4), because both S-S and C-S bonds in TD and BDC or XDC can dissociate photochemically into radicals.

From the analysis of end groups of the polymer chain, it was confirmed that the number of the carbamate groups were two during polymerization of St with TD, suggesting that TD gives a difunctional polymeric photoiniferter, i.e. macroiniferter. However, the terminal carbamate groups of the both chain ends are not identical, i.e., the carbamate group bond to head and tail positions of the terminal St monomer units of the polymer, as shown in eq. (5) [16].

$$\mathrm{TD} \xrightarrow[h\nu]{n\ \mathrm{St}} (C_2H_5)_2N\overset{C}{\downarrow}C(=S)-\overset{B}{\downarrow}S-CH_2-CH(C_6H_5)-(CH_2-CH(C_6H_5))_{n-2}-CH_2-CH(C_6H_5)\overset{A}{\downarrow}-SC(=S)N(C_2H_5)_2 \quad (5)$$

TD-Poly(St)

In the head end, the dissociation always occurs at the bond which is indicated by the arrow A, giving a more reactive carbon-centered radical and a less reactive thiyl radical which leads to termination of the active chain ends. In the case of the tail end, there is a possibility that the bond at the C position dissociates to producc the $(C_2H_5)_2NC(=S)^\bullet$ and thiyl radical [17] in addition to the preferable bond scission at B. Such a dissociation at C may not induce a controlled living radical polymerization.

The time-conversion and time-*M* relationships in the photopolymerizations of St with BDC and XDC are shown in Figure 3, in which the carbamate group concentration in these iniferters was identical, i.e.,[BDC]/2 = [XDC]. The both time-conversion curves are superimposable each other, indicating that the C-S bonds in BDC and the resulting poly(St) (BDC-poly(St)) may dissociate with the same probability as those in XDC and XDC-poly(St) into benzylic and thiyl radicals which show the identical reactivities. This result may also suggest that the polymerizations by BDC and XDC proceed via living radical mechanisms, and that BDC-poly(St) and XDC-poly(St) serve as mono- and difunctional macro-photoiniferters, respectively. The yield and *M* are observed to

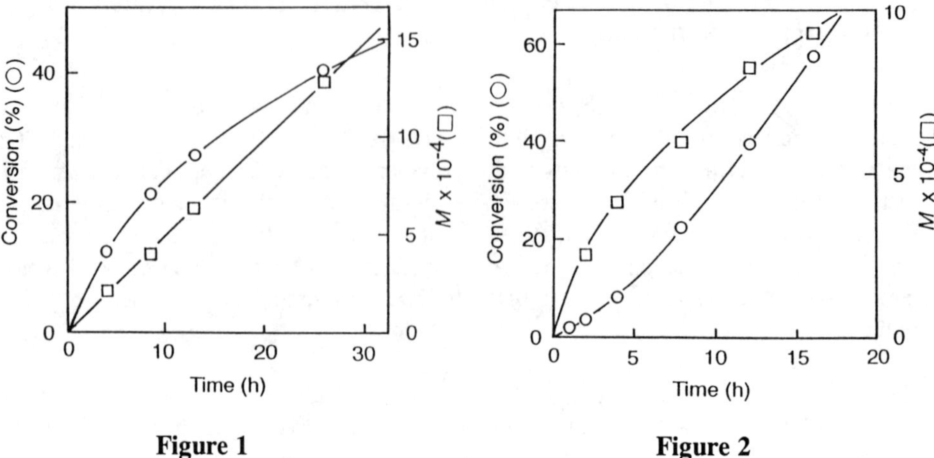


**Figure 1.** Time-conversion and time-$M$ relationship for photopolymerization of St in bulk with TD at 30°C; [TD] = $7.7 \times 10^{-3}$ mol/L.

**Figure 2.** Time-conversion and time-$M$ relationship for photopolymerization of MMA with TD in benzene at 30°C; [MMA] = 4.7 mol/L, [TD] = $4.6 \times 10^{-3}$ mol/L.

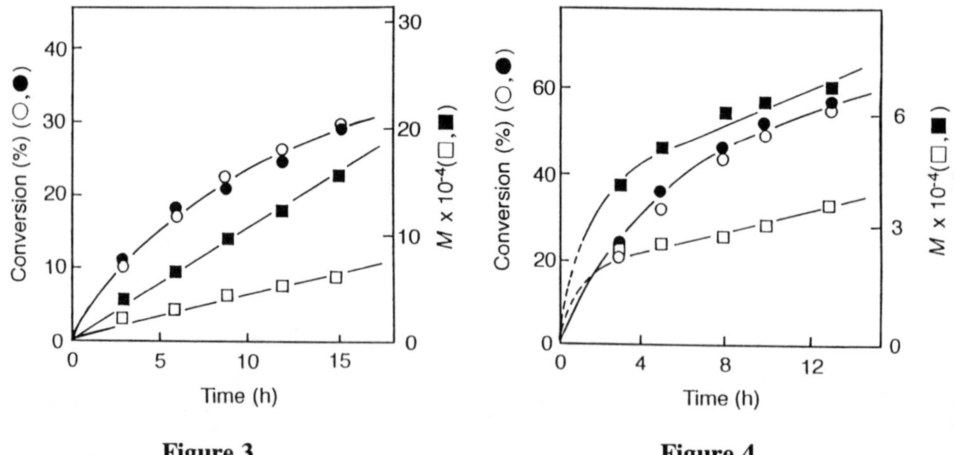


**Figure 3.** Time-conversion and time-$M$ relationship for photopolymerization of St with BDC and XDC in benzene at 30°C; [St] = 6.9 mol/L, [BDC] = $7.8 \times 10^{-3}$ mol/L, [XDC] = $3.8 \times 10^{-3}$ mol/L: ( ○ , □ ) BDC, ( ● , ■ ) XDC.

**Figure 4.** Time-conversion and time-$M$ relationship for photopolymerization of MMA with BDC and XDC in benzene at 30°C; [MMA] = 2.8 mol/L, [BDC] = $7.8 \times 10^{-3}$ mol/L, [XDC] = $3.8 \times 10^{-3}$ mol/L: ( ○ , □ ) BDC, ( ● , ■ ) XDC.

increase as a function of the polymerization time (i.e., conversion). The slope of the curves of the molecular weight relationships for XDC was twice as large as that with BDC, i.e., *M* of the polymer obtained with XDC are large twice as high as that with BDC. Similar results were also found for radical polymerizations of MMA with BDC and XDC, and of St and MMA with other sulfides such as benzyl *N*-ethyldithio - carbamate, *p*- xylylene bis(*N*-ethyldithiocarbamate) [14] and diphenyl sulfide [11] as photoiniferters. The number of the end group per one polymer chain are almost constant, i.e., 1 and 2 for BDC and XDC, respectively, independent of the polymerization time, strongly supporting that these polymerizations are performed according to eq. (4), and the polymers with the carbamate end groups are always produced.

The changes in *M* and its dispersity ($M_w/M_n$) of the poly(St) were further determined as a function of the reaction time. The results are shown in **Figure 5**, from which it is noted that $M_w/M_n$ of the polymers increases as well as *M* with the reaction time (conversion), contrary to living ionic polymerizations, which give polymers with a narrow molecular weight distribution, i.e., $M_w/M_n$ is close to unity. This observation is understandable from the facts that the radical-producing step is slow (i.e., rate-determining step) in the radical polymerization, and in the case using TD the dissociation of the S-S bond in TD and the C-S bond in the polymer chain end may occur with different rates.

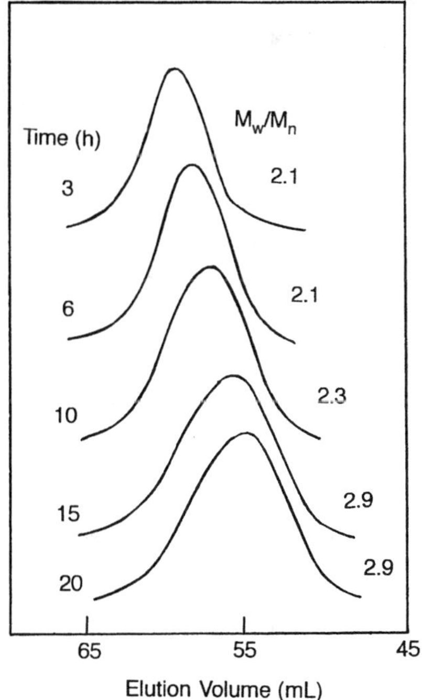


**Figure 5.** GPC elution curves of polymers obtained from photopolymerization of St in bulk with BDC at 30°C; [BDC] = 7.3 x $10^{-3}$ mol/L.

From the observed slope and linearity of the conversion-*M* relationships, the living nature in the polymerization of St or MMA with TD, BDC, and XDC as photoiniferters was fairly high. However, the efficiency of the block copolymer formation was about 70-90 %, as describe below, indicating that some undesirable side reactions leading to deactivation of the iniferter site as a dormant propagating radical species disturbing the ideal living radical polymerization (eq. (4)) might occur during polymerization, e.g., ordinary bimolecular termination, initiation by the carbamate radicals, and photo-dissociation of the bonds other than the specified bond leading to living radical polymerization, etc.

As described above, the polymers obtained by BDC and XDC further serve as mono- and difunctional macroiniferters, respectively [12]. If these BDC-poly(St) and XDC-poly(St) are used for the polymerization of MMA as a second monomer, AB and ABA type block copolymers, respectively, would be synthesized, as shown in eqs. (6) and (7).

$$\text{BDC} \xrightarrow{n\ \text{St}} \text{C}_6\text{H}_5\text{-CH}_2\text{-(St)}_n\text{-SC(=S)N(C}_2\text{H}_5)_2 \xrightarrow{m\ \text{MMA}}$$

BDC-Poly(St)

$$\text{C}_6\text{H}_5\text{-CH}_2\text{-(St)}_n\text{-(MMA)}_m\text{-SC(=S)N(C}_2\text{H}_5)_2 \quad (6)$$

$$\text{XDC} \xrightarrow{n\ \text{St}} (\text{C}_2\text{H}_5)_2\text{NC(=S)S-(St)}_n\text{-CH}_2\text{-C}_6\text{H}_4\text{-CH}_2\text{-(St)}_n\text{-SC(=S)N(C}_2\text{H}_5)_2 \xrightarrow{m\ \text{MMA}}$$

XDC-Poly(St)

$$(\text{C}_2\text{H}_5)_2\text{NC(=S)S-(MMA)}_m\text{-(St)}_n\text{-CH}_2\text{-C}_6\text{H}_4\text{-CH}_2\text{-(St)}_n\text{-(MMA)}_m\text{-SC(=S)N(C}_2\text{H}_5)_2 \quad (7)$$

The results of block copolymerizations of MMA with macroiniferters obtained by BDC and XDC are shown in Table 1. When the *M* of BDC-poly(St) and XDC-poly(St) increased, i.e.,the concentration of the iniferter-end group decreased, the total yield of MMA polymerization decreased. The yields of the AB- and ABA-type block copolymers with BDC-poly(St) and XDC-poly(St) were 70-90 %.

These block copolymer syntheses are advantageous for the synthesis of the polymer consisting of many kinds of vinyl monomer units, especially polar and functional monomers, such as acrylic acid (AA) and vinyl acetate (VAc). Similar technique was also applied for the synthesis of AB and ABA block copolymers containing random and alternating copolymer sequences as shown in Table 2. Poly(St-*co*-MMA)-*b*-poly(VAc), poly(VAc)-*b*-poly(St-*co*-MMA)-*b*-poly(VAc), poly(St)-*b*-poly(DiPF-*alt*-IBVE), poly(IBVE-*alt*-MAn)-*b*-poly(St)-*b*-poly(IBVE-*alt*-MAn), in which DiPF, IBVE, and MAn are diisopropyl fumarate, isobutyl vinylether, and maleic anhydride, respectively, were synthesized [18].

Table 1 Block Copolymerization of MMA with BDC-poly(St) and XDC-poly(St)[a)]

| Macro-iniferter | $M \times 10^{-4}$ | Total Yield (g) | Fraction extracted (%) Unreacted iniferter | $M_2$ homo-polymer | Block copolymer |
|---|---|---|---|---|---|
| BDC-poly(St) | 3.2 | 0.78 | 7 | 15 | 78 |
| | 4.3 | 0.64 | 8 | 16 | 76 |
| | 5.5 | 0.56 | 9 | 17 | 74 |
| | 6.3 | 0.50 | 13 | 16 | 71 |
| | 8.6 | 0.42 | 16 | 20 | 64 |
| XDC-poly(St) | 3.8 | 1.10 | 0 | 8 | 92 |
| | 6.3 | 0.99 | 2 | 9 | 89 |
| | 12.0 | 0.58 | 6 | 6 | 88 |
| | 15.0 | 0.49 | 0 | 8 | 92 |
| | 28.0 | 0.45 | 9 | 1 | 90 |

a) Poymerization at 30°C under irradiation of UV light from a distance of 10 cm; macroiniferter 0.2 g, [MMA] = 4.7 mol/L in benzene.

Table 2 Block Copolymer Synthesis Containing Random and Alternating Copolymer Sequences by Using Macroiniferter

| Macroiniferter | Monomer ($M_2$) | Time (h) | Fraction extracted (%) Unreacted iniferter | $M_2$ homo-polymer | Block copolymer |
|---|---|---|---|---|---|
| XDC-poly(St) | MMA | 7 | 2 | 9 | 89 |
| XDC-[poly(St)-*co*-poly(MMA)] | VAc | 8 | 5 | 13 | 82 |
| XDC-[poly(St)-*co*-poly(MMA)] | AA | 0.3 | 3 | 9 | 88 |
| XDC-[poly(St)-*co*-poly(IB)] | MMA | 3 | 15 | 19 | 66 |
| XDC-[poly(St)-*co*-poly(VAc)] | MMA | 3 | 14 | 3 | 83 |
| XDC-[poly(St)-*co*-poly(VC)] | MMA | 3 | 9 | 12 | 79 |
| XDC-poly(St) | MAn/IBVE | 6 | 0 | 31 | 69 |
| XDC-[poly(St)-*co*-poly(DiPF)] | MAn/IBVE | 4 | 6 | 24 | 70 |

# 4. SYNTHESIS OF MULTI-COMPONENT BLOCK, STAR AND GRAFT COPOLYMERS

The synthesis of some multi-block copolymers was attempted by successive polymerization using this iniferter technique. However, pure tri- or tetra-block copolymers free from homopolymers were not isolated by solvent extraction, because no suitable solvent for separation was found.

In 1963, Merrifield reported [19] a brilliant solid-phase peptide synthesis using a reagent attached to the polymer support. If a similar idea can be applied to the iniferter technique, pure block copolymer could be synthesized by radical polymerization. The dithiocarbamate group attached to a polystyrene-gel (PSG) through a hydrolyzable ester spacer was prepared and used as a PSG photoiniferter (eq. (8)) [20]. After photopolymerization of a monomer was carried out for a given time, the polymerization mixture was poured into excess precipitant to isolate the polymer, which was then extracted with a solvent to separate polymer grafted onto PSG and homopolymer.

$$\mathrm{PSG{-}CH_2OC(=O)CH_2Cl} \xrightarrow{\mathrm{NaSCSN(C_2H_5)_2}}$$

$$\mathrm{PSG{-}CH_2OC(=O)CH_2{-}SC(=S)N(C_2H_5)_2} \xrightarrow{n\ \mathrm{M_1}}$$

$$\mathrm{PSG{-}CH_2OC(=O)CH_2{-}\!\left(M_1\right)_{n}\!{-}SC(=S)N(C_2H_5)_2} \xrightarrow{m\ \mathrm{M_2}}$$

$$\mathrm{PSG{-}CH_2OC(=O)CH_2{-}\!\left(M_1\right)_{n}\!{-}\!\left(M_2\right)_{m}\!{-}SC(=S)N(C_2H_5)_2} \xrightarrow{l\ \mathrm{M_3}}$$

$$\mathrm{PSG{-}CH_2OC(=O)CH_2{-}\!\left(M_1\right)_{n}\!{-}\!\left(M_2\right)_{m}\!{-}\!\left(M_3\right)_{l}\!{-}SC(=S)N(C_2H_5)_2} \xrightarrow{\mathrm{Hydrolysis}}$$

$$\mathrm{PSG{-}CH_2OH} + \mathrm{HOC(=O)CH_2{-}\!\left(M_1\right)_{n}\!{-}\!\left(M_2\right)_{m}\!{-}\!\left(M_3\right)_{l}\!{-}SC(=S)N(C_2H_5)_2} \quad (8)$$

This grafted poly(St) onto PSG acted as a macroiniferter for radical polymerization of MMA. After extraction of homopolymer of MMA and hydrolysis of the resulting graft-block copolymer attached to PSG, a block copolymer of poly(St) with poly(MMA) was isolated by benzene extraction. The yield and $M$ increased with the reaction time, but the block copolymer thus obtained contain 10-15 % homopoly(St), indicating that a part of the iniferter site at the grafted poly(St) chain end was destroyed during the polymerization of St as a result of deviations from eq. (4).

To avoid homopoly(St) formation, TD, which can dissociate to *N,N*-diethyldithiocarbamate radicals, was added to the photopolymerization of St. When the polymer grafted onto PSG thus obtained was used as a PSG photoiniferter for the polymerization of MMA, followed by extraction and hydrolysis of the resulting graft-block copolymer, a block copolymer of poly(St) with poly(MMA) which contained only a trace of homopoly(St) was isolated. The polymer consisting of St-MMA block attached to PSG obtained was confirmed to induce further photopolymerization, leading to pure poly(St)-*b*-poly(MMA)-*b*-poly(St) triblock copolymers. In a similar way, poly(St)-*b*-poly(MMA)-*b*-poly(ClSt) triblock copolymer, poly(St)-*b*-poly(MMA)-*b*-poly(St)-*b*-poly(MMA) and poly(St)-*b*-poly(MMA)-*b*-poly(EMA)-*b*-poly(MOSt) tetrablock copolymers were also prepared, where ClSt, EMA, and MOSt are *p*-chlorostyrene, ethyl methacrylate, and *p*-methoxystyrene, respectively. These block copolymer syntheses were confirmed by GPC and NMR measurements of the polymers which were isolated at the each steps, as shown in **Figure 6**.

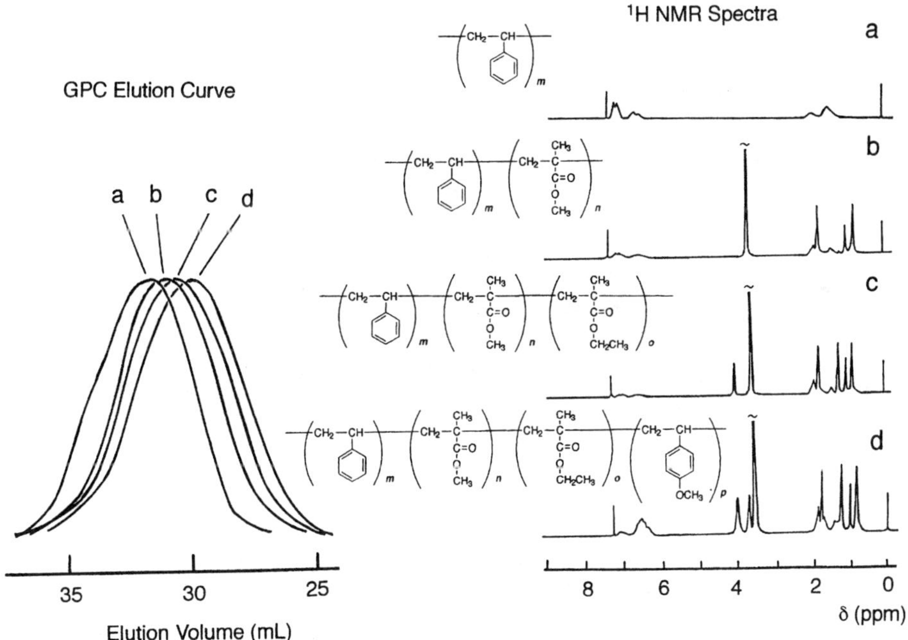


**Figure 6.** GPC elution curves and $^1$H NMR spectra of block copolymers: (a) poly(St), (b) poly(St)-*b*-poly(MMA), (c) poly(St)-*b*-poly(MMA)-*b*-poly(EMA), and (d) poly(St)-*b*-poly(MMA)-*b*-poly(EMA)-*b*-poly(MOSt).

1,2,4,5-Tetrakis(*N,N*-diethyldithiocarbamyl)benzene (DDC) was prepared as a tetrafunctional photoiniferter [21].

**DDC**

When the St polymerization was carried out with DDC under the conditions identical to those in Figure 3, i.e., [BDC]/4 = [XDC]/2 = [DDC], the formation of benzene-insoluble polymers was observed from the initial stage of the polymerization. The results at a lower concentration of these photoiniferters are shown in **Figure 7**. Although BDC and XDC induced living mono- and diradical polymerization similarly as mentioned above, benzene-insoluble polymers were formed in the polymerization with DDC, and *M* of the soluble polymers separated decreased with the reaction time. It suggests that a part of the propagating polymer radicals underwent ordinary bimolecular termination by recombination, leading to the formation of the cross-linked polymer, which was prevented by the addition of TD.

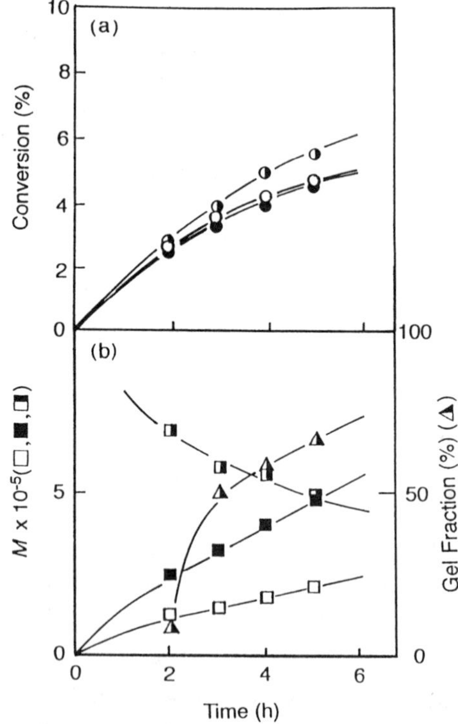


**Figure 7.** Time-conversion (a) and time-*M* (b) relationships for photopolymerization of St with BDC, XDC, and DDC in benzene at 30°C; [St] = 6.9 mol/L, [BDC] = 5.4 $\times 10^{-4}$ mol/L, [XDC] = 2.6 $\times 10^{-4}$ mol/L, [DDC] = 1.3 $\times 10^{-3}$ mol/L: (○,□) BDC, (●,■) XDC, ( ◐ , ◨ , ▲ ) DDC.

Similar polymerization was also carried out using MMA. The results are shown in **Figure 8**. The plots of the polymer yield with the reaction time are on a same straight line, and no benzene-insoluble polymers are produced, being different from that for St. Although *M* of the poly(MMA) obtained increase with the time, the relative ratio of increasing molecular weight was 1.0, 1.7, and 2.5 for BDC, XDC, and DDC, respectively. These indicate that the polymerization of MMA by BDC, XDC, and DDC do not proceed according to the ideal living mono-, di-, and tetra- radical mechanisms owing to unfavourable side reactions such as self-termination as was pointed out for diradical propagation in thermal polymerization. It may also be attributed to the uncertainty of the molecular weight determination by viscometric and GPC measurements because of the alteration of the polymer shape from linear to star. This polymer obtained from DDC could further act as a macroiniferter and initiate photopolymerization of methacrylic esters to give a star-block copolymer.

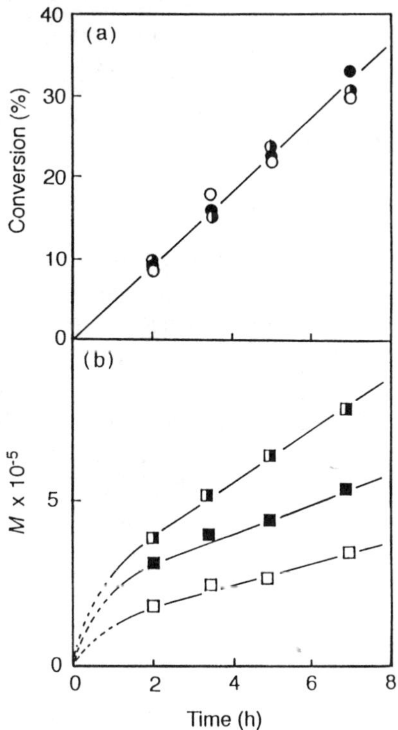


**Figure 8.** Time-conversion (a) and time-*M* (b) relationships for photopolymerization of MMA with BDC, XDC, and DDC in benzene at 30°C; [MMA] = 7.5 mol/L, [BDC] = 5.4 x$10^{-4}$ mol/L, [XDC] = 2.6 x $10^{-4}$ mol/L, [DDC] = 1.3 x $10^{-3}$ mol/L: (○ , □) BDC, (● , ■) XDC, (◑ , ◧) DDC.

If the dithiocarbamate photoiniferter having a polymerizable double bond, i.e., a monomer iniferter, was synthesized and used successively as both monomer and iniferter, macromonomers and graft copolymers would be obtained according to eq. (9) [22].

$CH_2{=}CH$–$C_6H_4$–$CH_2$–$SC(=S)N(C_2H_5)_2$ (VBDC) $\xrightarrow{+\ M_1}$ ---$CH_2$–$CH$(–$C_6H_4$–$CH_2$–$SC(=S)N(C_2H_5)_2$)----$M_1$---

$\downarrow + n\ M_2$ $\qquad\qquad$ $\downarrow + n\ M_2$

$CH_2{=}CH$–$C_6H_4$–$CH_2(M_2)_n$–$SC(=S)N(C_2H_5)_2$ $\xrightarrow{+\ M_1}$ ---$CH_2$–$CH$(–$C_6H_4$–$CH_2(M_2)_n$–$SC(=S)N(C_2H_5)_2$)----$M_1$--- **(9)**

4-Vinylbenzyl *N,N*-diethyldithiocarbamate (VBDC) was used for the synthesis and design of graft and cross-linked polymers. In the absence of light, VBDC easily polymerized with azo initiators. The resulting poly(VBDC) and its copolymers with St were found to act as polyfunctional macroiniferters of the living radical polymerization leading to graft copolymers consisting of benzene-soluble and insoluble fractions, in which the amount of the latter increased when the VBDC units increased. The photopolymerization of St with a catalytic amount of VBDC photoiniferter gave a benzene-soluble polymer that contains a styryl double bond and a dithiocarbamate group at polymer chain ends. When this macromonomer-iniferter was copolymerized with a second monomer in the presence of an azo initiator, and the formation of a high molecular weight graft copolymer was confirmed by GPC analysis.

## 5. CONCLUSION

As described before, the iniferter technique provides a novel synthetic method for designing the chain end structure of the resulting polymers in the radical polymerization. Especially, some compounds having *N,N*-diethyldithiocarbamate group serve as excellent photoiniferters and induce the living radical polymerization of St and MMA in a homogeneous system. Macroiniferters are applied to synthesis of various tailor-made polymers such as functional, telechelic, block, star, and graft polymers. However, there are some disadvantages, e.g., we can not obtain any polymer with a narrow molecular weight distribution close to monodisperse because the propagation is fast and chain reactions in the polymerization which proceeds according to eq. (4).

Recently several reports [23-32] on the polymer syntheses with some iniferters and the living radical polymerizations have been published, but none exceeds the ideal model in eq. (4). In the limiting case of eq. (4), if radical dissociation of the iniferter bond, addition of one monomer molecule, and reproduce of the iniferter bond by primary radical termination or chain transfer would occur, the polymerization could proceed in a stepwise manner. The polymerization system in which propagation consists of the homolytic insertion of a monomer molecule into the iniferter bond might be a new model for a living radical polymerization, be comparable to the living systems in ionic polymerizations.

## REFERENCES

1. O. W. Webster, *Science*, 251, 887 (1991).
2. M. Szwarc, M. Levy, and R. Milkovich, *J. Am. Chem. Soc.*, 78, 2656 (1956).
3. T. Otsu and M. Yoshida, *Makromol. Chem., Rapid Commun.*, 3, 127 (1982).
4. T. Otsu, M. Yoshida, and T. Tazaki, *Makromol. Chem., Rapid Commun.*, 3, 133 (1982).
5. T. Otsu, *J. Polym. Sci.*, 21, 559 (1956); T. Otsu, *J. Polym. Sci.*, 26, 113 (1957); K. Nayatani, I. Muto, and M. Imai, *Makromol. Chem.*, 27, 142 (1958); T. Otsu and K. Nayatani, *Makromol. Chem.*, 27, 149 (1958).
6. T. Otsu, A. Matsumoto, and T. Tazaki, *Polym. Bull.*, 17, 323 (1987).
7. T. Tazaki and T. Otsu, *Polym. Bull.*, 17, 127 (1987).
8. A. Bledzki and D. Braun, *Makromol. Chem.*, 182, 1047 (1981); A. Bledzki, H. Balard, and D. Braun, *Makromol. Chem.*, 182, 1057 (1981); H. Balard, A. Bledzki, and D. Braun, *Makromol. Chem.*, 182, 1063 (1981); A. Bledzki, H. Balard, and D. Braun, *Makromol. Chem.*, 182, 3195 (1981).
9. T. Otsu, M. Yoshida, and A. Kuriyama, *Polym. Bull.*, 7, 45 (1982).
10. T. Otsu and M. Yoshida, *Polym. Bull.*, 7, 197 (1982).
11. T. Otsu, A. Kuriyama, and M. Yoshida, *Kobunshi Ronbunshu*, 40, 583 (1983).
12. T. Otsu and A. Kuriyama, *Polym. Bull.*, 11, 135 (1984).
13. T. Otsu and A. Kuriyama, *J. Macromol. Sci.,-Chem.*, A21, 961 (1984).
14. T. Otsu, Y. Kobayashi, and A. Kuriyama, *Kobunshi Ronbunshu*, 46, 253 (1989).
15. T. Otsu and T. Tazaki, *Polym. Bull.*, 16, 277 (1986).
16. T. Otsu, T.Matsunaga, A. Kuriyama, and M. Yoshioka, *Eur. Polym. J.*, 25, 643 (1989).
17. M. Okawara, T. Nakai, K. Morishita, and E. Imoto, *Kogyo Kagaku Zasshi*, 67, 2108 (1964).
18. T. Otsu and A. Kuriyama, *Polym. J.*, 17, 97 (1985).
19. R. B. Merrifield, *J. Am. Chem. Soc.*, 85, 2149 (1963).
20. T. Otsu, T. Ogawa, and T. Yamamoto, *Macromolecules*, 19, 2087 (1986).
21. A. Kuriyama and T. Otsu, *Polym. J.*, 16, 511 (1984).
22. T. Otsu, K. Yamashita, and K. Tsuda, *Macromolecules*, 19, 287 (1986).

23. A. Bledzki, D. Braun, W. Menzel, and K. Titzchkau, *Makromol. Chem.*, 184, 287 (1983); A. Bledzki, D. Braun, and K. Titzschkau, *Makromol. Chem.*, 184, 745 (1983); A. Bledzki, D. Braun, and H. Tretner, *Makromol. Chem.*, 186, 2491 (1985); A. Bledzki, D. Braun, *Makromol. Chem.*, 187, 2599 (1986); A. Bledzki and D. Braun, *Polym. Bull.*, 16, 19 (1986); A. Bledzki, D. Braun, and K. Titzschkau, *Makromol. Chem.*, 188, 2061 (1987); A. Bledzki, H. Balard, and D. Braun, *Makromol. Chem.*, 189, 2807 (1988); D. Braun, H. J. Lindner, and H. Tretner, *Eur. Polym. J.*, 25, 725 (1989).
24. J. V. Crivello, D. A. Conlon, and J. L. Lee, *J. Polym. Sci., Part A, Polym. Chem.*, 24, 1197 (1986); J. V. Crivello, J. L. Lee, and D. A. Conlon, *J. Polym. Sci., Part A, Polym. Chem.*, 24, 1251 (1986); D. Feng, G. L. Wilkes, and J. V. Crivello, *Polymer*, 30, 1800 (1989).
25. C. P. R. Nair, G. Clouet, and J. Brossas, *J. Macromol. Sci.,-Chem.*, A25, 1089 (1988); C. P. R. Nair and G. Clouet, *Polymer*, 29, 1909 (1988); C. P. R. Nair, G. Clouet, and P. Chaumont, *J. Polym. Sci., Part A, Polym. Chem.*, 27, 1795 (1989); C. P. R. Nair and G. Clouet, *Macromolecules*, 23, 1361 (1990).
26. S. R. Turner and R. W. Blevins, *Macromolecules*, 23, 1856 (1990).
27. P. Lambrinos, M. Tardi, A. Polton, and P. Sigwalt, *Eur. Polym. J.*, 26. 1125 (1990).
28. C. V. Krechhoven, H. Van den Broeck, G. Smets, and J. Huybrechts, *Makromol. Chem.*, 192, 101 (1991).
29. Z. Liu, D. Yan, and J. Shen, *Makromol. Chem., Rapid Commun.*, 9, 27 (1988).
30. D. H. Solomon, E. Rizzardo, and P. Cacioli, *Eur. Pat. Appl.*, EP135280; *Chem. Abstr.*, 102, 221335q (1985).
31. M. H. Acar and Y. Yagci, *Macromol. Reports*, A28(Suppl), 177 (1991).
32. J. D. Druliner, *Macromolecules*, 24, 6079 (1991).

# Macroiniferters Chapter 13

## POLYMERIC PHOTOINIFERTERS

**Ramesh C. Kumar, Ronald R. Dueltgen, and Milton H. Andrus, Jr.**

Specialty Adhesives and Chemicals Laboratory
Specialty Materials Division
3M Center,
St Paul, MN 55144 -1000
USA





Macromolecular Design: Concept and Practice (Macromonomers, Macroinitiators, Macroiniferters, Macroinimers, Macroiniters, and Macroinifers) Pages 487 - 497

# 1. INTRODUCTION

This chapter presents a review of recent work in our laboratories and elsewhere on the production of polymeric photoiniferters ("macroiniferters") and their use in producing block and graft copolymers. Otsu et al. [1] have provided the initial framework of the study of iniferters ( initiator - transfer agent - terminator) delineating their ability to dissociate when irradiated to form highly-active initiating radicals A* and much less active terminating radicals A'* (**Scheme 1**). Reaction of the initiating radical with available reactive vinyl monomer, such as an acrylate or styrene, produces a growing polymer chain which is reversibly terminated by reaction with A'*. Since the new C-A' bond is approximately as labile as the A-A' bond, the new polymer chain, terminated by the A' group, has become a polymeric photoiniferter, capable of further reaction with a new monomer to produce a block copolymer.

$$A - A' \xrightarrow{h\nu} A^* + A'^*$$

$$A^* + n\,CH_2{=}\underset{Y}{\underset{|}{CH}} \longrightarrow A(CH_2-\underset{Y}{\underset{|}{CH}})_n{}^*$$

$$A(CH_2-\underset{Y}{\underset{|}{CH}})_n{}^* + A'^* \longrightarrow A(CH_2-\underset{Y}{\underset{|}{CH}})_n A'$$

$$A(CH_2-\underset{Y}{\underset{|}{CH}})_n A' + m\,CH_2{=}\underset{X}{\underset{|}{CH}} \xrightarrow{h\nu} A(CH_2-\underset{Y}{\underset{|}{CH}})_n(CH_2-\underset{X}{\underset{|}{CH}})_m A'$$

**Scheme 1**

In the earliest discussions of block copolymer formation, poly(acrylate) and poly(styrene) macroiniferters are clearly present. That is, in order to produce block copolymers, a homopolymer bearing an iniferter end-group must first be obtained, and that homopolymer becomes, by definition, a macroiniferter. This chapter reviews recent reports of the advantageous use of macroiniferters to produce mixed block and graft copolymers of poly(acrylates) and non- poly(acrylates), notably condensation-type polymers such as polyurethanes and siloxanes.

Initial investigations of photoiniferters by Otsu et al. [2,3] were directed mainly toward polymers in which vinyl-type monomers formed addition polymers such as poly(acrylates) and polystyrene. More recent studies, using macroiniferters, have expanded their scope to include siloxanes [4 - 6], polyurethanes [7], polyolefins [8, 9], poly(vinyl chloride) [10 - 13], and various macromonomers [14 - 16].

Siloxane and polyurethane macroiniferters are covered in greater detail later in this chapter. Polyolefin macroiniferters have been produced [8, 9] by reaction of halogen-terminated polyolefins with the sodium salt of a dialkyl dithiocarbamate. Irradiation in the presence of a reactive acrylate monomer produced, e.g., a polypropylene-*b*-poly(glycidylmethacrylate) copolymer. Alternatively, dithiocarbamate-terminal polyolefins can be obtained by direct reaction of a thiuram disulfide, such as N,N,N',N'-tetraethylthiuram disulfide, with the crude Ziegler-Natta olefin polymerization reaction mixture. Nucleophilic displacement of chlorine from poly(vinyl chloride) with e.g.,sodium diethyldithiocarbamate has been shown [10 - 13] to produce a dithiocarbamated PVC to which polyacrylates can be readily photo-grafted under UV irradiation. Macromonomer photoiniferters based on butyl acrylate and styrene have been described [14 - 16], for purposes of producing block copolymers having a specified molecular weight or domain size contributed by the macromonomer iniferter. In the report by Imoto et al. [16] photopolymerization of vinyl acetate with a styrene macroiniferter gave a styrene-b-vinyl acetate block copolymer in good yield, whereas these two monomers will not copolymerize under other polymerization conditions.

Grafting of dithiocarbamate-functional species onto silica gel [17 - 19], glass [20], and solvent- swollen polystyrene gel [21] has provided an even wider range of macroiniferters. In the case of silica gel, the macroiniferters are produced [17 - 19] by treatment of silica gel hydroxyls with a chlorine-functional silane followed by chlorine displacement by, e.g., sodium N,N-diethyldithiocarbamate and subsequent photografting of a vinyl monomer onto the silica gel core. Production of grafted block copolymers has also been described, where vinyl or acrylic block copolymers are produced from successive monomer addition to the original dithiocarbamate -substituted silica gel. Dithiocarbamate-substituted trimethoxypropyl silane (DATMS) has been shown [20] to be a macroiniferter when bound to a glass surface via hydrolysis of the methoxy silane moiety in the presence of a glass plate. Photo-graft-polymerization of sodium styrene sulfonate (SSS) onto the glass surface was easily accomplished by irradiating the glass plate in a dilute solution of SSS. Otsu et al. [21] have described the chloromethylation of solvent-swollen poly(styrene-divinylbenzene) beads followed by treatment with potassium (N,N-diethyldithio-carbamyl)acetate to produce beads containing >20% dithiocarbamate functionality. UV irradiation in the presence of styrene and methyl methacrylate monomers gave polymer-grafted beads in good yield.

Block copolymers derived from siloxane macroiniferters have also been described [22 -24]. In these cases, polysiloxanes having chlorine atoms at one or both ends of the molecular chain are converted to dithiocarbamate photoiniferters, as described in more detail below. Macroiniferters based on polyesters have also been described [25]. In this case, halogen-terminated polyesters were converted to the corresponding bis-dithiocarbamates by well-known methods, and block copolymers of polyesters and vinyl monomers were obtained.

While most studies of photoiniferters have been concerned with the production of block copolymers, Yamashita and coworkers [26 - 29], and the present authors, have produced graft copolymers from macroiniferters (as opposed to grafting of polymers onto other surfaces, as above) by incorporating dithiocarbamate functionality into a polymer backbone followed by photo-grafting of an added monomer.

Polymerization of 4-vinylbenzyl N,N- diethyldithiocarbamate (VBDC) [26 - 29] by standard (dark) radical polymerization methods provides polystyrene having a photolabile N,N-diethyldithiocarbamate group at every repeating unit.

Photopolymerization of styrene and methyl methacrylate in the presence of poly(VBDC) produced polymers whose yield and intrinsic viscosity increased with time, indicating a living radical mechanism. Likewise, random copolymers of VBDC and a number of vinyl monomers, including styrene, methyl methacrylate, N-vinylpyrrolidone, vinyl acetate and acrylonitrile, were shown to be macroiniferters for graft polymerization of a wide variety of vinyl monomers. Chloromethylation of polystyrene gave [30] polystyrene bearing aliphatic chlorines on some, but not all, aromatic rings. Displacement of chlorine by sodium diethyldithiocarbamate gave a polymer which was subsequently photo-grafted with vinyl monomers such as styrene and methyl methacrylate.

Incorporation of a dithiocarbamate photoiniferter moiety into the side-chain of a polymer other than polystyrene, whether via copolymerization or a substitution reaction of an already-formed polymer, is less well known. Kumar and Ohkubo [31] prepared a hydroxy-functional polyurethane, after which a number of free hydroxyls were taken up by reaction with 2-chloropropionyl chloride. The pendant chloride was displaced with sodium N,N- diethyldithiocarbamate, and the resulting photoiniferter-grafted polymer was easily converted into a number of side- chain poly(vinyl) grafted copolymers.

Crosslinkable block copolymers based on macroiniferters have also been described. In the first instance [32], radiochemical polymerization of S-vinyl diethyldithiocarbamate gave a higher yield of a polymer having a much higher molecular weight than that obtained by radical thermal polymerization using azobis(isobutyronitrile). Crosslinking via the dithiocarbamate is assumed. In the second instance [33], divinyl benzene was copolymerized with methyl methacrylate into the A-block of an ABA copolymer based on a butyl acrylate macroiniferter. Improved elongation and chemical resistance of the crosslinked copolymer was noted.

## 2. SILOXANE INIFERTERS

Recently Himori [23] has described the preparation of polysiloxanes containing dithiocarbamate groups which can be used as siloxane- iniferters for the preparation of vinyl/silicone copolymers. The first step in the synthesis is the preparation of a halogenated polysiloxane, the halogen atoms of which can then be displaced with dithiocarbamate groups. In considering the rates of such nucleophilic substitutions, the halogen atoms are preferably chlorine or bromine. Polysiloxanes halogenated at one or both ends of the polymer chain are obtained by known methods in which living polymerization of either hexamethylcyclotrisiloxane, "$D_3$," or octamethylcyclotetrasiloxane, "$D_4$,"is terminated with dimethylchlorosilane or tetramethyl-siloxane to give polymers with one or two hydrogens bound to silicones at the termini, respectively. Platinum-catalyzed hydrosilylation with p- chloromethylstyrene yielded polysiloxan es with either one or two terminal aliphatic chlorine atoms. Himori also points out that polysiloxanes bearing non-terminal chloromethyl groups can be obtained by ring opening copolymerizations of $D_3$ or $D_4$ with chlorocyclosiloxanes or by hydrosilylation of polysiloxanes containing hydrogens bound to non-terminal silicones such as are commercially available (e.g., SH-1107 from Toray Silicone or KF-99 from Shinetsu Silicone).

Kumar, Andrus and Mazurek [22] have described a similar synthesis of terminally-halogenated polysiloxanes by reacting $D_3$ with lithium trimethylsilanoate, $Me_3SiOLi$, at 0°C in THF. When all of the $D_3$ is consumed, 1-(dimethylchlorosilyl)- 2-(p-chloromethylphenyl)ethane (II) was added and the resulting mixture was stirred at room temperature, terminating the anionic polymerization. Removal of precipitated lithium chloride and any unreacted end-capper gave a polysiloxane with a terminal chloromethyl group (III), as per **Scheme 2**.

$$(CH_3)_3SiOLi + nD_3 \longrightarrow \underset{\text{I}}{Me_3SiO\,[Si(CH_3)_2O]_n^-\,Li^+}$$

$$\text{I} + \underset{\text{II}}{ClCH_2-C_6H_4-CH_2CH_2Si(CH_3)_2Cl} \longrightarrow \underset{\text{III}}{Me_3SiO[Si(CH_3)_2O]_nSiCH_2CH_2-C_6H_4-CH_2Cl}$$

$$\text{III} + Et_2N\overset{S}{\overset{\|}{C}}S^- \longrightarrow \underset{\text{IV}}{Me_3SiO[Si(CH_3)_2O]_nSiCH_2CH_2-C_6H_4-CH_2S\overset{S}{\overset{\|}{C}}NEt_2}$$

**Scheme 2**

Substitution of chlorine by dithiocarbamate takes place at temperatures between about 50° to 90°C in a solvent such as toluene, xylenes, or ethyl acetate under nitrogen. A phase transfer catalyst such as trioctylmethylammonium chloride or tetrabutylammonium chloride is often used advantageously.

Kumar et al. [32] have also described the synthesis of a polysiloxane, halogenated at both chain ends, using the symmetrical bis- chloro siloxane V, in **Scheme 3**. Hydrolytic coupling of chlorosilane II gave V, which also acts as an endcapper in the acid-catalyzed polymerization of cyclosiloxanes such as $D_4$. In this case, $D_4$, the endcapper V, and a small amount of sulfuric acid catalyst were stirred and heated under nitrogen at 90°C in the presence of carbon black (as an acid adsorbent). Filtration and stripping gave the desired bis-chlorinated polysiloxaneVI. Displacement of chlorine with N,N-diethyldithiocarbamate gave the macro di-iniferter VII, in good yield.

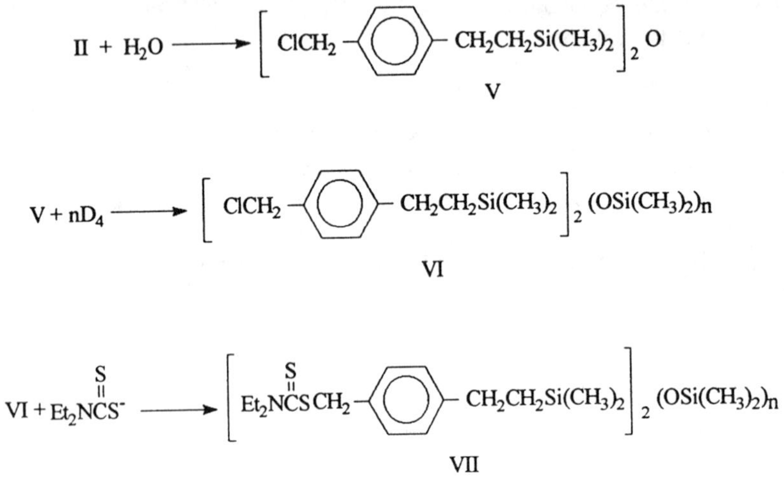

$$\mathrm{II + H_2O \longrightarrow \left[ ClCH_2-C_6H_4-CH_2CH_2Si(CH_3)_2 \right]_2 O}$$

V

$$\mathrm{V + nD_4 \longrightarrow \left[ ClCH_2-C_6H_4-CH_2CH_2Si(CH_3)_2 \right]_2 (OSi(CH_3)_2)_n}$$

VI

$$\mathrm{VI + Et_2N\overset{S}{\overset{\|}{C}}S^- \longrightarrow \left[ Et_2N\overset{S}{\overset{\|}{C}}SCH_2-C_6H_4-CH_2CH_2Si(CH_3)_2 \right]_2 (OSi(CH_3)_2)_n}$$

VII

**Scheme 3**

Siloxane iniferters are used to make block and graft copolymers by irradiating them, usually with UV, in the presence of a free-radically copolymerizable monomer. Typically, such monomers are substituted vinyl (i.e., ethylenically-unsaturated) compounds which include the esters, amides, nitriles, etc., of (meth)acrylic acids, styrene derivatives, vinyl esters, allylic compounds, and the like. Copolymerization of the monomer and the polysiloxane iniferter takes place in non-reactive UV-transparent solvent such as methyl ethyl ketone or ethyl acetate. Polysiloxane monoiniferters such as those described in **Scheme 2** form block copolymers having AB, CAB, etc., architecture. Diiniferters (**Scheme 3**) form block copolymers of the ABA, CABAC, etc., type. In both cases, "B" represents the polydimethylsiloxane block, and "A," "C," etc., represent blocks from the vinyl monomers. A typical reaction, in this case with methyl methacrylate, is shown in **Scheme 4**. **Table 1** shows some composition and the physical properties of various polysiloxane block copolymers [22].

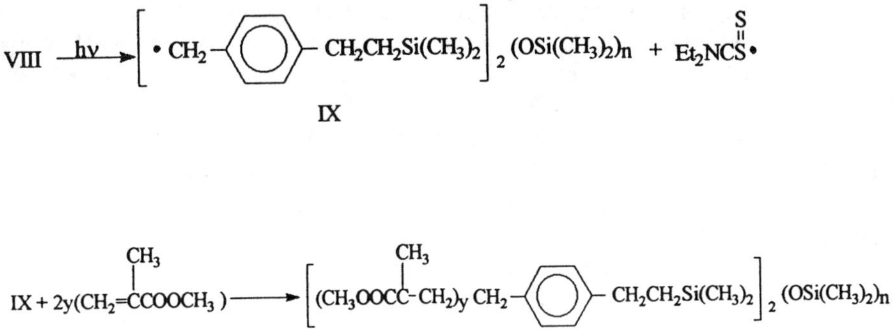

$$\text{VIII} \xrightarrow{h\nu} \left[ \bullet CH_2-C_6H_4-CH_2CH_2Si(CH_3)_2 \right]_2 (OSi(CH_3)_2)_n + Et_2NC(=S)S\bullet$$

IX

$$\text{IX} + 2y(CH_2{=}C(CH_3)COOCH_3) \longrightarrow \left[ (CH_3OOCC(CH_3)-CH_2)_y\, CH_2-C_6H_4-CH_2CH_2Si(CH_3)_2 \right]_2 (OSi(CH_3)_2)_n$$

X

MMA - b - PDMS - b - MMA

**Scheme 4**

**Table 1.** *Composition and physical properties of various polysiloxane block copolymers*

| Copolymer Composition | $M_n$ | $M_w$ | $M_w/M_n$ | $T_g$, °C |
|---|---|---|---|---|
| MMA-b-PDMS-b-MMA | 17,150 | 30,850 | 1.76 | -124<br>+105 |
| (MA/AA)x-b-PDMS-b-(MA/AA)x | 22,200 | 53,000 | 2.39 | -126<br>+60 |
| HEA-b-MMA-b PDMS-b-MMA-b-HEA | 13,160 | 20,650 | 1.56 | -120<br>+107 |
| MMA-b-PDMS | 15,400 | 20,400 | 1.33 | -124<br>+106 |

The vinyl - polysiloxane block and graft copolymers are easily prepared from the polysiloxane iniferter. Control of the ratio of vinyl to polysiloxane blocks allows rather precise tailoring of copolymer composition. Thus, unique materials that exhibit the combined properties of dissimilar polymers are possible, which may find utility in a wide range of applications, such as low adhesion backsizes, impact- resistant materials, fabric protectors, etc.

Polydimethylsiloxane (PDMS) photoiniferters have also been described by Inoue and Kohama [34] for the purpose of photografting vinyl monomers onto PDMS surfaces. In this case, chloromethylheptamethylcyclotetrasiloxane was polymerized using Amberlyt™ 15 ion exchange resin as a catalyst and precipitated from methanol to give poly(chloromethylheptamethylcyclotetrasiloxane) (C-PDMS), which was coated onto a glass slide and cured under a high- pressure mercury lamp. This crosslinked C-PDMS was reacted with sodium diethyldithiocarbamate in a DMF-THF solution to attach diethyldithiocarbamate (DDC) groups by a displacement reaction. UV-irradiation of the resultant DDC-PDMS in solvent in the presence of hydrophilic vinyl monomers such as 2-hydroxyethyl methacrylate and acrylamide gave a number of vinyl polymer-surface grafted PDMS materials. Each of the surface-grafted materials showed a decrease in water contact angle, indicating increased hydrophilicity of the grafted PDMS surface.

## 3. POLYURETHANE INIFERTERS

Kumar and Ohkubo [7] have described polyurethane iniferters useful as starting materials to produce stand-alone (i.e., not blended) high Tg, high modulus, high resilience resins for dispersing and binding ferromagnetic pigments for magnetic recording media. Commercial magnetic binders are blends of polyurethane and vinyl (usually polyvinyl chloride) copolymers, blended to optimize Tg, modulus and coatability of the resin. In theory, single resins would incorporate both the softer polyurethane component required for good coatability and conformance and the harder vinyl polymeric component required for abrasion resistance. In addition, the controllable nature of photoiniferter polymerizations would allow tailoring of the polymer make-up and, therefore, its properties.

The first step in obtaining polyurethane iniferters for magnetic binders was the preparation of hydroxy- functional polyurethane (HPU) and sulfonated hydroxy-functional polyurethane (SHPU) pre-polymers. Three key components of SHPU resins are 1) a sulfonated diol, 2) a polyisocyanate, and 3) a triol. The reaction of 2 - 4 equivalents of a diol such as a polyester diol, a polyether diol or a polycarbonate diol (TONE™ 0201, from Union Carbide) with one equivalent of dimethyl sodium sulfoisophthalate (DMSSIP) leads to an extended polyester diol which also contains an aromatic sulfonate group. Esterification catalysts such as tetrabutyltitanate are used to facilitate the reaction. A wide variety of polyisocyanates is useful, particularly diisocyanates such as diphenylmethane diisocyanate, isophorone diisocyanate, tetramethylene diisocyanate and p-phenylene diisocyanate, either singly or as a mixture. Likewise, a number of useful triols can be used, particularly those with molecular weights between 200 and about 1000. The list includes polycaprolactone triols (e.g., TONE™ 0305, from Union Carbide), polypropylene oxide triol, and polyester triols such as polybutylene adipate triol. HPU polymers contain the same basic materials without, of course, DMSSIP.

A typical SHPU preparation proceeds via co-reaction of a sulfonated diol monomer, neopentyl glycol and diphenylmethane diisocyanate in methyl ethyl ketone (MEK) in a ratio of about two isocyanate groups for each hydroxyl. When all hydroxy groups are consumed, the prepolymer is further reacted with an excess of a hydrophobic triol such that the ratio of isocyanate groups to hydroxy groups is in the range of 0.3 to 0.6. The resultant polyurethane has hydroxy groups pendant from the hydrophobic polymer chain segments arising from the triol.

Preparation of carbamate-functional macroiniferters (CSHPU and CHPU, respectively) requires two more steps. First, a functional group bearing an easily-displaced moiety is attached to the polymer backbone via the pendant hydroxyls. Examples of such functional groups include 2-chloropropionate, 1-chloropropionate and 1,2-dichloropropionate, each of which is available from its respective acid chloride. Displaceable groups other than chloride include bromide, tosylate and mesylate. Acyl groups of chain length greater than 3 are also useful as long as the leaving group is relatively readily accessible. A typical acylation takes place in tetrahydrofuran (THF) or MEK by dissolving the polymer to about 25 - 45% solids, adding a small amount of organic base (triethylamine, etc.), then dropwise addition of the acid chloride in THF. Appearance of precipitated triethylamine hydrochloride signifies onset of the reaction, which can be followed by conventional means. In the final step, the stirred slurry is treated with sodium diethyl dithiocarbamate in THF at 30° -35°C for 5 - 6 hours to obtain the CSHPU or CHPU macroiniferter. **Scheme 5** illustrates these reactions.

UV irradiation of the macroiniferter in the presence of a high Tg monomer such as methyl methacrylate or styrene produced the expected graft copolymer, which was shown to have a favorable combination of high modulus and pigment dispersibility for magnetic pigments such as Fe, $CrO_2$, $Fe_2O_3$, BaFe, etc. in high-density recording applications.

## 4. CONCLUSION

The pioneering work of Otsu described a unique method to prepare a wide variety of block copolymers. The intermediates in these block copolymer syntheses are macromolecular iniferters or macroiniferters. This chapter described the extension of this chemistry to block copolymers combining block segments often of widely different chemistry, e.g., polyolefins with vinyls, polyurethanes with vinyls and acrylics, and silicones with acrylics. These polymers can be synthesized with high conversions (>90%) and high molecular weights.

The combination of block segments with widely different chemistry and properties provides routes to polymeric materials which combine the best features of each block without loss of other desired properties. Thus rigid polymers can be flexibilized or toughened without loss of modulus and polymers with desired modulus can be prepared with widely different surface properties.

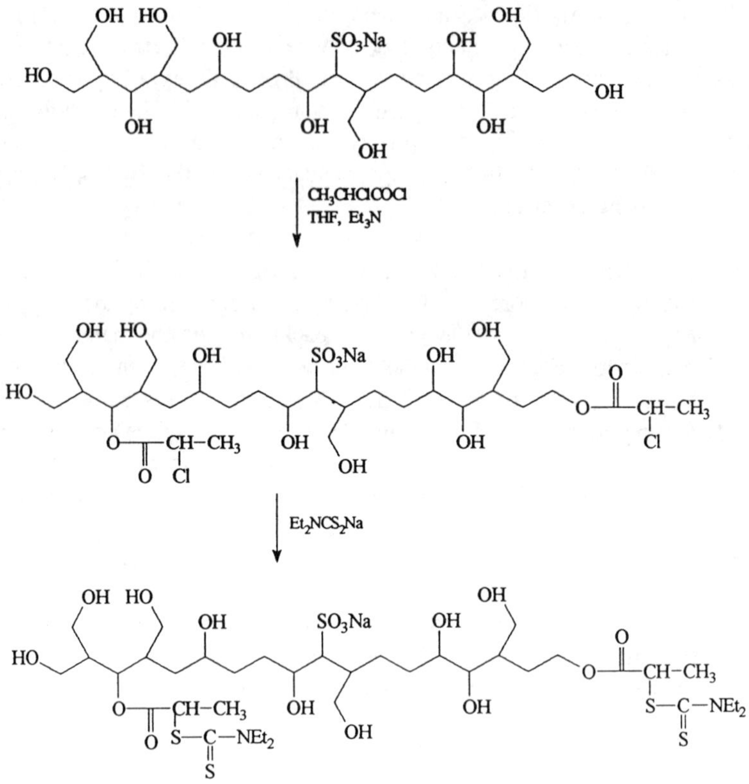

$CH_3CHClCOCl$
THF, $Et_3N$

$Et_2NCS_2Na$

**Scheme 5**

In addition to block copolymers, macroiniferters were shown to be useful in the synthesis of graft copolymers, also with interesting properties. As with the block copolymers, polymer segments with widely differing chemistry and properties can be combined to give materials exhibiting the desired properties of each segment. The work presented herein represents advances in the preparation of novel polymeric materials. Much room is left for continued development using macroiniferters.

## REFERENCES


1. Otsu ,T. and Yoshida, M., *Makromol. Chem., Rapid Commun.* **3,** 127-132(1982)
2. Otsu, T. and Yoshida, M., *Polym. Bull.*, **7,** 197 - 203 (1982)
3. Otsu, T. and Kuriyama, A., *Polym. Bull.*, **11**, 135-142 (1984)
4. Inoue, H. and Kohama, S., *J. Appl. Polym. Sci.*, **29**, 877 -889 (1984)
5. Inoue, H. and Kohama, S., *Kagaku to Kogyo* (Osaka), **60,** 81-90 (1986), C.A.105:173534
6. Kumar, R., and Kantner, S., U.S. Patent 5,089,336 (Feb. 18, 1992)

7. Kumar, R., and Ohkubo, T., U. S. Patent 5,118,580(June 2, 1992)
8. Yanagihara, H. and Tamano, H., Japan Patent Application 01 201,303(Aug. 14, 1989) C.A. 112:57006
9. Yanigahara, H. and Tamano, H., Japan Patent Application 63 43,906(Feb. 25, 1988) C.A. 109:38448.
10. Miyama, H., Fuji, N.,Shimazaki, Y., Ikeda, K., *Polymer Photochemistry,* **3,** 445-461 (1983)
11. Martiniz, G., Gomez-Daza, M., Mijangos, C., Millan, *J., Rev. Plast. Mod.*, **58,** 213-220 (1989) C.A. 112:78109
12. Fujii, N., Shimazaki, Y., Ikeda, K., and Miyama, H., *Kagaku Kogaku Ronbunshu*, **9**, 427-433 (1983) C.A. 99:123072
13. Nagaoka, S., Shiota, M., Mori, Y., and Kikuchi, T., *Kobunshi Ronbunshu,* **38**, 571-576 (1981) C.A. 95:169900
14. Ikeda, Y. and Kawanami, H., *Kobunshi Ronbunshu,* **47**, 659- 666 (1990) C.A. 13:172889
15. Konishi, H., Shinagawa, Y., Azuma, A. and Okano, T., *Makromol. Chem.,* **183,** 2941 - 2947 (1982)
16. Imoto, M., Otsu, T. and Yonezawa, J., Makromol. Chem., 36, 93 - 101 ( 1960).
17. Zaremskii, M. and Olenin, A., Zh. Prikl. Khim. (S.- Petersburg), 64, 2145- 2149 (1991) C. A. 116:236230
18. Olenin, A., Zaremskii, M. and Zubov, V., U.S.S.R. Patent No. 1,467,083(Mar. 23, 1989) C.A. 111:8049
19. Andrianov, A., Ukolova, E., Olenin, A., Garina, E. and Zubov, V., *Vyskomol. Soedine Ser. A.*, **29**, 588-592 (1987) C.A. 106:214438
20. Kobayaski, T., Takahashi, S., Nosaka, Y. and Fujii, N., *Chem. Lett.*, 1321-1324 (1992)
21. Otsu, T., Ogawa, T., and Yamamoto, T., Macromolecules, 19, 2087-2089 (1986)
22. Kumar, R. C., Andrus, M. H., and Mazurek, M. H. U. S. Patent 5,057,619 (October 15, 1991)
23. Himori, S., U. S. Patent 5,071,936 (December 10, 1991)
24. Inoue, H. and Kohama, S., *J. Appl. Poly. Sci.*, **29**, 877- 889 (1984).
25. Himori, S., U. S. Patent No. 5,097,007 (March 17, 1992).
26. Yamashita, K., Ito, K., Tsuboi, H., Takahama, S., and Tsuda, K., *J. Appl. Poly. Sci.*, **40**, 1445-1452 (1990)
27. Yamashita, K., Kanamori, T., and Tsuda, K., *J. Macromol. Sci.- Chem.* **A27,** 897-909 (1990)
28. Yamashita, K., Nakano, A., and Tsuda, K., *J. Appl. Polym. Sci.*, **35,** 465-473 (1988)
29. Otsu, T., Yamashita, K. and Tsuda, K., *Macromolecules,* **19,** 287-290 (1986)
30. Okawara, M and Nakai, T., *Bull. Tokyo Inst. Tech.*, **78,** 1 -16 (1966).
31. Kumar, R. C. and Ohkubo, T., U. S. Patent 5,118,580 (June 2, 1992).
32. Fujiwara, Y. and Nakagawa, T., Japan Patent Application 80 3,362 (Jan. 24, 1980) C.A. 93:205325.
33. Himori, S., Japan Patent Application 03 9,914 (Jan. 17, 1991) C.A. 114:249101.
34. Inoue, H. and Kohama, K., *J. Appl. Polym. Sci.*, **29**, 877-889 (1984)

# Chapter 14

## AN INTRODUCTION TO MACROINIMER, MACROINITER AND MACROINIFER

**Munmaya K. Mishra** and **Yusuf Yagci**

Polymers Research
Texaco, R & D Department
P.O.Box 509
Beacon, NY 12508, USA

Department of Chemistry
Istanbul Technical University
Maslak, Istanbul - 80626
TURKEY





Macromolecular Design: Concepts and Practice (Macromonomers, Macroinitiators, Macroiniferters, Macroinimers, Macroiniters, and Macroinifers) Pages 499 - 506

# 1. INTRODUCTION

Recently macromolecular engineering is playing a major role in the design of block and graft copolymers. The precise control of various parameters with respect to structure, molecular weight, degree of polymerization, and uniformity (polydispersity) have been achieved using various synthetic techniques and some of the details have been described in previous chapters. The synthesis, application, and properties of many macro-intermediates such as macroinitiators, macromonomers, macroiniferters have also been discussed in previous chapters. This chapter presents an overview of recent developments of macroinimers, macroiniters, and macroinifers.

# 2. MACROINIMER

Macroinimer is the abbreviation for macromolecular-initiator-monomer. A macroinimer is generally referred to as a macromolecule that can act simultaneously as a monomer and an initiator in a polymerization reaction. Macroinimers behave as macroinitiator, macromonomer, and macrocrosslinker in thermal polymerizations by themselves or copolymerization with a vinyl monomer/s [1]. In most cases the polymerizable group is vinylic. Macroinimers are used to synthesize block and graft copolymers however, the published literature in this subject is very scanty [1-3].

The synthesis of a macroinimer having two vinyl (MIM-2) groups [2] was carried out by reacting polyazoesters and isocyanato ethyl methacrylate (IEM) in the presence of dibutyl tin dilaurate (DBTDL). Similarly the macroinimer having one vinyl (MIM-1) terminal group was synthesized by reacting IEM at one end of a polyazoester and capping the other hydroxyl end group of the polyazoester with phenyl isocyanate. The reactions are presented in **Scheme 1**.

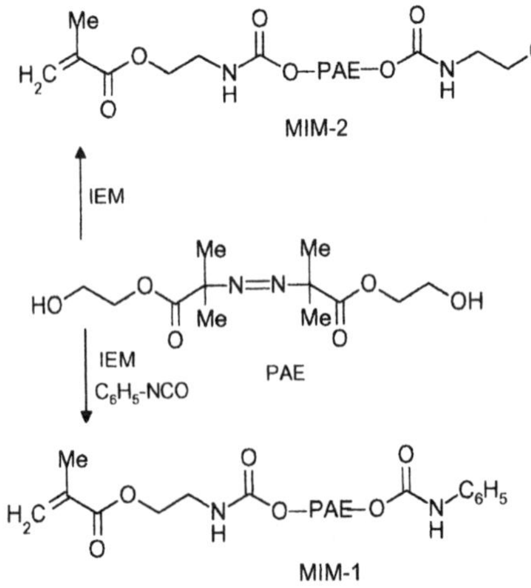

MIM-2

IEM

PAE

IEM
$C_6H_5$-NCO

MIM-1

**Scheme 1**

Polymerization of styrene or methyl methacrylate (MMA) by macroinimers having one vinyl terminal group yielded the corresponding block copolymer [2]. However, the use of macroinimers having two vinyl terminal groups to polymerize styrene or MMA resulted only crosslinked polymer/block copolymer. The reactivity of the macroinimers depend on the molecular weights and/or the content of vinyl group. The decrease in vinyl group and increase in the molecular weight of macroinimers made them less reactive.

The ester type macroinimers obtained by the reaction of 4, 4'-dicyano-4,4'-azovaleryl chloride [2], polyethylene glycol, and methacryloyl chloride was described by Hazer [1]. Macroinimers having different molecular weights containing polyethylene glycol of different molecular weights were synthesized. The synthesis of macroinimers (**Scheme 2**) was carried out at room temperature and the side reactions were minimized by decreasing reaction time and using a slight excess of methacryloyl chloride. Similarly, macroinimers were prepared from polyazoester with 4-vinylbenzoyl chloride [4].

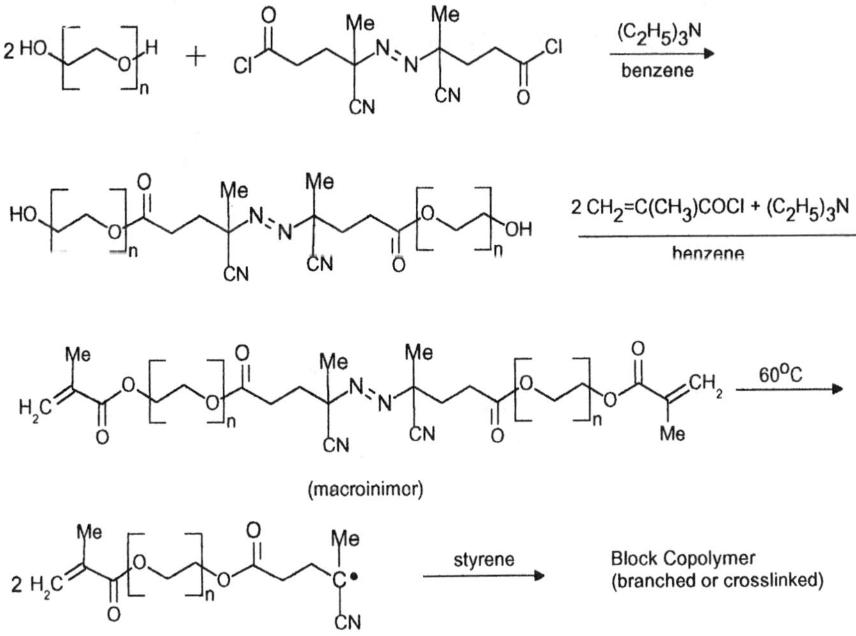

$(C_2H_5)_3N$, benzene

$2\ CH_2{=}C(CH_3)COCl + (C_2H_5)_3N$, benzene

60°C

(macroinimer)

styrene → Block Copolymer (branched or crosslinked)

**Scheme 2**

The thermal homo polymerization of macroinimers at 60°C containing PEG of 200, 400, 1000 molecular weights produced crosslinked PEG in yields of 38.3, 23.1 and 21.9% respectively. However, the polymerization of styrene with macroinimers gave branched and crosslinked styrene-PEG block copolymer.

# 3. MACROINITERS

Macroiniter is the abbreviation for macromolecular-initiator-terminator. These polymeric compounds have the ability to initiate the polymerization and take part in termination reactions. The research in this area has not been well identified. The use of this term is not common in the literature and the published literature using this concept is also limited. But one would not ignore the idea that macroiniferters (described in previous chapters) are also macroiniters. However, in this section one example [5] of possible macroiniter is presented.

It has been shown that radical transfer polymerization can be used successfully for the preparation of polymers with functional end groups. Yamashita et al. recently prepared carboxy [6] and crown-ether [7] terminated poly(methyl methacrylate) using the appropriate mercapto compound. Recently Yagci and his coworkers [5] have synthesized polymers with terminal trityl (triphenylmethyl) groups, using the corresponding chain transfer compound triphenylmethyl mercaptan (TPMM). They have described that these polymers are capable of acting as a thermal "initer" (initiator-terminator) and can be used for the preparation of block copolymers. However, depending upon the reaction condition/mechanism and how one would consider, the polymer bearing thio-trityl groups could be classified as a macroinifer.

Styrene (St) and methyl methacrylate (MMA) were polymerized [5] in the presence of 2,2'-azobisisobutyronitrile (AIBN) and TPMM as initiator and chain transfer agent, respectively, yielding polymers with trityl end groups. For the MMA polymerization the reaction scheme may be represented as follows (**Scheme 3**).

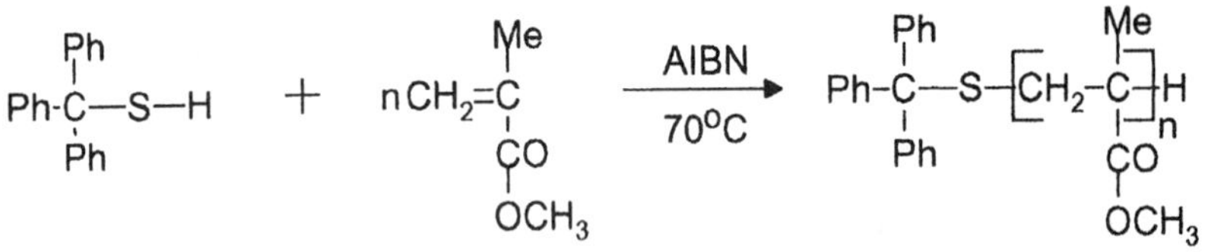

$$Ph_3C{-}S{-}H + n\,CH_2{=}C(Me)(COOCH_3) \xrightarrow[70^oC]{AIBN} Ph_3C{-}S{-}[CH_2{-}C(Me)(COOCH_3)]_n{-}H$$

**Scheme 3**

The polymer obtained has a trityl moiety at one end. Trityl terminated polystyrene (TPM- PSt) was used in the polymerization of styrene. The number average molecular weight $M_n$ of the polymers increased linearly with conversion indicating that polymerization proceeds in a quasi-living radical mechanism in a manner proposed by Otsu et al. [8].

Polymerization involves dissociation of trityl terminated polymer into radicals to give chain extended high molecular weight polymer. Bond dissociation is expected to occur preferentially at the S-$CPh_3$ bond. The thiyl radical reacts with the monomer, where as the trityl radical reacts only with the propagating radical (**Scheme 4**).

$$-\!\!\left[CH_2-\underset{Ph}{CH}\right]_n\!\!-S-\underset{Ph}{\overset{Ph}{C}}-Ph \longrightarrow -\!\!\left[CH_2-\underset{Ph}{CH}\right]_n\!\!-S^{\bullet} + {}^{\bullet}\underset{Ph}{\overset{Ph}{C}}-Ph$$

$$-\!\!\left[CH_2-\underset{Ph}{CH}\right]_n\!\!-S^{\bullet} + mCH_2{=}\underset{Ph}{CH} \longrightarrow -\!\!\left[CH_2-\underset{Ph}{CH}\right]_n\!\!-S-\!\!\left[CH_2-\underset{Ph}{CH}\right]_{m-1}\!\!-CH_2-\underset{Ph}{CH^{\bullet}} \xrightarrow{{}^{\bullet}CPh_3}$$

$$-\!\!\left[CH_2-\underset{Ph}{CH}\right]_n\!\!-S-\!\!\left[CH_2-\underset{Ph}{CH}\right]_m\!\!-\underset{Ph}{\overset{Ph}{C}}-Ph$$

**Scheme 4**

The trityl terminated polymers are used as polymeric initers in the synthesis of block copolymers. The blocking efficiency is rather low. Although the trityl terminated polymers are used as polymeric initiators to produce block copolymer, they can be used as polymeric inifers in specific reaction conditions.

# 4. MACROINIFER

Telechelics are oligomers with well defined functional end groups and can be synthesized using radical, anionic, or condensation polymerization techniques. Classical cationic polymerization of vinyl monomers is generally unsuitable for synthesizing telechelics because of predominance of transfer, isomerization, and termination reactions which led to mixtures of head and tail groups in the product. A breakthrough in the synthesis of telechelic polymers was the inifer technique developed by Kennedy and his coworkers [9,10]. The subsequent major breakthrough in the area of carbocationic living polymerization of olefins by Kennedy and Faust [11], Kennedy and Mishra [12], Mishra and Osman [13,14], and also by Higashimura and Sawamoto et al. [15] spurred the understanding of macromolecular engineering and led to the synthesis of functional/telechelic polymers.

Although 5-7% of all scientific publications and patents in the field of polymerization concern cationic polymer and oligomer synthesis, modifications, derivatization and degradation [16], there was little hope until 1979 that it would be possible to convert olefins into telechelics via a cationic polymerization.

In order to synthesize tailor-made polymers, e.g., telechelics, it is essential to control the initiation, the propagation, and the termination (or transfer) reactions. Thus, control of initiation allows introduction of head groups, and control of propagation and termination or transfer allow control of molecular weight and terminal groups, respectively (**Scheme 5**).

$$RX + MtX_n \longrightarrow {}^{+}R + MtX_{n+1}^{-}$$

$$R^{+} + nM \longrightarrow R\sim\sim\sim^{+}$$

$$R\sim\sim\sim^{+} + MtX_{n+1}^{-} \longrightarrow R\sim\sim\sim X + MtX_n$$

$$R\sim\sim\sim^{+} + RX \longrightarrow R\sim\sim\sim X + R^{+}$$

**Scheme 5**

The **Scheme 5** emphasizes that both termination and transfer can be utilized to introduce the desired endgroups. The transfer reaction involved here is a halogen interchange which is facilitated by the fact that carbocation (R+) formed is thermodynamically more stable than the growing polymer chain. The discovery in this field " Inifer Technique" by Kennedy et al. has been described elsewhere [9,10,16] (**Scheme 6**).

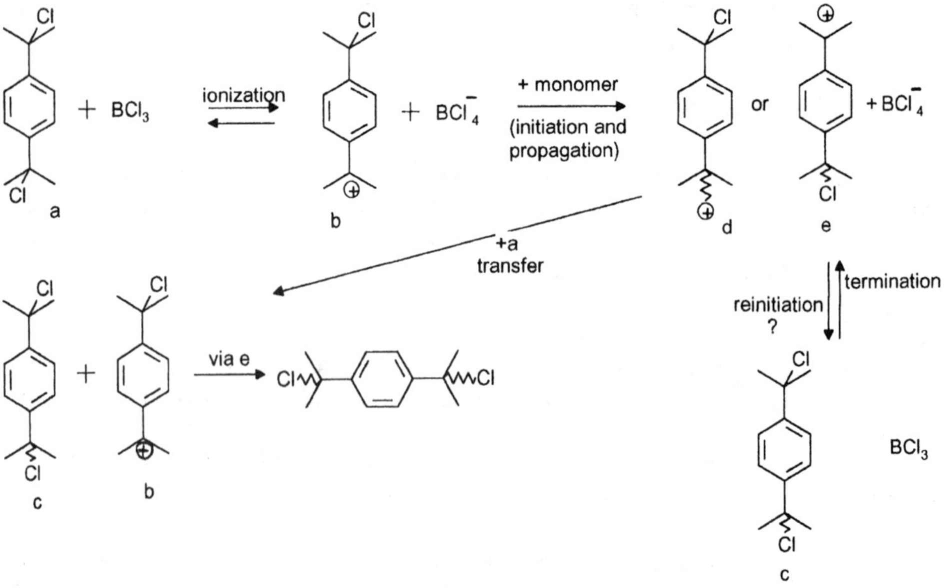

**Scheme 6**

The three basic inifers are 1. 1-chloro-1-methylethyl benzene, 2. 1,4-bis( 1-chloro-1-methyl ethyl benzene), and 3. 1,3,5-tris(1-chloro- 1-methyl ethyl benzene). These are called (1) mono- (2) bi- (3) trifunctional "Inifers". These inifers in conjunction with $BCl_3$ initiate the polymerization of isobutylene to produce polyisobutylene with well defined -$C(CH_2)Cl$ endgroups. They are presented in **Scheme 7.**

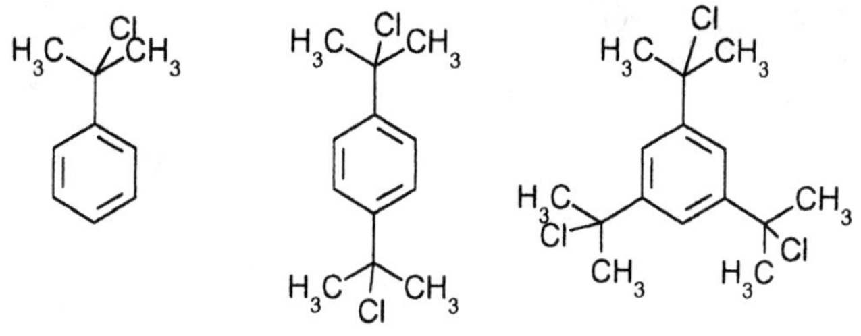

1 2 3

**Scheme 7**

Nuyken and his coworkers [17] have described the synthesis of polyfunctional ω-tert.-chloro-polyisobutylene graft copolymers by using macroinifers (macromolecular-initiator-transfer). The macroinifer was synthesized by the anionic polymerization of 1,3- or 1,4-diisopropenyl benzene followed by hydrochlorination according to the following reaction shown in **Scheme 8**. Nuyken et al. have also used this macroinifer for the cationic polymerization of isobutylene (**Scheme 9**). This method of graft copolymerization [17] is another example of multi-mode polymerization that has been described in a previous chapter of this book by Yagci and Mishra [18].

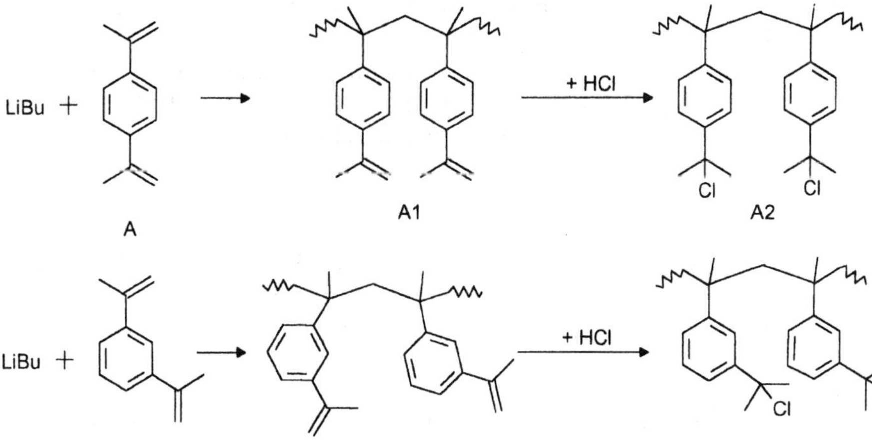

LiBu + → + HCl →

A A1 A2

LiBu + → + HCl →

B B1 B2

**Scheme 8**

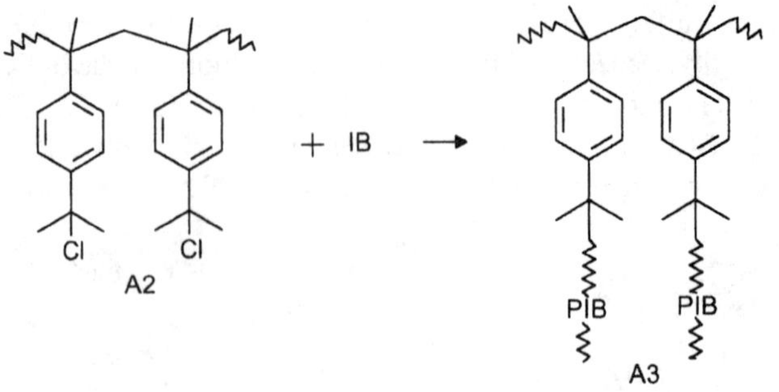

analogous B2 —(different amounts of IB used)→ B3, B4, B5 .. etc.

**Scheme 9**

## REFERENCES


1. B. Hazer, *Makromol. Chem.*, **193**, 1081 (1992)
2. B. Hazer, *Macromol. Rept.*, **A28**(suppl. 1), 47 (1991)
3. B. Hazer, *J. Appl. Polym. Sci.*, **47**(6), 1097 (1993)
4. B. Hazer, unpublished results.
5. P. Demircioglu, M. H. Acar and Y. Yagci, *J. Appl. Polym. Sci.*, **46**, 1639 (1992)
6. Y. Yamashita, Y. Chujo, H. Kobayashi, and Y. Kawakami, *Polym. Bull.*, **5**, 361 (1981)
7. Y. Chujo, T. Nakamura and Y. Yamashita, *J. Polym. Sci., Polym. Chem. Ed.*, **28**, 59 (1990)
8. T. Otsu, T. Matsunaga, A. Kuriyama and M. Yoshioka, *Eur. Polym. J.*, **25**, 643 (1989) and refs.
9. J. P. Kennedy and R. A. Smith, *Polymer Preprints*, **20**, 316 (1979)
10. J. P. Kennedy and R. A. Smith, *J. Polym. Sci., Polym. Chem.*, **18**, 1523 (1980)
11. J. P. Kennedy and R. Faust, US Pat., 4,910,321(1990)
12. J. P. Kennedy and M. K. Mishra, US Pat., 4,929,683(1990)
13. M. K. Mishra and A. Osman, US Pat., 4,908,421(1990)
14. M. K. Mishra and A. Osman, US Pat., 4,943,616(1990)
15. For a review, see: T. Higashimura, S. Aoshima and M. Sawamoto, *Makromol. Chem., Makromol. Symp.*, **13/14** 457 (1988).
16. J. P. Kennedy and E. Marechal, "Carbocationic Polymerization", J. Wiley, New York (1982)
17. H. -P. Muhlbauer and O. Nuyken, Presented at the 9th International Symposium in Ionic Polymerization and Related Processes, Strasburg, France (1989)
18. Y. Yagci and M. K. Mishra in "Macromolecular Design :Concept and Practice," M. K. Mishra, Ed., Polym. Front. Int'l, Inc., New York, Chapter 10 (1994)

# Index

## (Selected Subject Index)